Springer Series on

ATOMIC, OPTICAL, AND PLASMA PHYSICS

The Springer Series on Atomic, Optical, and Plasma Physics covers in a comprehensive manner theory and experiment in the entire field of atoms and molecules and their interaction with electromagnetic radiation. Books in the series provide a rich source of new ideas and techniques with wide applications in fields such as chemistry, materials science, astrophysics, surface science, plasma technology, advanced optics, aeronomy, and engineering. Laser physics is a particular connecting theme that has provided much of the continuing impetus for new developments in the field. The purpose of the series is to cover the gap between standard undergraduate textbooks and the research literature with emphasis on the fundamental ideas, methods, techniques, and results in the field.

Please view available titles in *Springer Series on Atomic, Optical, and Plasma Physics* on series homepage http://www.springer.com/series/411

Michael Bonitz
Norman Horing
Patrick Ludwig
(Editors)

Introduction to Complex Plasmas

With 226 Figures

Springer

Editors
Professor Dr. Michael Bonitz
Dr. Patrick Ludwig
Christian-Albrechts Universität Kiel
Institut für Theoretische Physik und Astrophysik
Leibnizstr. 15, 24098 Kiel, Germany
E-mail: bonitz@physik.uni-kiel.de, ludwig@theo-physik.uni-kiel.de

Professor Dr. Norman Horing
Stevens Institute of Technology, Deptartment of Physics
Castle Point on Hudson, Hoboken, NJ 07030, USA
E-mail: nhoring@stevens.edu

Springer Series on Atomic, Optical, and Plasma Physics ISSN 1615-5653

ISBN 978-3-642-10591-3 e-ISBN 978-3-642-10592-0

DOI 10.1007/978-3-642-10592-0

Springer Heidelberg Dordrecht London New York

Library of Congress Control Number: 2010930285

Typesetting and production: SPi
Cover concept: eStudio Calmar Steinen
Cover design: SPI Publisher Services

SPIN 12587322 57/3180/spi
Printed on acid-free paper

9 8 7 6 5 4 3 2 1

Springer is a part of Springer Science+Business Media (springer.com).

Preface

Within the wide and important field of plasma research, this tutorial book focuses on modern developments in the field of *particle-containing plasmas*. A central issue is the inherent overlap of three key scientific problems of complex plasma physics: *correlations*, *dynamics*, and *reactivity*. Examples include:

1. Coupling effects of highly charged dust particles in plasma traps giving rise to strongly correlated plasma states
2. Dynamics of multispecies plasmas and plasma–surface interaction
3. Chemical processes in plasmas and on plasma boundaries

In this book, these fundamental problems are approached using complementary experimental, computational, and theoretical methods which combine the authors' expertise from plasma physics, surface and solid-state physics, chemical physics, and materials science.

The central goal of this book is to provide graduate students and young researchers with the necessary knowledge base in the fast-growing field of complex plasma research. The style of each chapter is review-like, that is, the authors do not focus only on their own work but also give a survey of the state of the art. For easy access to the various aspects of complex plasmas by newcomers, each chapter opens with an introduction and overview of the particular topic, and also the basics – which are typically not covered in scientific journal publications – are explained in great detail. Furthermore, the chapters are enriched with much valuable background information, which should be of interest to a broad readership.

Part I of this book briefly introduces the very fundamentals of complex plasma physics. This part addresses the key questions and hot topics in modern complex plasma research and links them to the other chapters of this book. Part II is devoted to the field of quantum plasmas and their description with modern simulation techniques. In this part, graphene – the rising star of condensed-matter physics – is introduced as a very recent and promising example for the broad applicability of (quantum) plasma physics. Part III covers strong correlation effects and order phenomena occurring in complex plasmas in traps and introduces powerful numerical methods used for a "first-principle" simulation of dusty plasmas. Finally, Part IV deals with the issue of reactivity and surface processes, which have strong impact for nanotechnological applications.

This book is based on tutorial lectures given at the Graduate Summer Institute on "Complex Plasmas" at the Stevens Institute of Technology, Hoboken, NJ (USA) from July 30 to August 8, 2008. The workshop was jointly organized by the SFB-TR24 "Fundamentals of Complex Plasmas" Greifswald/Kiel (Germany) and Stevens. The school was attended by about 110 participants – scientists and graduate students. The chapters in this book take account of the lively discussions at this summer school and should serve as valuable introductory material for the active field of complex plasmas.

Our thanks are to all authors who contributed their knowledge to this tutorial book. We gratefully acknowledge for financial support by the Deutsche Forschungsgemeinschaft via SFB-TR24, the Air Force Office of Scientific Research (AFOSR), Army Research Office (ARO), National Science Foundation (NSF), the Polytechnic University New York, and the Princeton Plasma Physics Lab and the help of many people who have made the workshop and this tutorial book possible.

Kiel and Hoboken
May 2010

M. Bonitz
N. Horing
P. Ludwig

Contents

Part I Introduction

1 Complex Plasmas ... 3
Patrick Ludwig, Michael Bonitz, and Jürgen Meichsner
1.1 Plasmas in Nature and in the Laboratory ... 3
1.2 Complex Plasmas ... 7
1.3 Low-Temperature Plasmas and Technological Applications ... 9
1.4 Outline of this book ... 12
References ... 13

Part II Classical and Quantum Plasmas

2 Principles of Transport in Multicomponent Plasmas ... 17
Igor D. Kaganovich, Raoul N. Franklin, and Vladimir I. Demidov
2.1 Introduction ... 18
2.1.1 Production and Destruction Mechanisms of Negative Ions ... 19
2.1.2 The Drift–Diffusion Approximation for the Description of Plasma Transport ... 20
2.2 Ambipolar Diffusion ... 21
2.3 Temporal Dynamics of Negative Ion Flows in Multicomponent Plasmas ... 23
2.4 Afterglow in Multicomponent Plasmas and Consequent Wall Fluxes of Negative Ions ... 27
2.5 Steady-State Profiles of Plasmas with Negative Ions ... 30
2.6 The Sheath in Strongly Electronegative Gases ... 33
2.7 The Connection Between Plasmas with Negative Ions, Dusty Plasmas, and Ball Lightning ... 35
References ... 38

3 Introduction to Quantum Plasmas ... 41
Michael Bonitz, Alexei Filinov, Jens Böning, and James W. Dufty
3.1 Introduction ... 41
3.2 Relevant Parameters of Quantum Plasmas ... 43
3.3 Different States of Quantum Plasmas ... 46
3.4 Occurrences of Quantum Plasmas ... 48
3.4.1 Astrophysical Plasmas ... 48
3.4.2 Dense Laboratory Plasmas ... 48
3.4.3 Laser Plasmas ... 49
3.4.4 Plasmas in Condensed Matter Systems ... 49
3.4.5 Highly Compressed Two-Component Plasmas: Mott Effect ... 50
3.4.6 Ultra-Dense Plasmas in Nuclear Matter: Quark–Gluon Plasma and the Big Bang ... 52
3.5 Theoretical Description of Quantum Plasmas ... 53
3.5.1 Basic Equations ... 54
3.5.2 Thermodynamics of Partially Ionized Plasmas ... 54
3.5.3 Spin Effects in Quantum Plasmas ... 58
3.5.4 Bose Plasmas ... 60
3.5.5 Plasmas of Particles Having Fermi Statistics ... 64
3.5.6 Quantum Kinetic Theory ... 66
3.5.7 More Advanced Approach: The Method of Second Quantization ... 68
3.5.8 Other Approaches to Quantum Plasmas ... 71
3.6 Conclusions ... 75
References ... 75

4 Introduction to Quantum Plasma Simulations ... 79
Sebastian Bauch, Karsten Balzer, Patrick Ludwig, and Michael Bonitz
4.1 Introduction ... 79
4.2 Time-Dependent Schrödinger Equation ... 80
4.2.1 1D Crank–Nicolson Method ... 81
4.2.2 TDSE Solution in Basis Representation ... 86
4.2.3 Computational Example: Electron Scattering in a Laser Field ... 88
4.3 Hartree–Fock Method ... 89
4.3.1 Standard Approach ... 90
4.3.2 NEGF Approach ... 92
4.3.3 Example ... 94
4.4 Quantum Monte Carlo Methods ... 97
4.4.1 Metropolis Monte Carlo Method ... 98
4.4.2 Path-Integral Monte Carlo ... 100
4.5 Summary ... 105
References ... 106

5 Quantum Effects in Plasma Dielectric Response: Plasmons and Shielding in Normal Systems and Graphene 109
Norman J.M. Horing
5.1 Introduction 109
5.1.1 Background 109
5.1.2 Quantum Theory of Dielectric Response 111
5.2 Quantum Effects in Normal Solid-State Plasmas 113
5.2.1 Three-Dimensional Quantum Plasma 113
5.2.2 Dielectric Properties of Low-Dimensional Systems 115
5.2.3 Dielectric Function of a Magnetized Quantum Plasma 117
5.3 Graphene 121
5.3.1 Introduction 121
5.3.2 Graphene Hamiltonian, Green's Function, and RPA Dielectric Function 123
5.3.3 Some Physical Features of Graphene 127
5.4 Summary 130
References 131

Part III Strongly Coupled and Dusty Plasmas

6 Imaging Diagnostics in Dusty Plasmas 135
Dietmar Block and André Melzer
6.1 Introduction 135
6.2 Imaging 2D Systems 136
6.2.1 Imaging Particles 136
6.2.2 Image Analysis 138
6.3 Imaging 3D Systems 141
6.3.1 Scanning Video Microscopy 141
6.3.2 Color Gradient Method 142
6.3.3 Stereoscopy 143
6.3.4 Digital Holography 146
6.4 Summary and Outlook 152
References 152

7 Structure and Dynamics of Finite Dust Clusters 155
André Melzer and Dietmar Block
7.1 Introduction 155
7.2 Trapping of Dust Clouds 156
7.3 Formation of Finite Dust Clusters 158
7.4 Structural Transitions in 1D Dust Clusters 158
7.5 Structure of 2D Dust Clusters 160
7.6 Normal Mode Dynamics of Dust Clusters 161
7.7 Formation of 3D Dust Clusters 163
7.8 Structure of 3D Dust Clusters 165
7.9 Metastable Configurations of Yukawa Balls 167

7.10 Shell Transitions in Yukawa Balls 170
7.11 Dynamical Properties of Yukawa Balls 171
7.12 Summary 172
References 173

8 Statistical Theory of Spherically Confined Dust Crystals 175
Christian Henning and Michael Bonitz
8.1 Introduction 175
8.2 Variational Problem of the Energy Functional 176
8.3 Ground-State Density Profile Within Mean-Field Approximation ... 180
8.3.1 The Coulomb Limit and Electrostatics 180
8.3.2 General Solution 181
8.3.3 Density Profile for Harmonic Confinement 183
8.3.4 Force Equilibrium Within Yukawa Electrostatics 185
8.4 Simulation Results of Spatially Confined Dust Crystals 187
8.4.1 Ground-State Simulations 188
8.4.2 Comparison of Simulation and Mean-Field Results 190
8.5 Inclusion of Correlations by Using the Local Density Approximation 191
8.5.1 LDA Without Correlations 192
8.5.2 LDA with Correlations 195
8.5.3 Comparison of Simulation and LDA Results 197
8.6 Shell Models of Spherical Dust Crystals 198
8.7 Summary and Discussion 200
References 201

9 PIC–MCC Simulations of Capacitive High-Frequency Discharge Dynamics with Nanoparticles 203
Irina V. Schweigert
9.1 Introduction 203
9.2 Combined PIC–MCC Approach for Fast Simulation of a Radio-Frequency Discharge at Low Gas Pressure 205
9.2.1 Combined PIC–MCC Approach 206
9.2.2 Description of the Algorithm 207
9.2.3 How Many Simulation Particles We Need? 210
9.2.4 Simulation Results of a CCRF-Discharge in Helium and Argon 211
9.3 Physical Model of Discharge Plasma with Movable Dust 217
9.3.1 Algorithm of Calculation 218
9.3.2 Ion Drag Force 220
9.3.3 Transition Between Different Modes 222
9.3.4 Dust Motion Effect 225
9.4 Conclusion 228
References 230

10 Molecular Dynamics Simulation of Strongly Correlated Dusty Plasmas 231
Torben Ott, Patrick Ludwig, Hanno Kählert, and Michael Bonitz
10.1 Introduction 231
10.2 Basics of Molecular Dynamics Simulation 232
10.2.1 Simulation Model of Strongly Coupled Dusty Plasmas 234
10.2.2 Equations of Motion of a One-Component Plasma 236
10.2.3 Velocity Verlet Integration Scheme 238
10.2.4 Runge–Kutta Integration Scheme 239
10.3 Equilibrium Simulations: Thermodynamic Ensembles 240
10.3.1 Velocity Scaling 241
10.3.2 Stochastic Thermostats 241
10.3.3 Nosé–Hoover Thermostat 242
10.3.4 Langevin Dynamics Simulation 242
10.3.5 Dimensionless System of Units 244
10.4 Simulation of Macroscopic Systems 245
10.4.1 Potential Truncation 245
10.4.2 Electrostatic Interactions 246
10.4.3 Finding of Neighboring Particles 246
10.4.4 Periodic Boundary Conditions 247
10.5 Input and Output Quantities 249
10.5.1 Pair Distribution Function and Static Structure Factor 249
10.5.2 Transport Properties 251
10.6 Applications I: Mesoscopic Systems in Traps 251
10.6.1 Simulated Annealing 252
10.6.2 Effect of Screening 254
10.6.3 Effect of Friction 255
10.7 Applications II: Macroscopic Systems 258
10.7.1 Simulation Results 259
10.8 Conclusion 261
References 262

Part IV Reactive Plasmas, Plasma–Surface Interaction, and Technological Applications

11 Nonthermal Reactive Plasmas 267
Jürgen Meichsner
11.1 Introduction 267
11.2 Nonthermal Plasma Conditions 271
11.3 Plasma Kinetics and Plasma Chemical Reactions 272
11.3.1 Boltzmann Equation 272
11.3.2 Reaction Rate Coefficient 274

11.4 Plasma–Surface Interaction 276
11.4.1 Plasma Sheath 276
11.4.2 Surface on Floating Potential 277
11.4.3 High-Voltage Plasma Sheath, Radio-Frequency Plasma Sheath 278
11.5 Low-Pressure Oxygen rf-Plasma 280
11.5.1 Plasma Characterization 281
11.5.2 Interaction of Oxygen Plasma with Polymers 291
References 296

12 Formation and Deposition of Nanosize Particles on Surfaces 299
Rainer Hippler, Satya R. Bhattacharyya, and Boris M. Smirnov
12.1 Introduction 299
12.2 Magnetron Discharge 300
12.3 Nucleation Processes in a Magnetron Plasma 302
12.4 Nanosize Cluster Deposition 305
12.5 Melting Temperature and Lattice Parameters of Ag Clusters 307
12.6 Rapid-Thermal Annealing (RTA) of Deposited Cluster Films 308
12.7 Evaporation of Clusters 312
12.8 Conclusions 313
References 313

13 Kinetic and Diagnostic Studies of Molecular Plasmas Using Laser Absorption Techniques 315
Jürgen Röpcke, Richard Engeln, Daan Schram, Antoine Rousseau, and Paul B. Davies
13.1 Introduction 316
13.2 Plasma Chemistry and Reaction Kinetics 319
13.2.1 Studies of $Ar/H_2/N_2/O_2$ Microwave Plasmas 319
13.2.2 On the Importance of Surface Association to the Formation of Molecules in a Recombining N_2/O_2 Plasma 322
13.3 Kinetic Studies and Molecular Spectroscopy of Radicals 326
13.3.1 Line Strengths and Transition Dipole Moment of CH_3 326
13.3.2 Molecular Spectroscopy of the CN Radical 330
13.4 Quantum Cascade Laser Absorption Spectroscopy for Plasma Diagnostics and Control 331
13.4.1 General Considerations 331
13.4.2 Time-Resolved Study of a Pulsed DC Discharge: NO and Gas Temperature Kinetics 333

13.4.3 Trace Gas Measurements Using Optically Resonant Cavities ... 335
13.4.4 In Situ Monitoring of Plasma Etch Processes with a QCL Arrangement in Semiconductor Industrial Environment ... 338
13.5 Summary and Conclusions ... 340
References ... 340

14 X-Ray Diagnostics of Plasma-Deposited Thin Layers ... 345
Harm Wulff
14.1 Introduction ... 345
14.2 X-Ray Analytical Methods ... 347
14.2.1 Grazing Incidence X-Ray Diffractometry, Asymmetric Bragg Case ... 347
14.2.2 GIXD, Bragg Case, Specular Reflected ... 348
14.2.3 X-Ray Reflectometry ... 349
14.3 Examples ... 350
14.3.1 Characterization of ITO Films ... 350
14.3.2 Study of Al_2O_3 Formation During Microwave Plasma Treatment of Al Films in Ar–O_2 Gas Mixtures ... 358
14.4 Summary ... 365
References ... 365

15 The Use of Nonthermal Plasmas in Environmental Applications ... 367
Kurt H. Becker
15.1 Introduction ... 367
15.2 Commercially Viable, Large-Scale Plasma-Based Environmental Applications ... 369
15.2.1 Ozonizers ... 369
15.2.2 Electrostatic Precipitation ... 373
15.3 Decomposition of Volatile Organic Compounds in Microplasmas ... 377
15.3.1 Experimental Details ... 377
15.3.2 VOC Destruction Efficiency ... 379
15.3.3 Byproduct Formation ... 381
15.3.4 Kinetic Studies ... 382
15.3.5 Summary ... 385
15.4 Pulsed Electrical Discharges in Water ... 386
15.4.1 Background ... 386
15.4.2 Experimental Systems ... 387
15.4.3 Selected Experimental Results ... 389
15.4.4 Summary ... 391
15.5 Conclusion ... 391
References ... 392

16 Complex (Dusty) Plasmas: Application in Material Processing and Tools for Plasma Diagnostics 395
Holger Kersten and Matthias Wolter
16.1 Introduction 395
16.2 Disturbing Side Effects of Dust Particles in Plasma Processing 396
16.3 Formation and Modification of Powder Particles in Plasmas for Various Industrial Applications 398
16.3.1 Coating of Powder Particles in a Magnetron Discharge 402
16.3.2 Deposition of Protective Coatings on Individual Phosphor Particles 407
16.3.3 Particles as Microsubstrates 410
16.4 Particles as Electrostatic Probes 413
16.4.1 Dust Particles in Front of an Adaptive Electrode 417
16.4.2 Interaction Between Dust Particles and Ion Beams 426
16.5 Particles as Thermal Probes 434
References 439

Index 443

Contributors

Karsten Balzer Institut für Theoretische Physik und Astrophysik, Christian-Albrechts Universität zu Kiel, 24098 Kiel, Germany, balzer@theo-physik.uni-kiel.de

Sebastian Bauch Institut für Theoretische Physik und Astrophysik, Christian-Albrechts Universität zu Kiel, 24098 Kiel, Germany, bauch@theo-physik.uni-kiel.de

Kurt H. Becker Department of Physics, Polytechnic Institute of New York University, Six MetroTech Center, Brooklyn, NY 11201, USA, kbecker@poly.edu

Satya R. Bhattacharyya Surface Physics Division, Saha Institute of Nuclear Physics, 1/AF Bidhan Nagar, Kolkata 700 064, India, satyar.bhattacharyya@saha.ac.in

Dietmar Block Institut für Experimentelle und Angewandte Physik, Christian-Albrechts-Universität Kiel, Olshausenstr. 40-60, 24098 Kiel, Germany, block@physik.uni-kiel.de

Jens Böning Institut für Theoretische Physik und Astrophysik, Christian-Albrechts Universität zu Kiel, 24098 Kiel, Germany, boening@theo-physik.uni-kiel.de

Michael Bonitz Institut für Theoretische Physik und Astrophysik, Christian-Albrechts Universität zu Kiel, 24098 Kiel, Germany, bonitz@physik.uni-kiel.de

Paul B. Davies University of Cambridge, Cambridge CB2 1EW, UK, pbd2@cam.ac.uk

Vladimir I. Demidov UES, Inc., Dayton-Xenia Rd., Beavercreek, OH 45432, USA, demidov@wvu.edu

James W. Dufty Department of Physics, University of Florida, Gainesville, FL 32611, USA, dufty@phys.ufl.edu

R. Engeln Eindhoven University of Technology, P.O. Box 513, 5600 MB, Eindhoven, The Netherlands,
r.a.h.engeln@tue.nl

Alexei Filinov Institut für Theoretische Physik und Astrophysik, Christian-Albrechts Universität zu Kiel, 24098 Kiel, Germany,
filinov@theo-physik.uni-kiel.de

Raoul N. Franklin Department of Physics and Astronomy, The Open University, Milton Keynes MK7 6AA, UK,
r.n.franklin@open.ac.uk

Christian Henning Institut für Theoretische Physik und Astrophysik, Christian-Albrechts Universität zu Kiel, 24098 Kiel, Germany,
henning@theo-physik.uni-kiel.de

Rainer Hippler Institut für Physik, Ernst-Moritz-Arndt-Universität Greifswald, Felix-Hausdorff-Str. 6, 17489 Greifswald, Germany,
hippler@physik.uni-greifswald.de

Norman J.M. Horing Department of Physics and Engineering Physics, Stevens Institute of Technology, Hoboken, NJ 07030, USA,
nhoring@stevens.edu

Igor D. Kaganovich Plasma Physics Laboratory, Princeton University, Princeton, NJ 08543, USA,
ikaganov@pppl.gov

Hanno Kählert Institut für Theoretische Physik und Astrophysik, Christian-Albrechts Universität zu Kiel, 24098 Kiel, Germany,
kaehlert@theo-physik.uni-kiel.de

Holger Kersten Institut für Experimentelle und Angewandte Physik, Christian-Albrechts-Universität Kiel, Leibnizstr.19, 24098 Kiel, Germany,
kersten@physik.uni-kiel.de

Patrick Ludwig Institut für Theoretische Physik und Astrophysik, Christian-Albrechts Universität zu Kiel, 24098 Kiel, Germany,
ludwig@theo-physik.uni-kiel.de

Jürgen Meichsner Institut für Physik, Ernst-Moritz-Arndt-Universität Greifswald, Felix- Hausdorff-Str. 6, 17489 Greifswald, Germany,
meichsner@physik.uni-greifswald.de

André Melzer Institut für Physik, Ernst-Moritz-Arndt-Universität Greifswald, Felix-Hausdorff-Str. 6, 17489 Greifswald, Germany,
melzer@physik.uni-greifswald.de

Torben Ott Institut für Theoretische Physik und Astrophysik, Christian-Albrechts Universität zu Kiel, 24098 Kiel, Germany,
ott@theo-physik.uni-kiel.de

Jürgen Röpcke INP-Greifswald, Felix-Hausdorff-Str. 2, 17489 Greifswald, Germany, roepcke@inp-greifswald.de

Antoine Rousseau LPTP, Ecole Polytechnique, CNRS, 91128 Palaiseau Cedex, France,
Antoine.Rousseau@lpp.polytechnique.fr

Daan Schram Eindhoven University of Technology, P.O. Box 513, 5600 MB, Eindhoven, The Netherlands,
d.c.schram@tue.nl

Irina V. Schweigert Institute of Theoretical and Applied Mechanics, Russian Academy of Sciences, Novosibirsk 630090, Russia,
ischweig@itam.nsc.ru

Boris M. Smirnov Joint Institute for High Temperatures, Russian Academy of Sciences, Izhorskaya 13/19, Moscow 125412, Russia,
bmsmirnov@gmail.com

Matthias Wolter Institut für Experimentelle und Angewandte Physik, Christian-Albrechts-Universität Kiel, Leibnizstr.19, 24098 Kiel, Germany,
wolter@physik.uni-kiel.de

Harm Wulff Institut für Biochemie, Ernst-Moritz-Arndt-Universität Greifswald, Felix-Hausdorff-Str. 4, 17487 Greifswald, Germany,
wulff@chemie.uni-greifswald.de

Part I
Introduction

Chapter 1
Complex Plasmas

Patrick Ludwig, Michael Bonitz, and Jürgen Meichsner

Abstract Complex plasmas are a modern research area in plasma science. Such plasmas differ from conventional high-temperature plasmas in several ways (1) they may contain additional species, including nanometer- to micrometer-sized particles, negative ions, molecules, and radicals and (2) they exhibit strong correlations (e.g., dusty plasmas) or quantum effects. Numerous applications of particle-containing plasmas and microplasmas are already emerging. This chapter provides an introduction into the field of complex plasmas and an outline on the chapters of this book.

1.1 Plasmas in Nature and in the Laboratory

Among all objects observed in nature, about 90% exist in ionized form. Plasmas, often considered the fourth state of matter, span a huge diversity of parameter scales and exist throughout the universe, in laboratories and numerous technological applications. Examples include astrophysical plasmas such as in dilute interstellar gas clouds or the highly energetic and dense matter in the strongly compressed interior of stars or giant gas planets like Jupiter (see Fig. 1.1). On earth, examples include hot plasma ions in the magnetosphere surrounding our planet, the aurora borealis, lightning bolts, or the plasma of a candle flame. Besides these *natural plasmas*, plasmas find technological application in many modern industrial processes, in plasma chemistry, in nanoparticle sources and a variety of surface processing technologies for the treatment of metals, optical components, and plastic materials (functionalization, sterilization, etc.). Furthermore, they are widely used in the field

P. Ludwig (✉) and M. Bonitz
Institut für Theoretische Physik und Astrophysik, Christian-Albrechts-Universität zu Kiel, 24098 Kiel, Germany
e-mail: ludwig@theo-physik.uni-kiel.de; bonitz@theo-physik.uni-kiel.de

J. Meichsner
Institut für Physik, Ernst-Moritz-Arndt-Universität Greifswald, Felix-Hausdorff-Str. 6, 17489 Greifswald, Germany
e-mail: meichsner@physik.uni-greifswald.de

M. Bonitz et al. (eds.), *Introduction to Complex Plasmas*, Springer Series on Atomic, Optical, and Plasma Physics 59, DOI 10.1007/978-3-642-10592-0_1,

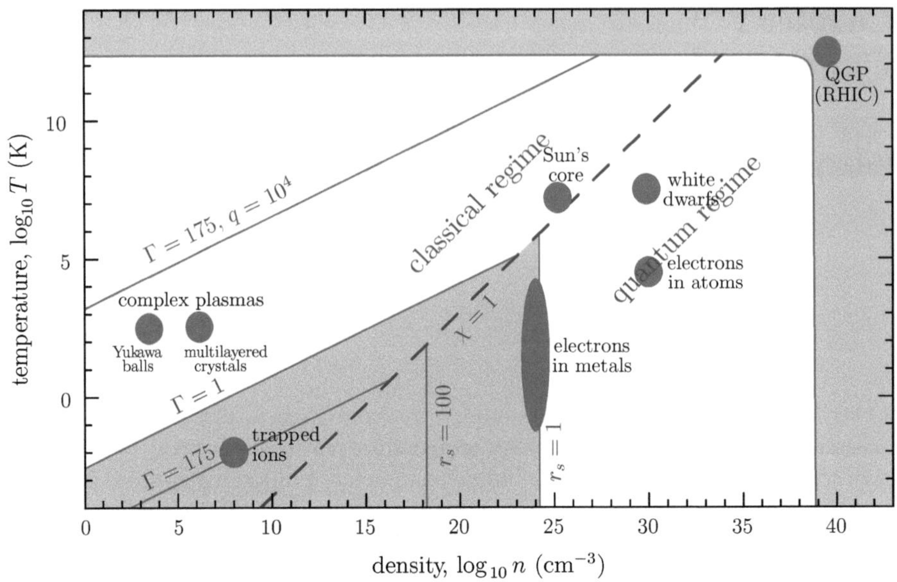

Fig. 1.1 Density–temperature phase diagram of ionized matter in nature and laboratory. The *dashed line* $\chi = 1$ divides the plane into the regimes of classical and quantum systems. In the case of singly charged particles, $q = 1$, the regime of strong correlations and structure formation is observed in the *shaded area* enclosed by the *lines* $\Gamma = 1$ and $r_s = 1$. Higher charged species, $q > 1$, such as in complex plasmas, widely extend the area where strong correlation effects are present. The *outer shaded area* indicates the deconfined state of hadronic matter, that is, the quark–gluon plasma (QGP). Examples of particle-containing complex plasmas are also shown

of nanotechnologies which includes plasma-assisted deposition or etching in the semiconductor industry. Plasmas play a central role in the development of improved light sources, display technology, lasers, and solar cells. Also, promising medical applications on living tissues are already emerging. Other novel fields of plasma research are laser-produced plasmas, particle acceleration in plasma wake fields, the generation of ultra-dense plasmas (so-called *warm dense matter*) by focusing of intense laser beams on small targets, as well as energy research in large-scale nuclear fusion experiments such as ITER in France or the National Ignition Facility in the US.

A conventional plasma consists of freely moving charged particles which are typically electrons and ions. This implies that high thermal energies on the order of several electron volts (10^5 K) are involved to break neutral atoms and molecules into free electrons and ions. At high temperatures and low densities, thermal energy dominates and the essentially classical particles are not affected by each other and move in an uncorrelated manner.[1] The transition from the classical to the quantum

[1] Correlated means that the particle properties (e.g., motion) depend not only on its position, but also on the positions and velocities of the other particles in the plasma.

plasma regime is quantified by the degeneracy parameter χ. Exceeding $\chi = 1$ (to the right of the dashed line in Fig. 1.1), that is, at densities at which the mean interparticle distance equals the spatial extension of the particle wave function (its thermal wavelength), quantum mechanics becomes essential.

Quantum plasmas are known to exist under high-density conditions such as created in many modern experiments with short laser pulses or ion beams. But also in various astrophysical objects such as white dwarf stars or neutron stars, the plasma electrons behave fully quantum mechanically – even at temperatures comparable to hot fusion plasma (10^8 K). Another example of naturally existing quantum plasma is the *Fermi gas* of delocalized electrons in metals whose detailed understanding is of high technological importance. An in-depth introduction on the topic of quantum plasmas is given in Chap. 3.

Correlated behavior arises from the long-range nature of electromagnetic Coulomb forces and plays a significant role in the particle dynamics. Interactions and collective many-particle behavior (which causes the structure of atoms, solid matter, and all classical plasma correlation effects) come into play, when the Coulomb interaction energy dominates over kinetic energy, which counteracts the formation of correlated states of matter. Even though many-particle behavior is strongly affected by the respective type of the potential energy and the strength of quantum effects, similar correlation phenomena can be observed in very different charged many-particle systems. In fact, due to universal scaling laws, many plasma properties are not only related to a specific system, but rather are of fundamental nature (Fig. 1.2).

The strength of many-particle correlations in classical interacting Coulomb systems can be quantified by the coupling (correlation) parameter Γ. This plasma parameter is defined as the ratio of the mean (nearest neighbor) *interaction energy* to the average *thermal energy* of the system:

$$\Gamma = \frac{|E_{\mathrm{int}}|}{E_{\mathrm{therm}}}. \tag{1.1}$$

By means of this parameter, universal trends in classical plasmas can be very generally quantified as ranging from *ideal gas-like behavior* (for $\Gamma \ll 1$) *strongly*

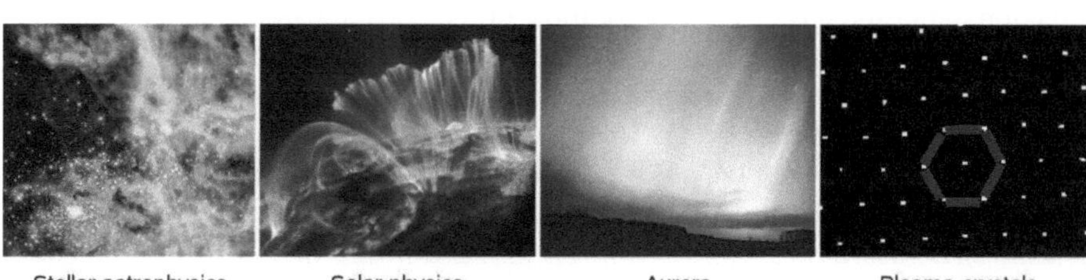

Fig. 1.2 Plasma as a nonlinear system encompasses a wide field of diverse appearances ranging from astrophysical plasmas, fusion plasmas, the aurora borealis to particle-containing dusty plasmas in the laboratory

coupled (liquid-like) systems with short-range order (for $\Gamma \gtrsim 1$) to *crystalline long-range order*, when the dimensionless parameter exceeds a critical value that is typically around $\Gamma_{\mathrm{cr}} \approx 100$. Hence, the phenomenon of spatial correlations and structure formation expresses the particles' tendency to minimize their potential energy by avoiding the close neighborship to other particles.

A second key quantity which describes particle correlations in quantum systems is the Brueckner parameter:

$$r_{\mathrm{s}} = \frac{a}{a_{\mathrm{B}}}, \tag{1.2}$$

which is given by the ratio of the mean interparticle distance a and the effective Bohr radius $a_{\mathrm{B}} = 4\pi\varepsilon\hbar^2/(mq^2)$, where m denotes the particle's mass and q its charge. The quantum coupling parameter r_{s} takes into account the increase of *quantum kinetic energy* at high densities[2] and reflects the impact of nonclassical effects due to wave function overlap. The parameter expresses the transition from a weakly coupled, ideal quantum system ($r_{\mathrm{s}} \ll 1$) to a purely classical one ($r_{\mathrm{s}} \to \infty$).

It is remarkable that, despite the greatly different ranges and nature of plasmas, the occurrence of cooperative many-particle effects such as structure formation is completely captured by these two dimensionless parameters alone. Considering the phase diagram (Fig. 1.1), crystalline states of matter are found only in a relatively small range at low temperatures and moderate particle densities, where $\Gamma \geq \Gamma_{\mathrm{cr}} \approx 100$ and $r_{\mathrm{s}} \geq r_{\mathrm{s_{cr}}} \approx 100$, respectively. In contrast, both limits $\Gamma \ll 1$ and $r_{\mathrm{s}} \ll 1$ are fully structureless.

The experimental measurement and analysis of *strongly correlated systems* require in most cases experimentally challenging low temperatures which are, for example, in the case of Doppler laser-cooled ion crystals in electromagnetical Paul and Penning traps on the mK scale. Only at mK temperatures, the electrostatic energy of the mutual Coulomb interactions between the ions becomes sufficiently larger than the thermal energy, that is, $\Gamma > 100$. However, strongly coupled Coulomb systems can be observed also at room temperature. A prominent representative are so-called *complex (dusty) plasmas*. At suitable plasma conditions, a complex plasma containing highly charged "dust particles" can spontaneously self-organize into a highly ordered state, where the detailed spatial arrangement strongly affects all many-particle features of the system and leads to many fascinating plasma properties. The outline of this unique state of *soft matter* and novel direction in plasma physics is a major topic of the book at hand.

It is interesting to note that even under the extreme conditions of temperature and density of the quark–gluon plasma, a strong *color-Coulomb interaction* allows for strong correlation effects [1]. The plasma consisting of deconfined quarks and gluons plays a key role in the description of the early universe and of neutron stars or quark stars and is briefly discussed in Chap. 3.

---

[2] Even at $T = 0$, zero-point quantum fluctuations are present which increase with density.

1.2 Complex Plasmas

Complex plasmas are a young field of plasma research and among the most promising classical systems showing strongly correlated many-body behavior. The title of this book defines the name "complex plasma" in a wider sense than the present usage of this name in the dusty plasma community.[3] A "complex plasma" is, in our terms, a multicomponent low-temperature plasma containing, besides electrons and positive ions, additional species such as negative ions, and charged nano- to micrometer-sized solid (dust) particles or clusters, as well as reactive atoms or molecules strongly interacting with surfaces (see Fig. 1.3). Like *complex fluids*, complex plasmas belong to the group of so-called *soft matter*. Pierre-Gilles de Gennes, Nobel Prize laureate in 1991, defined soft matter as:

> Supramolecular substances which exhibit special properties such as macroscopic softness or elasticity, which have an internal equilibrium structure that is sensitive to external forces, which process excited metastable states and where the relevant physics is far above the quantum level [3].

Dust and dusty plasmas are ubiquitous in nature, occurring in interplanetary and interstellar clouds, dust rings around planets like Saturn, on the surface of the moon,

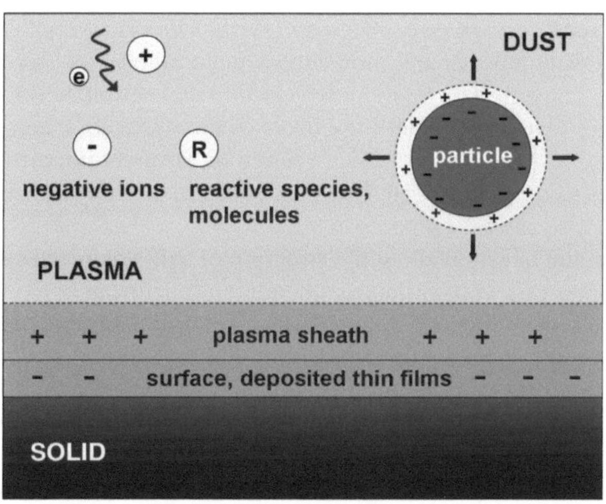

Fig. 1.3 The complexity of the plasma (electrons "e" and positive ions "+") arises on the one hand from the embedding of dust particles, which may in addition be strongly coupled and lead to order phenomena and many modified plasma properties. On the other hand negative ions "−" and reactive species "R" as well as the plasma sheath and the interaction with solid surfaces further increase the complexity of the plasma from [4]

[3] Originally, the name "complex plasma" was chosen in analogy to "complex fluids"; since complex plasmas can be regarded as the fourth state of soft matter, very much like ordinary plasmas can be regarded as the fourth state of ordinary matter [2].

noctilucent clouds in the mesosphere, or thunder clouds. They are responsible for fundamental astrophysical processes such as the formation of the solar systems and planets and are involved in many industrial processes, for example, the production of nanopowders. In the case of a "dusty plasma" the additional plasma species consist of small polymer microspheres which are dispersed in a (partially) ionized gas of electrons, positive ions, and neutral atoms. By electron and ion interaction, the dust particles acquire, depending on their size, high (in experiments typically negative) charges of several thousand of elementary charges and interact amongst themselves by a (screened) Coulomb repulsion. The interacting dust grains in the low-pressure discharge plasma define a *single-species subsystem*, which is only weakly damped by neutral gas friction. As a main difference to complex fluids, where the (charged) colloidal particles in the fluid are heavily damped, complex plasma research is not restricted to equilibrium studies. Rather, in high-precision experiments, all dynamical and collective processes of dust particles in the optically thin plasma medium can be fully resolved in *both* real time and space! This is due to the fact that the grains are large enough to be observed and manipulated individually.

As a crucial consequence of the high electrical charges on the individual dust particles, complex plasmas can be strongly coupled even at room temperature (see the $\Gamma = 175$-line for $q = 10^4$ in Fig. 1.1). In fact, the strong dust–dust interaction is connected to one of the most fascinating properties of complex plasmas: the spontaneous self-organization of plasma constituents into a strongly coupled crystalline plasma state – the *plasma (Wigner) crystal*. Its discovery in 1994 marked a milestone in plasma physics and motivated a huge number of scientific investigations, aiming at characterizing and understanding this new state (of matter). In this book, we expose that strong coupling effects of the dust subsystem make complex plasmas an ideal test ground for basic research on strongly interacting matter with increasing links to other fields, such as condensed matter physics, nuclear physics, and ultra-cold Bose and Fermi gases in traps.

In particular, the question about the structure of self-organized systems emerges as a key issue for a variety of collective phenomena occurring in physical systems at completely different energy and length scales. A unique example for such universal scaling behavior are the common features of finite Coulomb systems in confined geometries, where the strong coupling enables the emergence of crystal-like structures[4] even in small (nonthermodynamic) systems consisting of only 100 or less charged particles. A striking feature of such finite systems is that their structure and properties are very sensitive to the *exact* particle number. Interestingly enough, even without change of density or temperature, just by adding or removing a single charged particle qualitative transformations of the collective interplay can be achieved, resulting in drastically different physical properties (structural, electronic, magnetic, transport, or optical). Taking the dusty plasma as a unique representative, in the book at hand, many novel features of three dimensionally confined dust

[4] To differentiate between the ordered state of finite systems from the thermodynamic solid phase in macroscopically extended systems, we speak of a crystal-like state instead of a crystal.

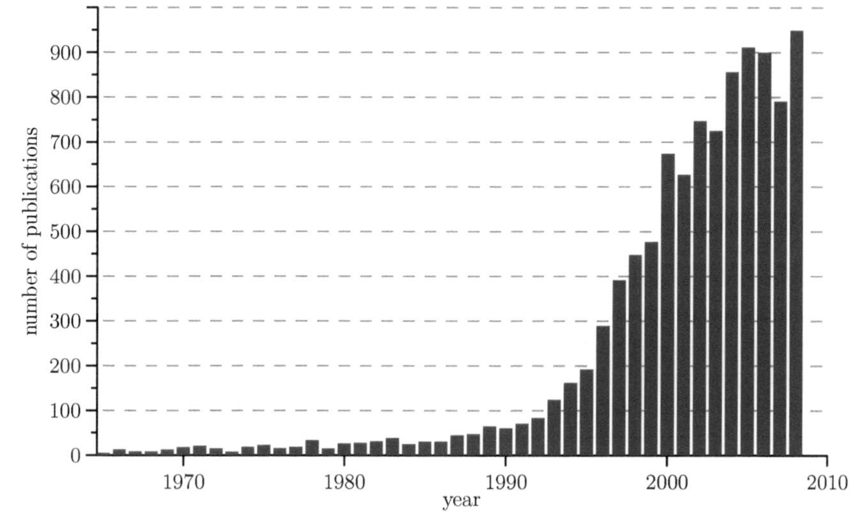

Fig. 1.4 Estimated number of publications on "complex plasmas" per year. The strong growth reflects the recent breakthrough of this young research field [5]

systems (so-called *Yukawa balls*) are revealed by use of complementary experimental analysis (Chap. 7), theoretical analysis (Chap. 8), and modern computational methods (Chap. 10).

Besides the strong correlations and strong modifications of the plasma parameters by high local charges, the complexity of a particle-containing plasma is further increased by processes connected to the existence of *negative ions* and *reactive species*, the *plasma sheath*, as well as the *plasma interaction with surfaces*.[5] The fast-growing interdisciplinary research combining plasma physics with surface physics, solid-state physics, materials science, and physical chemistry results in a strongly increasing number of publications over a broad spectrum of scientific journals in the last two decades (see Fig. 1.4). Today, complex plasma research has evolved into a mature state and has become the second most important plasma research field next to fusion.

1.3 Low-Temperature Plasmas and Technological Applications

Complex plasmas, as defined above, belong to the group of cold (and in most cases dilute) plasmas. Various kinds of multispecies plasmas play an exceptionally important role in technological plasma applications, such as computer chip etching,

---

[5] An introduction to the transport properties of multicomponent plasmas, in particular plasmas containing negative ions, is given in Chap. 2.

or nanopowder production. Dusty plasmas are also technologically relevant for deposition processes, for example, in nanocrystalline solar cells or in polymer coatings with embedded nanoparticles. Hence, a quantitative understanding of plasma processes and plasma surface interaction represents the basis for progress in new plasma technologies and future process control. In this context, some fundamental aspects, like the growth mechanisms of particles in hydrocarbon plasmas, the deposition of metal clusters on surfaces, and the deposition of nanocomposite materials are of particular interest. Such experiments have a potential for further development on the one hand toward the study of chemical reactions or catalytic processes involving dust particles, and on the other hand toward tailoring the physical and chemical properties of metal–polymer nanocomposites.

Low-temperature and nonequilibrium plasmas share scientific challenges with other branches of plasma research. For instance, the principles underlying plasma heating, stability, and control in the low-temperature regime are the same that govern processes in high-temperature plasmas of magnetic fusion devices, just as the emergence of collective behavior is shared with many other areas of plasma science and physics in general.

Another cross-cutting topic are plasma interactions with surfaces (see Chaps. 12 and 14). These interactions are often the desired outcome of certain low-temperature engineering procedures, whereas in fusion devices, the goal is to control and minimize these interactions. Situated at the border of plasma physics and solid-state physics, many fundamental questions including chemical reactions and electronic quantum effects are far from being understood. Related to this open questions, a deep understand of quantum plasmas is inevitable. An introduction is given in Chap. 3.

The recent review on plasma research and its perspectives in the next decade, given in "Plasma Science – Advancing Knowledge in the National Interest" by the National Research Council, USA (2007), has concluded that the expanding scope of plasma research provides new scientific opportunities and challenges: "Plasma science is on the cusp of a new era. It is poised to make significant breakthroughs in the next decade that will transform the field" [6]. In view of the NRC report, the essential fields of importance for fundamental research and applications cover an extremely wide range of temperatures, densities, and magnetic fields including relativistic, classical, and highly correlated plasmas. A special challenge for low-temperature plasma research is the *large parameter space* and the *diversity of physical conditions* which are encountered:

<u>*Plasma size*</u>: from large and stable plasmas to *micro- and nanoplasmas*

<u>*Plasma pressure*</u>: from low pressure to *atmospheric and higher pressure*

<u>*Plasma chemistry*</u>: from rather simple rare-gas plasmas to more complex and *reactive molecular plasmas* (e.g., oxygen, hydrocarbons, fluorocarbons) and their interaction with condensed matter

<u>*Time scales*</u>: from electron and ion dynamics to *chemical reactions* and *collective behavior* of massive dust particles

Because of the multitude of species, and collective interactions at strongly different scales, the exploration of complex plasmas is still far from exhaustive. Due to the fundamental problems, the way of gaining insight in industry was often characterized by empirical methods.

The recent growth of predictive capability in the field of reactive and dusty plasmas is characterized by the progress from fundamental understanding to useful science-based models. It has been driven by advances in diagnostics which can probe the internal dynamic of the plasma and yield a quantitative understanding of plasma processes and plasma surface interaction. For example, plasma crystallization, pattern formation, or phase transitions are systematically studied in experiments by using high-speed video microscopy, stereoscopic imaging, and digital holography for tracing particles in dusty plasmas (see Chap. 6). These novel diagnostic techniques in complex plasma research allow for the investigation of fundamental (collective) mechanisms in *strongly coupled Coulomb systems* with an unprecedented experimental resolution. Processes can be followed at the kinetic level of individual particles and the system lends itself to one-to-one comparisons with particle simulations and is thus predestinated for the investigation of phase transitions, structure formation, and the intricate interplay between plasma and embedded dust particles. For Yukawa balls, the influence of system size or dimensionality on crystal structure, vibrational spectra, thermodynamic properties, or the competition of various forms of order are of high interest. More fundamentally, the forces on individual particles or the linear and nonlinear waves or instabilities need to be addressed. Another kind of modern diagnostics which uses particles as probes is discussed in Chap. 16. A particular experimental challenge is the diagnostics of chemical reactions in complex plasmas. Here, modern techniques had to be developed which range from laser absorption to X-ray diffraction. These methods are discussed in Chaps. 13 and 14.

On the other hand, theoretical and computational advances have led to models that can make more accurate predictions of plasma behavior. In particular, the combination of different modern simulation techniques can tackle the challenging multiscale and the correlation problem arising in complex plasmas and have progressed our knowledge about plasma chemical processes in plasma bulk and on surfaces over the last years. At the most fundamental scale, quantum-mechanical techniques are necessary to describe elementary electronic processes in the bulk of the plasma as well as on its boundary. Central methods of modern quantum plasma simulations such as the exact diagonalization techniques, Hartree–Fock and quantum path integral Monte Carlo are explained in Chap. 4. On the intermediate scale, where typical plasma phenomena such as screening and sheath formation occur, semiclassical kinetic equations and PIC–MCC simulations are the method of choice. A new combined particle-in-cell Monte Carlo collisions (PIC–MCC) approach is discussed in Chap. 9. Finally, global models must be employed to describe the physics of the discharge on the largest scale, the scale of the discharge vessel. Here, an introduction is given in Chap. 2. While the ultimate goal may be the integration of all the different levels into a self-consistent multiscale simulation, also simplified models can yield an adequate agreement with experiment and teach us the essential mechanisms which are responsible for the experimental observations. This is demonstrated in Chaps. 8 and 10. In the latter, the correlation problem is approached with first-principle classical Langevin dynamics and molecular dynamics simulations. With these techniques, not only ground-state and thermodynamic quantities, but also exact (classical) results for the strongly nonlinear many-particle dust dynamics can be obtained. The introduced methods may also be a valuable starting point for many related problems such as the ion dynamics in response to electromagnetic fields.

The systematic exploration of the fundamentals of complex plasmas is a timely task with valuable contributions to basic science and impact on many branches in plasma technology. The relevant questions are addressed using modern methods of plasma physics, atomic and molecular physics, and solid-state physics. Research on complex plasmas also integrates overlapping questions from neighboring disciplines into plasma physics, such as surface science, condensed matter physics, materials science, and chemistry. For example, plasmas at *atmospheric pressure* and *small dimensions* (microplasmas) are significantly controlled by the *discharge boundaries*. Thus, the plasma synthesis of thin films, particles with well-defined physical or chemical properties, or the nano- and microstructuring of surfaces require a deep understanding of the interaction of plasmas with surfaces and of the behavior of particles and chemically active species. But also, ordinary plasma processes like ionization and recombination, or reactive processes inside the plasma bulk, the plasma sheath and at walls require further substantial investigations due to the multitude of involved species, the wide spatial and temporal scales, and the complexity of the system.

Qualitatively novel insights into the fundamental issues are achieved by a complementary use of experimental, computational, and theoretical methods. Driven by the innovative progress over the recent years, complex plasmas have proven to be systems well-suited for investigating order phenomena and phase transitions on different spatial and temporal scales, for studying waves, transport and ion kinetics as well as for generating radicals, molecules, and nanoclusters and for forming thin functional surface layers.

1.4 Outline of this book

This book covers a broad spectrum of topics, from basic physics to chemistry. Technological aspects are devoted to the understanding of plasma chemistry, particle growth, and transport. This book addresses various fundamental aspects such as self-organization, kinetics of phase transitions and comprises – on the one hand – experimental and theoretical investigations of fundamental questions, such as plasma crystallization in strongly coupled dusty plasmas (see Chaps. 6 and 10), and – on the other hand – applied topics, for example, particle growth and catalytic reactions (which are covered in Chaps. 11 and 12). In this book, in-depth introductions to fundamental theoretical concepts and simulation methods in the field of quantum plasmas are given in Chaps. 3 and 4. Chapter 5 is devoted to the theory of quantum plasmas in solid-state physics, with focus on the dynamic and nonlocal features of the dielectric response. In particular, it covers one of the most spectacular recent developments: the properties of two-dimensional carbon layers – graphene. This chapter shows that knowledge and application of basic plasma principles is extremely important in many modern fields of physics.

Finally, modern technological applications of complex plasmas emerge in many fields, including environmental applications (Chap. 15), surface and particle coating (Chap. 16), and creation of novel surface structures (Chap. 12). But this is only a small part of the applications of complex plasmas which are expected to grow rapidly during the coming years.

References

1. U. Heinz, J. Phys. A: Math. Theor. **42**, 214003 (2009)
2. G.E. Morfill et al., Appl. Phys. B: Lasers Opt. **89**, 527 (2007). See also M. Bonitz, C. Henning, D. Block, Reports Prog. Phys. 73, 066501 (2010)
3. Pierre-Gilles de Gennes cited by G. Morfill et al., in: Europhys. News **39**, 30–32 (2008)
4. J. Meichsner (ed.), *Fundamentals of Complex Plasmas*, Report and Research Plan of the SFB-TR24, 2009
5. Google Scholar (http://www.scholar.google.com) search for papers containing the phrases: `''dusty plasma'' OR ''dusty plasmas'' OR ''complex plasma'' OR ''complex plasmas'' OR ''Coulomb crystal'' OR ''Coulomb crystals'' OR ''Yukawa crystal'' OR ''Yukawa crystals'' OR ''Coulomb ball'' OR ''Coulomb balls'' OR ''Yukawa ball'' OR ''Yukawa balls'' OR ''plasma crystal'' OR ''plasma crystals'' OR ''dust crystal'' OR ''dust crystals''`, May 2009
6. National Research Council, *Plasma Science: Advancing Knowledge in the National Interest* (National Academies, Washington, DC, 2007)

Part II
Classical and Quantum Plasmas

Chapter 2
Principles of Transport in Multicomponent Plasmas

Igor D. Kaganovich, Raoul N. Franklin, and Vladimir I. Demidov

Abstract The main principles of transport in multicomponent plasmas are described. Because the bulk plasma is charged positively to keep electrons together with positive ions, negative ions are confined by electrostatic fields inside the plasma and they flow from the plasma periphery toward the center. It is shown that the flow velocity of negative ions is a nonlinear function of the negative ion density. Increasing the negative ion density makes the electron density profile flatter and leads to a decrease of the electric field. Such a nonlinear dependence of the negative ion flow velocity on their density results in the formation of steep gradients of negative ion density, or negative ion fronts. Addition of negative ions makes the plasma afterglow a complex process as well. Typically, two stages of afterglow appear. In the first stage, the negative ions are trapped inside the plasma and only electrons and positive ions can reach the walls. However, at a later time, electrons quickly leave the plasma, and the second stage of afterglow begins, in which electrons are totally absent and an ion–ion plasma forms. During this stage, only the negative and positive ions contribute to the wall fluxes. The complex structure of the radio frequency sheath in strongly electronegative gases is also reviewed. Similar phenomena are observed in dusty plasmas. A possible relevance to ball lightning is discussed.

I.D. Kaganovich (✉)
Plasma Physics Laboratory, Princeton University, NJ 08543, USA
e-mail: ikaganov@pppl.gov

R.N. Franklin
Department of Physics and Astronomy, The Open University,
Milton Keynes MK7 6AA, UK
e-mail: r.n.franklin@open.ac.uk

V.I. Demidov
UES, Inc., Dayton-Xenia Rd., Beavercreek, Ohio 45432, USA
and
Department of Physics, West Virginia University, Morgantown, WV 26506
e-mail: demidov@wvu.edu

M. Bonitz et al. (eds.), *Introduction to Complex Plasmas*, Springer Series on Atomic, Optical, and Plasma Physics 59, DOI 10.1007/978-3-642-10592-0_2,

2.1 Introduction

Most elements of the periodic table produce negative ions. Table 2.1 shows the electron affinity of elements for the formation of negative ions. The larger the electron affinity, the easier it is to form a negative ion. From Table 2.1, it is evident that elements of the groups 6 and 7 have a large electron affinity of the order of a few electron volts and thus most readily produce negative ions. Halogen gases of the group 7, most importantly chlorine and fluorine, are frequently used in the semiconductor industry for dry-etching processes [3] and elimination of notching [4]. The creation of negative ions in oxygen plasmas is an important process in atmospheric electricity and in the formation of the ionospheric D-layer [5]. The production and acceleration of hydrogen or deuterium negative ions are utilized for generating powerful neutral beams to heat and drive current in magnetically confined plasmas for fusion energy research [6]. Negative ion beams made from halogens were also proposed as potential drivers for heavy ion fusion [7].

This review aims at introducing a general reader to the qualitative effects in plasma transport arising from the addition of negative ions. Given the fact that the chapter is limited in size, only the most important and robust effects are described. Further details can be found in cited literature. The organization of the chapter is as follows. Section 2.1.1 describes the production and destruction mechanisms of negative ions. Section 2.1.2 derives the drift–diffusion approximation describing plasma transport. Section 2.2 briefly reminds the reader about ambipolar diffusion to prepare for Sect. 2.3 on the more complicated temporal dynamics of negative ion flows in multicomponent plasmas. Section 2.4 describes the various stages of afterglow in multicomponent plasmas and wall fluxes of negative ions. General properties of steady-state profiles of plasmas with negative ions are reviewed in Sect. 2.5. Section 2.6 is devoted to a very interesting example of influence of negative ions on the rf-sheath structure in strongly electronegative gases. And last, but not least, Sect. 2.7 shows the relevance of the described phenomena associated with negative ions to dusty plasmas and ball lightning.

Table 2.1 Electron affinity, E_f, for formation of negative ions (from [1, 2])

Group period	I	II	III	IV	V	VI	VII	VIII
1	H 0.75							He 0.075[a]
2	Li 0.62	Be 0.004[a]	B 0.28	C 1.2	N 0.2[a]	O 1.45	F 3.4	Ne 0.095[a]
3	Na 0.54	Mg 0.004[a]	Al 0.44	Si 1.4	P 0.75	S 2.08	Cl 3.61	Ar 0.17[a]
4	K 0.5	Ca 0.02	Sc 0.19	Ge 1.2	As 0.81	Se 2.02	Br 3.36	Kr 0.65[a]
5	Rb 0.49	Ba 0.15	Sr 0.11	Sn 1.1	Sb 1.1	Te 1.97	I 3.06	Xe 1.25[a]
6	Cs 0.47	Ra	Ba 0.15	Pb 0.36	Bi 0.95	Po 1.9	At 2.8	Rn

[a]Metastable negative ions; their lifetime is typically a fraction of ms.

2.1.1 Production and Destruction Mechanisms of Negative Ions

Production of negative ions occurs in a number of processes [1–3]:

Electron attachment.

$$\mathrm{e} + \mathrm{A} \rightarrow \mathrm{A}^-$$

An electron attaches to an atom A and the released energy of the electron affinity, E_f, transfers to photons or to a third body.

Dissociative attachment.

$$\mathrm{e} + \mathrm{AB}(^*) \rightarrow \mathrm{A}^- + \mathrm{B}$$

An electron attaches to a molecule AB, often in an excited state and the energy of the electron affinity and excitation, $E_\mathrm{f} - E^*$, is absorbed in dissociation of the molecule.

Charge transfer.

$$\mathrm{A}^- + \mathrm{B} \rightarrow \mathrm{A} + \mathrm{B}^-$$

An electron is transferred from a negative ion A^- to another atom B.

Clustering reactions.

$$\mathrm{A}^- + \mathrm{B} + \mathrm{C} \rightarrow \mathrm{AB}^- + \mathrm{C}$$

Negative ions often play an important role in the formation of a cluster, as an initial seed of a clustering reaction (see, e.g., [8]).

Destruction of negative ions occurs due to following processes [1–3]:

Electron detachment.

$$\mathrm{e} + \mathrm{A}^- \rightarrow \mathrm{A} + 2\mathrm{e}; \quad \mathrm{e} + \mathrm{AB}^- \rightarrow \mathrm{A} + \mathrm{B} + 2\mathrm{e}$$

An electron impacts a negative ion, which leads to "ionization" of the negative ion, in which a loosely bound electron detaches from the negative ion.

Associative detachment.

$$\mathrm{A}^- + \mathrm{B} \rightarrow \mathrm{AB} + \mathrm{e}$$

A collision of a negative ion with an atom yields the formation of a molecule; the difference between the dissociation energy and the electron affinity is transferred to a free electron.

Charge transfer.

$$\mathrm{A}^- + \mathrm{B} \rightarrow \mathrm{A} + \mathrm{B}^-$$

An electron is transferred from a negative ion A^- to another atom B.

Positive ion–negative ion recombination.

$$\mathrm{A}^- + \mathrm{B}^+ \rightarrow \mathrm{A} + \mathrm{B}^*$$

A collision of a negative ion with a positive ion leads to recombination of ions.

At high pressures, three-body recombination between electron and ion in the presence of a third atom can become important.

A detailed description of the formation of negative ions must therefore include a set of several reactions between neutrals, positive and negative ions, and molecules. In many cases, the formation of metastable states of atoms or even vibrationally excited molecules makes it important to take these processes into account for a correct description of negative ion creation. For commonly used gases such as oxygen, chlorine, silane, and sulfur hexafluoride (SF_6), such sets of participating reactions have been developed and tested against experimental measurements [3,9]. For many other gas mixtures, it remains a difficult task to determine all of the possible routes for negative ion formation and destruction.

2.1.2 *The Drift–Diffusion Approximation for the Description of Plasma Transport*

Most plasma devices utilizing negative ions operate in discharge chambers and at pressures large enough to be able to produce a sufficient amount of negative ions. Therefore, the ion mean free path is typically small compared with the discharge chamber dimensions and ion transport can be described making use of the drift–diffusion equations [3, 10–12]. For discussion of effects in collisionless plasma, see, for example, [13–16]. The momentum balance for the positive ions with density, n_p, and temperature, T_i, reads

$$-\nabla n_p T_i + e n_p \mathbf{E} - M_{pa} n_p \nu_{ia} \mathbf{u}_p = 0. \tag{2.1}$$

Here, we have neglected the ion inertia term, which is small compared to the ion friction term, $M_{pa} n_p \nu_{pa} \mathbf{u}_p$, based on the assumption of small mean free path. Here, M_{pa} is the reduced mass of the positive ion and gas atom [10]. If the mean ion–neutral atom collision frequency, ν_{pa}, can be assumed independent of the mean ion flow velocity, $\mathbf{u}_p$, the ion flux can be expressed as

$$\Gamma_p = n_p \mathbf{u}_p = \frac{-\nabla n_p T_i + e n_p \mathbf{E}}{M_{pa} n_p \nu_{pa}} = \frac{-T_i \nabla n_p + e n_p \mathbf{E}}{M_{pa} n_p \nu_{pa}}. \tag{2.2}$$

The assumption of the collision frequency, ν_{pa}, being independent of the mean ion flow velocity, $\mathbf{u}_p$, requires that the ion mean flow velocity be small compared with the ion thermal velocity, which may fail in the limit of low pressures [3]. Nevertheless, accounting for the variable collision frequency, $\nu_{pa}(u_p)$ does not change results qualitatively, but makes analytic results less transparent. Corresponding results can be easily generalized for the case of the variable collision frequency (see, e.g., [3, 17]). Therefore, in the following, we assume that ν_{pa} is independent of the mean ion flow velocity and utilize drift–diffusion equations. Similarly, in (2.2) we assumed that the ion temperature is constant. This is not accurate due to gas heating by the discharge current (see, e.g., [11, 18]). Likewise, it is not important to take this effect into account to examine qualitative effects described in this chapter.

Equation (2.2) is traditionally rewritten in the form of drift–diffusion equation:

$$\Gamma_{\mathrm{p}} = -D_{\mathrm{p}} \nabla n_{\mathrm{p}} + e n_{\mathrm{p}} \mu_{\mathrm{p}} E, \tag{2.3}$$

where

$$D_{\mathrm{p}} = \frac{T_{\mathrm{i}}}{M_{\mathrm{pa}} \nu_{\mathrm{pa}}} = \frac{v_{T_{\mathrm{i}}}^2}{\nu_{\mathrm{pa}}} = \frac{v_{T_{\mathrm{i}}}^2}{\nu_{\mathrm{pa}}^2} \nu_{\mathrm{pa}} = \lambda_{\mathrm{pa}}^2 \nu_{\mathrm{pa}}$$

is the ion diffusion coefficient and $\mu_{\mathrm{p}} = e/M_{\mathrm{pa}}\nu_{\mathrm{pa}}$ is the positive ion mobility. Here, $v_{T_{\mathrm{i}}} = \sqrt{T_{\mathrm{i}}/M_{\mathrm{pa}}}$ is the ion thermal velocity.

Similarly, the negative ion flux can be expressed as a function of negative ion density and its gradients, n_{n}, as

$$\Gamma_{\mathrm{n}} = -D_{\mathrm{n}} \nabla n_{\mathrm{n}} - e n_{\mathrm{n}} \mu_{\mathrm{n}} E, \tag{2.4}$$

where $D_{\mathrm{n}} = T_{\mathrm{i}}/M_{\mathrm{n}}\nu_{\mathrm{na}}$ is the ion diffusion coefficient and $\mu_{\mathrm{n}} = e/M_{\mathrm{n}}\nu_{\mathrm{na}}$ is the negative ion mobility. Here, M_{na} is the reduced mass of the negative ion gas and atom and ν_{pa} is the mean negative ion–neutral atom collision frequency.

Finally, the electron flux is determined from

$$\Gamma_{\mathrm{e}} = -D_{\mathrm{e}} \nabla n_{\mathrm{e}} - e n_{\mathrm{e}} \mu_{\mathrm{e}} E, \tag{2.5}$$

where $D_{\mathrm{e}} = T_{\mathrm{e}}/m\nu_{\mathrm{ea}}$ is the electron diffusion coefficient and $\mu_{\mathrm{e}} = e/m\nu_{\mathrm{ea}}$ is the electron mobility, m is the electron mass and T_{e} is the electron temperature. Note that in accord with the Einstein relation for each species k, the diffusion coefficient is equal to the product of the mobility and the species temperature $D_k = \mu_k T_k$ [10].

2.2 Ambipolar Diffusion

Let us start by briefly describing the transport of a two-component plasma without negative ions. The electron density is determined from the continuity equation:

$$\frac{\partial n_{\mathrm{e}}}{\partial t} = -\frac{\partial}{\partial x}\left(-D_{\mathrm{e}} \frac{\partial n_{\mathrm{e}}}{\partial x} - e n_{\mathrm{e}} \mu_{\mathrm{e}} E\right) + S_{\mathrm{e}}, \tag{2.6}$$

where S_{e} is the total volumetric electron production rate, which is sum of the source and loss terms for electron production. Typically, the term associated with electron diffusion is much larger than any source terms and electrons diffuse away from the positive ions until the plasma polarizes and the electric field stops the electron flux, as shown in Fig. 2.1 [3, 10–12]

$$\Gamma_{\mathrm{e}} = -D_{\mathrm{e}} \frac{\partial n_{\mathrm{e}}}{\partial x} - e n_{\mathrm{e}} \mu_{\mathrm{e}} E \approx 0. \tag{2.7}$$

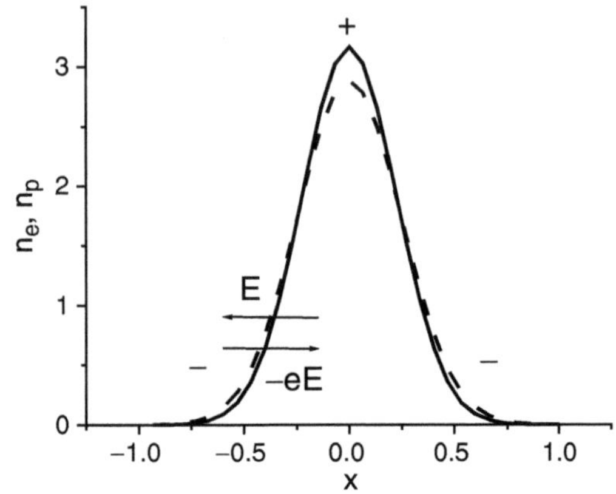

Fig. 2.1 Plasma polarization during ambipolar diffusion

Phenomena associated with current-carrying plasma are described in [12]. A zero electron flux corresponds to the electrons obeying a Boltzmann relation, that is, the electric field is determined by

$$eE = \frac{-D_e \nabla n_e}{\mu_e n_e} \approx -T_e \frac{\nabla n_e}{n_e}. \tag{2.8}$$

Substituting (2.8) into the continuity equation for the positive ion density

$$\frac{\partial n_p}{\partial t} = -\frac{\partial}{\partial x}\left(-D_p \frac{\partial n_p}{\partial x} + e n_p \mu_p E\right) + S_p \tag{2.9}$$

yields the ambipolar diffusion equation

$$\frac{\partial n_p}{\partial t} = \frac{\partial}{\partial x}\left(\mu_p (T_e + T_p) \frac{\partial n_p}{\partial x}\right) + S_p. \tag{2.10}$$

Note that we have used the quasineutrality condition, $n_e = n_p$, to derive (2.10). The ambipolar diffusion equation (2.10) shows that the ambipolar diffusion rate is increased by a factor $(T_e/T_i + 1)$ compared with that of the ion diffusion due to plasma polarization and ion acceleration in the self-consistent electric field. The important property of the ambipolar diffusion equation (2.10) is that it is linear, thus describing relatively simple linear dynamics of the initially complex, nonlinear drift–diffusion equations (2.6) and (2.9). Any change of the indicated assumptions will result in restoring nonlinear effects contained initially in the nonlinear drift–diffusion description of (2.6) and (2.9) [12, 19]. For example, addition of negative ions results in a significant change of plasma transport, in which the simple concept of ambipolar diffusion is no longer valid [12, 20, 21].

2.3 Temporal Dynamics of Negative Ion Flows in Multicomponent Plasmas

As shown in the previous section, electrons achieve a Boltzmann equilibrium distribution as long as electron diffusion is much larger than the electron production terms, $|\partial \Gamma_e/\partial x| \gg |S_e|$. Because electrons are trapped by the self-consistent electric field, any negatively charged particle is pulled by the electric field from the discharge periphery into the plasma center. Substituting (2.8) into the continuity equation for the negative ions yields

$$\frac{\partial n_n}{\partial t} = -\frac{\partial}{\partial x}\left(-D_n \frac{\partial n_n}{\partial x} + e\mu_n T_e \frac{n_n}{n_e}\frac{\partial n_e}{\partial x}\right) + S_n. \tag{2.11}$$

In typical discharge conditions, the ion diffusion coefficient, $D_n = \mu_n T_i$, is small compared with $\mu_n T_e$, because the electron temperature is large compared with the ion temperature:

$$T_e \gg T_i \tag{2.12}$$

(ions are cooled in collisions with neutral atoms much faster than electrons, which in contrast to ions only lose a small portion of their energy, proportional to m/M). Therefore, a convective term, $\mu_n e E n_n$, dominates the dynamics of negative ions, unless the electron profile becomes very flat. Figure 2.2 shows the evolution of plasma profiles at the beginning of the active phase of the pulsed discharge in oxygen [22, 23]. The initial condition of the plasma at the beginning of the active glow corresponds to the final condition at the end of the afterglow of the previous pulse. For the chosen set of parameters, this corresponds to an ion–ion plasma with smooth

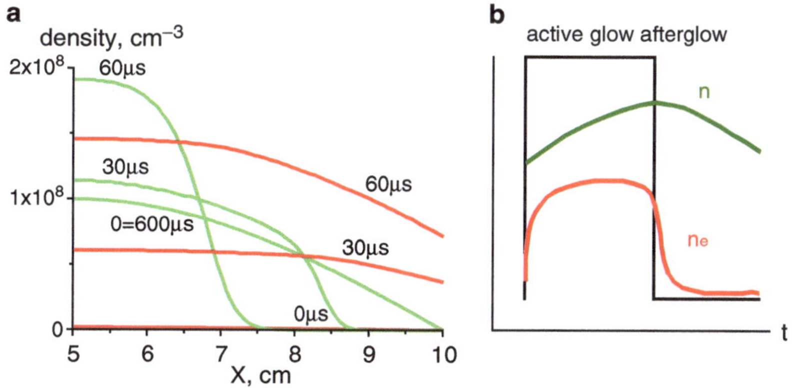

Fig. 2.2 Evolution of electron and negative ion densities in early active glow of a pulsed discharge. (**a**) Negative ion (*green lines*) and electron density (*red lines*) in the early active glow of Ar and 3% O_2 plasma. Interelectrode gap: 10 cm; total pressure: 5 mTorr; averaged power density: 1.0 mW/cm^3; pulse duration: 600 ms; duty ratio: 0.5; and *numbers* denote time in the active glow in µs. (**b**) Time evolution of the electron and negative ion density in the center of the discharge

cosine-like profiles of charged species densities (lines for 0 ms). Once the power is switched on (at time $t = 0$), the electron temperature rises to several eV. Because the electric field increases with the electron temperature, negative ions are squeezed toward the center of the plasma. However, as is evident from Fig. 2.2, the electron density gradient is much larger at the periphery of the discharge than in the center, indicating that negative ions move much faster at the periphery than in the center. Because negative ions slow down more rapidly in the densest part of the discharge, their density rises and can eventually lead to formation of large negative ion density gradients or negative ion density fronts. At the same time, because the negative ions are swept from the periphery region quickly, their density is low near discharge walls, where an electropositive plasma of mostly electrons and positive ions forms. *Thus, plasmas with negative ions tend to self-separate into two distinct regions: a region of electropositive plasma where the electric field is high and the negative ion density is small compared to the electron density, and an electronegative region where the negative ion density is large compared to the electron density and the electric field is weak* [24–31]. Such separation is completely contrary to the prediction of the second law of thermodynamics where mixing of all species is expected. This occurs due to the large difference between the electron and ion temperatures in low-temperature plasmas.

The formal description of the separation effect is not straightforward. To describe the flattening of the electron density gradient in the self-consistent evolution of plasma profiles in the presence of negative ions, we derive an effective equation for the electrons by subtracting the equation for the negative ions (2.11) from the equation for the positive ions (2.10), neglecting source and ion diffusion terms, where

$$\frac{\partial n_e}{\partial t} = \frac{\partial}{\partial x}\left(eT_e \frac{\mu_n n_n + \mu_p n_p}{n_e} \frac{\partial n_e}{\partial x}\right). \tag{2.13}$$

From (2.13), it is evident that the effective coefficient of electron diffusion,

$$D_{\text{eff}}\left(\frac{n}{n_e}\right) = eT_e \frac{\mu_n n_n + \mu_p n_p}{n_e} = \mu_p eT_e + eT_e(\mu_n + \mu_p)\frac{n_n}{n_e}, \tag{2.14}$$

is a strongly nonlinear function of the negative ion density and electronegativity, n_n/n_e. The larger the ratio of the negative ion density to the electron density, the larger is the effective diffusion coefficient. Given that changes in the electron flux, $\Gamma_e = -D_{\text{eff}}(\partial n_e/\partial x)$, are of the order of unity in the discharge, whereas the electronegativity, n_n/n_e, can change by large factors, the increase of the electronegativity eventually leads to a flattening of the electron density profile and a reduction of the electric field. Substituting the electron flux, $\Gamma_e = -D_{\text{eff}}(\partial n_e/\partial x)$, instead of the electron density into the equation for negative ions, (2.11) yields

$$\frac{\partial n_n}{\partial t} = -\frac{\partial}{\partial x}\Gamma_n \approx -u_{\text{eff}}\left(\frac{n_n}{n_e}\right)\frac{\partial}{\partial x}n_n, \tag{2.15}$$

where we have neglected the variation in the electron flux and density compared with those of the negative ions. Here, $\Gamma_n = -\Gamma_e[\mu_n n_n/(\mu_n n_n + \mu_p n_p)]$ and $u_{eff} = \partial\Gamma_n/\partial n_n$ are the negative ion signal propagation velocity [32]:

$$u_{eff} = \frac{\mu_n \mu_p T_e}{\mu_n n_n + \mu_p n_p}\frac{\partial n_e}{\partial x} = \frac{\mu_p n_e}{\mu_n n_n + \mu_p n_p} u_n. \tag{2.16}$$

From (2.16), it is evident that the velocity of propagation of the negative ion density perturbation is different from the velocity of negative ions, $u_n = \mu_n eE$. In the case when positive and negative ion mobilities coincide, the perturbation velocity is a factor of $2n_n/n_e + 1$, smaller than the negative ion velocity:

$$u_{eff} = \frac{1}{2n_n/n_e + 1} u_n. \tag{2.17}$$

If the negative ion density is small compared with the electron density, both velocities coincide, $u_{eff} = u_n$. However, if the negative ion density is large compared with the electron density, the signal propagation velocity is much smaller than the velocity of the negative ions, because the electron density gradient is strongly affected by the negative ion density changes. This effect was verified in numerical simulations [33] and is shown in Fig. 2.3, where the propagation of a small perturbation of negative ion density for three different cases of electronegativity and the same profile of initial electron density are shown. As is evident from Fig. 2.3, the propagation velocity for the case (b) $n_n/n_e \cong 1$ is about three times smaller than in case $n_n/n_e = 0$, in accord with (2.17).

The nonlinear evolution of a large negative ion density perturbation is shown in Fig. 2.4. The general theory of one-dimensional flows [34] predicts that each point of initial profile $n_n(x)$ moves with its own velocity $u_{eff}(n_n/n_e)$. According to

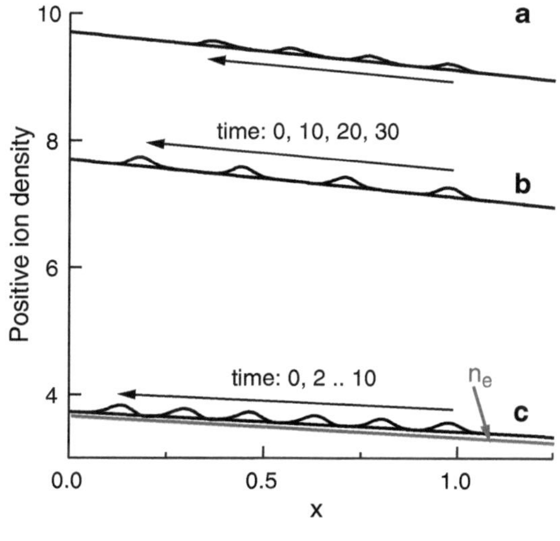

Fig. 2.3 Propagation of small perturbation of negative ion density for three different values of electronegativity n_n/n_e: (**a**) $n_n/n_e \cong 1.5$, (**b**) $n_n/n_e \cong 1$, and (**c**) $n_n/n_e = 0$, with the same profile of the initial electron density, $n_e = n_0(3.7 - 0.3x)$. All variables are dimensionless, normalized to reference values; density n/n_0, coordinate x/L, and time $t\mu_n T_e/L^2$. Ion diffusion is neglected and ion mobilities were assumed to be the same $\mu_n = \mu_p$

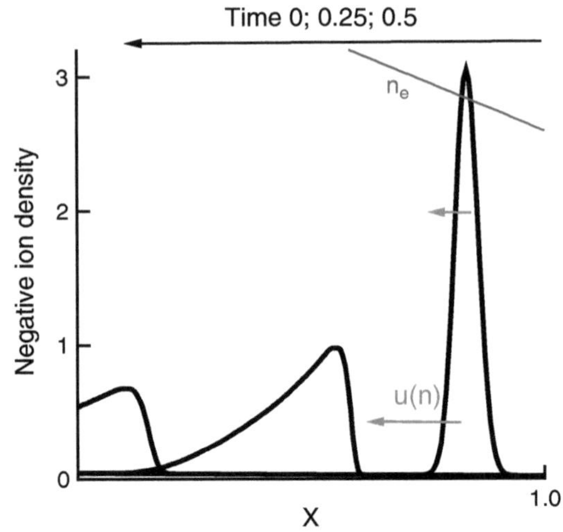

Fig. 2.4 Formation of negative ion density fronts during propagation of a large perturbation of negative ion density for the conditions of Fig. 2.3, but with $T_i/T_e = 10^{-3}$ and $n_e = n_0(6.2 - 3.6x)$. The negative ion density profiles are plotted three times every 0.25 units of dimensionless time $tL^2/\mu_n T_e$ [33]

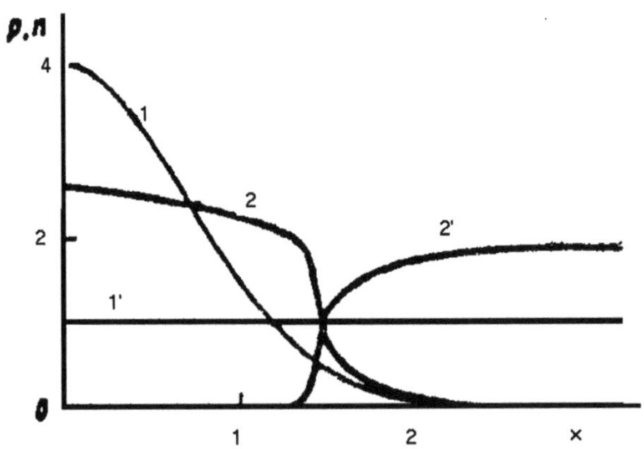

Fig. 2.5 Temporal evolution of the two positive species [32]. The ratio of ion mobilities is 0.1. Initially, the density profile of the plasma species with a large mobility was uniform (denoted by $1'$) and the localized addition of a less mobile species was added (denoted by 1). After evolution in the self-consistent electric field, the mobile species (denoted by $2'$) flows away from the less mobile species (denoted by 2) and they self-separate

the theoretical predictions, $u_{\rm eff}(n_n/n_e)$ is inversely proportional to electronegativity (see (2.17)) and the regions of small negative ion density move faster than regions of large negative ion density. As a result, the front of the profile spreads out, and the back of the negative ion density profile steepens, leading to profile breaking and the formation of an ion density discontinuity – *ion density fronts*. Note that negative ion density fronts are formed at the back of the profile in contrast to gas dynamic shocks which are formed at the front of the profile. This is again due to the fact that $u_{\rm eff}(n_n/n_e)$ is a decreasing function of electronegativity, n_n/n_e.

The effect of self-separation can also occur in multicomponent plasmas with two species of positive ions, as shown in Fig. 2.5. For this effect to occur, the two species should have very different ion mobilities and the electron temperature should be much greater than the ion temperature. In this case ion diffusion does not spread out the two species significantly. If a large ion plasma perturbation with low mobility is added on top of a uniform two-component plasma of mobile ions, the resulting self-consistent electric field accelerates both ions away from the perturbation – but the most mobile species flows outwards faster than the less mobile species and they self-separate as shown in Fig. 2.5 [32].

2.4 Afterglow in Multicomponent Plasmas and Consequent Wall Fluxes of Negative Ions

In the previous section, it was shown that negative ions are trapped in a plasma by the self-consistent electric field. An important practical question is whether there is a way to extract negative ions from the plasma. One way to do so is to apply an external magnetic field to reduce the electron mobility near the extracting electrodes, as is done in negative ion beam sources [7]. Another approach is to extract negative ions during the afterglow, when electrons eventually leave the plasma. We pose and answer the following question: when do negative ions start reaching the walls and how big are the wall fluxes in the afterglow. The analysis of this problem was performed in [22, 23, 35–37]. When the discharge current is switched off, the electron temperature rapidly drops to room temperature and, consequently, the electron and ion temperatures equilibrate, $T_e = T_i$. Subsequently, ion diffusion becomes the dominant process. At the end of the afterglow phase of a pulsed discharge, the negative ions that accumulated in the discharge center diffuse toward the wall (see evolution of the negative ion profile at 60–600 μs in Fig. 2.2). Interestingly, if volumetric losses due to ion–ion recombination are small compared to wall losses, the solution becomes self-similar, in which all profiles are geometrically similar [38, 39]:

$$\frac{n_n(x,t)}{n_n(0,t)} = \frac{n_p(x,t)}{n_p(0,t)} = \frac{n_e(x,t)}{n_e(0,t)} = f\left(\frac{x}{\Lambda}\right), \tag{2.18}$$

where $f(x/\Lambda) = \cos(x/\Lambda)$ for slab geometry and $f(r/\Lambda) = J_0(r/\Lambda)$ for cylindrical geometry, J_0 is the Bessel function, and $x = 0$ corresponds to the center of the discharge.

Substituting the self-similar solution for plasma profiles (2.18) into the equation for the positive ion density, (2.9) yields

$$\frac{\partial n_p}{\partial t} = -\frac{\partial}{\partial x}\left(-2D_p\frac{\partial n_n}{\partial x}\right), \tag{2.19}$$

that is, the positive ions diffuse to the wall with the ambipolar diffusion coefficient, which is twice as large as that of free ion diffusion for the case of equal ion and electron temperatures. The solution of (2.19) is

$$n_p(x,t) = n_p(0,t) f(x/\Lambda) e^{-2t/\tau_d}, \tag{2.20}$$

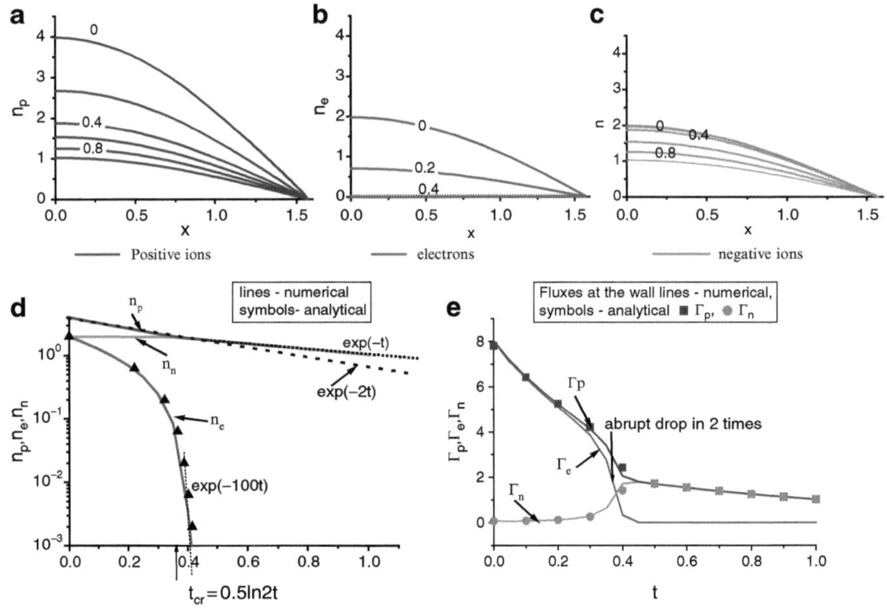

Fig. 2.6 Evolution of plasma profiles in the afterglow after the electron and ion temperatures equilibrate and establish a self-similar solution. Positive ion density (**a**), electron density (**b**), and negative ion density (**c**) profiles are plotted every 0.2 units of dimensionless time $\Lambda^2/\mu_n T_e$ [22,23]. Ion mobilities were taken to be the same. The initial profiles at $t = 0$ were taken according in accordance with to the fundamental diffusion mode of (2.18), and plasma chemical processes were neglected in the afterglow. Plasma parameters versus dimensionless time: *solid lines* are numerical results and *symbols* are analytic estimates

where $\tau_d = \Lambda^2/\mu_p T$, as shown in Fig. 2.6d. Note that such diffusion is independent of the electron density and occurs even if the electron density is much smaller than the positive ion density. Substituting the self-similar solution for plasma profiles (2.18) into the equation for the negative ion density (2.11), one obtains the result that the negative ion flux exactly vanishes, $\Gamma_n = 0$, and the negative ion density profile remains unchanged as is evident in Fig. 2.6c while electrons are present. Because the negative ions are trapped inside the plasma, the quasineutrality condition yields

$$n_e(x,t) = [n_p(0,t)e^{-2t/\tau_d} - n_n(0)]f(x/\Lambda). \tag{2.21}$$

Equation (2.21) indicates that the electron density vanishes completely at certain finite time given by

$$t_{ii} = \frac{1}{2}\tau_d \ln\left(\frac{n_p(0)}{n_n(0)}\right). \tag{2.22}$$

Electrons and positive ions are lost together to the walls; and the positive and negative ion fluxes to the walls are equal, as is evident in Fig. 2.6e. Since the electron density becomes small compared with the positive ion density, the rate of electron loss strongly increases as the electron density tends to zero, as shown in Fig. 2.6d. Eventually the electron density becomes so small that it cannot support the electric field sufficiently to stop free electron diffusion to the walls; and the remaining electrons leave the plasma, thus a purely ion–ion plasma develops. If the negative ion and positive ion mobilities are the same, both the positive and negative ion free diffusion fluxes are automatically the same, and the ambipolar electric field is absent for a purely ion–ion plasma. Therefore, further plasma decay occurs with the free ion diffusion rate:

$$n_{\mathrm{p}}(x,t) = n_{\mathrm{n}}(x,t) = n_{\mathrm{n}}(0,t) f(x/\Lambda)\mathrm{e}^{-t/\tau_{\mathrm{d}}}. \tag{2.23}$$

Experimental measurements of the wall fluxes [40] shown in Fig. 2.6e agree well with the analytic description outlined earlier.

In the description of temporal plasma decay in an afterglow we neglected volumetric processes in the plasma. Generalization of these results taking into account attachment and detachment processes is straightforward and is described in [22, 23, 35]. The most important qualitative effect occurs due to detachment and production of electrons from negative ions. If the detachment frequency, γ_{d}, is faster than the ambipolar wall loss frequency (see (2.20)):

$$\gamma_{\mathrm{d}} > 2/\tau_{\mathrm{d}}. \tag{2.24}$$

The attached electrons are produced quickly enough by this mechanism so that electrons are always present in the afterglow. Therefore, the negative ions are always trapped in this case. Because the detachment rate is proportional to the gas pressure, but the diffusion coefficient is inversely proportional to the gas pressure, increasing pressure can lead to a sudden transition from the regime in which detachment is small to the regime in which the detachment frequency dominates the wall losses and the criterion condition in (2.24) is satisfied. The trapping of negative ions throughout afterglow is demonstrated in Fig. 2.7 for pressures higher than 3.5 mTorr.

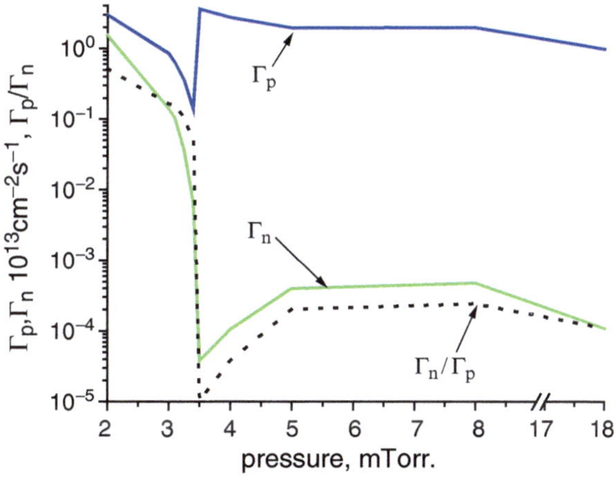

Fig. 2.7 Maximum negative ion flux in the oxygen afterglow as a function of pressure [35]. The discharge interelectrode gap is 10 cm and average power density is 10 mW/cm^3

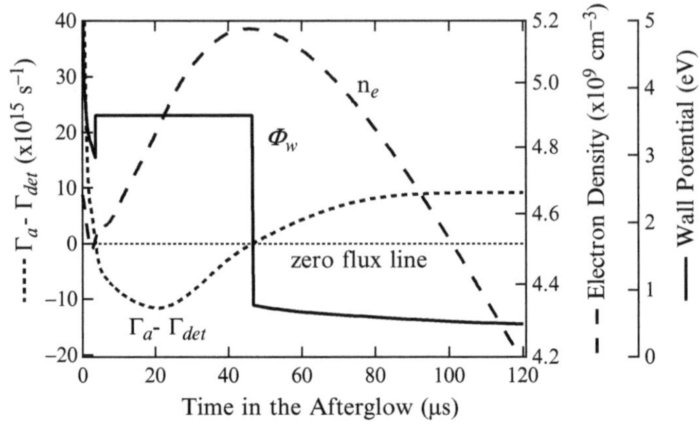

Fig. 2.8 The electron density, the wall potential, and the electron temperature evolution in the afterglow of discharge in oxygen [42]. Calculated difference between the ion flux Γ_a and the electron flux available from negative ion detachment Γ_{det} is shown on the *left*; the electron density and the calculated wall potential are shown on the *right*

Note also that associative detachment of electrons can lead to creation of free electrons with energies much larger than the representative average energies of electrons and ions in afterglow (which can be typically close to the room temperatures). For instance, in oxygen reaction $O^- + O \rightarrow O_2 + e$, the energy of generated electrons can be up to 3.6 eV. Due to this process, walls of the plasma volume can be charged negatively up to −3.6 V in the above example. Without the presence of the fast electrons, one would normally expect the electron density, the wall potential and the electron temperature to monotonically decrease. However, during the afterglow, a sharp change in sheath voltage drop is possible due to generation of the fast electrons, as described in [41]. For an example, Fig. 2.8 shows the electron density, the wall potential and the electron temperature evolve non-monotonically in the afterglow of discharge in oxygen. It is evident from Fig. 2.8 that the electron density initially decreases, then increases for a time between 3 and 45 μs, and then decreases again. This increase corresponds to the time interval in which electrons are trapped by the high wall potential, which is also shown in the figure. The wall potential is determined by the difference between the ion flux to the wall and the available flux of the fast electrons generated due to associative detachment in oxygen [42]. We can see that during the time between 3 and 45 μs, the difference in those fluxes is negative, that is, the available flux of fast electrons is large compared with the flux of ions. As a result, all thermal electrons and a portion of the fast group are trapped during this part of the afterglow to satisfy ambipolarity condition.

2.5 Steady-State Profiles of Plasmas with Negative Ions

Examples of measured and simulated density positional profiles in plasmas with the addition of negative ions are shown in Figs. 2.9–2.11.

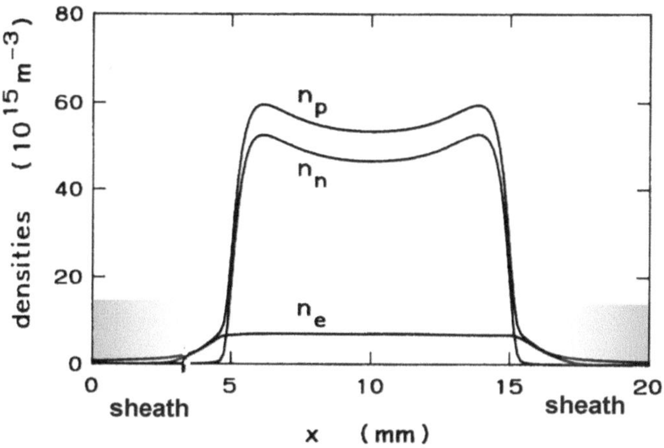

Fig. 2.9 Plasma profiles in rf-discharge, 0.5 Torr CF4, 13.6 MHz [43]. Shaded region marks sheath

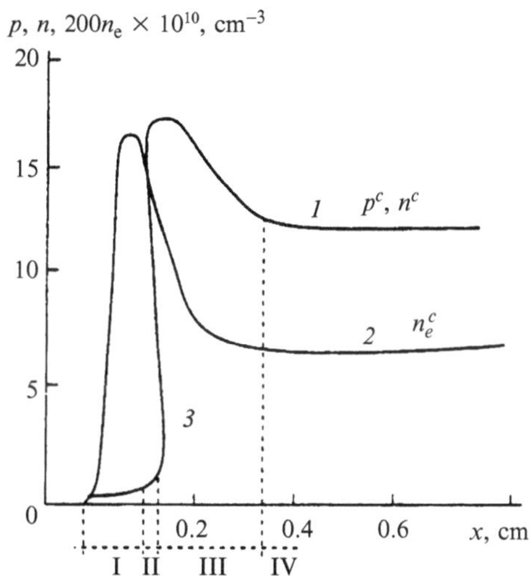

Fig. 2.10 Profiles of the charged particle densities obtained in simulations of rf-discharge in SF_6 [45]. The trace 1 corresponds to the positive and negative ion densities, and trace 2 to the time-averaged electron density. The discharge parameters are the gas pressure 0.13 Torr, rf-frequency 13.6 MHz, and current density 2 mA/cm^2. Region I corresponds to the sheath, region III to the ion density perturbation associated with the sheath, region IV is a region of nearly uniform ion density profile in the plasma bulk, and region II is the narrow transition region between regions I and III where negative ion density jumps

The common feature of all profiles is that the plasma tends to stratify into a bulk plasma region where the negative ion density is larger than the electron density and a peripheral region near the sheath where the electron density is large compared with the negative ion density. (A notable exception is the case of a very strongly

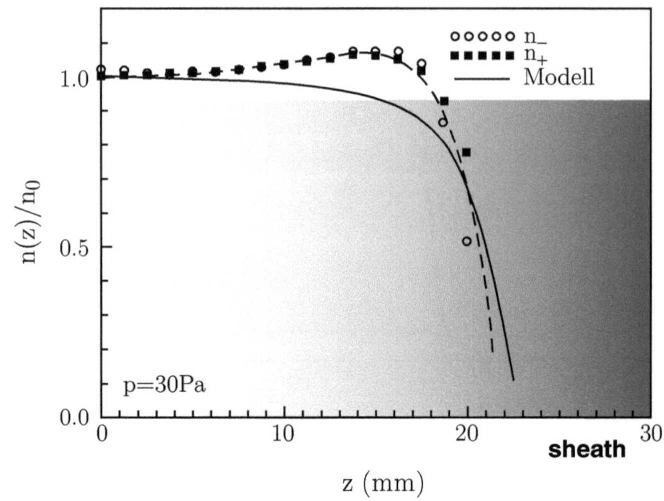

Fig. 2.11 Experimentally measured and simulated negative ion density profiles in an rf-discharge in oxygen, pressure 0.21 mTorr [44]. The central ion density is $3.5 \times 10^9\,\mathrm{cm}^{-3}$ and the electron density profile is nearly uniform with density $1.5 \times 10^8\,\mathrm{cm}^{-3}$. Shaded region marks the sheath

electronegative gas, SF_6, shown in Fig. 2.10.) In the sheath region, the electric field accelerates negative ions toward the plasma center. The negative ion flux increases toward the plasma center due to attachment. As shown in Sect. 2.3, the negative ion flow velocity decreases as soon as the negative ion density becomes large compared with the electron density. At this point the negative ion density increases sharply forming a distinct boundary between the electropositive peripheral region, in which $n_\mathrm{n}/n_\mathrm{e} < 1$ and the electronegative core with $n_\mathrm{n}/n_\mathrm{e} \gg 1$, as shown, for example, in Fig. 2.9.

An analytic calculation of the exact width of the electropositive peripheral region requires an accurate calculation of the negative ion flux and ion velocity, which can be affected by the number of processes included in simulations of realistic plasma profiles: for example, accounting for finite ion mean free path, effects of ion diffusion and width of the sheath region. These will not be described here; see [3, 17, 20, 21, 25–30, 33, 44, 46] for details. However the central region should satisfy a simple relation between ion and electron densities. In the bulk plasma ion diffusion can be neglected and the positive and negative ion fluxes become mostly convective $\Gamma_\mathrm{p} \approx \mu_\mathrm{p} e n_\mathrm{p} E$ and $\Gamma_\mathrm{n} \approx -e\mu_\mathrm{n} n_\mathrm{n} E$. As the negative ion density becomes large, $n_\mathrm{n}/n_\mathrm{e} \gg 1$, the ion fluxes should be nearly equal, $\mu_\mathrm{n}\Gamma_\mathrm{p} + \mu_\mathrm{p}\Gamma_\mathrm{n} \approx 0$ as shown in Fig. 2.12. Because of this $\mu_\mathrm{n}\partial\Gamma_\mathrm{p}/\partial x + \mu_\mathrm{p}\partial\Gamma_\mathrm{n}/\partial x \approx 0$, and the sources of positive and negative ions have to satisfy the relation

$$\mu_\mathrm{n} S_\mathrm{p} + \mu_\mathrm{p} S_\mathrm{n} \approx 0. \tag{2.25}$$

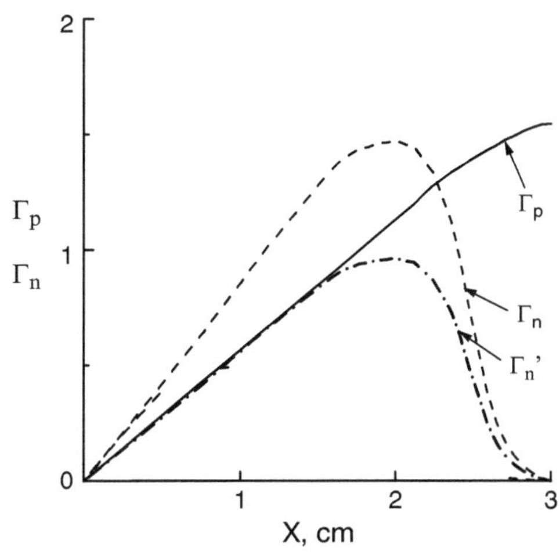

Fig. 2.12 Calculated ion fluxes in units of 10^{13} cm^2/s for the same parameters as in Fig. 2.11. *Dashed lines* represent negative ion fluxes. The *line* marked as Γ_n' corresponds to the negative ion flux divided by $D_n/D_p = 1.4$

For a typical electronegative gas, positive ion production is due to ionization and loss is due to ion–ion recombination, $S_p = Z_{iz}n_e - \beta_{ii}n_n n_p$; negative ion production is due to electron attachment and loss due to ion–ion recombination and detachment, $S_n = \alpha_{at}n_e - \gamma_d n_n - \beta_{ii}n_n n_p$. Substituting these equations for the positive and negative ion production rates into (2.25) yields [12, 20, 21, 24–30, 44, 47]

$$\left(\frac{\mu_n}{\mu_p}Z_{iz} + \alpha_{at}\right)n_e - \gamma_d n_n \approx \left(1 + \frac{\mu_n}{\mu_p}\right)\beta_{ii}n_n n_p. \qquad (2.26)$$

This relation was verified for the positive and negative ion and electron densities for the conditions of Figs. 2.8–2.10.

2.6 The Sheath in Strongly Electronegative Gases

The properties of a sheath in strongly electronegative plasmas can be even more complicated than the properties of quasineutral plasmas. Walls and rf-electrodes are charged negatively by the plasma to provide an ambipolarity condition for the electron and ion fluxes to the walls. Therefore, negative ions are driven toward the plasma center inside the sheath region. In an rf-discharge the electric field at the electrode is modulated in time and the plasma sheath boundary moves toward and away from the electrode in accord with the electric field variations. When the electric field decreases, electrons move closer to the electrode and during this phase they can produce negative ions due to attachment inside the rf-sheath. Ions are much slower compared with the electrons and cannot respond to the instantaneous electric field, and thus respond to a time-averaged electric field, which always pushes them toward the plasma center. In this averaged electric field negative ions gain velocity and drift toward the center. Their flux increases due to electron attachment and their velocity decreases toward the plasma sheath boundary. If the product of the frequency of attachment, ν_{at} times the time for a negative ion to traverse the sheath region, $\tau_i \sim L_{sh}/V_i$, is small compared with unity, $\nu_{at}\tau_i < 1$, the negative ion density in the sheath is small compared with the electron density and the sheath structure is not affected by the negative ions. In the opposite case, $\nu_{at}\tau_i \gg 1$, negative ion production is so strong that the presence of negative ions strongly modifies the sheath structure. As the attachment frequency increases, the negative ion density also increases inside the sheath region and so does the positive ion density. Analysis of the transport equations shows that in this case the electron density inside the sheath can greatly exceed the electron density in the bulk plasma, as is evident in Fig. 2.10. As a result, the sheath width is greatly reduced and, as shown in [48], the sheath region self-consistently adjusts itself to satisfy the condition $\nu_{at}\tau_i \approx 1$. The positive ion density jumps at the end of the sheath due to fast ion deceleration. Under certain conditions, such a rapid increase of ion density can also produce sharp peaks in ion density as shown in Fig. 2.13. For a detailed description of these effects, see [18,48].

The presence of negative ions in rf-discharges can result in a number of other complicated nonlinear phenomena, like the generation of low-frequency oscillations corresponding to the sheath restructuring from cathode-like to anode-like structures [49] and the generation of $f/3$ and $2f/3$ harmonics [45], where f is the discharge frequency.

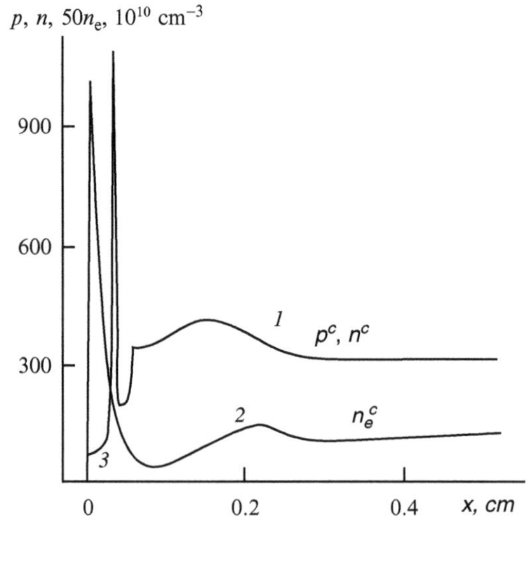

Fig. 2.13 The formation of ion density peaks in the simulation [45]. The conditions are the same as those of Fig. 2.10 except for higher pressure (1.33 Torr) and current (100 mA/cm^2)

2.7 The Connection Between Plasmas with Negative Ions, Dusty Plasmas, and Ball Lightning

Particles are often formed in processing plasmas and they affect plasma properties. Deliberate synthesis of nanoparticles is used in applications ranging from photovoltaics to cancer treatment. As will be described below, the presence of small particles may also explain the properties of ball lightning. When particles are very small, about a nm in diameter, most of them have charge zero or ± 1, with the particle charging process dominated by stochastic fluctuations. Larger particles are always charged negatively due to the large incoming electron flux incident on the particle. Negatively charged particles act as negative ions and are pushed to the center of the plasma by the electric field. Charged species' profiles in dusty plasmas are shown in Fig. 2.14 [50,51]. The formation of large ion and particle density gradients is apparent inside the plasma, similar to the negative ion density fronts described in Sect. 2.5. For larger particles ion drag has to be taken into account as well as the electric field. When the particle size increases even further the ion drag overcomes the force due to the electric field and pushes the particles out of the center, producing a void. This void has been observed to undergo a "heartbeat instability," in which it periodically expands and contracts [52]. Further details about particle transport and dynamics can be found in Chaps. 7 and 9.

Here, we only wish to point out the connection of negative ions and negatively charged microparticles to ball lighting, photographs of which are shown in Fig. 2.15. A remarkable feature of ball lightning reported by observers is that it resumes a rounded shape after colliding with an obstacle and even in passing through an

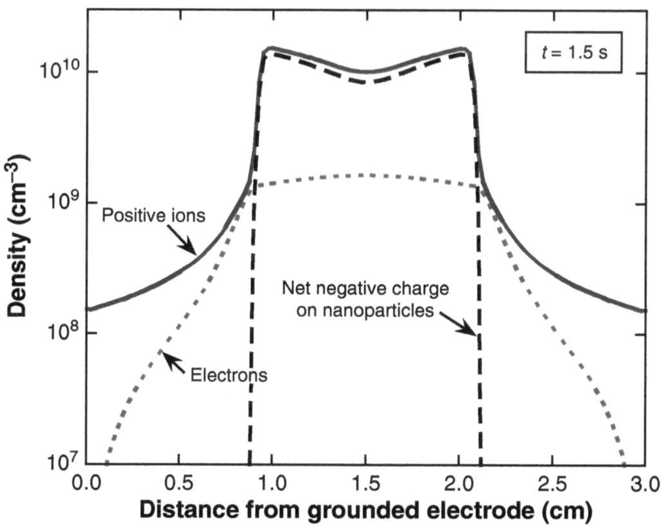

Fig. 2.14 Predicted profiles of charge carrier densities at 1.5 s following onset of nucleation, with particle size about 20 nm [48]

Fig. 2.15 Photographs of ball lightning (from http://www.zeh.ru/shm/index_e.php)

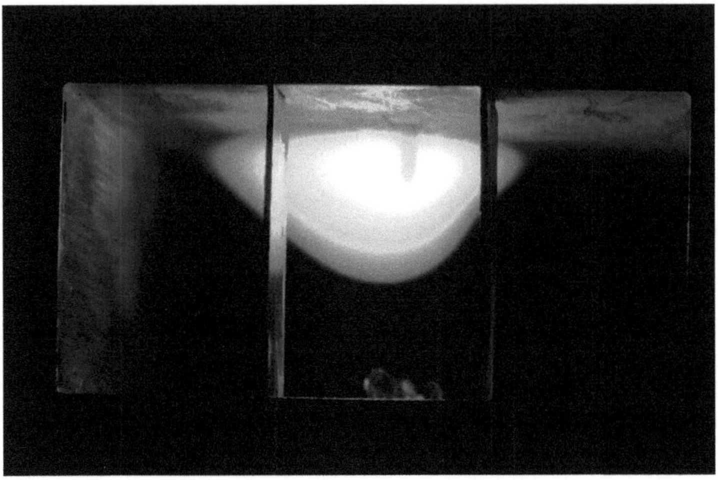

Fig. 2.16 Photograph of a fireball in a microwave drill (from [54])

aperture. There were many attempts to reproduce ball lightning in the laboratory, most notably the recent experimental studies of [53, 54]. Jerby and Dikhtyar [53] produced a fireball resembling ball lightning in an industrial microwave drill. The *Nature* article [55] comments that "Such fire balls (pictured in Fig. 2.16) mimic two of the most perplexing aspects of ball lightning – they persist after the initial source of energy is removed and they float in air. This supports previous suggestions that ball lightning could be driven by the oxidation of particles in a cloud generated by an energetic event, such as a conventional lightning strike." In previous experiments including that of [53], fireballs extinguished almost immediately after the discharge power was switched off (Fig. 2.16). In contrast to these previous experiments, Paiva et al. [54] performed electric arc discharges in pure silicon and were able to generate luminous balls with a lifetime of the order of seconds. The source of energy was chemical – burning of small silicon particles produced in the arc, shown in Fig. 2.17.

The presence of a large number of silicon particles in the plasma of a fireball was verified recently by small-angle X-ray scattering [56]. The authors claim that "The results show that the fireballs contain particles with a mean size of ~50 nm with

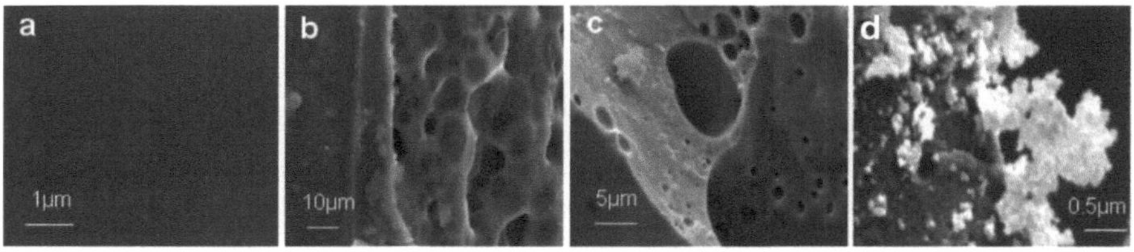

Fig. 2.17 SEM of the Si wafer before (**a**) and after electrical discharge (**b**)–(**d**). The surface of the samples subjected to electrical discharges shows holes (**b**, **c**) and chains of micrometer-sized particles (**d**) (from [54])

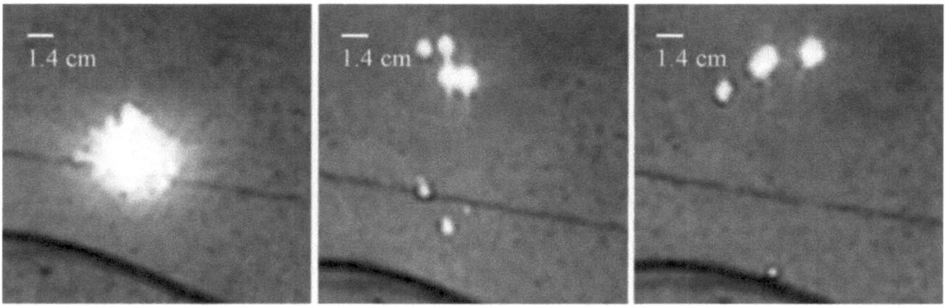

Fig. 2.18 Successive video frames showing the luminous balls bouncing off the ground. Time interval between the frames is 80 ms (from [54]). See also the supplementary video 2 (from [58])

average number densities on the order of $\sim 10^9$. Hence, fireballs can be considered as a dusty plasma which consists of an ensemble of charged nanoparticles in the plasma volume" [56]. All these findings give support to the Abrahamson–Dinniss theory for the formation of ball lightning [57] based on the generation of an oxidizing silicon particle network liberated by lightning striking the ground.

As explained in Sects. 2.4 and 2.5, the self-consistent electric field in a plasma $\vec{E} = -T_e \vec{\nabla} \ln n_e$ pushes negatively charged particles into the plasma center. If ball lightning or fireballs are supported by the burning of dust particles, such coupling between plasma production and transport may explain the tendency of these objects to resume a rounded shape, because all the negative ions move toward the plasma density maximum, thus acting as an effective surface tension. An example of such an event restoring the rounded shape of a fireball is shown in Fig. 2.18 (taken from [54]).

In summary, the presence of negative ions makes plasma transport much more complicated and interesting. Many complex nonlinear phenomena remain to be explained.

Acknowledgments The authors are thankful for stimulating discussions with Norman J.M. Horing, Larry Grisham, Lev D. Tsendin, and Alan J. Lichtenberg. This work was partially supported by the Air Force Office of Scientific Research through STTR Phase 2 contract.

References

1. H.S.W. Massey, *Negative Ions* (Cambridge University Press, Cambridge, 1974)
2. B.M. Smirnov, *Negative Ions* (McGraw-Hill, New York, 1982)
3. M.A. Lieberman, A.J. Lichtenberg, *Principles of Plasma Discharges and Materials Processing, 2nd edn.* (Wiley, New York, 2005)
4. G.S. Hwang, K.P. Giapis, Phys. Rev. Lett. **79**, 845 (1997)
5. W. Swider, *Ionospheric Modeling* (Birkhauser, Basel, 1988)
6. I M. Kuriyama et al., Fusion Eng. Des. **39–40**, 115 (1998)
7. L.R. Grisham, Nucl. Instrum. Methods Phys. Res. A **464**, 315 (2001)
8. A.A. Howling, I. Sansonnens, J.-L. Dorier, C. Hollenstein, J. Phys. D Appl. Phys. **26**, 1003 (1993)
9. N.L. Aleksandrov, A.P. Napartovich, Sov. Phys. Uspekhi **36**, 107 (1993)
10. B.M. Smirnov, *Physics of Ionized Gases* (Wiley-IEEE, New York, 2001), p. 165
11. P.P. Raizer, *Gas Discharge Physics* (Springer, Berlin, 1991)
12. V.A. Rozhansky, L.D. Tsendin, *Transport Phenomena in Partially Ionized Plasma* (CRC, Boca Raton, 2001)
13. I.G. Kouznetsov, A. Lichtenberg, M. Lieberman, Plasma Sci. Technol. **5**, 662 (1996)
14. I.G. Kouznetsov, A. Lichtenberg, M. Lieberman, J. Appl. Phys. **86**, 4142 (1999)
15. V.I. Kolobov, D.J. Economou, Appl. Phys. Lett. **72**, 656 (1998)
16. P. Chabert, A.J. Lichtenberg, M.A. Lieberman, Phys. Plasmas **14**, 093502 (2007)
17. I.D. Kaganovich, L.D. Tsendin, IEEE Trans. Plasma Sci. **20**, 66 (1992)
18. I.D. Kaganovich, L.D. Tsendin, N.A. Yatsenko, Sov. Phys. Tech. Phys. **39**, 1215 (1994)
19. A.P. Dmitriev, V.A. Rozhansky, L.D. Tsendin. Sov. Phys. Uspekhi **28**, 467 (1985)
20. R.N. Franklin, Plasma Sour. Sci. Technol. **11**, A31 (2002)
21. R.N. Franklin, J. Phys. D Appl. Phys. **36**, 828 (2003)
22. I.D. Kaganovich, D.J. Economou, B. Ramamutri, V. Midha, Phys. Rev. Lett. **84**, 1918 (2000)
23. I.D. Kaganovich, B. Ramamutri, D.J. Economou, Phys. Rev. E **64**, 036402 (2001)
24. L.D. Tsendin, Sov. Phys. Tech. Phys. **34**, 11 (1989)
25. C. Ferreira, G. Gousset, M. Touseau, J. Phys. D Appl. Phys. **21**, 1403 (1988)
26. C. Ferreira, G. Gousset, J. Phys. D Appl. Phys. **24**, 775 (1991)
27. P.G. Daniels, R.N. Franklin, J. Phys. D Appl. Phys. **22**, 780 (1989)
28. P.G. Daniels, R.N. Franklin, J. Snell, J. Phys. D Appl. Phys. **23**, 823 (1990)
29. R.N. Franklin, P.G. Daniels, J. Snell, J. Phys. D Appl. Phys. **26**, 1638 (1993)
30. R.N. Franklin, J. Snell, J. Phys. D Appl. Phys. **27**, 21823 (1990)
31. I.D. Kaganovich, Phys. Plasmas **8**, 2540 (2001)
32. I.D. Kaganovich, L.D. Tsendin, Plasma Phys. Rep. **19**, 645 (1993)
33. I. Kaganovich, S.V. Berezhnoi, C.B. Shin, Phys. Plasmas **8**, 719 (2001)
34. G.B. Whitham, *Linear and Nonlinear Waves* (Wiley, New York, 1974)
35. I.D. Kaganovich, B.N. Ramamurthi, D.J. Economou, Appl. Phys. Lett. **76**, 284 (2000)
36. E.A. Bogdanov, A.A. Kudryavtsev, L.D. Tsendin, Tech. Phys. Lett. **27**, 652 (2001)
37. E.A. Bogdanov, A.A. Kudryavtsev, L.D. Tsendin, Tech. Phys. **46**, 404 (2001)
38. M.V. Koniukov, Sov. Phys. JEPT **12**, 629 (1958)
39. A.V. Phelps, J. Res. Natl. Inst. Stand. Technol. **95**, 407 (1990)
40. D. Smith, A.G. Dean, N.G. Adams, J. Phys. D Appl. Phys. **7**, 1944 (1974)
41. V.I. Demidov et al., Phys. Rev. Lett. **95**, 215002 (2005)
42. C.A. DeJoseph Jr. et al., Phys. Plasmas **14**, 057101 (2007)
43. J.D. Passchier, W.J. Goedheer, J. Appl. Phys. **73**, 1073 (1993)
44. S.V. Berezhnoi, U. Buddemeier, I. Kaganovich, C.B. Shin, Appl. Phys. Lett. **77**, 800 (2000)
45. V.A. Schweigert, Plasma Phys. Rep. **17**, 844 (1991)
46. A.J. Lichtenberg, I.G. Kouznetsov, Y.T. Lee, M.A. Lieberrman, I.D. Kaganovich, L.D. Tsendin, Plasma Sour. Sci. Technol. **6**, 437 (1997)
47. M. Lampe, W.M. Manheimer, R.F. Fernsler, S.P. Slinker, G. Joyce, Plasma Sour. Sci. Technol. **13**, 15–26 (2004)

48. I.D. Kaganovich, Plasma Phys. Rep. **21**, 434 (1995)
49. Y.P. Raizer, M.N. Shneider, N.A. Yatsenko, *Radio Frequency Capacitive Discharges* (CRC, Boca Raton, 1995)
50. L. Ravi, S.L. Girshick, IEEE Trans. Plasma Sci. **36**, 1022 (2008)
51. S.J. Warthesen, S.L. Girshick, Plasma Chem. Plasma Process. **27**, 292 (2007)
52. M. Mikikian, L. Boufendi, Phys. Plasmas **11**, 3733 (2004)
53. E. Jerby, V. Dikhtyar, Phys. Rev. Lett. **96**, 045002 (2006)
54. G.S. Paiva et al., Phys. Rev. Lett. **98**, 048501 (2007)
55. Nat. Phys. 2 March 2006 Research Highlights Great balls of fire!
56. J.B.A. Mitchell, J.L. LeGarrec, M. Sztucki, T. Narayanan, V. Dikhtyar, E. Jerby, Phys. Rev. Lett. **100**, 065001 (2008)
57. J. Abrahamson, J. Dinniss, Nature (Lond.) **403**, 519 (2000)
58. See EPAPS Document No. E-PRLTAO-98-047705 for video clips showing the experiment and the luminous balls. For more information on EPAPS, see http://www.aip.org/pubservs/epaps.html

Chapter 3
Introduction to Quantum Plasmas

Michael Bonitz, Alexei Filinov, Jens Böning, and James W. Dufty

Abstract Plasmas are generally associated with a hot gas of charged particles which behave classically. However, when the temperature is lowered and/or the density is increased sufficiently, the plasma particles (most importantly, electrons) become quantum degenerate, that is, the extension of their wave functions becomes comparable to the distance between neighboring particles.

This is the case in many astrophysical plasmas, such as those occurring in the interior of giant planets or dwarf and neutron stars, but also in various modern laboratory setups where charged particles are compressed by very intense ion or laser beams to multi-megabar pressures. Furthermore, quantum plasmas exist in solids – examples are the electron gas in metals and the electron–hole plasma in semiconductors. Finally, the exotic state of the Universe immediately after the Big Bang is believed to have been a quantum plasma consisting of electrons, quarks, photons, and gluons. In all these situations, a description in terms of classical mechanics, thermodynamics, or classical kinetic theory fails.

In this chapter, an overview of quantum plasma features and their occurrence is given. The conditions for the relevance of quantum effects are formulated and discussed. The key concepts for a theoretical description of quantum plasmas are developed and illustrated by simple examples.

3.1 Introduction

When the term "plasma" was introduced by Tonks and Langmuir in the late 1920s, they had in mind the gas of electrons and ions existing in the ionosphere. They wanted to distinguish this system from a "normal" gas consisting of neutral atoms

M. Bonitz (✉), A. Filinov, and J. Böning
Institut für Theoretische Physik und Astrophysik, Christian-Albrechts Universität zu Kiel, 24098 Kiel, Germany
e-mail: bonitz@physik.uni-kiel.de; filinov@theo-physik.uni-kiel.de; boening@theo-physik.uni-kiel.de

J.W. Dufty
Department of Physics, University of Florida, Gainesville, FL 32611
e-mail: dufty@phys.ufl.edu

M. Bonitz et al. (eds.), *Introduction to Complex Plasmas*, Springer Series on Atomic, Optical, and Plasma Physics 59, DOI 10.1007/978-3-642-10592-0_3,

or molecules. Of course, upon strong heating, this neutral gas will ionize (electrons gain kinetic energy and escape the atom) and transform into a plasma. In this ionized gas, electrons and ions interact under classical dynamical laws subject to Newtonian mechanics and Maxwellian statistical physics.

In the 1930s, it became clear that gas-like many-particle systems of electrons also exist in entirely different systems: electrons in metals and semiconductors exhibit very similar properties, behaving like an *electron gas* or *electron–hole plasma*, respectively. There is, however, a fundamental difference: In the solid-state environment, the quantum nature of the electrons becomes relevant, leading to wave properties (interference) and Fermi statistics effects (Pauli exclusion principle).[1]

But it soon became clear that quantum effects in plasmas are not confined to condensed matter. Observations of astrophysical objects led to the conclusion that many planets and stars contain ionized matter in their interior at very high pressure, where the behavior of the electrons is governed by quantum mechanics. Finally, since the 1980s, researchers have been able to produce similar plasmas on Earth by applying high-intensity laser pulses or ion beams to solid-state targets, which leads to ionization and strong compression and the emergence (for an intermediate period of time) of quantum behavior of the released electrons. Such experiments are of rapidly increasing importance, as they promise to produce new states of matter (there is even a new term *warm dense matter* – WDM) and, in particular, they are of relevance for inertial confinement fusion.

Finally, the exotic state of the Universe immediately after the Big Bang is believed to have been a quantum plasma as well, being at an extremely high density and temperature, consisting of charged electrons, positrons, photons, quarks, antiquarks, and gluons. This *quark–gluon plasma* (QGP) is of fundamental importance for our understanding of the structure of matter and it is now being produced in large accelerators at Brookhaven and CERN; more details will be given in Sect. 3.4.

Thus, basic understanding of the properties of quantum plasmas is of key importance for many fields of modern physics. For a theoretical description, one has to leave the fields of classical physics and use a quantum description. This is based on the Schrödinger equation and quantum many-body theory involving methods such as quantum statistics, quantum kinetic theory, Hartree–Fock, nonequilibrium Green's functions (NEGF), and quantum Monte Carlo. These methods have been the subject of a number of excellent textbooks, for example, [1–5] and review articles [6–9] which are well suited for starting a thorough study of this rapidly evolving field.

This chapter does not duplicate these works but aims to give a compact introduction into and provide basic understanding of the main concepts in quantum plasma theory, and also includes a discussion of new developments which were not covered before. We start by introducing the main parameters of quantum plasmas (Sect. 3.2) and discuss the main examples of quantum plasmas and where they are located on a temperature–density plane in Sect. 3.4. Basic theoretical concepts are

[1] The particular properties of solid-state plasmas are discussed in detail in Chap. 5.

then introduced in Sect. 3.5 and include the aspects of bound state formation, Fermi and Bose statistics, quantum kinetic theory, and second quantization. The interested reader will find proper references to more detailed papers and textbooks on special issues. Those who are interested in numerical methods for quantum plasmas will find an excellent introduction in Chap. 4.

3.2 Relevant Parameters of Quantum Plasmas

Quantum plasmas, like conventional plasmas, are governed by the Coulomb interaction between the charged particles which – due to its long range – is responsible for most of the plasma properties, including collective excitations and instabilities. The quantum nature of the particles requires a quantum-mechanical approach which starts from the Hamilton operator. For a system of particles with mass m_i and charge e_i interacting via a Coulomb potential, it is given by

$$\hat{H} = \sum_{i=1}^{N} \left[-\frac{\hbar^2}{2m_i} \nabla_i^2 + V(\boldsymbol{r}_i) + \sum_{j<i}^{N} \frac{e_i e_j}{\epsilon r_{ij}} \right], \tag{3.1}$$

where $r_{ij} = \left| \boldsymbol{r}_i - \boldsymbol{r}_j \right|$ and ϵ denotes a static background dielectric constant, which equals 1 in vacuum and in a gas plasma but is of the order of 10 in case of an electron–hole plasma in a semiconductor.

Despite their apparently different forms, all Coulomb systems exhibit similar fundamental properties governed by the strength of the Coulomb interaction, which is measured by dimensionless control parameters: the coupling parameters Γ_a, and r_{sa} of particle species a and the quantum degeneracy parameter χ_a. These parameters are determined by the ratio of characteristic energy and length scales, respectively [3, 8, 9]:

Length scales.

1. $\bar{r}$ – the average interparticle distance, $\bar{r} \sim n^{-1/d}$ (n and $d \in \{1, 2, 3\}$ denote the density and dimensionality of the system, respectively).
2. Λ – the quantum-mechanical extension of the particles. For free particles, we have the De Broglie wavelength $\Lambda_a(p) = h/p$, where p is the momentum. In case of a plasma in thermodynamic equilibrium, we may replace p by the thermal momentum and obtain a wavelength which depends only on temperature and the mass, $\Lambda_a^{\text{free}}(T_a; m_a) = h/\sqrt{2\pi m_a k_B T_a}$. On the other hand, for particles bound in an atom, Λ is given by the extension of the bound state wave function. In particular, for electrons in the ground state of an atom, $\Lambda_a^{\text{bound}} = 2\pi a_B$.
3. a_B – the relevant Bohr radius of an electron–ion pair with charges e_k and masses m_k, $k \in \{\text{e}, \text{i}\}$: $a_B = (\epsilon/e_a e_b)(\hbar^2/m_{ab})$, where $m_{ab}^{-1} = m_a^{-1} + m_b^{-1}$.

4. a_{Ba} – the effective Bohr radius of a one-component plasma (OCP)[2] is sometimes used and defined by analogy to the hydrogen-type Bohr radius a_B according to $a_{Ba} = (\epsilon/e_a^2)(\hbar^2/m_a)$.

Energy scales.

1. $\langle K \rangle$ – the mean kinetic energy, which in a classical system is given by the thermal energy $\langle K_a \rangle_{cl} = (d/2)k_B T_a$. In contrast, in a highly degenerate Fermi system, $\langle K_a \rangle_{qm} = (3/5)E_{Fa}$ holds. Here, we introduced the Fermi energy, $E_F(n) = \hbar^2(3\pi^2 n)^{2/3}/2m$, which is a function of density but is independent of temperature.
2. $\langle U_C^{ab} \rangle$ – the mean Coulomb energy, given for free and bound particles by $\langle U_C^{ab} \rangle^{free} = (e_a e_b/4\pi\epsilon)(1/\bar{r})$ and $\langle U_C^{ab} \rangle^{bound} = (e_a e_b/4\pi\epsilon)(1/2a_B) \equiv E_R$ (Rydberg), respectively.

Dimensionless control parameters. Combining various length or energy scales allows us to distinguish parameter regions of qualitatively different plasma behavior. These regions are shown in the phase diagram (Fig. 3.1):

1. The quantum *degeneracy parameter*

$$\chi_a \equiv n_a \Lambda_a^d \propto \left(\frac{\Lambda_a}{\bar{r}_a} \right)^d \tag{3.2}$$

provides a reasonable estimate about the strength of quantum effects in many-body systems. If the thermal De Broglie wavelength is small compared to the average interparticle distance ($\chi \ll 1$), the system can be described classically, whereas in the opposite case ($\chi \gg 1$) it has to be treated quantum mechanically. Thus, the line $\chi_a \approx 1$ roughly separates the phase diagram into the classical and quantum regimes for the species a.
2. The ratio of average potential to kinetic energy $|\langle U_C \rangle| / \langle K \rangle$ is the relevant parameter determining the phase of the system. Depending on its nature (see earlier), we can define the *Coulomb coupling parameter*

$$\Gamma_a \equiv \frac{|\langle U_C^{aa} \rangle|}{k_B T_a} \tag{3.3}$$

for classical systems. When replacing the classical kinetic energy by its quantum counterpart, we obtain *quantum Coulomb coupling parameter* or *Brueckner parameter*:

$$r_{sa} \equiv \frac{\bar{r}_a}{a_{Ba}} \propto \frac{|\langle U_C^{aa} \rangle|}{E_{Fa}}. \tag{3.4}$$

---

[2] The OCP is a standard theoretical model of plasmas and condensed matter systems (where it is called *jellium model*) to simplify the description of multicomponent plasma systems to a single "active" component – normally the heavy particles. It assumes charge neutrality which is assured by the light component which is considered to act as a homogeneous background.

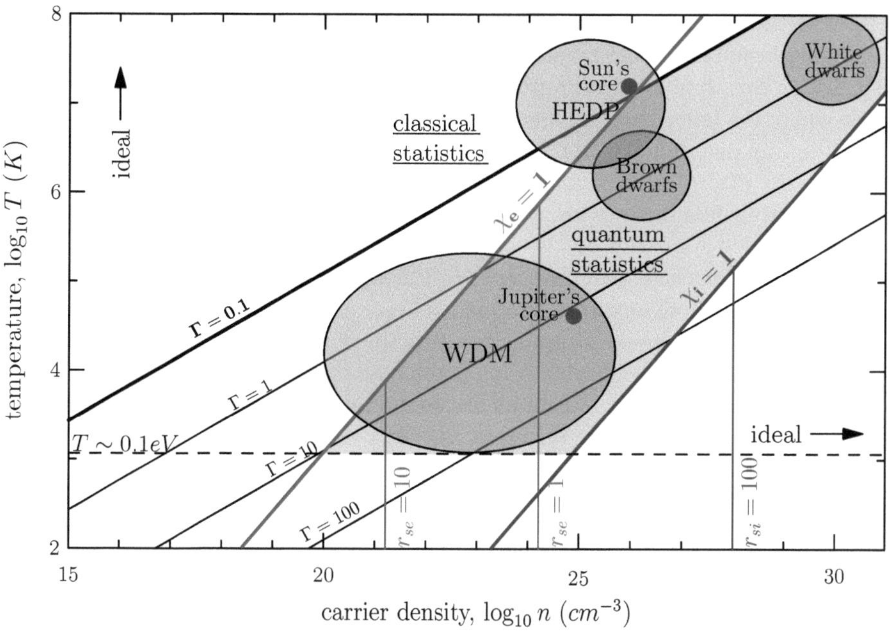

Fig. 3.1 Phase diagram of a two-component plasma of electrons and singly charged ions in thermodynamic equilibrium. The *line* $\chi_e = 1$ ($\chi_i = 1$) separates the region of classical (*upper left*) and quantum (*lower right*) behavior of the electrons (ions). Also, the *lines* of constant classical (Γ) and quantum (r_s) coupling strength are shown. For $\Gamma \ll 1$, the plasma is ideal and classical; for $r_{se} \ll 1$ ($r_{si} \ll 1$), an ideal quantum behavior of electrons (ions) is observed. The *lines* $\Gamma = 1$ and $r_s = 1$ enclose the region of strong coupling which may contain regions of liquid-like and crystalline behavior and (in a TCP) bound states. The *line* $r_{si} = 1$ is located outside the figure at higher densities. For singly charged ions, the classical coupling parameters for electrons and ions coincide. Also, a few typical examples of plasmas in space and in the laboratory (warm dense matter, WDM and high-energy density plasmas, HEDP) are included

This parameter is related to the parameter, $r_s = \bar{r}/a_B$, familiar from atomic units, $r_s = r_{sa}[1 - (m_a/(m_a + m_b))]$. One can introduce similar coupling parameters for different species and other pair interactions.

Parameters of two-component plasmas (TCPs). In a plasma with differing masses and/or charges of the species, the coupling and quantum degeneracy parameters of the species may be unequal (we now go beyond the OCP model). In particular, in a dense electron–ion plasma, classical ions and quantum electrons may coexist. Analogously, ions may be strongly coupled while the electrons are only weakly coupled. The ratio of the degeneracy parameters scales as $\chi_a/\chi_b = (m_b/m_a)^{1/2}$, whereas the ratios of the coupling parameters are given by $\Gamma_a/\Gamma_b = (e_a/e_b)^{2-1/d}$ and $r_{sa}/r_{sb} = (m_a/m_b)(e_a/e_b)^{2+1/d}$, where local charge neutrality, $n_a e_a = n_b e_b$, has been assumed.

Let us now examine the appearance of these characteristic parameters of a quantum plasma in thermodynamic equilibrium on the density/temperature plane of Fig. 3.1. In this double-logarithmic plot, the lines of constant Γ and χ are straight lines where the latter ones are steeper than the former. The lines of constant quantum coupling parameter r_s are straight vertical lines. Thus, to the right of the line $\chi_e = 1$, the electrons are quantum degenerate and have to be described quantum mechanically while the ions are still classical. As one can see, this is observed even at moderate densities well below solid-state densities. Only to the right of the line $\chi_i = 1$ are the ions also quantum, and a full quantum-mechanical description has to be performed for both electrons and ions.

A quantum description alone is not difficult if the particles are ideal, that is, interaction effects are negligible. This is the case to the right of the line $r_s = 0.1$, where the plasma behaves like an ideal quantum gas obeying either Fermi or Bose statistics. While electrons are fermions and obey the Pauli principle, ions can be of either statistics. The different behaviors of fermions and bosons will be discussed in Sect. 3.5.3.

3.3 Different States of Quantum Plasmas

Using the above-discussed dimensionless parameters, we can easily understand the qualitatively different behaviors which may occur in dense plasmas in thermodynamic equilibrium. Let us first consider plasma behavior at low density and high temperature (top left part of Fig. 3.1). Here, the plasma is classical and ideal. Reduction of temperature leads to a successive crossing of the lines of increasing Γ values. For $\Gamma > 1$, the plasma is moderately coupled, that is, Coulomb interaction effects start to dominate the behavior. Instead of random gaseous particle motion, the plasma particles become spatially correlated as in a liquid. Upon further temperature reduction, we eventually reach very strong coupling, $\Gamma \gg 1$ (bottom left part of Fig. 3.1). Here, kinetic energy of the particles plays only a minor role and the plasma behavior is governed by the total potential (sum of external potential V and all Coulomb pair potentials). For the particles, it is now energetically favorable to settle in the local minima of this potential. This behavior resembles particles confined to the lattice sites of a crystal. Indeed, the formation of a Coulomb crystal was predicted as early as 1934 by Wigner [10] for the electrons in a metal. Since then, it has been observed experimentally (see below). Also, theoretical analysis and, in particular, computer simulations have clearly confirmed this prediction and given accurate critical data: a classical Coulomb crystal forms if Γ exceeds a value of about 175 in 3D and 137 in a 2D system (for details, see [9]).

Before proceeding, we note that reaching high values of Γ is easily possible only in a OCP. In contrast, in a TCP containing electrons and positive ions, temperature reduction leads to recombination of electrons and ions, that is, to the reformation of atoms and molecules. These neutral entities interact much more weakly with each other than do charged particles; thus, the coupling strength is greatly reduced. Having mentioned bound state formation (which is, of course, an entirely quantum-mechanical feature in itself), we postpone a detailed discussion of this effect to Sect. 3.4.5. Thus, to reach very strong coupling, bound state formation has to be avoided. One way to do this consists in choosing plasma parameters such that bound states are unstable. This will be discussed in Sect. 3.4.5.

Another approach to reach strong coupling is very simple: create a plasma which contains only a single component, that is, a non-neutral plasma. To maintain a stable plasma state, obviously, the Coulomb repulsion of the particles has to be balanced by other forces – normally, one applies an external "confinement potential" which can be, for example, an electrostatic potential. Such spatial confinement of a pure ion plasma has been achieved with Penning or Paul traps [11, 12] and in storage rings [13, 14] (for an overview, see [15]). Wigner crystals have also been predicted to occur in semiconductor quantum dots containing a finite number of electrons confined by an intrinsic or external electrostatic potential [16]. Finally, Coulomb crystals have been produced in dusty plasmas [17–19] which is discussed in detail in Chaps. 7 and 8.

So far, we have discussed states where the plasma is at low temperature and low density such that it behaves classically. Let us now consider the changes occurring when the plasma is being compressed at low temperature (cf. lower bottom of Fig. 3.1). When we cross the line $\chi_e = 1$, the electron behavior starts to be dominated by quantum mechanics. Thus electrons can no longer be regarded as point particles, but they have a finite extension of their wave functions given by the length scale Λ. This means that the degree of nonideality of the electrons is no longer given by the classical parameter Γ but by the Brueckner parameter r_{se}. In this case too, small values of the coupling parameter, $r_{se} \ll 1$, correspond to negligibility of kinetic energy and to crystal formation. The theoretical predictions for the critical values are still somewhat uncertain being around $r_s^{cr} \approx 100$ for fermions and 160 for bosons in 3D, and $r_s^{cr} \approx 37$ in 2D (see [9] and references therein).

Further increase of the density, at constant low temperature, leads to a continuous reduction of r_s. Below the critical value for crystallization but above $r_s = 1$, the plasma shows liquid-like behavior. Depending on the spin statistics, one observes either a Fermi liquid or a Bose liquid. In the latter case, the liquid may be partially superfluid – which means that the system loses part of its viscosity due to quantum coherence effects of the particles. Finally, at $r_s < 1$, the particles behave as a weakly nonideal fluid and, at $r_s \ll 1$, all interaction effects are much weaker than the quantum kinetic energy, and the system behaves as an ideal quantum (Bose or Fermi) gas.

While all these discussions have been concentrated on a OCP, the results are readily generalized to multicomponent systems. Then, depending on the mass ratio of the components, quantum degeneracy and nonideal behavior are shifted in density, as indicated in Fig. 3.1 (for the example of hydrogen) by the corresponding lines of constant χ and r_s of electrons and protons, respectively.

3.4 Occurrences of Quantum Plasmas

Let us now turn to the various occurrences of quantum plasmas in nature and in the laboratory. The phase diagram (Fig. 3.1) contains a variety of the most important examples and indicates where they are located in the density/temperature plane. The figure includes dense plasmas in the core of the giant planets, of the Sun and in dwarf stars as well as WDM and HEDP. Some of these examples will be discussed more in detail below.

3.4.1 Astrophysical Plasmas

There are an enormous variety of plasmas in space. The interstellar medium contains electrons and ions but its density is so low that they behave classically. The same is true for dust and gas clouds filling the space between stars and planets. On the other hand, quantum effects are expected to occur in many dense astrophysical plasmas [7], in particular in the core of the giant planets such as Jupiter, Saturn, and Neptune, due to the high density caused by gravitation. Theoretical models predict that the plasma density there exceeds the density of solid matter and pressure reaches the multi-megabar range [20–24].

In stars like our Sun, matter is usually also fully ionized due to strong heat production in the core. Again, gravity produces a radial density profile which is peaked in the center. As a result, also quantum degeneracy and nonideality increase toward the center. Figure 3.1 indicates that in the core of the Sun, electrons are expected to be quantum degenerate and weakly nonideal (e.g., [25]). At the same time, the ions – mostly protons and α-particles (hydrogen and helium nuclei) – are classical (for more details, see Sect. 3.5.2). Other representatives of quantum plasmas include compact stars in the late stages of their evolution: brown or white dwarf stars (see, e.g., [26]). These are very massive objects with a core density exceeding solid densities by many orders of magnitude, exhibiting unusual behavior such as ion crystals in the presence of a Fermi gas of electrons. This will be discussed in Sect. 3.4.5. The last astrophysical object we mention are neutron stars. The plasma inside is even denser than those in dwarf stars [27, 28], so that even the heavy particles – protons and heavier atomic nuclei – are quantum degenerate (see Sect. 3.4.5).

3.4.2 Dense Laboratory Plasmas

For several decades, researchers have tried to compress plasmas in the laboratory to study new states of matter and also to understand the behavior of compact astrophysical objects. Plasma compression has been achieved by many groups worldwide using a variety of static and dynamic techniques, including diamond anvils [29], gas guns [30, 31], explosive devices [32, 33], pinches in high-current-carrying plasmas [34, 35], and shock waves [36]. While the first three have been the main methods employed in recent decades, the latter two are currently actively used, together with lasers and ion beams [37] (see below). In recent years, the densities and pressures were increased steadily and now typical solid-state conditions of $10^{23}\,\text{cm}^{-3}$ have been achieved. At these densities, it is possible to produce degenerate electrons as long as the temperatures are moderate. Usually, this is the case in the initial stage of the compression, while later on temperature increases steadily, and the plasma becomes classical again. In many cases, it also expands hydrodynamically with the same consequence of reducing the electron degeneracy. The corresponding density–temperature range is indicated by the area named "WDM" in Fig. 3.1.

3.4.3 Laser Plasmas

With the advent of high-power lasers (e.g., [38]), there exists a new tool to compress matter by simply exploiting the pressure produced by an electromagnetic wave [39–41]. Laser beam ionization of metal or plastic foils is now routine, and the ionized matter – a mixture of electrons and ions – is subsequently compressed well beyond solid-state densities. The most spectacular application of this method is aimed at thermonuclear (inertial confinement) fusion, where a small capsule of deuterium and tritium is compressed and heated so strongly that it is hoped that fusion of two hydrogen atoms into one helium atom will occur (e.g., [39–41]). Here, quantum effects are not the main goal; the thrust is to achieve high pressure and temperature for a sufficiently long period to fulfill the Lawson criterion for fusion. The corresponding experiments are now set up at the National Ignition Facility (NIF) at Lawrence Livermore National Lab, and are expected to yield first results in 2010 (e.g., [42]). Similar possibilities exist with the use of ion beams instead of laser beams or in a combination of both [37]. In recent years, substantial activities have been initiated in various countries, in particular at GSI in Darmstadt, Germany and at LMJ in France.

3.4.4 Plasmas in Condensed Matter Systems

As discussed in Sect. 3.1, a different kind of quantum plasma exists in solid-state systems. In metals, some of the valence electrons of the atoms are quasi-free, behaving like a weakly nonideal Fermi gas. This ionized gas has all properties of a plasma: it contains charged particles interacting by Coulomb potentials and it is quasineutral. The main differences compared to a "normal" plasma are: the electrons are not classical point particles but are delocalized in space; they are described by Bloch waves, which solve the Schrödinger equation. At the same time, the positive charges are not mobile but are fixed in the lattice of the solid.

A different kind of quantum plasma exists in semiconductors. At room temperature, these materials do not have free mobile electrons. It requires an external excitation by heat, a voltage, or a laser field to excite electrons from the valence band into the formerly empty conduction band, and this gives rise to vacancies in the valence band which act like free positive charges called *holes*. If the excitation is sufficiently strong, the number of electron–hole pairs can be large with densities reaching $10^{20}\,\mathrm{cm}^{-3}$, behaving very much like a TCP which is commonly called an *electron–hole plasma*. There are several differences compared to gas plasmas: the masses of electrons and holes differ from that of a free electron, they are determined by the band structure of the material, and they may be anisotropic. Further, electrons and holes are affected by the crystal lattice and by lattice vibrations (phonons) which gives rise to an effective dielectric constant of the order of 10, thus weakening (dynamic screening) the Coulomb interaction. These effects are described in detail in Chap. 5.

3.4.5 Highly Compressed Two-Component Plasmas: Mott Effect

We now focus attention on dense TCPs in the range of solid-state densities. As discussed in Sect. 3.3, a TCP at low temperature, that is, $k_B T < E_R$, tends to form bound states, so that the plasma behavior is dominated by neutral atoms and molecules. While these bound states can exist only at low temperature, we may question if there is also a restriction with respect to density. Consider compressing the material so strongly that the average distance between two atoms becomes comparable to the average extension of the electron wave function, which is given by the Bohr radius a_B. Then the electronic wave functions of two electrons of neighboring atoms will overlap, allowing one electron to move away from its parent ion to the next one. Such electrons are no longer tightly bound inside the atom, that is, we observe tunnel ionization of the atom. This is also called *pressure ionization* or *Mott effect* and it is a purely quantum effect which occurs even at zero temperature, at densities corresponding to $r_{se} \approx 1$.

Thus, strong compression of a neutral gas has the same effect as strong heating – it transforms matter from a neutral state of atoms and molecules into a fully ionized plasma consisting of electrons and ions. This is discussed more in detail for the example of hydrogen in Sect. 3.5.2. The corresponding region of existence of bound states is located in the lower left corner of the density–temperature plane (see Fig. 3.2). Now consider that we pressure ionize atoms with charge number Z. Then we are left with N-free electrons and N-singly charged ions, each still containing $Z-1$ electrons. Upon continued compression, eventually, the wave functions of the bound electrons will overlap, giving rise again to a Mott effect. This procedure may be repeated until all electrons become ionized, resulting in a plasma of $N \times Z$ electrons and N-bare Z-fold charged nuclei.

What are the properties of such a dense plasma of electrons and nuclei? It turns out that very different states of matter are possible, depending on temperature,

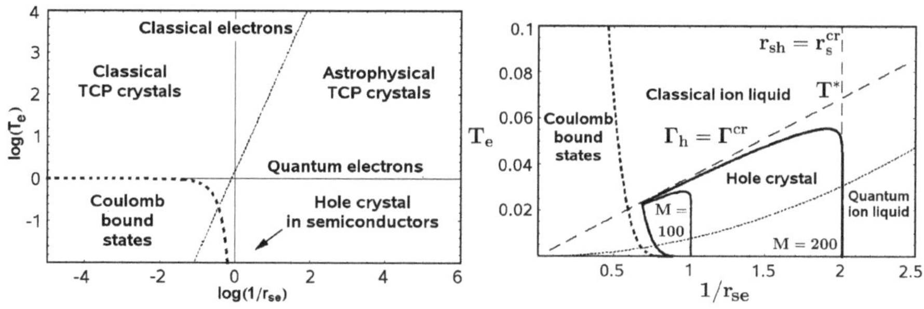

Fig. 3.2 *Left*. Phase diagram of a dense two-component plasma of electrons and ions (or holes). Density increases toward the right. The *dotted line* separates the region of classical electron behavior (*upper left*) from the quantum domain (*lower right*). The *dashed line* surrounds the area of Coulomb bound states (atoms and molecules). *Right*. Enlarged part of left figure, in the vicinity of the Mott point. Note the linear scale

density, and the mass ratio M of ions and electrons (see the phase diagram in Fig. 3.2). Consider first intermediate densities such that the ions are still classical, that is, $\chi_e \gg 1$ and $\chi_i < 1$. At high temperature, the ions form a classical gas or liquid, whereas the electrons are in a Fermi gas state. When the system is cooled, the ions may freeze into a classical crystal. This crystal is stable up to a maximum density corresponding to $r_{sh} \approx r_s^{cr}$. Beyond this density (the ions have already become quantum degenerate), quantum melting of the crystal occurs and the ions form a quantum liquid (Fermi or Bose liquid, depending on the spin of the ions), and further compression gives rise to a quantum gas. Is this behavior universal? Clearly, the coexistence of an ion crystal with a Fermi gas of electrons is only possible at densities and temperatures where $\chi_e \gg 1 \gg \chi_i$. It is obvious that this is only possible if the mass ratio M is sufficiently large.

The solution of this problem has been given recently [43]: there exists a critical mass ratio, $M^{cr} \approx 83$, above which crystallization of the heavy component is possible.[3] For smaller mass ratios, the ions will form only a liquid-like state. Systems with mass ratios above 83 exist in some semiconductors – there the analysis predicts hole crystallization [45–47] which has been confirmed by computer simulations [43, 48]. Other obvious candidates are hydrogen or helium plasmas where crystallization of protons or α-particles is expected at a high density exceeding 10^{25} cm^{-3} [43]. With increasing mass ratio, the stability region of the crystal grows further. Therefore, likely candidates to exhibit ion crystals are brown dwarf or white dwarf stars (cf. Sect. 3.4 and Fig. 3.1), which are expected to contain fully ionized oxygen and carbon ions in their core. Also, some neutron stars which contain fully ionized iron might exhibit an ion crystal in their outer crust [7, 28].

[3] We mention that heavy particle crystallization is possible at much lower M in other geometries. For example if electrons and ions (holes) are spatially separated in a double layer, crystallization is predicted already for $M \approx 4$ [44].

3.4.6 *Ultra-Dense Plasmas in Nuclear Matter: Quark–Gluon Plasma and the Big Bang*

We now consider continued compression of the quantum plasma of electrons and ions (nuclei). The crystal of nuclei can exist only up to densities of $r_{si} \geq r_s^{cr}$. At higher densities, it exhibits quantum melting, giving rise to a quantum fluid of nuclei embedded into a Fermi gas of electrons. Suppose we further compress this plasma. Although nuclei are extremely small entities, we have to expect that their internal structure – i.e. their composition of protons and neutrons – will become relevant at high densities. Indeed, if the distance between two nuclei becomes comparable to the quantum extension of a proton or a neutron we again have the situation of Mott transition: a proton (neutron) can tunnel from one nucleus to the next. As a consequence, nuclei break up giving rise to a plasma of free electrons, protons, and neutrons. This Mott transition is expected to occur at the nuclear density $\rho_N = 0.17\,\text{fm}^{-3}$, where $1\,\text{fm} = 10^{-15}\,\text{m}$.

But this is still not the end of the compression road: it is easy to imagine what would happen if the density is so large that two neighboring nucleons overlap. This is expected to happen at a density of about $\rho^{\text{deconf}} \approx 10\rho_N \approx 1.7 \times 10^{39}\,\text{cm}^{-3}$ (see, e.g., [49]): particles which are bound inside one nucleon would become free to tunnel out.[4] In fact, the standard model of elementary particles predicts that protons and neutrons are not elementary particles themselves but each is composed of three quarks. Then at this density quarks would become free (the so-called *deconfinement transition*), and the nuclear matter undergoes another Mott effect and transforms into the *QGP*. This is a very special kind of quantum plasma consisting of charged particles (electrons and quarks), their antiparticles, and their interaction quanta (photons and gluons). While details of the interactions in this plasma are quite complex, recent lattice gauge theory calculations indicate that the QGP is also dominated by Coulomb interaction, not only between normal charges but also between "color" charges [50–52].

How long can this compression procedure and sequence of Mott transitions continue, is there an end to this road? Most present theories agree that there exist true elementary particles – electrons, quarks, and their antiparticles – which are not composed of smaller ones, so no further Mott transition occurs. But why does not all matter collapse into a single point in space such as a black hole? The reason that this does not happen is one of the key properties of quantum plasmas – it is the quantum statistical nature of these elementary particles. Electrons and quarks obey Fermi statistics and thus, the Pauli principle applies (for details, see Sect. 3.5.3). While compression would lower the Coulomb interaction energy[5] (which is negative and

[4] This estimate is for low temperature. Alternatively, the breakup of hadrons can be achieved at lower densities by heating. The current estimate of the energy needed is of the order of 150–200 MeV, corresponding to $1.8–2.4 \times 10^{12}$ K.

[5] Here, we give the estimates for the energy per particle. Alternatively, to have the energy per volume, one has to multiply with the density n.

its magnitude grows with density as $n^{1/3}$), the quantum kinetic energy (which is positive) grows like $n^{2/3}$. Thus, a collapse would increase the total energy which is in contradiction to the stability of matter.[6]

Are these exorbitant densities not only pure fantasy? There are strong reasons to believe that this is a very realistic scenario. Since our Universe is presently expanding increasingly fast, one can extrapolate back and estimate when all matter must have been so close together that the density of matter would have equaled ρ^{deconf}. The present estimate for this time is 13.7 billion years, setting the current estimate for the age of the Universe. While we do not know what might have caused such extreme compression at this time, we know that it did happen and that it started with an giant explosion – the Big Bang. The driving force of this explosion was the Pauli principle of the fermionic quantum QGP, giving rise to a continuing expansion of the Universe still observed today.[7]

Finally, we mention that similar plasma conditions have been observed in various laboratory experiments. Researchers at the Relativistic Heavy Ion Collider (RHIC) in Brookhaven have generated collisions of very energetic heavy nuclei reaching similar densities and temperatures. Their recent experiments do indeed indicate the creation of a QGP. It is expected that these plasma conditions will soon be reached and even exceeded at the Large Hadron Collider at CERN, allowing for a detailed experimental study of this very special quantum plasma.

3.5 Theoretical Description of Quantum Plasmas

The theoretical description of a quantum plasma is lodged in the application of quantum mechanics applied to many-particle systems with Coulomb interactions. Here, we focus on nonrelativistic systems which are described by the theories of Schrödinger and Heisenberg.

---

[6] This estimate for the kinetic energy per particle applies to nonrelativistic fermions. At very high temperature, however, one has to apply relativistic quantum theory. In the limit of ultrarelativistic particles, the kinetic energy per particle scales like $n^{1/3}$ rather than $n^{2/3}$, that is, exactly like the Coulomb energy, and the above stability argument does not hold anymore. A similar situation is observed in astrophysics where the role of the attractive Coulomb interaction is taken over by gravity which has the same distance dependence. There it was discovered by Chandrasekhar [53] who showed that stars slightly more massive than our Sun would not be stabilized by the Pauli principle due to relativistic effects. They could collapse into neutron stars, black holes, or the hypothetic quark stars.

[7] As discussed before, today it is an open question whether in an ultrarelativistic quark–gluon plasma the Pauli principle alone would be able to trigger this explosion. There are many theories and speculations about additional factors. Updated overviews can be found, for example, on Wikipedia.

3.5.1 Basic Equations

The central equation governing nonrelativistic quantum dynamics is the *Schrödinger equation*:

$$i\hbar \frac{\partial}{\partial t} |\Psi(t)\rangle = \hat{H} |\Psi(t)\rangle, \quad |\Psi(0)\rangle = |\Psi_0\rangle, \tag{3.5}$$

which describes the time evolution of a general quantum-mechanical state $|\Psi(t)\rangle$ starting from the initial state $|\Psi_0\rangle$. The dynamics of a specific quantum system are thus governed by its *Hamiltonian*, which is defined, as in (3.1), for plasmas,

$$\hat{H} = \sum_{i=1}^{N} \left[-\frac{\hbar^2}{2m_i} \nabla_i^2 + V(\boldsymbol{r}_i) + \sum_{j<i}^{N} W(r_{ij}) \right]. \tag{3.6}$$

The solutions of the Schrödinger equation are known as *wave functions*. If the Hamiltonian is not a function of time, that is, its potentials do not explicitly change over time, the solutions are readily expressed by the stationary states or a linear combination of such. These states can be obtained from the simpler *stationary Schrödinger equation*:

$$\hat{H} |\psi_j\rangle = E_j |\psi_j\rangle, \quad |\Psi(t)\rangle = e^{-(i/\hbar)E_j t} |\psi_j\rangle, \tag{3.7}$$

as the time dependence can be separated out. The first part of (3.7) has the mathematical form of an eigenvalue problem, where the energy eigenvalues, E_j, and the corresponding eigenvectors, $|\psi_j\rangle$, are to be determined self-consistently. Note that the equations are written for N particle systems and the wave functions depend on the coordinates (and spin projections) of all N particles. This makes solution extremely difficult. With modern computers, the time-dependent Schrödinger equation can now be solved exactly for two or three particles. For larger particle numbers, alternative approaches based on many-body theory are used. One commonly used approximation is the Hartree–Fock approximation (for more details, see Chap. 4).

A different problem arises for N particle systems at *finite temperatures*, as is typically the case for plasmas. In these systems, no precise ("pure") initial state $|\Psi_0\rangle$ is known. In contrast, these systems are in a "mixed" state and as the Schrödinger equation only yields solutions for "pure" states, such as the ground state, we need additional input indicating how to mix these states to obtain a physically meaningful N particle quantum state. For macroscopic systems, this can be solved by statistical means provided by the density operator, which will be discussed below.

3.5.2 Thermodynamics of Partially Ionized Plasmas

The statistical mixture of pure states in a closed system is described by the *density operator* in its most general form (e.g., [3])

$$\hat{\rho} = \sum_j p_j |j\rangle \langle j|, \tag{3.8}$$

where the coefficients p_j are probabilities which are nonnegative ($0 \leq p_j \leq 1$) and add up to 1 ($\sum_j p_j = 1$). We can regard the mixed state as representing a statistical ensemble of systems, that is, large number of copies of the system in question, where p_j is the fraction of the ensemble being in the state $|j\rangle$.[8] The equation of motion of the density operator is given by the *von Neumann equation*[9]:

$$\mathrm{i}\hbar \frac{\partial \hat{\rho}}{\partial t} = \left[\hat{H}, \hat{\rho}\right], \tag{3.9}$$

which describes how the density operator evolves in time, just as the Schrödinger equation does for pure states. Both formalisms are equivalent.

Thermodynamic equilibrium is described by the *canonical ensemble*[10] for particle number N, volume V, and temperature T, with the probability of the ensemble to be in state $|j\rangle$ given by the *Boltzmann factor*, $e^{-\beta E_j}/Z$, where E_j is the corresponding energy eigenvalue, $\beta = 1/k_\mathrm{B}T$ is the inverse temperature, and Z is the partition function. The density operator $\hat{\rho}$ then takes a form similar to the distribution of states, $P(E)$, for classical systems:

$$P(E) = \frac{1}{Z} e^{-\beta E}, \quad \hat{\rho} = \frac{1}{Z} \sum_j e^{-\beta E_j} |j\rangle\langle j| = \frac{1}{Z} e^{-\beta \hat{H}}. \tag{3.10}$$

The partition function assures normalization, that is, $\mathrm{Tr}\,\hat{\rho} = 1$, so $\hat{\rho}$ plays the role of a quantum-mechanical probability density, in place of the probability density $|\psi|^2$ of the quantum mechanics of pure states.[11] Correspondingly, for any *observable* $\hat{A}$ of the system, the *expectation value* of its measurement can be calculated using the density operator according to

$$\langle \hat{A} \rangle = \mathrm{Tr}(\hat{\rho}\hat{A}). \tag{3.11}$$

Examples of such observables which are of interest for plasmas in thermodynamic equilibrium are the total and potential energy of the system, the free energy, the pressure (equation of state), the pair distribution function and static structure factor, and so on.

---

[8] The limit of a pure state follows as a special case, if every copy of the system in that ensemble is in the same state, and we recover the description based on the Schrödinger equation.

[9] This is derived straightforwardly from the Schrödinger equation and the definition (3.8).

[10] Thermodynamic equilibrium may also be described by other statistical ensembles like the *grand canonical ensemble*, for the chemical potential, μ, instead of the particle number. For this ensemble, the relations given in (3.10) gain an additional factor, the fugacity $e^{\beta\mu N}$.

[11] Here Tr denotes the trace, that is, the sum over all diagonal matrix elements.

3.5.2.1 Weakly Coupled Quantum Plasmas

As for the case of classical plasmas, analytical results for the density operator can be obtained only when correlations are weak, that is, for $r_s < 1$, in analogy to the classical weak coupling limit, $\Gamma < 1$. Then one can construct a perturbation theory with respect to r_s, taking the ideal quantum gas limit ($r_s = 0$) as the basis, and any quantity has the form

$$\langle \hat{A} \rangle = \langle \hat{A} \rangle^{\text{id}} + \langle \hat{A} \rangle^{\text{int}}, \tag{3.12}$$

where the first term is the ideal contribution, obtained for $r_s = 0$, and the second the interaction correction. In this limit, the ideal quantum plasma is described by a Fermi or Bose distribution function (which replaces the classical Maxwell distribution), depending on the spin statistics of the particles (see Sect. 3.5.3):

$$f^{\text{F/B}}(E_j; \mu, T) = \frac{1}{e^{\beta(E_j - \mu)} \pm 1}, \tag{3.13}$$

where μ is the chemical potential. The corresponding thermodynamic properties are well known from statistical mechanics, and we refer the reader to standard textbooks showing how to compute thermodynamic quantities and how to evaluate integrals involving Fermi or Bose distributions. However, when $r_s > 1$, correlation effects cannot be treated perturbatively and the thermodynamic quantities differ strongly from those of an ideal Fermi or Bose gas. Here, more sophisticated analytical approaches or computational tools have to be applied.

3.5.2.2 Chemically Reacting Quantum Plasma

If the plasma contains positive and negative charges, strong Coulomb attraction will eventually lead to the formation of bound electron–ion pairs, that is, to atoms and molecules, if the temperature and density are sufficiently low (cf. Fig. 3.2). In general, part of the particles will be bound and others will be free, so we have a partially ionized plasma. As in classical thermodynamics, eventually an equilibrium state is established in which the two fractions (i.e., the degree of ionization) have reached stationary values. This chemical equilibrium follows from the equality of the chemical potentials of the reacting species.

We consider, as an example, the ionization equilibrium in a dense hydrogen plasma at temperatures of about one tenth of the atomic binding energy (1 Ryd $\approx$ 155.000 K). There, electrons (e) can bind to protons (p) and form a hydrogen atom (H). Furthermore, two atoms can recombine to a molecule (H_2). The corresponding chemical equilibrium is determined by the coupled equations:

$$\mu_{\text{e}} + \mu_{\text{p}} = \mu_{\text{H}}, \quad 2\mu_{\text{H}} = \mu_{\text{H}_2}, \tag{3.14}$$

where the chemical potentials are of the form (3.12) and depend on temperature and all densities $\mu_a = \mu_a(T, n_{\text{e}}, n_{\text{H}}, n_{\text{H}_2})$, where we assume charge neutrality, $n_{\text{e}} = n_{\text{p}}$, and conservation of the total number of electrons, $n = n_{\text{e}} + n_{\text{H}} + 2n_{\text{H}_2} = \text{const}$.

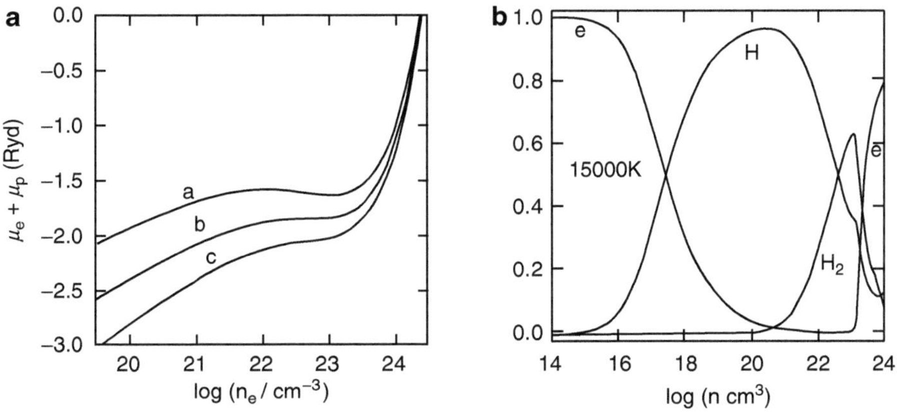

Fig. 3.3 Chemical composition of partially ionized hydrogen (from [20]). (**a**) Correlated equilibrium chemical potentials of electrons and protons versus free electron density for $T = 14{,}000$ K (*a*), $T = 17{,}000$ K (*b*), and $T = 20{,}000$ K (*c*). The minimum is due to Coulomb correlations and bound states (1 Ryd $= E_H = 13.6$ eV). (**b**) Fractions of electrons which are free (e) or bound in atoms (H) or molecules (H_2) versus total electron density. The region of bound states indicates the influence of correlations. At high densities, the *effective binding energy* I^{eff} of atoms and molecules decreases due to screening and quantum effects, and bound states are no longer stable, they vanish due to pressure ionization (Mott effect)

The results for the chemical potential isotherms of electrons and protons are shown in Fig. 3.3a. One sees the crossover from the ideal chemical potential of a classical plasma (low density) $\mu^{\text{id}}_{\text{cl},a} = k_B T \ln \chi_a$ (where χ_a is the degeneracy parameter; Sect. 3.2) to the ideal chemical potential of a Fermi gas (high density). In both limits, the chemical potential is a monotonically increasing function of the electron density. However, at intermediate densities, the curves show a decrease which becomes more pronounced as the temperature is lowered. This is a consequence of the correlation correction μ^{int} due to the Coulomb interaction between the particles, which is negative. In the range of this minimum, the plasma is strongly correlated, $\Gamma > 1$ and (for the electrons) $r_{se} > 1$, which is manifest in strong Coulomb attraction of electrons and protons – this interaction energy exceeds the kinetic energy of the particles, giving rise to formation of atoms and molecules. This is shown in Fig. 3.3b: the fraction of free electrons is high at low density and then decreases almost to zero, in favor of hydrogen atoms. At high densities, the atoms vanish in favor of molecules, but the latter exist only in a narrow density range and are replaced by free electrons (and protons) at densities exceeding $10^{23}\,\text{cm}^{-3}$. This is nothing but pressure dissociation (Mott effect), which was discussed in Sect. 3.4.5.

Finally, we mention that a nonmonotonic behavior of the chemical potential may give rise to instability of the homogeneous state, that is, a phase transition, similar to the liquid–gas transition. In fact, there have been many predictions of a *plasma phase transition* in dense hydrogen (e.g., [20, 22, 54]). We do not discuss here how the interaction parts of the chemical potentials were computed – it is a very challenging task of quantum statistical theory; the present results are from Schlanges et al. [20]. Details of the theory have been presented in various textbooks (see, e.g., [1,2]).

The present approach to chemically reacting quantum plasmas is called the *chemical picture* because one postulates the existence of different chemical species – such as free electrons, atoms, and molecules – and studies the chemical reactions between them (see, e.g., [1,23]). This subdivision, of course, loses meaning at the points where the degree of ionization changes substantially, such as the Mott point, at which other approaches are required. Further, we have to mention that in the strongly correlated regime, all analytical approximations become inaccurate or even qualitatively wrong and questionable. The alternative is computer simulations, such as quantum Monte Carlo or quantum molecular dynamics (QMD) which are discussed in Chap. 4.

3.5.3 Spin Effects in Quantum Plasmas

Up to this point, we have not considered the restrictions imposed on quantum many-particle states due to the indistinguishability of particles. We may attempt to write the full many-particle state of the system as a superposition of product states $|j\rangle$, where each is written as tensor product of single-particle states:

$$|j\rangle \equiv |j_1 j_2 \cdots j_N\rangle_{\mathrm{D}} = |j_1\rangle \otimes |j_2\rangle \otimes \cdots \otimes |j_N\rangle\,, \tag{3.15}$$

where $|j_i\rangle$ denotes the single-particle state occupied by particle i. This is simply the canonical way of constructing a basis for a tensor product space from the individual spaces. However, the indices given to individual particles imply that we are able to distinguish between them, which is *not* the case. Any permutation of labels cannot alter the measurement. Formally, such a permutation of two single-particle states $|j_i\rangle$ and $|j_k\rangle$ may be introduced with the *permutation operator* $\hat{P}_{ik}$:

$$\hat{P}_{ik}\,|\ldots j_i \ldots j_k \ldots\rangle = |\ldots j_k \ldots j_i \ldots\rangle = \pm\,|\ldots j_i \ldots j_k \ldots\rangle\,, \tag{3.16}$$

which is a Hermitian and unitary operator. Acting twice with $\hat{P}_{ik}$ on the same state changes nothing, that is, $\hat{P}_{ik}^2 = 1$, thus its eigenvalues are $+1$ and -1. We call states with $+1$ *symmetric* and states with -1 *antisymmetric*. As the operator is Hermitian, this symmetry is indeed measurable.

In nature exist two species of particles, *bosons* and *fermions*, which differ in their symmetry. Bosons occupy a *totally symmetric state*, which is symmetric under the exchange of *any two* particle labels. In the same vein, fermions occupy *totally antisymmetric states*. It can easily be shown that a product state constructed by using (3.15) does *not* satisfy (3.16), neither for the bosonic case nor for the fermionic one. Instead, we provide alternative construction formulae for the two cases, which automatically assure the correct symmetry of every many-particle basis state $|j\rangle$.

The (anti)symmetrization requires that we superimpose all possible product states which can be generated by exchanging from two to N particles (there is a total of $N!$ possibilities) with the result

$$|j_1 j_2 \cdots j_N\rangle_{\mathrm{S}} = \sqrt{\frac{\prod_j N_j!}{N!}} \sum_{p \in S_N} |j_{p(1)}\rangle \otimes |j_{p(2)}\rangle \otimes \cdots \otimes |j_{p(N)}\rangle , \tag{3.17}$$

$$|j_1 j_2 \cdots j_N\rangle_{\mathrm{A}} = \sqrt{\frac{1}{N!}} \sum_{p \in S_N} \mathrm{sgn}(p)\, |j_{p(1)}\rangle \otimes |j_{p(2)}\rangle \otimes \cdots \otimes |j_{p(N)}\rangle . \tag{3.18}$$

The sum is taken over all possible N particle permutations, S_N, and $\mathrm{sgn}(p)$ is the signature[12] or parity of each permutation p. The square root on the right-hand side is a normalization constant. The quantity N_j stands for the number of times each of the single-particle states appears in the N particle state. This factor can be omitted in the second case (fermions) since an important consequence of fermionic symmetry is the exclusion of states where the same single-particle state is occupied more than once, that is, for fermions, all N_j can equal only 0 or 1.

The exchange symmetry must also be taken into account for mixed states. This can easily be achieved when using the coordinate representation of the density operator introduced above which reads

$$\rho(R, R'; \beta) \equiv \langle R|\hat{\rho}|R'\rangle , \tag{3.19}$$

with $R = (\boldsymbol{r}_1, \ldots, \boldsymbol{r}_N)$, where $\boldsymbol{r}_i$ denotes the position of the ith particle. As Feynman [55] has shown, the density matrices for bosonic and fermionic systems can be obtained as follows:

$$\rho^{\mathrm{S}}(R, R'; \beta) = \frac{1}{N!} \sum_p \rho(R, p(R'); \beta), \tag{3.20}$$

$$\rho^{\mathrm{A}}(R, R'; \beta) = \frac{1}{N!} \sum_p \mathrm{sgn}(p) \rho(R, p(R'); \beta). \tag{3.21}$$

The nature of (anti)symmetric states has important consequences for the statistical properties of many-particle systems in the quantum regime in general and for quantum plasmas in particular. The *Pauli (exclusion) principle* forbids identical fermions from sharing the same single-particle state. As a result, adding Fermi particles to a system at low temperature is only possible if they occupy higher energy levels (assuming that all lower ones are occupied), which leads to a drastic increase of the total energy (per particle) with density as $n^{2/3}$. We have already discussed

---

[12] $\mathrm{sgn}(p)$ is always positive for bosons, and for fermions it is positive (negative) if the N particle permutation can be decomposed into an even (odd) number of pair permutations.

in Sect. 3.4 that this effect prevents the collapse of matter in our Universe since it would be energetically more costly than a state where all particles remain at a lower density.

Such a restriction does not exist for bosons which in fact may collapse into a *Bose–Einstein condensate* (BEC), in which a finite fraction of bosons occupies the same single-particle quantum state. These statistical properties are described as Bose–Einstein statistics. Fortunately, our Universe contains much less of bosons than of fermions. Further, in many cases, bosons are composed of several elementary particles (e.g., the hydrogen atom is a boson which is composed of two fermions). In such a case, the spin statistics of the composite particles depend on the number of constituents. For example, an even number of fermions always gives rise to a composite particle obeying (approximately) Bose statistics. Now, if such a system of Bose particles collapses, obviously, again a Mott-type transition would occur leaving us with fermions which resist further compression.

3.5.4 Bose Plasmas

As an example, we consider few bosons in a harmonic trapping potential. This will allow us to illustrate the effects of coupling strength and quantum statistics. We will focus on a system of charged particles (e.g., bosonic ions) which can be called a *bosonic nanoplasma*. Such systems are as interesting as trapped fermions, although the densities necessary to reach strong quantum degeneracy are not yet experimentally accessible.

Consider the Hamiltonian (3.6):

$$\hat{H} = \sum_{i=1}^{N} \left[-\frac{\hbar^2}{2m_i} \nabla_i^2 + \frac{1}{2} m\omega^2 \mathbf{r}_i^2 + \sum_{j<i}^{N} \frac{q^2}{r_{ij}} \right], \tag{3.22}$$

where ω is the trapping frequency and q the charge per ion. The thermodynamic state of the nanoplasma is characterized by three parameters: particle number N, temperature, and the dimensionless coupling parameter,[13] $\lambda = e^2/(l_0 \hbar\omega)$, which is the ratio of the Coulomb interaction to the trap energy scale (with the square of the oscillator length given by $l_0^2 = \hbar/m\omega$). λ characterizes the confinement strength and can be externally controlled by the trap frequency ω. Reducing ω increases the coupling and lowers the density. In the limit $\lambda \to \infty$, the system approaches a classical crystal of point-like charges.

In the following, we present simulation results obtained using the path-integral Monte Carlo (PIMC) method (this is explained in Chap. 4). The results correspond to a fixed temperature, $k_B T/E_0 = 1/2{,}000$, which is given in units of the average

[13] In a finite system, the density is set by the trap frequency and λ plays the role of the coupling parameter r_s of a macroscopic system.

Coulomb interaction energy, $E_0 = m\omega^2 r_0^2 = e^2/r_0$, of two classical particles having ground-state separation r_0. The choice of this low temperature allows us to eliminate the influence of classical thermal fluctuations on the liquid–solid transition and to enhance the effect of zero-point motion and quantum statistics.

The general behavior of the system of bosonic particles in the crystalline state is similar to that of the case of fermions [16]. At low density, the particles are well localized and arranged in concentric shells. This shell structure is very robust and is observed not only in the crystalline state but also in the mesoscopic liquid-like state. As density increases, a two-stage melting process is observed. The first is orientational melting (OM), that is, the system transforms from complete spatial ordering to a second crystal phase which is only radially ordered (RO) – the particles are still localized on shells, but the shells can rotate relative to each other. When the density is increased further, a second transition – radial melting (RM) – sets in and the radial order gradually vanishes, giving rise to a liquid-like state. RM is typically observed at significantly higher densities than OM, corresponding to $\lambda_{RM} \approx 28$ and $\lambda_{OM} > 60$.

One of the most striking phenomena of Bose liquids is their ability to form a BEC when cooled to very low temperatures. Under such conditions, a large fraction of the atoms collapse into the lowest quantum state of the external potential, at which point quantum effects become apparent on a macroscopic scale. Experimentally, the first BEC was created by Eric Cornell, Carl Wieman, and coworkers at JILA in June 1995. They accomplished this by cooling a dilute vapor consisting of approximately 2,000 rubidium-87 atoms to below 170 nK using a combination of laser cooling and magnetic evaporative cooling [56]. About 4 months later, an independent effort led by Wolfgang Ketterle at MIT created a condensate consisting of sodium-23 atoms. Ketterle's condensate had about a hundred times more atoms, allowing him to obtain several important results such as the observation of quantum-mechanical interference between two different condensates [57]. Cornell, Wieman, and Ketterle won the 2001 Nobel Prize in Physics for their achievement [58].

An advantage of a trapped BEC is that it can be observed in both position and momentum space as a sharply spiked Gaussian against the flat distribution of the thermal cloud of noncondensate particles. This effect is hinted in Fig. 3.4 showing the density distribution of five charged bosons in a harmonic trap. At low coupling strength, $\lambda \ll 1$, the system is nearly ideal and all particles settle in the trap center occupying the lowest oscillator eigenstate. This indicates the tendency toward BEC (which, in fact, is a macroscopic phenomenon). As coupling increases, Coulomb repulsion increases and particles are pushed away from the trap bottom, that is, the ground state becomes depopulated, and the condensate is depleted. At strong coupling, $\lambda \sim 10$, the trap bottom is even emptied since the strong Coulomb repulsion favors a ring-shaped density distribution, until at $\lambda \gtrsim 30$ five individual density peaks emerge indicating the trend toward classical behavior resembling a Wigner crystal. The figure also shows the effect of temperature increase (see the right columns): the additional kinetic energy leads to a delocalization of the particles which start to oscillate around their local potential minimum. At large λ, the temperature increase leads to melting of the crystal.

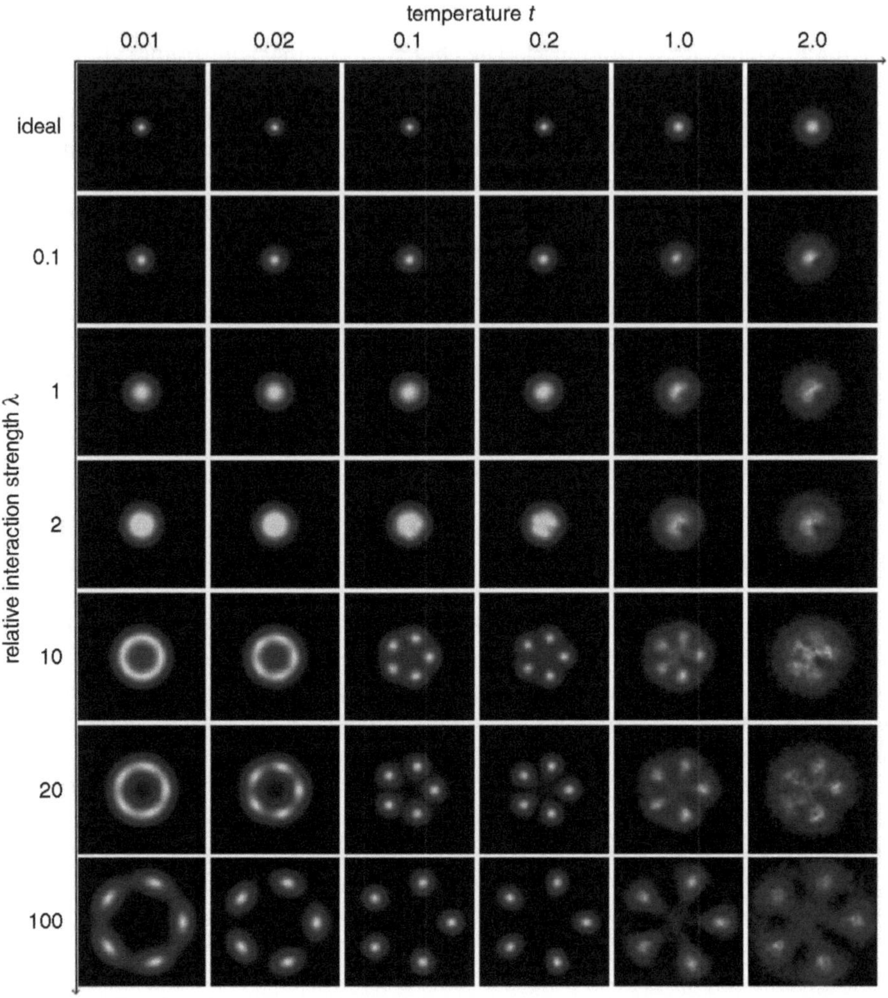

Fig. 3.4 Snapshots of a Bose "nanoplasma" of five particles in a harmonic trap for various values of the coupling parameter λ and temperature (from [59])

There is a second interesting effect in Bose systems – superfluidity. This phenomenon exists only in nonideal Bose liquids, that is, it requires substantial pair interaction. Superfluidity describes a phase of matter in which the liquid overcomes friction and its viscosity drops to zero. The temperature below which this occurs is known as the *lambda point*, and was first observed for helium-4. Such a liquid can be adequately described by a two-fluid model where the total density is decomposed into a normal and a superfluid part, $\rho = \rho_{\mathrm{n}} + \rho_{\mathrm{s}}$, with the normalization $\int \rho(\boldsymbol{r})\mathrm{d}^2r = N$. The superfluid part of a rotating liquid can be detected as *missing moment of inertia (MMI)* [60] since only the normal fluid part produces a moment of inertia. This allows one to determine the superfluid fraction γ_s experimentally or by numerical simulations as the deviation of the measured moment of inertia I from the familiar classical moment ($I^{cl} = \int \rho(\boldsymbol{r}) m r^2 d^2 r$) with the whole mass is rotating.

Here, we consider that the whole system is set into rotation around an axis perpendicular to the 2D plane, in our case, passing through the origin of the trapping potential, $\boldsymbol{r} = 0$. In path-integral simulations, the superfluid fraction γ_s can be computed with the *area formula* [61]:

$$\gamma_s = \frac{\rho_s}{\rho} = \frac{I^{cl} - I}{I^{cl}} = \frac{4m^2 k_B T \langle A^2 \rangle}{\hbar^2 I^{cl}}, \tag{3.23}$$

where A is the area enclosed by all particle trajectories in the 2D plane and $\langle \ldots \rangle$ denotes the thermodynamical average. The results for γ_s for $N = 19$ are shown in Fig. 3.5 [62]. We observe that the superfluid fraction decreases exponentially when approaching the crystallization point. To test the reliability of the results for γ_s, we also performed calculations for spinless "distinguishable" quantum particles. For this system, in the thermodynamic limit ($N \to \infty$), the RHS of (3.23) asymptotically goes to zero, $\gamma_s \to 0$, for any value of λ. This is not the case for simulations for a finite number of trapped particles, which yield small finite values, $\gamma_s \approx 0.1$–1%. The obtained values of γ_s for the spinless particles can be regarded as a noise level in our computer experiment causing a slight overestimate of the true superfluid fraction.

The systematic deviation of our results from the blue curve in Fig. 3.5 confirms that we indeed observe a collective quantum response, that is, a finite superfluid fraction, in the radially ordered solid phase, $28 \lesssim \lambda \lesssim 45$. For $\lambda \gtrsim 45$, the superfluid

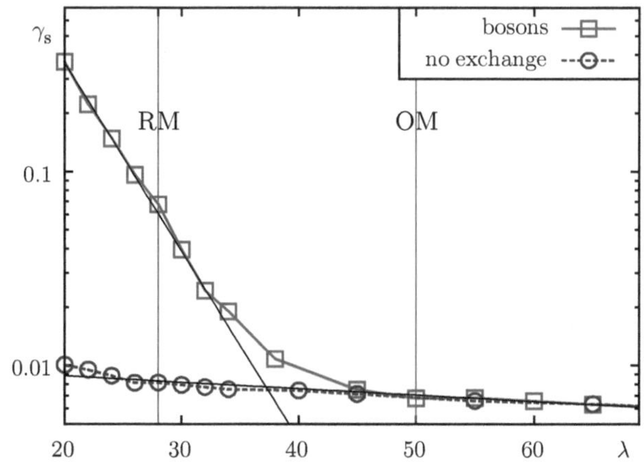

Fig. 3.5 Superfluid fraction in a strongly correlated Bose "nanoplasma" of 19 particles in a harmonic trap for various values of the coupling parameter λ and temperature. The *dashed line* shows the comparison to a quantum plasma without Bose statistics ("boltzmannons") (from [62])

fraction decays to the noise level and is negligible in the orientationally ordered crystal phase. A question concerning the possible coexistence of crystalline and superfluid behavior in a nonideal Bose system (a so-called *supersolid*) has been raised in the late 1960s [63–65]. It is presently very actively studied both experimentally and theoretically (e.g., [60] for an overview). The present mesoscopic Bose systems indicate that the crystalline phase may in fact contain a small fraction of a few percent of superfluidity. However, this is caused by the trapping potential, which breaks the translational symmetry, and the effect vanishes with increasing particle number [62].

Superfluidity is presently very actively studied in neutral Bose systems in traps and optical lattices (for a recent overview, see [66]). Finally, we note that superfluidity is also expected to occur in dense astrophysical objects, such as in the core of neutron stars. There, the density is so large (cf. Sect. 3.4.5) that nuclei are unstable giving rise to a fluid of protons and neutrons. It is expected that the latter are able to form bosonic pairs which can exhibit superfluidity [67, 68].

3.5.5 Plasmas of Particles Having Fermi Statistics

Consider now a "normal" plasma consisting of fermions, such as electrons. We already know that this system obeys Fermi–Dirac statistics and fulfills the Pauli principle. To illustrate its typical behavior, we consider the same Hamiltonian (3.22) as before – a few particles in a harmonic trap – but now fermions instead of bosons. The density distribution of two to seven fermions in the trap is shown in Fig. 3.6. It corresponds to strong coupling, so the Coulomb repulsion again leads to spatial separation of the individual particles. At first sight, there is no difference from the case of bosons. Differences appear only at lower coupling when neighboring particles come in contact with each other and "feel" their spin statistics. This is shown in Fig. 3.7 (top row) where we show the density distribution for six fermions in different approximations. Consider first the upper solid (green) curves which are computed for an ideal system in which all interactions are neglected, that is, $\lambda = 0$. There is a clear density modulation although particles do not repel each other. On energetic grounds alone, each particle would tend to occupy the bottom of the trap, $x = 0$, which is exactly what we saw in the case of bosons at $\lambda = 0$ (see Fig. 3.4). Thus, the difference must come from the differing spin statistics. In fact, in the case of ideal fermions, we observe an effective repulsion of the particles which can only be attributed to the Pauli principle. Indeed, only one particle can occupy the trap bottom (more precisely, the oscillator ground-state orbital), and the others have to move to other orbitals with higher energies.

This density modulation is progressively washed out at higher temperatures because quantum effects are reduced. Consider now the other curves corresponding to the inclusion of interaction effects, in which we see the combined action of "Pauli repulsion" and Coulomb repulsion which further enhances the density modulation.

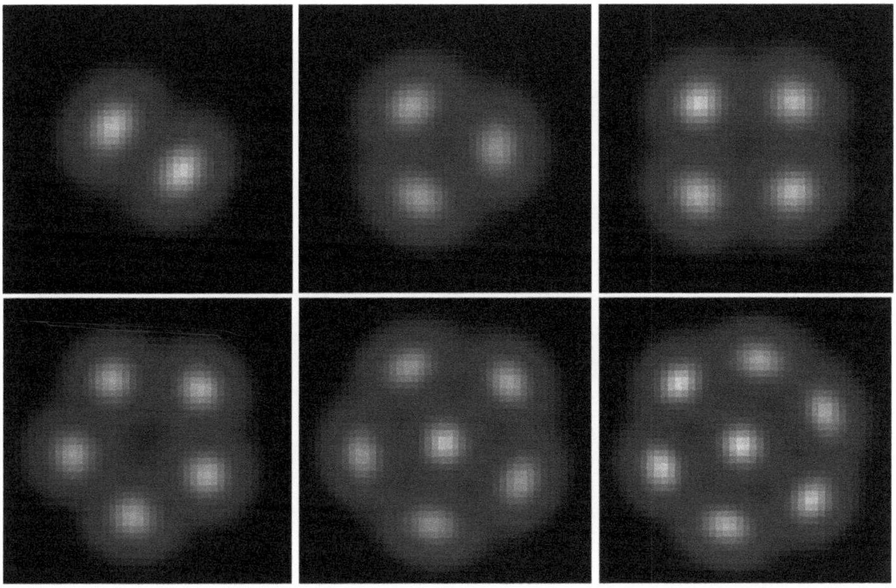

Fig. 3.6 Snapshots of the ground-state density of a Fermi "nanoplasma" consisting of two to seven particles in a harmonic trap at $\lambda = 5$ and $\beta = 100$ (from [69])

The two curves correspond to two different approximations – the Hartree–Fock (mean field) and the second Born approximation which takes into account Coulomb correlations.

That the above energy interpretation is correct becomes clear from the lower row of Fig. 3.7, in which we show the average occupation (probability) of various energy levels in the trap. Clearly, all states are occupied by no more than one particle, again confirming the Pauli principle. If there were no interactions, the energy distribution would be given by the Fermi function (3.13). This is just the full (green) curve which is close to a step function at low temperature (left figure). At higher temperatures, the step becomes rounded, and some particles from low-lying energy states move to higher states, due to thermal excitation: We can also see the effect of the Coulomb interaction, see, for example, the red line (circles) in the center figure: it also reduces the occupation of lower states (close to the sharp decrease of the function) and increases the occupation of higher states.

The present results have been computed with the use of the nonequilibrium Green's function formalism, and we provide some explanation in Sect. 3.5.7 (see also Chap. 4).

Finally, we mention that there are situations where fermions manage to transform into bosons, in order to achieve a lower energy, such as the trap bottom in the example above. This effect is known as *Cooper pair* formation where two electrons form a weakly bound pair which obeys Bose statistics. This is possible in some solids and it requires a strong attractive interaction of electrons with lattice

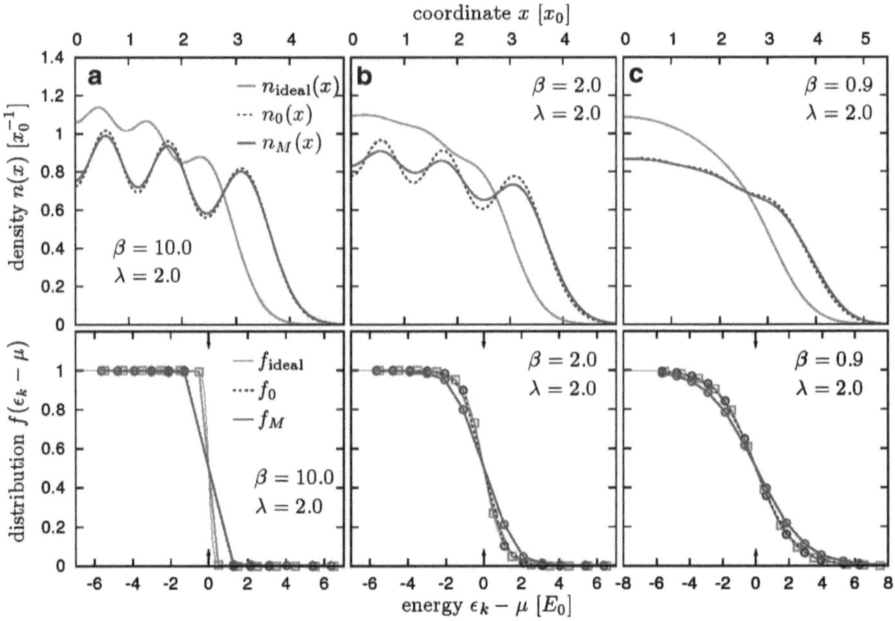

Fig. 3.7 Six Fermi particles in a harmonic trap at $\lambda = 2$ for three temperatures – *left column*: $k_B T = \hbar\omega/10$, *middle column*: $k_B T = \hbar\omega/2$, and *right column*: $k_B T = 1.1\hbar\omega$. *Upper row*: density distribution in the trap for an ideal Fermi system (*full green line*), in Hartree–Fock approximation (*dotted line*) and including correlations (*full red line*). Due to the trap symmetry only positive coordinates are shown, $n(-x) = n(x)$. *Lower row*: energy distribution functions for the same three cases (from [70])

vibrations (phonons) sufficient to overcome their Coulomb repulsion. Such Cooper pairs have in fact been found – they are responsible for the loss of electrical resistance, that is, superconductivity, in a variety of materials. This effect is analogous to superfluidity of bosons. This phenomenon is also expected to occur in dense quantum plasmas, such as in dwarf stars or in the core of neutron stars. But this is beyond the scope of our chapter. For more details, the interested reader is referred to textbooks on condensed matter theory.

3.5.6 Quantum Kinetic Theory

Having discussed the properties of quantum plasmas in equilibrium, we now move on to time-dependent phenomena, in particular, to the response to external fields. The standard description of *nonequilibrium plasmas* is based on kinetic theory, in which the central quantity is the phase-space distribution function $f_a(\boldsymbol{r}, \boldsymbol{p}, t)$ of electrons, ions, and neutral atoms, describing the probability density with normalization $\int d^3r d^3p\, f_a(\boldsymbol{r}, \boldsymbol{p}, t) = N_a(t)$ (e.g., [3, 71–73]). This function obeys a kinetic equation describing the time evolution of f_a under the action of external fields as well as its evolution toward the equilibrium state, due to scattering effects. A typical kinetic equation has the form[14]

$$\frac{\partial f}{\partial t} + \boldsymbol{v} \cdot \frac{\partial f}{\partial \boldsymbol{r}} - \frac{\mathrm{d}U}{\mathrm{d}\boldsymbol{r}} \cdot \frac{\partial f}{\partial \boldsymbol{v}} = \int \mathrm{d}^3 r_2 \, \mathrm{d}^3 p_2 \, \frac{\mathrm{d}V(\boldsymbol{r} - \boldsymbol{r}_2)}{\mathrm{d}\boldsymbol{r}} \frac{\partial f_{12}}{\partial \boldsymbol{v}_1} = I(\boldsymbol{r}, \boldsymbol{p}, t), \quad (3.24)$$

where U is the total potential energy appearing in the Hamiltonian (3.6) and I is the collision integral involving the binary distribution function f_{12}.

We now discuss how to generalize the concepts of kinetic theory to quantum plasmas. As is known from basic quantum mechanics, coordinate and momentum do not commute but obey Heisenberg's uncertainty relation, that is, they are not measurable at the same time. This immediately implies that a phase-space function, $f(\boldsymbol{r}, \boldsymbol{p}, t)$, may be ill-defined for a quantum system. We will see below that this problem, in fact, appears in the resulting quantum kinetic equation, giving rise to a more complex (nonlocal) dependence of the distribution function on momentum. We start the derivation using the nonequilibrium density operator $\hat{\rho}$ which contains full information about time dependence of the quantum many-particle system, as given by the von Neumann equation (3.9). Its classical analog is the Liouville equation for the N particle distribution function, $f_N(\boldsymbol{r}_1, \boldsymbol{p}_1, \ldots, \boldsymbol{r}_N, \boldsymbol{p}_N, t)$.

The single-particle distribution function, $f(\boldsymbol{r}_1, \boldsymbol{p}_1, t)$, is obtained from f_N by integrating over the additional variables, that is, $r_s, \ldots, r_N$ and $p_2, \ldots, p_N$. In full analogy, we define the one- and two-particle density operators ($s \in \{1, 2\}$) via a partial trace:

$$F_{1,\ldots,s} = \mathcal{V}^s \, \mathrm{Tr}_{s+1,\ldots,N} \, \rho_{1,\ldots,N}, \quad \frac{1}{\mathcal{V}^s} \mathrm{Tr}_{1,\ldots,s} F_{1,\ldots,s} = 1, \quad (3.25)$$

where we used the traditional normalization to the volume $\mathcal{V}$. It is straightforward to derive, from (3.9), the equation of motion of F_1 [3]:

$$\mathrm{i}\hbar \frac{\partial}{\partial t} F_1 - [H_1, F_1] = n \mathrm{Tr}_2 \, [V_{12}, F_{12}], \quad (3.26)$$

which couples F_1 to the two-particle density operator F_{12}. Similarly, the equation for F_{12} couples to F_{123} giving rise to a hierarchy of equations, the BBGKY hierarchy (for Bogolyubov, Born, Green, Kirkwood, and Yvon) exactly as in the classical case. Here, for simplicity, we neglect quantum exchange effects which will be restored in Sect. 3.5.7.

Thus, the operator equation (3.26) has similar structure to that of the classical kinetic equation (3.24). What remains is to transform to phase-space variables. In quantum mechanics, this is done using coordinate representation and transforming

[14] For simplicity, we consider a single-component plasma.

to center of mass and relative coordinates and applying the Wigner transform. Here, we present the final result (for details, see [3]):

$$\begin{aligned}
&\left\{\frac{\partial}{\partial t}+\frac{p_1}{m_1}\nabla_{R_1}\right\} f(R_1,p_1,t)-\frac{1}{\mathrm{i}\hbar}\int\frac{\mathrm{d}r_1\mathrm{d}\bar{p}_1}{(2\pi\hbar)^3}\,\mathrm{e}^{-(\mathrm{i}/\hbar)(p_1-\bar{p}_1)r_1}\\
&\quad\times\left\{U\left(R_1+\frac{r_1}{2}\right)-U\left(R_1-\frac{r_1}{2}\right)\right\} f(R_1,\bar{p}_1,t)\\
&=\frac{1}{\mathrm{i}\hbar}\int\frac{\mathrm{d}r_1\mathrm{d}\bar{p}_1}{(2\pi\hbar)^3}\,\mathrm{d}R_2\,\mathrm{d}p_2\,\mathrm{e}^{-(\mathrm{i}/\hbar)(p_1-\bar{p}_1)r_1}\\
&\quad\times\left\{V\left(R_1-R_2+\frac{r_1}{2}\right)-V\left(R_1-R_2-\frac{r_1}{2}\right)\right\} f_{12}(R_1,\bar{p}_1,R_2,p_2,t).
\end{aligned} \tag{3.27}$$

This equation can be directly compared to the classical kinetic equation. Differences appear only in the explicit form of the external potential U and the interaction potential V. Whereas in the classical case the equation contains the gradient (corresponding to the associated force), in the quantum case there appears a difference of potentials with shifted arguments. This is a pure quantum effect, and the classical limit is recovered in the limit of vanishing difference of arguments, that is, $r_1 \to 0$. This nonlocal coupling is a consequence of the Heisenberg's uncertainty principle: the value of f at a given phase-space point (R_1, p_1) depends not only on the value of the potential at the same point R_1 but also on its values at all other points. The reason is the finite spatial extension of quantum particles. The analysis shows that, in general, the main contribution comes from the point R_1, and the contribution of neighboring points decreases with distance. This can be accurately analyzed by performing a Taylor expansion, which yields quantum corrections to the classical expression. This is the basis for various semiclassical approximations and numerical schemes of *QMD* (for details, see the chapter on QMD in [5]).

There have been considerable activities over the last several decades to develop solutions of the quantum kinetic equation (3.27), mostly in solid-state physics and quantum plasma physics [74, 75]. However, this still remains a challenging problem which is handled much less efficiently than in the classical limit where highly optimized particle-in-cell methods exist (see Chap. 9).

Finally, we mention that a closed kinetic equation for F_1 requires the introduction of a decoupling approximation for the hierarchy which yields F_{12} as a functional of F_1. Typical approximations which are standard in many-body theory are the mean field (Vlasov or Hartree) approximation, the second Born approximation, the polarization (dynamically screened second Born or Lenard–Balescu) approximation, or the Boltzmann (or T-matrix) approximation, and so on, see e.g. [3].

3.5.7 More Advanced Approach: The Method of Second Quantization

In the preceding section, we have exhibited the derivation of a quantum kinetic equation starting from the *N* particle density operator. So far, this equation omits a key property of quantum particles – the correct spin statistics. This can be corrected within the density operator formalism by performing an (anti)symmetrization of the reduced density operators (e.g., [3, 76]), but there exists a simpler approach – the method of *second quantization* which we now discuss.

The key idea is to introduce the bosonic (fermionic) (anti)symmetry of an N particle state at the lowest level, when adding a single quantum particle to a previously empty system given by the vacuum state $|0\rangle$. To this end, one defines a *creation operator* $\hat{\psi}^\dagger$ with the property

$$\hat{\psi}^\dagger \, |0\rangle = |1\rangle \,. \tag{3.28}$$

This can be generalized to an arbitrary state $|n\rangle$ which contains n particles,[15] and we can also define an *annihilation operator*, $\hat{\psi}$, which removes a particle

$$\hat{\psi}^\dagger \, |n\rangle = C_n^+ \, |n+1\rangle \,, \tag{3.29}$$

$$\hat{\psi} \, |n\rangle = C_n^- \, |n-1\rangle \,. \tag{3.30}$$

Obviously, the action of the latter on the vacuum state gives nothing, $\hat{\psi}\,|0\rangle = 0$, by definition of the vacuum. The normalization constants C_n^- and C_n^+ differ for bosons and fermions. They are chosen in such way that we can act on a state successively with both operators obtaining

$$\hat{n}\,|n\rangle \equiv \hat{\psi}^\dagger \hat{\psi}\,|n\rangle = n\,|n\rangle \,. \tag{3.31}$$

$\hat{n} = \hat{\psi}^\dagger \hat{\psi}$ is called the *particle number operator*, the action of which on a state $|n\rangle$ gives just the number of particles in this state. Now we can easily build into this approach the Pauli principle. We only need to require that we cannot add a second particle to any state, that is, $\hat{\psi}^\dagger\,|n\rangle = 0$ for all $n \geq 1$. Naturally, this applies only to the case of fermions, whereas for bosons no such restriction exists.

With these introductory remarks, we can formulate the basic expressions of the method of second quantization. First, we generalize the creation and annihilation operators corresponding to a specific quantum state to the case of many such available states. We do this by adding the position argument $\boldsymbol{r}$ to the operators which now describe particles propagating in the entire space with different arguments referring to different positional states. This means $\hat{\Psi}^\dagger(\boldsymbol{r},t)$ ($\hat{\Psi}(\boldsymbol{r},t)$) creates (annihilates) a particle at space point $\boldsymbol{r}$ at time t. With these arguments, the operators are usually called *field operators*. The fermionic/bosonic character of the operators is guaranteed by imposing the following equal time (anti)commutation relations:

$$\left[\hat{\Psi}^{(\dagger)}(\boldsymbol{r}_1), \hat{\Psi}^{(\dagger)}(\boldsymbol{r}_2)\right]_{\mp} = 0, \quad \left[\hat{\Psi}(\boldsymbol{r}_1), \hat{\Psi}^{\dagger}(\boldsymbol{r}_2)\right]_{\mp} = \delta(\boldsymbol{r}_1 - \boldsymbol{r}_2), \tag{3.32}$$

---

[15] The space of all symmetric or antisymmetric states with 0, 1, 2, ... particles is called *Fock space*.

where $[a, b]_\pm = ab \pm ba$. Here, the minus (plus) sign refers to bosons (fermions). These equations state that two annihilators or two creators of bosons (fermions) always commute (anticommute), whereas a creator and an annihilator do not commute (anticommute).

From these definitions, a new representation of quantum mechanics of many particles is possible. Indeed, one can show that all quantum-mechanical operators can be expressed in terms of the field operators. For example, a straightforward calculation yields the Hamilton operator (3.6) in the form

$$\hat{H}(t) = \int d^3r\, \hat{\Psi}^\dagger(\boldsymbol{r}) \left[-\frac{\hbar^2}{2m}\nabla^2 + V(\boldsymbol{r}, t) \right] \hat{\Psi}(\boldsymbol{r}) + \frac{1}{2} \iint d^3r\, d^3\bar{r}\, \hat{\Psi}^\dagger(\boldsymbol{r}) \hat{\Psi}^\dagger(\bar{\boldsymbol{r}}) w(\boldsymbol{r} - \bar{\boldsymbol{r}}) \hat{\Psi}(\bar{\boldsymbol{r}}) \hat{\Psi}(\boldsymbol{r}), \tag{3.33}$$

and analogously for all other observables.

Now, for nonequilibrium (as well as equilibrium) applications, we are interested in the time evolution of all observables. To this end, we need the time dependence of the field operators, which may be obtained from the Heisenberg's picture of quantum mechanics, which states that the time evolution of any operator is given by its commutator with the Hamiltonian (we assume that the operator itself is time independent):

$$\mathrm{i}\hbar \frac{\partial}{\partial t} \hat{A} = \left[\hat{A}, \hat{H} \right]. \tag{3.34}$$

Using this equation, we can obtain the equations governing the dynamics of $\hat{\Psi}^\dagger$ and $\hat{\Psi}$ and any of their combinations by simple algebra.

Finally, we discuss how to extract physically useful results from this formalism. As always in quantum mechanics, the operators alone are random quantities with a spectrum of possible realizations (eigenvalues). To compare with experiments, we need information about the probability of a certain value, that is, about the expectation value of the operator. The same is true here. The probability of different realizations is given by the density operator $\hat{\rho}$, so as discussed earlier (see (3.11)) the ensemble average $\langle \hat{A} \rangle$ is given by the trace $\mathrm{Tr}\hat{\rho}\hat{A}$. For example, as we have seen earlier, the product of two field operators is related to the particle number operator. Generalizing this to coordinate-dependent operators, we can determine the particle density at position $\boldsymbol{r}$ at time t, via the expectation value:

$$n(\boldsymbol{r}, t) = \langle \hat{\Psi}^\dagger(\boldsymbol{r}, t) \hat{\Psi}(\boldsymbol{r}, t) \rangle \,. \tag{3.35}$$

This is an important but rather specialized observable, and we generally wish to extract the maximum possible information from the field operators. This can be achieved by generalizing (3.35) to the case of different space and time arguments which we denote compactly $1 \equiv \boldsymbol{r}_1, t_1$, etc.:

$$G^<(1,\bar{1}) = \pm\frac{1}{\mathrm{i}\hbar}\left\langle \hat{\Psi}^\dagger(\bar{1})\,\hat{\Psi}(1)\right\rangle, \qquad (3.36)$$
$$G^>(1,\bar{1}) = \frac{1}{\mathrm{i}\hbar}\left\langle \hat{\Psi}(1)\,\hat{\Psi}^\dagger(\bar{1})\right\rangle,$$

where the plus (minus) sign refers to bosons (fermions). Note that the ordering of the operators is important since they do not commute (anticommute) at different times; therefore, the functions $G^>$ and $G^<$ are, in general, independent. These functions are called *two-time correlation functions* (or *nonequilibrium Green's functions*) and contain the full information about the many-body system. Here, we only note their relation to the density[16] and to the single-particle density matrix discussed earlier:

$$\rho(\boldsymbol{r}_1, \boldsymbol{r}_{\bar{1}}; T) = \pm\mathrm{i}\hbar G^<(1,\bar{1})\big|_{t_1=t_{\bar{1}}=T}\,, \qquad (3.37)$$

$$n(\boldsymbol{R}, T) = \pm\mathrm{i}\hbar G^<(1,\bar{1})\big|_{t_1=t_{\bar{1}}=T,\,\boldsymbol{r}_1=\boldsymbol{r}_{\bar{1}}=\boldsymbol{R}}\,. \qquad (3.38)$$

From the density matrix, we immediately obtain the phase-space distribution (Wigner function) via a Wigner transform as explained in Sect. 3.5.6. Then, from the equations of motion of the correlation functions $G^{\gtrless}$, one directly recovers the quantum kinetic equations discussed before, but now generalized to the inclusion of spin (exchange) effects.

The method of NEGF has been developed in the late 1950s by Schwinger, Martin, Keldysh, Baym, Kadanoff, and others (see, e.g., [77, 78] and references therein). It generalizes the approach of quantum kinetic theory and has become a powerful tool of many-particle physics in many fields, including dense quantum plasmas [1–3]. The central equations of this theory are the Keldysh/Kadanoff–Baym equations (KKBE), that is, the equations of motion for the Green's function $G^{\gtrless}$ (cf. (3.36), and they are successfully being solved numerically [3, 5, 70]). In fact, our results shown earlier in Figs. 3.6 and 3.7 for a nonideal fermion nanoplasma were obtained by solving the KKBE. The interested reader is referred to Chap. 4 for more details. We also note that some applications of the NEGF theory to solid-state plasmas, in particular to the treatment of dielectric properties and dynamical screening, are given in Chap. 5.

3.5.8 Other Approaches to Quantum Plasmas

Several practical methods for the theoretical description and simulation of quantum systems have been developed from semiclassical representations of the quantum problem of interest. These include *Bohmian trajectories*, in which the system

---

[16] Compare to (3.35).

particles obey an effective classical Newtonian dynamics, and the closely related field of *quantum hydrodynamics*, in which the Wigner distribution is represented in terms of equations for its momentum moments. Both are exact in principle, and form the basis for the introduction of practical approximate descriptions of quantum effects.

3.5.8.1 Bohmian Quantum Mechanics

In an attempt to address the conceptual paradoxes of "reality" in quantum theory, Bohm [79, 80] proposed a reinterpretation of the solution to the Schrödinger equation and associated measurements. Consider a system of N identical particles in the quantum state $\Psi(\{\boldsymbol{r}_i\}, t)$, where $\{\boldsymbol{r}_i\} \equiv \{\boldsymbol{r}_1, \ldots, \boldsymbol{r}_N\}$ denotes a configuration for the system. Bohm showed that the particles could be considered "real" in the classical sense of having their trajectories determined by Newtonian mechanics with specific positions and momenta. However, quantum effects occur through additional forces determined by Ψ, and a statistical distribution of the trajectories determined from $|\Psi|^2$. In this sense, the physical features of a quantum system are determined by an ensemble of classical trajectories. A brief overview of this remarkable interpretation follows.

The wave function $\Psi(\{\boldsymbol{r}_i\}, t)$ can be expressed in terms of its amplitude $A(\{\boldsymbol{r}_i\}, t)$ and phase $S(\{\boldsymbol{r}_i\}, t)$ as

$$\Psi(\{\boldsymbol{r}_i\}, t) = A(\{\boldsymbol{r}_i\}, t)\, \mathrm{e}^{(i/\hbar)S(\{\boldsymbol{r}_i,\}t)}. \tag{3.39}$$

These two real functions obey equations that follow directly from the real and imaginary parts of the time-dependent Schrödinger equation:

$$\partial_t A = -\frac{1}{2m}\sum_{i=1}^{N}\left[2\left(\nabla_{\boldsymbol{r}_i} S\right)\cdot\nabla_{\boldsymbol{r}_i} A + A\nabla^2_{\boldsymbol{r}_i} S\right], \tag{3.40}$$

$$\partial_t S = -\sum_{i=1}^{N}\left[\frac{1}{2m}\left(\nabla_{\boldsymbol{r}_i} S\right)^2 + U + Q\right], \tag{3.41}$$

where U is the potential energy of the Hamiltonian and Q is an *effective quantum potential* given by

$$Q(\{\boldsymbol{r}_i\}, t) = \frac{\hbar^2}{2m}\frac{\nabla^2_{\boldsymbol{r}_i} A(\{\boldsymbol{r}_i\}, t)}{A(\{\boldsymbol{r}_i\}, t)}. \tag{3.42}$$

Equation (3.40) implies the conservation of probability for the probability density field, $P = A^2$. The notion of this probability density "flow" as being associated with a configuration space trajectory of the system's particles, $\{\boldsymbol{r}_i(t)\}$, follows from Bohm's observation that the second equation (3.41) has the form of the Hamilton–Jacobi equation with $S(\{\boldsymbol{r}_i\}, t)$ being the generating function

$$\partial_t S(\{\boldsymbol{r}_i\}, t) + \mathcal{H}(\{\boldsymbol{r}_i\}, t; \{\nabla_{\boldsymbol{r}_i} S\}) = 0, \tag{3.43}$$

where the classical Hamiltonian $\mathcal{H}$ is identified as

$$\mathcal{H}(\{\boldsymbol{r}_i\};\{\nabla_{\boldsymbol{r}_i}S\}) = \sum_{i=1}^{N}\frac{p_i^2}{2m} + U(\{\boldsymbol{r}_i\}) + Q(\{\boldsymbol{r}_i\},t). \tag{3.44}$$

The canonical momenta in this context are obtained from the generating function according to $\boldsymbol{p}_i = \nabla_{\boldsymbol{r}_i}S\,(\{\boldsymbol{r}_i\},t)$. This identifies a classical Hamiltonian dynamics for the variables $\{\boldsymbol{r}_i(t),\boldsymbol{p}_i(t)\}$. In particular, this definition of the momenta provides the configuration space trajectories from the solutions of

$$m\dot{\boldsymbol{r}}_i = \nabla_{\boldsymbol{r}_i}S\,(\{\boldsymbol{r}_i\},t)\,. \tag{3.45}$$

It follows from the continuity equation in configuration space that $P(\{\boldsymbol{r}_i\},t) = P(\{\boldsymbol{r}_i(t)\},0)$, where $\boldsymbol{r}_i(t)$ are determined from (3.45) for specified initial values of $\{\boldsymbol{r}_i\}$ and $\Psi(\{\boldsymbol{r}_i\},0)$. Therefore, the time evolution of the probability density can be calculated from the trajectories. However, the presence of the quantum potential Q in the Hamiltonian $\mathcal{H}$, or S in (3.45), requires the evolution of A and S as well. Consequently, the trajectories are entirely determined from the coupled set of (3.40), (3.41), and (3.45). Numerical integration of these equations is at the heart of current applications of Bohmian quantum mechanics. For a brief review see [81], or for more details of the implementation and references see the book by Wyatt [82]. The objective of application of this approach is an effective reconstruction of the wave function, and some of its advantages over direct solution of the Schrödinger equation are discussed in the last two references.

The above description is restricted to pure states, that is, systems with a well-defined wave function. More generally, a system may be specified only statistically, with probabilities $\{p_\alpha\}$ given for a set of possible states $\{\Psi_\alpha\}$. This is then a mixed state given by the density operator (3.8), which may be written in coordinate representation as

$$\rho(\{\boldsymbol{r}_i\},\{\boldsymbol{r}_i'\};t) = \sum_\alpha p_\alpha \Psi_\alpha(\{\boldsymbol{r}_i\},t)\Psi_\alpha(\{\boldsymbol{r}_i'\},t), \quad \sum_\alpha p_\alpha = 1. \tag{3.46}$$

Each Ψ_α is a pure state as described earlier, but the trajectories in each case obey different equations of motion and the reconstruction of $\rho(\{\boldsymbol{r}_i\},\{\boldsymbol{r}_i'\};t)$ is therefore more difficult. Instead, for these more general states, a related approach known as *quantum hydrodynamics* provides a useful extension of the pure state equations.

3.5.8.2 Quantum Hydrodynamics

For simplicity of notation, only single-particle systems are considered in this section to illustrate the main ideas. The density operator in the Wigner representation then obeys (3.27) with the right-hand side equal to zero. The first two moments of $f(\boldsymbol{R},\boldsymbol{p};t)$ with respect to $\boldsymbol{p}$ are the probability density $P(\boldsymbol{r},t)$ and mass times its flux $\boldsymbol{J}(\boldsymbol{R},t)$. It is convenient to introduce an associated velocity field by $\boldsymbol{J}(\boldsymbol{R},t) \equiv P(\boldsymbol{R},t)\boldsymbol{v}(\boldsymbol{R},t)$. Equations for $P(\boldsymbol{R},t)$ and $\boldsymbol{v}(\boldsymbol{R},t)$ follow directly from the first two moments of the Wigner function equation:

$$0 = \partial_t P + \nabla_{\boldsymbol{R}} \cdot (P\boldsymbol{v}), \tag{3.47}$$

$$0 = (\partial_t + \boldsymbol{v} \cdot \nabla_{\boldsymbol{R}}) v_i + \frac{1}{m}\nabla_{R_i} U + \frac{1}{mP}\nabla_{R_j} \cdot P_{ij}, \quad i \in \{1,2,3\}. \tag{3.48}$$

These equations have the form of the hydrodynamic equations for continuity and flow velocity, with P_{ij} being analogous to the pressure tensor:

$$P_{ij}(\boldsymbol{R},t) = \int \mathrm{d}\boldsymbol{p}\, [p_i - mv_i(\boldsymbol{R},t)]\left[p_j - mv_i(\boldsymbol{R},t)\right] f(\boldsymbol{R},\boldsymbol{p};t). \tag{3.49}$$

Hence, they are referred to as quantum hydrodynamic equations. This interpretation was introduced very early by Madelung [83] and exploited by Bohm [84] for his interpretation of the trajectories (motion of an "impurity" perfectly convected by the underlying fluid).

Equations (3.47) and (3.48) apply for both pure and mixed states. In the former case, they become equivalent to (3.40) and (3.41) and hence are sufficient to determine completely the Wigner distribution $f(\boldsymbol{R},\boldsymbol{p};t)$. This follows from the fact that the pressure tensor becomes the unit tensor times the quantum potential (3.42) [85]. Thus, the trajectories of the previous section are characteristic of the underlying quantum hydrodynamics and may be determined from it. For mixed states, (3.47) and (3.48) still define a quantum hydrodynamics but the pressure tensor cannot be determined simply in terms of the hydrodynamic fields P and $\boldsymbol{v}$. Instead, a new equation for the second moment of the Wigner function equation must be developed, which, in turn, couples the second moment to the third moment of f. In this way, it is seen that for the mixed state, quantum hydrodynamic equations comprise an infinite hierarchy for all moments, which in general does not truncate at any finite order. Reconstruction of the Wigner function requires, in principle, all moments, and hence all the corresponding "hydrodynamic" equations.

Applications of quantum hydrodynamics therefore generally require a truncation of the hierarchy. For example, the Nth moment could be approximated (uncontrolled) by its form for a Gaussian f, thereby expressing it in terms of moments up to degree 2. A nontrivial closed set of N equations to determine the first N moments then follows and the Wigner distribution would thusly be reconstructed approximately using these moments (e.g., a correspondingly truncated Taylor series expansion, cumulant expansion, etc.).

The equation for the Wigner function is a quantum version of the classical Liouville equation for dynamics in phase space. For systems of many degrees of freedom, extensions of the moment representations have been described in which the moments are taken only with respect to a subset of those degrees of freedom [86]. Consequently, a mixed hydrodynamic and Liouville motion results. Various versions of quantum hydrodynamics have been applied to low-temperature bosons, the best known one is the mean-field equation of Pitaevskii and Gross [87, 88], together with improved approximations [89]. Similarly, there have been recent applications to fermions, such as electrons in metals [90], which have been applied to inhomogeneous dense quantum plasmas, although a rigorous justification based on the strict methods presented in this chapter is still missing. Further details of quantum hydrodynamics, recent developments, and applications can be found in [82].

3.6 Conclusions

In this chapter, we have provided an overview on quantum plasmas. We have discussed their main occurrences in astrophysical systems, such as giant planets, dwarf stars, or neutron stars and in dense laboratory plasmas and in solid-state systems. As for any plasma, these systems are governed by the long-range Coulomb interaction, but quantum plasmas are characterized by a number of additional features. Due to high density and/or low temperature, the electrons (in some ultra-dense astrophysical systems, also the ions or hadrons) are quantum degenerate – their wave functions overlap. This leads to the governing role of the laws of quantum mechanics and to the important impact of the spin statistics. We have seen, in simple examples of few-particle systems in a trapping potential, that fermions and bosons show qualitatively different behavior characterized by either the Pauli principle or Bose condensation and superfluidity, respectively. All these effects become increasingly important in laboratory experiments (in particular using intense laser or ions beams) that compress matter to higher densities, far exceeding solid densities.

In the second part of the chapter, we briefly discussed the main theoretical approaches to dense quantum plasmas – the Schrödinger equation for the wave function, the density operator, quantum kinetic theory, and the method of second quantization. Naturally, this was only an introduction intended to point out the main concepts and methods. We have provided detailed references suitable for an in-depth study of the subject for the interested reader. We have also briefly mentioned modern computational methods which have emerged from the theoretical concepts. More details about the numerical treatment of quantum plasmas are presented in Chap. 4.

Acknowledgments The authors acknowledge many stimulating discussions with K. Balzer, S. Bauch, V. Filinov, and C. Henning. This work is supported by the Deutsche Forschungsgemeinschaft via SFB-TR24 and by the US Department of Energy award DE-FG02-07ER54946.

References

1. W.D. Kraeft, D. Kremp, W. Ebeling, G. Röpke, *Quantum Statistics of Charged Particle Systems* (Akademie, Berlin, 1986)
2. D. Kremp, M. Schlanges, W.D. Kraeft, *Quantum Statistics of Nonideal Plasmas* (Springer, Berlin, 2005)

3. M. Bonitz, *Quantum Kinetic Theory* (Teubner, Stuttgart, 1998)
4. S. Ichimaru, *Statistical Plasma Physics, vol. II: Condensed Plasmas* (Westview, Boulder, 2004)
5. M. Bonitz, D. Semkat (eds.), *Introduction to Computational Methods for Many-Body Physics* (Rinton, Princeton, 2006)
6. S. Ichimaru, Rev. Mod. Phys. **54**, 1017 (1982)
7. H.M. Van Horn, Science **252**, 384 (1991)
8. M. Bonitz et al., J. Phys. A: Math. Gen. **36**, 5921 (2003)
9. M. Bonitz et al., Phys. Plasmas **15**, 055704 (2008)
10. E. Wigner, Phys. Rev. **46**, 1002 (1934)
11. D.J. Wineland, J.C. Bergquist, W.M. Itano, J.J. Bollinger, C.H. Manney, Phys. Rev. Lett. **59**, 2935 (1987)
12. M. Drewsen, C. Brodersen, L. Hornekær, J.S. Hangst, J.P. Schiffer, Phys. Rev. Lett. **81**, 2878 (1998)
13. T. Schätz, U. Schramm, D. Habs, Nature **412**, 717 (2001)
14. U. Schramm, T. Schätz, D. Habs, Phys. Rev. Lett. **87**, 184801 (2001) and Phys. Stat. Sol. (b), **221**, 231 (2000)
15. D.H.E. Dubin, T.M. O'Neill, Rev. Mod. Phys. **71**, 87 (1999)
16. A.V. Filinov, M. Bonitz, Yu.E. Lozovik, Phys. Rev. Lett. **86**, 3851 (2001)
17. J.H. Chu, I. Lin, Phys. Rev. Lett. **72**, 4009 (1994)
18. H. Thomas, G.E. Morfill, V. Demmel, J. Goree, B. Feuerbacher, D. Möhlmann, Phys. Rev. Lett. **73**, 652 (1994)
19. Y. Hayashi, K. Tachibana, Jpn. J. Appl. Phys. **33**, L804 (1994)
20. M. Schlanges, M. Bonitz, A. Tschttschjan, Contrib. Plasma Phys. **35**, 109 (1995)
21. V.S. Filinov, M. Bonitz, W. Ebeling, V.E. Fortov, Plasma Phys. Contr. Fusion **43**, 743 (2001)
22. W. Ebeling, G. Norman, J. Stat. Phys. **110**(3–6), 861 (2003)
23. B. Holst, N. Nettelmann, R. Redmer, Contrib. Plasma Phys. **47**, 368 (2007)
24. J. Vorberger, I. Tamblyn, S.A. Bonev, B. Militzer, Contrib. Plasma Phys. **47**, 375 (2007)
25. W. Däppen, A. Nayfonov, Astrophys. J. Suppl. Ser. **127**, 287 (2000)
26. L. Segretain, Astron. Astrophys. **310**, 485 (1996)
27. G. Chabrier, Astron. J. **414**, 695 (1993)
28. A.Y. Potekhin, G. Chabrier, D.G. Yakovlev, Contrib. Plasma Phys. **41**, 231 (2001)
29. P. Loubeyre et al., High Pressure Res. **24**, 25 (2004) and references therein
30. H.J. Kusch, J. Phys. E **18**, 654 (1985)
31. H. Hess, Contrib. Plasma Phys. **26**, 209 (1986)
32. V.B. Mintsev, V.E. Fortov, High Temp. **20**, 584 (1982)
33. V.B. Mintsev, V.K. Gryaznov, M.I. Kulish, V.E. Fortov, Contrib. Plasma Phys. **41**, 119 (2001)
34. E. Boggasch et al., Phys. Rev. Lett. **66**, 1705 (1991)
35. M.K. Matzen et al., Plasma Phys. Contr. Fusion **41**, A175 (1999)
36. V.B. Mintsev, V.E. Fortov, J. Phys. A: Math. Gen. **39**, 4319 (2006)
37. D.H.H. Hoffmann et al., Phys. Scripta **T123**, 1 (2006)
38. M.D. Perry, G. Mourrou, Science **264**, 917 (1994)
39. J. Nuckolls, L. Wood, A. Thiessen, G. Zimmerman, Nature **239**, 139 (1972)
40. R. Kodama et al., Nature **412**, 798 (2001)
41. M. Tabak et al., Phys. Plasmas **1**, 1626 (1994)
42. LLNL, *Science and Technology Review*, July/August 2007 (see also http://lasers.llnl.gov)
43. M. Bonitz, V.S. Filinov, V.E. Fortov. P.R. Levashov, H. Fehske, Phys. Rev. Lett. **95**, 235006 (2005)
44. P. Ludwig, A. Filinov, Yu. Lozovik, H. Stolz, M. Bonitz, Contrib. Plasma Phys. **47**, 335 (2007)
45. M. Bonitz, V.S. Filinov, V.E. Fortov, P.R. Levashov, H. Fehske, J. Phys. A: Math. Gen. **39**, 4717 (2006)
46. V. Filinov, H. Fehske, M. Bonitz, V.E. Fortov, P.R. Levashov, Phys. Rev. E **75**, 036401 (2007)
47. A.A. Abrikosov, J. Less-Common Met. **62**, 451 (1978)
48. V.S. Filinov, M. Bonitz, V.E. Fortov, JETP Lett. **72**, 245 (2000) [Pis'ma v ZhETF **72**, 361 (2000)]; V.S. Filinov, V.E. Fortov, M. Bonitz, D. Kremp, Phys. Lett. A **274**, 228 (2000)
49. M.H. Thoma, IEEE Trans. Plasma Sci. **32**, 738 (2004)

50. M. Hofmann et al., Phys. Lett. B **478**, 161 (2000)
51. A. Peshier, B. Kämpfer, G. Soff, Phys. Rev. D **66**, 094003 (2002)
52. B.A. Gelman, E.V. Shuryak, I. Zahed, Phys. Rev. C **74**, 044908 (2006)
53. For this discovery, Chandrasekhar was awarded the Nobel prize in 1983. His original paper is: S. Chandrasekhar, Mon. Not. R. Astr. Soc. **95**, 207 (1935)
54. V. Filinov, M. Bonitz, V.E. Fortov, P.R. Levashov, JETP Lett. **74**, 384 (2001) [Pis'ma v ZhETF **74**, 422 (2001)]
55. R.P. Feynman, *Statistical Mechanics – A Set of Lectures (Frontiers in Physics)* (Perseus, Reading, 1972)
56. M.H. Anderson, J.R. Ensher, M.R. Matthews, C.E. Wieman, E.A. Cornell, Science **269**, 198 (1995)
57. K.B. Davis et al., Phys. Rev. Lett. **75**, 3969 (1995)
58. http://nobelprize.org/nobel_prizes/physics/laureates/2001/ – Web page of the Nobel Foundation for the Nobel Lectures
59. J. Böning, Superfluidity in mesoscopic systems of charged bosons. Diploma Thesis, Kiel University (2007)
60. A.V. Balatsky et al., Phys. Rev. B **75**, 094201 (2007)
61. P. Sindzingre, M.L. Klein, D.M. Ceperley, Phys. Rev. Lett. **63**, 1601 (1989)
62. A. Filinov, J. Böning, M. Bonitz, Yu.E. Lozovik, Phys. Rev. B **77**, 214527 (2008)
63. G.V. Chester, L. Reatto, Phys. Rev. **155**, 88 (1967)
64. A.F. Andreev, I.M. Lifshitz, Sov. Phys. JETP **29**, 1107 (1969)
65. A.J. Leggett, Phys. Rev. Lett. **25**, 1543 (1970)
66. I. Bloch, J. Dalibard, W. Zwerger, Rev. Mod. Phys. **80**, 885 (2008)
67. A.B. Migdal, Zh. Eksp. Theor. Fiz. **37**, 249 (1959)
68. G. Baym, C. Pethick, D. Pines, Nature **224**, 674 (1969)
69. K. Balzer, M. Bonitz, J. Phys. A **42**, 214020 (2009) (preprint ArXive:0810.2633)
70. K. Balzer, Nonequilibrium Green's function approach to artificial atoms. Diploma Thesis, Kiel University (2007)
71. R. Balescu, *Statistical Mechanics of Charged Particles* (Wiley, London, 1963)
72. Yu.L. Klimontovich, *Kinetic Theory of Nonideal Gases and Nonideal Plasmas* (Pergamon, Oxford, 1982)
73. V.V. Belyi, Yu.A. Kukharenko, Contrib. Plasma Phys. **47**, 240 (2007)
74. D. Kremp, Th. Bornath, M. Bonitz, M. Schlanges, Phys. Rev. E **60**, 4725 (1999); M. Bonitz, Th. Bornath, D. Kremp, M. Schlanges, W.D. Kraeft, Contrib. Plasma Phys. **39**, 329 (1999)
75. H. Haberland, M. Bonitz, D. Kremp, Phys. Rev. E **64**, 026405 (2001)
76. D. Boercker, J.W. Dufty, Ann. Phys. (NY) **119**, 43 (1979)
77. L.P. Kadanoff, G. Baym, *Quantum Statistical Mechanics* (Benjamin, New York, 1962)
78. L.V. Keldysh, Zh. Eksp. Theor. Fiz. **47**, 1515 (1964) [Sov. Phys. JETP **20**, 235 (1965)]
79. D. Bohm, Phys. Rev. **85**, 166 (1952)
80. D. Bohm, Phys. Rev. **85**, 180 (1952)
81. D.-A. Deckert, D. Durr, P. Pickl, J. Phys. Chem. A **111**, 41, 10325 (2007)
82. R.E. Wyatt. *Quantum Dynamics with Trajectories* (Springer, Berlin, 2005)
83. E. Madelung, Z. Physik **40**, 332 (1926)
84. D. Bohm, J. Vigier, Phys. Rev. **96**, 208 (1954)
85. P. Holland, *The Quantum Theory of Motion* (Cambridge, New York, 1993)
86. I. Burghardt, K. Moller, G. Parlant, L. Cederbaum, E. Bittner, Int. J. Quant. Chem. **100**, 1153 (2004)
87. L.P. Pitaevskii, Zh. Eksp. Theor. Fiz. **40**, 646 (1961) [Sov. Phys. JETP **13**, 451 (1961)]
88. E.P. Gross, Nuovo Cimento **20**, 454 (1961)
89. E.B. Kolomeisky et al., Phys. Rev. Lett. **85**, 1146 (2000)
90. G. Manfredi, F. Haas, Phys. Rev. B **64**, 075316 (2001)

Chapter 4
Introduction to Quantum Plasma Simulations

Sebastian Bauch, Karsten Balzer, Patrick Ludwig, and Michael Bonitz

Abstract This chapter contains a brief introduction to the field of quantum simulations. Beginning with a numerical treatment of single-particle problems by exact numerical solution of the time-dependent Schrödinger equation, we demonstrate concepts useful in the computational treatment of quantum systems. These rather basic techniques are limited by the number of particles, N. Considering an increase of system size, approximation methods arising from many-particle theories are necessary. Here, we introduce two powerful approaches: the (time-dependent) Hartree–Fock method with improvements for inclusion of correlations based on nonequilibrium Green's functions and, for the calculation of time-independent phenomena, a rigorous quantum Monte Carlo technique. These computational tools complement each other and thus provide for a comprehensive theoretical modeling of quantum plasmas.

4.1 Introduction

Modern experimental techniques allow for the selective manipulation of small micro- and nanoscale systems (quantum plasmas) of even less than 100 particles. Although the fundamental physical laws which govern these measurements are well known, exact analytical solutions are available only for a very limited number of many-particle systems, such as ideal solids (i.e., highly periodic structures without any lattice defects or distortions) or noninteracting classical (i.e., $\Gamma \ll 1$) or quantum gases ($r_s \ll 1$). Consideration of interaction makes things much more interesting, but also more complex and theoretically challenging. In most practical cases, the fundamental many-body Hamiltonian (4.1) cannot be directly diagonalized and more efficient numerical methods are needed. In fact, even simple models

S. Bauch (✉), K. Balzer, P. Ludwig, and M. Bonitz
Institut für Theoretische Physik und Astrophysik, Christian-Albrechts Universität zu Kiel, 24098 Kiel, Germany
e-mail: bauch@theo-physik.uni-kiel.de; balzer@theo-physik.uni-kiel.de; ludwig@theo-physik.uni-kiel.de; bonitz@physik.uni-kiel.de

M. Bonitz et al. (eds.), *Introduction to Complex Plasmas*, Springer Series on Atomic, Optical, and Plasma Physics 59, DOI 10.1007/978-3-642-10592-0_4,

used to describe interacting quantum systems in the regime of strong particle correlations are computationally very demanding (see Chap. 3).

In the investigation of ground states and phase transitions, we utilize different classical and quantum *bottom-up approaches*. This means that the theoretical description starts at the microscopic level of individual particles and, thereby, takes full account of all microscopic many-particle interactions. The only simulation input data involved are the fundamental pair interaction potentials as well as the boundary (confinement) conditions. Hence, the theoretical framework of computational bottom-up methods on hand allows for highly flexible modeling with regard to the specific experimental setup (trap geometry, external fields, number of particles, etc.) and yields direct results that are free from any fitting parameters.

The numerical modeling of quantum systems starts from the fundamental many-body Hamiltonian introduced in Chap. 3:

$$\hat{H} = \sum_{i=1}^{N} \frac{\hat{p}_i^2}{2m_i} + \sum_{i=1}^{N} V_i(\boldsymbol{r}_i, t) + \frac{1}{2} \sum_{i \neq j}^{N} w(\boldsymbol{r}_i, \boldsymbol{r}_j). \tag{4.1}$$

Again, N is the particle number involved, m_i the mass of the ith particle, $V_i(\boldsymbol{r}_i, t)$ the single-particle potential (e.g., external confinement, time-dependent perturbations, etc.) of the ith particle, and $w(\boldsymbol{r}_i, \boldsymbol{r}_j)$ the binary interaction between the ith and jth particles, for example, the Coulomb interaction between charged particles $w(\boldsymbol{r}_i, \boldsymbol{r}_j) = e^2 / \left|\boldsymbol{r}_i - \boldsymbol{r}_j\right|$.

The Hamiltonian (4.1) fully describes the system of interest, its ground-state (equilibrium) properties, as well as its dynamical behavior following a perturbation, $V(t)$. The first part of this introduction deals with solution schemes for investigation of the temporal development of excited systems on the basis of the single-particle time-dependent Schrödinger equation (TDSE) (Sect. 4.2). Then, in Sect. 4.3, we discuss equilibrium and nonequilibrium properties of many-body systems by means of (time-dependent) Hartree–Fock (HF, TDHF) simulations and systematic improvement of the approximation with respect to the binary interparticle interaction w. This chapter continues with an overview of the path-integral Monte Carlo (PIMC) method, which allows for a finite temperature description of equilibrium properties of large quantum systems in Sect. 4.4.

4.2 Time-Dependent Schrödinger Equation

Time-dependent phenomena – such as ionization, scattering, and excitation – are accurately described within the framework of the TDSE, which reads as

$$i\hbar \frac{\partial}{\partial t} \Psi(\boldsymbol{x}_1, \ldots, \boldsymbol{x}_N, t) = \hat{H} \Psi(\boldsymbol{x}_1, \ldots, \boldsymbol{x}_N, t). \tag{4.2}$$

Here, $\boldsymbol{x}_i = (\boldsymbol{r}_i, \sigma_i)$ denotes the combination of the spatial coordinate vector, $\boldsymbol{r}_i$, and the spin variable σ_i.

Due to the great complexity of the time-dependent problem, the many-body TDSE (4.2) can only be solved in very few cases. Therefore, the following discussion of the numerical treatment involves two parts: the exact solution of the *one-particle* ($N = 1$) TDSE and the approximative solution of the many-body ($N > 1$) TDSE in Sect. 4.3.

In the atomic system of units, with $\hbar = m = e = 1$, (4.2) for one particle simplifies to

$$\mathrm{i}\frac{\partial}{\partial t}\Psi(\boldsymbol{r},t) = -\frac{1}{2}\Delta\Psi(\boldsymbol{r},t) + V(\boldsymbol{r},t)\Psi(\boldsymbol{r},t), \tag{4.3}$$

where the spin degree of freedom is neglected. The numerical solution of the TDSE is a widely studied subject. There exist many different approaches. Each computational technique has its own range of applicability and one has to choose carefully the most suitable investigative procedure as this can enormously affect the resulting efficiency and accuracy.

The formal solution of the TDSE for slow time variation of $V(\boldsymbol{r},t)$ is given by the time evolution operator

$$\hat{U}(t,t_0) = \exp\left[-\mathrm{i}\hat{H}(t-t_0)\right]\Psi_0(\boldsymbol{r}), \tag{4.4}$$

where the corresponding time evolution of the wave function $\Psi(\boldsymbol{r},t)$ takes the form

$$\Psi(\boldsymbol{r},t) = \hat{U}(t,t_0)\Psi_0(\boldsymbol{r}). \tag{4.5}$$

Here, $\Psi_0(\boldsymbol{r}) = \Psi(\boldsymbol{r}, t \equiv t_0)$ denotes the *initial condition*, that is, the state of the system at initial time t_0 of the time evolution. Since we are, for numerical reasons, interested in propagation over a small time step of duration Δt, we only consider the case of Hamilton operators which are not explicitly time dependent. The external potential, $\hat{V}(\mathbf{r},t)$, is taken to be slowly time-dependent, for example, by modeling an external perturbation (laser field, etc.). Otherwise, $\hat{U}$ would take a more complicated form, which is well known from textbook quantum mechanics. Here, the main idea is that $\hat{V}(\boldsymbol{r},t)$ is approximated to be constant during a certain, small time interval Δt. This can (always) be assured by the choice of a sufficiently small Δt. In the following, we will discuss two methods to solve (4.5) numerically, which have advantages for different types of systems.

4.2.1 1D Crank–Nicolson Method

In this tutorial, we concentrate on the solution of the one-dimensional (1D) form of (4.3), which reads as

$$\mathrm{i}\frac{\partial}{\partial t}\Psi(x,t) = -\frac{1}{2}\frac{\mathrm{d}^2}{\mathrm{d}x^2}\Psi(x,t) + V(x,t)\Psi(x,t). \tag{4.6}$$

The generalization of the described method to systems of higher dimensions (e.g., by *operator splitting*) can be found in the literature (e.g., [1]). Equation (4.6) is a complex diffusion-like initial value problem which has to be supplemented by boundary conditions.

A stable, implicit time evolution scheme which leads to the *Crank–Nicolson* procedure is based on Cayley's form of the time evolution operator $\mathrm{e}^{-\mathrm{i}\hat{H}\Delta t}$ [1], which is the lowest order of a Pade expansion of the exponential function:

$$\mathrm{e}^{-\mathrm{i}\hat{H}\Delta t} \approx \frac{1-\frac{1}{2}\mathrm{i}\hat{H}\,\Delta t}{1+\frac{1}{2}\mathrm{i}\hat{H}\,\Delta t}. \tag{4.7}$$

This expansion implies a *unitary* time evolution; hence, the normalization of the wave function,

$$n(x,t) = \int_{-\infty}^{\infty} |\Psi(x,t)|^2\,\mathrm{d}x, \tag{4.8}$$

is assured to be one for all times t, which one might easily show for our case by inserting (4.7) into (4.8). In contrast to *explicit* schemes, where the TDSE is solved for the wave function Ψ and then integrated with respect to t, *implicit* schemes are numerically more advanced. Generally, in such an implicit scheme, the wave function is not directly accessible, but has to be obtained by solving a system of linear equations.

Further, we discretize our spatial coordinate x by introducing a spacing Δx (cf. Fig. 4.1). Therefore, for $x \in [x_i, \ldots, x_{i+1}]$ with $\Delta x = |x_{i+1} - x_i|$, we write $\Psi_i^n \equiv \Psi(x,t)$. The index $i = 1, \ldots, N_x$ indicates the spatial discretization with the step size Δx, whereas the superscript n denotes the corresponding discretization in time. Hence, $n+1 \equiv t + \Delta t$ and $i+1 \equiv x + \Delta x$, for example.

Using (4.7) as an approximation for $\hat{U}$ in (4.5), one represents the propagation of Ψ_i^n to the state Ψ_i^{n+1} of later time as

$$\left(1+\frac{1}{2}\mathrm{i}\hat{H}\,\Delta t\right)\Psi_i^{n+1} = \left(1-\frac{1}{2}\mathrm{i}\hat{H}\,\Delta t\right)\Psi_i^n. \tag{4.9}$$

Now, the remaining task is to find a representation of the Hamilton operator $\hat{H}$. In our case, it is replaced by a finite difference approximation. We consider a second-order expression for the derivatives:

$$\frac{\mathrm{d}^2}{\mathrm{d}x^2}\Psi(x,t) \approx \frac{\Psi_{i+1}^n - 2\Psi_i^n + \Psi_{i-1}^n}{(\Delta x)^2}. \tag{4.10}$$

Of course, higher-order schemes can be implemented (for a recent adaption of the methods see, e.g., [2]).

Fig. 4.1 One-dimensional grid. The TDSE is solved within the marked region on a number N_x discrete grid points. $i = 1$ and $i = N_x$ are defined by the boundary conditions

Combining (4.9) and (4.10) with $\alpha = \mathrm{i}\Delta t/(2\Delta x)^2$ and

$$a_i = -\alpha = c_i, \quad b_i = 1 + 2\alpha + \frac{1}{2}\mathrm{i}\Delta t\, V_i^n, \quad i = 2, \dots, N_x - 1, \tag{4.11}$$

the LHS of (4.9) is given by

$$\left(1 + \frac{1}{2}\mathrm{i}\Delta t \hat{H}\right) \Psi_i^{n+1} = a_i \Psi_{i-1}^{n+1} + b_i \Psi_i^{n+1} + c_i \Psi_{i+1}^{n+1}. \tag{4.12}$$

Analogously, the RHS of (4.9) transforms into

$$\left(1 - \frac{1}{2}\mathrm{i}\hat{H}\Delta t\right) \Psi_i^n = \Psi_i^n \left(1 - 2\alpha - \frac{1}{2}\mathrm{i}\Delta t\, V_i^n\right) + \alpha \Psi_{i-1}^n + \alpha \Psi_{i+1}^n \equiv r_i^n. \tag{4.13}$$

Expressing the combination of (4.12) and (4.13) in matrix form, one finds a tridiagonal form of the system of linear equations for the N_x unknown variables Ψ_i^{n+1}, $i = 1, \dots, N_x$:

$$\begin{pmatrix} b_1 & c_1 & 0 & 0 \\ a_2 & b_2 & c_2 & 0 \\ \vdots & \vdots & \ddots & \vdots \\ 0 & 0 & a_{N_x} & b_{N_x} \end{pmatrix} \cdot \begin{pmatrix} \Psi_1^{n+1} \\ \Psi_2^{n+1} \\ \vdots \\ \Psi_{N_x}^{n+1} \end{pmatrix} = \begin{pmatrix} r_1^n \\ r_2^n \\ \vdots \\ r_{N_x}^n \end{pmatrix}. \tag{4.14}$$

The elements b_1, c_1, and r_1^n are defined by the boundary conditions at the left edge of the grid (an example is given below). Similarly, b_{N_x}, c_{N_x}, and $r_{N_x}^n$ are defined at the right edge. Since the matrix of coefficients in (4.14) is very sparse, it can be solved and stored very efficiently. Many numerical libraries have specialized routines implemented to handle such matrices.

4.2.1.1 Boundary Conditions

The initial value problem of the TDSE is supplemented by boundary conditions, which close the partial differential equation mathematically. Several possibilities are available, for example, Dirichlet, von Neumann, and absorbing conditions. They are chosen by physical observations and define the system of interest. In this introduction, only the first one (Dirichlet) is considered.

In this case, the wave function is taken to vanish at the boundaries of the system,

$$\lim_{x\to\infty} \Psi(x,t) \to 0 \quad \text{and} \quad \lim_{x\to-\infty} \Psi(x,t) \to 0, \tag{4.15}$$

and a simulation box with reflecting grid boundaries is created. This special type of *Dirichlet boundary conditions* assures conservation of the normalization of Ψ for all times if no dissipation is artificially included (e.g., by an absorbing potential, see below). Equation (4.15) can be fulfilled if the wave function is zero at the right-most and the left-most element on the grid, that is, $\Psi_0^n = \Psi_{N_x}^n = 0$ for all time steps n. This can be satisfied by modifying the first and the last elements in the tridiagonal system of equations (4.14). As an exercise, the reader is left with the calculation that gives $a_{1,N_x} = c_{1,N_x} = r_{1,N_x} = 0$ and $b_{1,N_x} = 1 + i$. Obviously, this type of boundary condition requires large spatial grids to allow for long simulation runs without influence of reflected parts of the wave function stemming from nonphysical reflection of Ψ at the end of the simulation box.

4.2.1.2 Absorbing Boundary Conditions

A technique of avoiding computationally expensive large grids is the introduction of absorbing boundary conditions. In fact, there are basically two methods to be found in the literature. The first one is especially important for 1D calculations. It is based on a mathematical theory which allows complete absorption of the wave function at a specific grid point [3]. Its computational implementation is difficult and for higher-dimensional systems it is not applicable [4]. We will describe a simpler, easy to implement approach, which uses additional potentials in the system but lacks of mathematical rigor.

If a spatially confined imaginary part is added to the one-particle potential $V(x,t)$, the wave function $\Psi(x,t)$ is damped during the time of propagation over this region. This can be rationalized in terms of the equation of continuity since this so-called *optical potential* acts like a dissipation term. One finds that this damping effect increases with higher energies of the propagated particle. Thus, the faster a particle moves the more efficiently it can be absorbed by the potential. It should be mentioned that such absorbing potentials have to be chosen carefully. Every change in the potential, no matter if real or imaginary, leads to a reflected part of the wave function. Roughly speaking, the smoother the spatial variation of the imaginary potential, the better it will work. The simplest conceptual version is, for example, given by a linear potential. More efficient potentials are available in the literature [5, 6].

Figure 4.2 shows the absorption of a 1D wave packet for such a linear, imaginary potential, indicated by the black line. Only a very small fraction (10^{-7}) survives the damping and is reflected. Such grid boundaries are of course not boundary conditions in a mathematical sense. The system of equations (4.14) has yet to be closed by Dirichlet conditions discussed earlier.

4.2.1.3 Initial Conditions

Finally, the time propagation of the TDSE needs an initial condition, $\Psi_0(x) = \Psi(x, t = t_0)$. The choice of this state reflects the physical motivation of the problem. In the following, we will discuss two possibilities (1) the construction of eigenstates and (2) the treatment of free particles by Gaussian wave packets:

1. *Imaginary time propagation (ITP)*. The above-described time propagation code can easily be used to calculate stationary states by replacing the time t by an

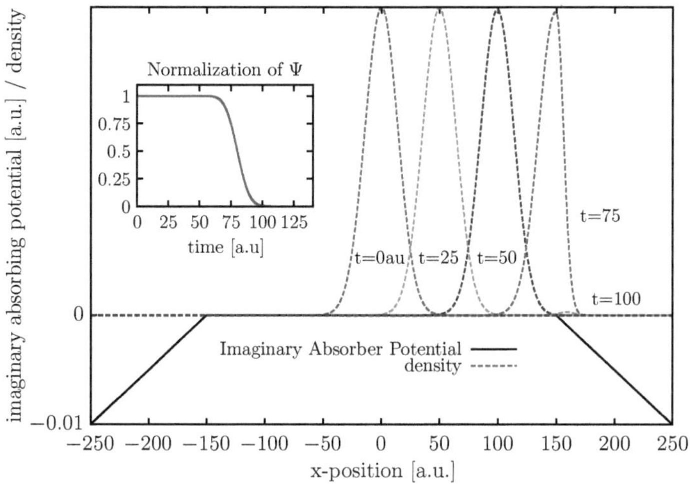

Fig. 4.2 Influence of a linear imaginary absorbing potential $\mathrm{i}V(x)$: absorption of a 1D wave packet [$k = 2.0$ a.u., (4.20)]. Almost the whole packet is absorbed and only a fraction smaller than 10^{-7} is reflected; see inset for normalization of Ψ. All quantities are given in atomic units (a.u.)

imaginary time it [4], which transforms the TDSE into a diffusion equation. An arbitrary state can be written as a superposition of eigenstates with expansion coefficient $c_j = \langle \psi_j | \Psi(t) \rangle$:

$$|\Psi(t)\rangle = \sum_j c_j \exp(-\mathrm{i}E_j t)\, |\psi_j\rangle\,, \tag{4.16}$$

with $|\psi_j\rangle$ describing the stationary states. Now, if the imaginary time is inserted, one obtains

$$|\Psi(t)\rangle = \sum_j c_j \exp(E_j t)\, |\psi_j\rangle \tag{4.17}$$

and the corresponding states are exponentially decaying or increasing during the TDSE propagation depending on the sign of the energy eigenvalue E_j. Only the ground state survives because it decays less or increases much faster than the other states. Of course, this scheme does not conserve the normalization of the Ψ. Therefore, the wave function has to be renormalized at each time step.

Excited states, $(n+1)$, where n denotes the highest, previously constructed state, are also accessible by this procedure: The Schmidt orthogonalization of $|\psi(t)\rangle$

$$\psi_{n+1}^{\perp}(\boldsymbol{r}) = \psi_{n+1}(\boldsymbol{r}) - \sum_{i=0}^{n} \left(\int_{-\infty}^{\infty} \mathrm{d}^3 r\, \psi_{n+1}^{*}(\boldsymbol{r}) \psi_i(\boldsymbol{r}) \right) \cdot \psi_i(\boldsymbol{r}) \tag{4.18}$$

at each time step will force the wave function to converge to the next unknown eigenfunction. The initial wave function for the ITP may be chosen to be completely random or, what is better for convergence reasons, as near to the ground (excited) state wave function as possible.

During the imaginary iteration procedure, several convergence indicators can be used. The total energy appears to be an improper criterion especially for higher excited states. Its convergence is very fast but not sensitive to density changes. Thus, a density-based quantity such as

$$\Delta\xi = \int_{-\infty}^{\infty} |\Psi(x,t) - \Psi(x,t+\Delta t)|^2 \,\mathrm{d}x \tag{4.19}$$

is found to be of higher accuracy. The ITP method is, in contrast to other methods like the *shooting* algorithm [7], applicable to arbitrary potentials in an arbitrary number of spatial dimensions. It may be understood as a powerful method for the diagonalization of $\hat{H}$ in the spatial coordinate basis representation.

2. *Gaussian wave packets.* For other interesting physical questions, for example scattering situations, it may be of interest, to model initially free electrons. One possibility is the usage of Gaussian wave packets (see, e.g., [8]):

$$\Psi(x,t_0) = \frac{1}{\sqrt{2\pi\sigma}} \exp\left(-\frac{(x-x_0)^2}{2\sigma^2}\right) \exp(\mathrm{i}k_0 x). \tag{4.20}$$

The initial momentum k_0 describes how fast the electron travels, whereas x_0 determines its initial position at time $t = t_0$ and σ its spatial spreading. Through Heisenberg's uncertainty law, σ also defines the momentum distribution, which corresponds to a smoothed energy distribution via the free particle dispersion law $E = k_0^2/2$. All these issues have to be kept in mind in order to make accurate simulations of quantum systems.

4.2.2 TDSE Solution in Basis Representation

In this section, we describe a different solution scheme for the one-particle TDSE (cf. (4.3)), which only relies on matrix multiplications and (at least one) diagonalization of the Hamilton matrix. Depending on the systems considered (e.g., confined systems), it is very efficient. The applied techniques described here are the basis for the following introduction to Hartree–Fock methods (Sect. 4.3); hence, study of this section is strongly suggested before continuing to the next part.

To start, we express the wave function $|\Psi(t)\rangle$ in a complete orthonormal basis set $\{\varphi_i\}_{i=1,\ldots,\infty}$, where $\langle\varphi_i|\varphi_j\rangle = \delta_{ij}$ holds:

$$|\Psi(t)\rangle = \sum_{i=0}^{\infty} c_i(t)\,|\varphi_i\rangle\,. \tag{4.21}$$

The expansion coefficients are given by $c_i(t) = \langle\varphi_i|\Psi(t)\rangle$. For numerical reasons it is necessary, analogously to the introduction of the finite spatial grid in the previous section, to truncate the sum in (4.21) at a finite number N_b. With this, the basis is no longer complete in the mathematical sense and one has to assure that the chosen basis functions reflect the final solutions as closely as possible, minimizing the required number of basis functions, N_b. This is a challenging task, especially in the case of highly dynamical behavior of the system considered. During the time evolution, many intermediate states may be accessed and all these states have to be described as accurately as possible by the basis.

As described earlier, in addition to the Hamilton operator and the corresponding time evolution operator, an initial condition $\Psi(t = t_0)$ is needed, which corresponds to an initial set of expansion coefficients $\{c_i(t_0)\}_{i=1,\ldots,N_b}$.

4.2.2.1 Deriving a Time Evolution Scheme

The determination of an expression for the time evolution of the vector of N_b coefficients $\boldsymbol{c}(t)$ can be achieved by applying the time evolution operator $\hat{U}$ to the initial state: $|\Psi(t)\rangle = \hat{U}\,|\psi_0\rangle$. To extract $c_i(t)$, we expand $|\Psi(t)\rangle$ and insert $\hat{1} = \sum_i^{N_b} |\varphi_i\rangle\,\langle\varphi_i|$:

$$|\Psi(t)\rangle = \sum_{j=0}^{N_b} c_j(t)\,|\varphi_j\rangle = \sum_i^{N_b} \hat{U}\,|\varphi_i\rangle \left\langle \varphi_i \middle| \sum_l^{N_b} c_l(t = t_0) \middle| \varphi_l \right\rangle. \tag{4.22}$$

Multiplying the whole equation from the left with $\langle\varphi_k|$ yields

$$\sum_{j=0}^{N_b} c_j(t)\,\langle\varphi_k|\varphi_j\rangle = \sum_{i=0}^{N_b} \langle\varphi_k|\hat{U}|\varphi_i\rangle \left\langle \varphi_i \middle| \sum_{l=0}^{N_b} c_l(t = t_0) \middle| \varphi_l \right\rangle. \tag{4.23}$$

Therefore, we finally obtain the time-dependent coefficients as

$$c_k(t) = \sum_{i=0}^{N_b} U_{ki} \sum_{l=0}^{N_b} c_l(t = t_0)\,\langle\varphi_i|\varphi_l\rangle = \sum_i U_{ki} c_i(t = t_0). \tag{4.24}$$

This is simply a matrix product and can be written in the form

$$\mathbf{c}(t) = \mathbf{U}\cdot\mathbf{c}(t = t_0), \tag{4.25}$$

with $\mathbf{U} = \{\,U_{ij}\,\}_{i,j\in\{1,\ldots,N_b\}}$ denoting an $N_b \times N_b$ matrix.

4.2.2.2 Computation of Matrix Elements of U_{ij}

The remaining problem is to find the basis representation of the time evolution operator $\hat{U} = \exp(\mathrm{i}\hat{H}\Delta t)$ (cf. (4.5)), that is, the N_b^2 complex matrix elements U_{ij} of $\mathbf{U}$.

Applying basic linear algebra leads to

$$U_{ij} = \langle \varphi_i | \hat{U} | \varphi_j \rangle = \sum_{k=0}^{\infty} \langle \varphi_i | \psi_k \rangle \langle \psi_k | \varphi_j \rangle \exp(\mathrm{i} E_k \Delta t). \tag{4.26}$$

Here, $|\psi_k\rangle$ are the eigenfunctions of $\hat{H}$ corresponding to the eigenvalue E_m. The first two factors on the right-hand side of eq. (4.26) are simply the eigenvectors of $\hat{H}$ in the basis representation $|\varphi_i\rangle_{i=1,\ldots,N_b}$, which can be obtained by a diagonalization of the Hamiltonian. This diagonalization of $\hat{H}$ has to be performed for each temporal change in $\hat{H}$ which leads to a computationally very efficient propagation scheme for given excitations. From the obtained (time-dependent) expansion coefficients $c_i(t)$, all expectation values can be computed.

Boundary conditions, as considered earlier, are not to be specified explicitly. They are embedded in the behavior of the chosen basis functions. For the initial moment of time propagation, one only has to specify a certain set of $c_i(t = t_0)$; hence, it is very easy to prepare a system in a bound state if the basis is chosen to be a set of associated eigenstates. Thus, it is clear that this method has advantages for localized systems in traps, atoms, etc., but may reach its limits in the consideration of situations where combinations of free particles and localized states are involved. For this case, grid methods perform better.

4.2.3 Computational Example: Electron Scattering in a Laser Field

In this section, we demonstrate the utility of the above-described algorithms by their application to a simple physical system, which is, due to its computational complexity, analytically not accessible. Let us consider a free electron, represented by a wave packet of Gaussian shape (cf. (4.20)), traveling with a momentum k_0 toward an ion. The whole system is radiated with a strong, linearly polarized laser field, modeled by the potential (in dipole approximation):

$$V_{\text{laser}}(x,t) = -E_0 x \cos(\omega t). \tag{4.27}$$

The electron may now absorb energy from the laser field during the scattering process, and due to the quantization character, only in amounts equal to ω (remember $\hbar = m = e = 1$ in our system of units). The setup of the system is schematically drawn in Fig. 4.3.

After propagating the TDSE with the Crank–Nicolson procedure described earlier, the resulting wave function outside the ion potential is transformed by a fast Fourier transform into momentum space. Using the free particle dispersion relation, $E = p^2/2m$, the energy distribution is calculated. Figure 4.4 shows the result of the simulation of such a scattering process with $k_0 = 4.0$ a.u. in a strong laser field

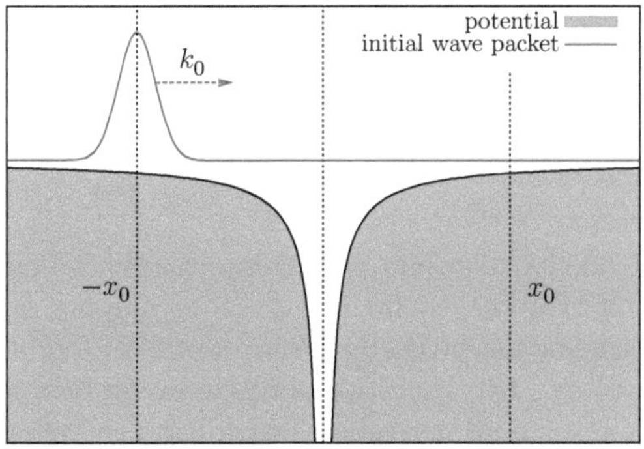

Fig. 4.3 Coulomb scattering process. The electronic wave packet is launched at a distance x_0 from the ion with a momentum k_0 directed toward the ion. The whole setup is placed in a strong linearly polarized laser field

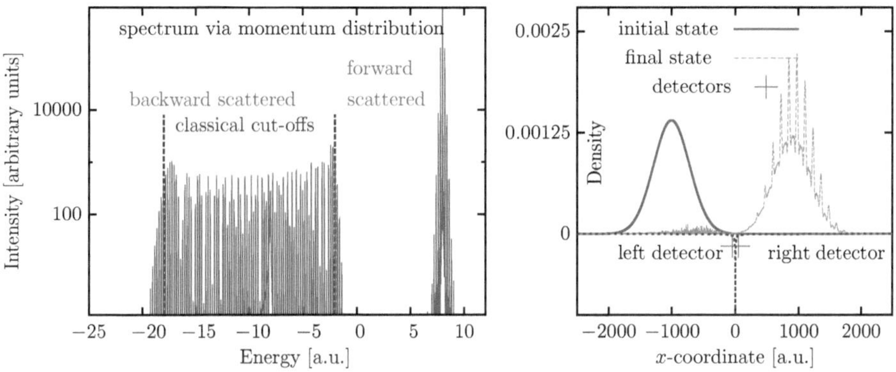

Fig. 4.4 Energy distribution (*left*) and initial + final electron density (*right*) of a Coulomb scattering process with $k_0 = 4.0$ a.u. in a strong laser field with $\omega = 0.2$ a.u. and $E_0 = 0.2$ a.u. Negative (positive) energies indicate backward (forward) scattering

with parameters given in the figure caption. The sign of the energy indicates forward (+) and backward (−) scattering, respectively. One easily identifies the peak of elastically forward scattered electrons with an energy of $E = k_0^2/2 = 8.0$ a.u. In backward direction (negative energies), a large plateau in the energy distribution is formed with two significant cutoff energies, which can easily be obtained using a simple classical theory [8, 9]. A closer look at the energy spectrum reveals a peak-like structure, where each individual peak is separated by the photon energy ω.

4.3 Hartree–Fock Method

The TDHF method aims at approximately solving the time-dependent N particle Schrödinger equation:

$$\mathrm{i}\frac{\partial}{\partial t}\Psi(\boldsymbol{x}_1, \ldots, \boldsymbol{x}_N, t) = \hat{H}\Psi(\boldsymbol{x}_1, \ldots, \boldsymbol{x}_N, t), \tag{4.28}$$

with $\boldsymbol{x}_i = (\boldsymbol{r}_i, \sigma_i)$ and Hamiltonian[1]

$$\hat{H} = \sum_{i=1}^{N} \left(-\frac{\nabla_{\boldsymbol{r}_i}^2}{2m} + \sum_{i=1}^{N} V(\boldsymbol{r}_i, t) \right) + \sum_{i<j}^{N} w(\boldsymbol{r}_i - \boldsymbol{r}_j). \tag{4.29}$$

Thereby, it offers partial information when the exact solution is not accessible, for example, by the numerical techniques described in the last section. The TDHF scheme can be derived from several different (but equivalent) aspects of many-body theory [10]. Here, we concentrate on two common approaches. The first outlines the standard derivation which handles the problem on the level of an approximate N particle wave function, $\Psi(\boldsymbol{x}_1, \ldots, \boldsymbol{x}_N, t)$. The main quantity of the second is based on a two-time generalization of the one-particle density matrix, $\varrho(\boldsymbol{x}, \bar{\boldsymbol{x}}; t)$ – the nonequilibrium Green's function (NEGF).

The idea of TDHF dates back to Dirac [11] and has had first applications in nuclear collisions and atomic physics [12]. More recent approaches – to name only a few – include the dynamics of electrons in molecules [13], atoms exposed to strong laser fields [14], as well as single- and double-ionization processes (see, e.g., [15] and references therein). Furthermore, in condensed matter physics and nonideal plasmas, the TDHF ansatz has been used successfully in the framework of NEGFs [16–20]. Moreover, the method is closely related to the random phase approximation (RPA), for example, as applied in the theory of dielectric functions [21].

4.3.1 Standard Approach

Considering fermions, the TDHF method starts with the TDSE (4.28) approximating the total many-body wave function to be a totally antisymmetrized product of one-particle orbitals $\phi_i(\boldsymbol{x}_i, t)$ $(i \in \{1, \ldots, N\})$ which depend on time [10]. Thus, $\Psi(\boldsymbol{x}_1, \ldots, \boldsymbol{x}_N, t)$ becomes a single Slater determinant $\mathcal{D}$ according to

$$\begin{aligned} \Psi(\boldsymbol{x}_1, \ldots, \boldsymbol{x}_N, t) &= \hat{A} \prod_{i=1}^{N} \phi_i(\boldsymbol{x}_i, t) = \frac{1}{\sqrt{N!}} \mathcal{D}(\phi_i(\boldsymbol{x}_j, t)) \\ &=: |\Psi[\phi_1, \ldots, \phi_N]\rangle, \end{aligned} \tag{4.30}$$

where $\hat{A}$ denotes the antisymmetrization operator. The equations of motion for the orbitals ϕ_i then follow from the action functional [15]:

$$A[\phi_1, \ldots, \phi_N] = \int_{t_0}^{t_1} \mathrm{d}t \left\langle \Psi[\phi_1, \ldots, \phi_N] \middle| \mathrm{i}\frac{\partial}{\partial_t} - \hat{H} \middle| \Psi[\phi_1, \ldots, \phi_N] \right\rangle, \tag{4.31}$$

[1] For simplicity, we suppress the Planck quantum $\hbar$ in (4.28) and (4.29) as well as in the remainder of this chapter.

which must be stationary under variations of the orbitals when the constraint $\delta\phi_i = 0\ \forall i$ is applied at $t = t_0$ and $t = t_1$. Using this action principle, the norm of the many-body wave function Ψ is conserved while inserting definition (4.30) directly into the TDSE and presuming all ϕ_i to be orthonormal, that is, $\langle\phi_i|\phi_j\rangle = \delta_{ij}\ \forall t$, leads to a set of equations by which the norm of Ψ varies. The latter approach is more straightforward and leads to the coupled system of equations

$$\mathrm{i}\frac{\partial}{\partial t}\phi_i(\boldsymbol{x},t) = \left(\hat{H}^0(\boldsymbol{x},t) + \hat{\Sigma}[\phi](\boldsymbol{x},t)\right)\phi_i(\boldsymbol{x},t), \tag{4.32}$$

with the single-particle energy operator $\hat{H}^0(\boldsymbol{x},t) = -\nabla_{\boldsymbol{x}}^2/(2m) + V(\boldsymbol{x},t)$, and the self-energy operator $\hat{\Sigma}(\boldsymbol{x},t)$, which accounts for particle–particle interactions. The notation $\hat{\Sigma}[\phi]$ indicates that the self-energy is a functional of all ϕ_i with $i = 1,\ldots,N$. Further, the sum $\hat{H}^0 + \hat{\Sigma}$ is often called *Fock operator* in the literature [14]. Explicitly, the action of the operator $\hat{\Sigma}$ on the orbital ϕ_i is given by

$$\begin{aligned}\hat{\Sigma}(\boldsymbol{x},t)\phi_i(\boldsymbol{x},t) = &\left(\sum_j \int \mathrm{d}\bar{x}\ |\phi_j(\bar{\boldsymbol{x}},t)|^2\, w(\boldsymbol{x}-\bar{\boldsymbol{x}})\right)\phi_i(\boldsymbol{x},t)\\ &- \sum_{i<j}{}' \left(\int \mathrm{d}\bar{x}\ \phi_i(\bar{\boldsymbol{x}},t)\phi_j(\bar{\boldsymbol{x}},t)w(\boldsymbol{x}-\bar{\boldsymbol{x}})\right)\phi_j(\boldsymbol{x},t).\end{aligned} \tag{4.33}$$

The first term constitutes the classical mean-field (Hartree) potential as an integral over the orbital-resolved density $n_i(\boldsymbol{x},t) = |\phi_i(\boldsymbol{x},t)|^2$. The second is the exchange potential (or Fock term) which is the indispensable quantum-mechanical correction accounting for antisymmetrization and the Pauli exclusion principle. In this sense, the ansatz (4.30) neglects any correlation effects which arise from terms of higher than first order in the interaction potential. Systematic improvements of TDHF results are possible, for example, by multiconfiguration time-dependent Hartree–Fock (MCTDH) theory which involves superpositions of several Slater determinants [22], time-dependent density functional theory (TDDFT) including exchange-correlation functionals [23], or nonequilibrium Green's function theory with higher-order self-energies [16, 24].

Of course, in regard to time evolution, (4.32) needs to be supplied with initial conditions which are just the stationary (ground-state) solutions with $\mathrm{i}\partial/\partial_t\phi_i(\boldsymbol{x},t)$ replaced by $\epsilon_i\phi_i^0(\boldsymbol{x})$. The energy eigenvalues ϵ_i and the corresponding orbitals $\phi_i^0(\boldsymbol{x})$ can be computed either by propagation of (4.32) in imaginary time – as described for the TDSE in the previous section – or by iterating the stationary equations until a self-consistent solution $\{\epsilon_i,\phi_i^0(\boldsymbol{x})\}$ is obtained (self-consistent field method, see, e.g., [25–28]). However, the latter procedure is, in contrast to the ITP technique, not conveniently performed on spatial grids, but it is instead transformed into a generalized eigenvalue problem by using a basis expansion for the initial set of orbitals:

$$\phi_i^0(\boldsymbol{x}) = \sum_{j=1}^{n_\mathrm{b}} c_{ji}\varphi_j(\boldsymbol{x}), \tag{4.34}$$

where the n_b – in principle – arbitrary (or adapted) functions $\varphi_j(\boldsymbol{x})$ have overlap $O_{ij} = \langle\varphi_i|\varphi_j\rangle$ and c_{ji} are the unknown coefficients. Using (4.34) in the stationary form of (4.32) and integrating over space yields the so-called Roothaan–Hall equations [27,28]:

$$\sum_{k=1}^{n_\mathrm{b}} \left(H^0_{ik} + \Sigma_{ik}[c] - \epsilon_j O_{ik}\right) c_{kj} = 0. \tag{4.35}$$

Here, the one-particle energy and the self-energy become matrices, $\hat{H}^0(\boldsymbol{x}) \to H^0_{ij}$ and $\hat{\Sigma}[\phi](\boldsymbol{x}) \to \Sigma_{ij}[c]$, which are defined by

$$H^0_{ij} = \int \mathrm{d}x\, \varphi_i^*(\boldsymbol{x}) \hat{H}^0(\boldsymbol{x}) \varphi_j(\boldsymbol{x}), \tag{4.36}$$

$$\Sigma_{ij}[c] = \sum_{kl} (w_{ij,kl} - w_{il,kj}) \varrho_{kl}[c], \tag{4.37}$$

with the *single-particle* density matrix, $\varrho_{ij}[c] = \sum_{k\leq N} c_{ik} c^*_{jk}$, and two-particle integrals given by

$$w_{ij,kl} = \iint \mathrm{d}x\, \mathrm{d}\bar{x}\, \varphi_i^*(\boldsymbol{x}) \varphi_k^*(\bar{\boldsymbol{x}}) w(\boldsymbol{x} - \bar{\boldsymbol{x}}) \varphi_j(\boldsymbol{x}) \varphi_l(\bar{\boldsymbol{x}}). \tag{4.38}$$

To solve the Roothaan–Hall equations iteratively, one starts from a random or trial state characterized by the density matrix ϱ_{ij} (provide either ϱ_{ij} or the expansion coefficients c_{ij}). Then, the self-energy is computed from (4.37) and the general eigenvalue problem (4.35) is solved for a new set of coefficients c_{ij}. The procedure is finally repeated until self-consistency is achieved, in which all elements ϱ_{ij} have reached a fixed point. However, the solution contains, in general, more than N-orbitals. The N-energetically lowest (occupied) orbitals, $\phi^0_{i\leq N}$, form the desired (pure) ground-state wave function, whereas the remaining $n_\mathrm{b} - N$ (virtual) orbitals are connected with the excited states of the system. Hence, the self-consistent field method also allows for the description of mixed states when a finite temperature (grand canonical) density matrix, $\varrho_{ij} \to \sum_{k=1}^{n_\mathrm{b}} c_{ik} f_\beta(\epsilon_k - \mu) c^*_{jk}$, is used, where $f_\beta(\epsilon_i - \mu)$ denotes the Fermi–Dirac distribution with inverse temperature $\beta = (k_\mathrm{B}T)^{-1}$ and μ is the chemical potential obtained from $\sum_{k=1}^{n_\mathrm{b}} f_\beta(\epsilon_i - \mu) = N$. However, note that the TDHF equations (4.32) do not provide the basis for the propagation of mixed states.

4.3.2 NEGF Approach

The theory of NEGFs [16,24] offers an alternative formulation of the TDHF method in terms of a generalized one-particle density matrix $G(\boldsymbol{x}, t; \bar{\boldsymbol{x}}, \bar{t})$, which is called Green's function and depends on two space–time variables – again $\boldsymbol{x} = (\boldsymbol{r}, \sigma)$. Utilizing NEGFs it is, particularly, possible to go beyond the ansatz (4.30) by systematically including correlations via higher-order self-energy contributions [24, 29]. Also, the approach is capable of handling the time dependence of mixed states.

Using $\hbar = 1$, the NEGF is defined by

$$G(\boldsymbol{x},t;\bar{\boldsymbol{x}},\bar{t}) = -\mathrm{i}\,\langle T_{\mathcal{C}}\hat{\psi}_H(\boldsymbol{x},t)\hat{\psi}^\dagger_H(\bar{\boldsymbol{x}},\bar{t})\rangle\,, \quad \langle\ldots\rangle = \mathrm{Tr}\{\varrho\ldots\}, \tag{4.39}$$

with fermionic field operators $\hat{\psi}_H(\boldsymbol{x},t)$ in the Heisenberg picture,[2] and time arguments t and $\bar{t}$ defined on the complex Schwinger/Keldysh contour $\mathcal{C}$ [30]. Further, the operator $T_{\mathcal{C}}$ ensures time ordering on $\mathcal{C}$. From G, the one-particle density matrix, and many other observables [18], are recovered in the limit of equal times:

$$\varrho(\boldsymbol{x},\bar{\boldsymbol{x}};t) = -\mathrm{i}G^<(\boldsymbol{x},t;\bar{\boldsymbol{x}},t), \tag{4.40}$$

with definition $G(1,\bar{1}) = \theta(t-\bar{t})G^>(1,\bar{1}) - \theta(\bar{t}-t)G^<(1,\bar{1})$ and notation $1 = (\boldsymbol{x},t)$ and $\bar{1} = (\bar{\boldsymbol{x}},\bar{t})$. The two-time Green's function G obeys generalized kinetic equations, the Keldysh/Kadanoff–Baym equations (KBE) [24]

$$\left[\mathrm{i}\frac{\partial}{\partial t} - \hat{H}^0(1)\right]G(1,\bar{1}) = \delta_{\mathcal{C}}(1-\bar{1}) - \int_{\mathcal{C}}\mathrm{d}2\;\Sigma[G](1,2)G(2,\bar{1}), \tag{4.41a}$$

$$\left[-\mathrm{i}\frac{\partial}{\partial \bar{t}} - \hat{H}^0(\bar{1})\right]G(1,\bar{1}) = \delta_{\mathcal{C}}(1-\bar{1}) - \int_{\mathcal{C}}\mathrm{d}2\;\Sigma[G](1,2)G(2,\bar{1}), \tag{4.41b}$$

with $\delta_{\mathcal{C}}(1-\bar{1}) = \delta_{\mathcal{C}}(t-\bar{t})\delta(\boldsymbol{x}-\bar{\boldsymbol{x}})$, $\int_{\mathcal{C}}\mathrm{d}2 := \int\mathrm{d}x_2\int_{\mathcal{C}}\mathrm{d}t_2$, and self-energy $\Sigma[G]$. The TDHF problem is then formulated via the approximation

$$\Sigma[G](1,\bar{1}) = \mathrm{i}\delta(1-\bar{1})\int_{\mathcal{C}}\mathrm{d}3\;W(1{-}3)G(3,3^+) - \mathrm{i}G(1,\bar{1})W(1^+{-}\bar{1}), \tag{4.42}$$

where $W(1-\bar{1}) = \delta_{\mathcal{C}}(t-\bar{t})w(\boldsymbol{x}-\bar{\boldsymbol{x}})$ and 1^+ indicates the limit $t \to t + \varepsilon_{>0}$. The explicit form of (4.42) is readily obtained from an irreducible diagrammatic expansion of the self-energy [29] retaining only first-order terms in W, and has the same structure as (4.33): the first term is the time-local Hartree potential and the second incorporates exchange features.

The equilibrium solution of the KBE including (4.42) is given by the Matsubara Green's function:

$$G^{\mathrm{M}}(\boldsymbol{x},\bar{\boldsymbol{x}};\tau) = \sum_{i,j=1}^{n_b}\phi_i^{0*}(\boldsymbol{x})\phi_j^0(\boldsymbol{x})G^{\mathrm{M}}_{ij}(\tau),$$
$$G^{\mathrm{M}}_{ij}(\tau) = \delta_{ij}\,f_\beta(\epsilon_i-\mu)\mathrm{e}^{-\tau(\epsilon_i-\mu)}, \tag{4.43}$$

[2] The field operators satisfy the equal-time anticommutation relations $[\hat{\psi}_H(\boldsymbol{x},t),\hat{\psi}^\dagger_H(\bar{\boldsymbol{x}},t)]_+ = \delta(\boldsymbol{x}-\bar{\boldsymbol{x}})$ and $[\hat{\psi}^{(\dagger)}_H(\boldsymbol{x},t),\hat{\psi}^{(\dagger)}_H(\bar{\boldsymbol{x}},t)]_+ = 0$, where $[\hat{A},\hat{B}]_+ = \hat{A}\hat{B}+\hat{B}\hat{A}$.

where $\tau \in [-\beta, 0]$. The orbitals ϕ_i^0 together with ϵ_i and μ are solutions of the self-consistent field method, and any (pure) ground state is obtained in the limit $\beta \to \infty$. More formally, (4.43) solves the Dyson equation (the KBE for $t - \bar{t} = \mathrm{i}\tau$) on the Hartree–Fock level (see, e.g., [29, 31] and references therein). Further, $G^{\mathrm{M}}(\boldsymbol{x}, \bar{x}; \tau)$ serves as initial condition for real-time propagation according to

$$G(\boldsymbol{x}, 0 - \mathrm{i}\tau; \bar{\boldsymbol{x}}, 0 - \mathrm{i}\bar{\tau}) = \mathrm{i}[G^{\mathrm{M}}(\boldsymbol{x}, \bar{x}; \tau) - G^{\mathrm{M}}(\boldsymbol{x}, \bar{x}; -\bar{\tau})]. \tag{4.44}$$

Instead of making the orbitals $\phi_i^0(\boldsymbol{x})$ time-dependent, it is advantageous to approach (4.41) using a static basis:

$$G(1, \bar{1}) = \sum_{i,j=1}^{n_b} \phi_i^{0*}(\boldsymbol{x}) \phi_j^0(\boldsymbol{x}) G_{ij}(t, \bar{t}). \tag{4.45}$$

Consequently, one is left with equations of motion for the elements $G_{ij}(t, \bar{t}) = -\mathrm{i}\, \langle T_{\mathcal{C}}\, \hat{a}_i(t)\, \hat{a}_j^{\dagger}(\bar{t}) \rangle$ which are just the Green's functions with respect to the creation (annihilation) operators $a_i^{\dagger}(t)$ $(a_i(t))$ of the steady state i [32, 33]. In addition, the products of H^0 $(\Sigma[G])$ and G in (4.41) become standard matrix multiplications, and the time-dependent self-energy is evaluated in the manner of (4.37) as

$$\Sigma_{ij}[G](t, \bar{t}) = -\mathrm{i}\delta_{\mathcal{C}}(t - \bar{t}) \sum_{k,l=1}^{n_b} (w_{ij,kl} - w_{il,kj}) G_{\bar{k}l}^{<}(t, \bar{t}). \tag{4.46}$$

We note that, using the Hartree–Fock approximation (4.46), it is fully sufficient to propagate the KBE for the center of mass (c.m.) time $t_{\mathrm{c.m.}} = (t + \bar{t})/2$ alone, instead of expanding G into the whole two-time plane $[0, t] \times [0, \bar{t}]$. However, this is no longer the case when higher-order self-energies $\Sigma[G]$ are involved – examples are second Born- or GW-type formulas [24, 34] which are nonlocal in time and lead to correlation and memory effects [35].

4.3.3 Example

Let us consider two fermions of mass m and charge q confined in a 1D parabolic trap $V(\boldsymbol{r}_i) = (m/2)\omega^2 \boldsymbol{r}_i^2$ with frequency ω and $\boldsymbol{r}_i \to (x_i, 0, 0)$, $i = 1, 2$. Due to the electrostatic charge, both fermions repel each other via the Coulomb potential $w(x_1 - x_2) = q^2/(4\pi\epsilon_0\, |x_1 - x_2|)$. However, in 1D it is convenient to apply a regularized Coulomb potential [36] with $|x_1 - x_2|$ replaced by $\sqrt{(x_1 - x_2)^2 + \kappa^2}$. The parameter $\kappa > 0$ keeps the integrals in (4.33), (4.38), and (4.42) finite and, in a physical interpretation, allows for a transversal spread of the total wave function.

Using dimensionless units $\{x_i \to x_i/x_0, E \to E/E_0\}$ with $x_0 = \sqrt{\hbar/(m\omega)}$ and $E_0 = \hbar\omega$, the Hamiltonian reads

$$\hat{H} = \frac{1}{2}\left(-\nabla^2_{x_1} + x_1^2\right) + \frac{1}{2}\left(-\nabla^2_{x_2} + \bar{x}_2^2\right) + \frac{\lambda}{\sqrt{(x_1 - x_2)^2 + \kappa^2}}. \tag{4.47}$$

Here, $\lambda = E_{\mathrm{C}}/E_0 = x_0/a_{\mathrm{B}}$ is the coupling parameter – the ratio between the characteristic Coulomb energy $E_{\mathrm{C}} = q^2/(4\pi\epsilon_0 x_0)$ (oscillator length x_0) and the confinement energy E_0 (Bohr radius $a_{\mathrm{B}} = (4\pi\epsilon_0\hbar\omega)/q^2$). For fixed κ, the coupling parameter solely controls the system behavior. For $\lambda \ll 1$ (corresponding to high density), the two fermions will be found in a Fermi gas- or liquid-like state whereas in the (low-density) limit $\lambda \to \infty$, with $x_0 \gg a_{\mathrm{B}}$, quantum effects vanish in favor of classical, interaction dominated behavior. For moderate coupling $\lambda \gtrsim 1$, states with well-localized density can be formed [18, 37]. However, such structure formation strongly depends on the spin configuration.

In the following, we examine the ground-state and nonequilibrium situations of the two-fermion system (4.47) using the TDHF ansatz (4.30). Thereby, we further assume the orbitals ϕ_1 and ϕ_2 to be equal for all times t, that is, we consider

$$\Psi(x_1, x_2, t) = \phi(x_1, t)\phi(x_2, t). \tag{4.48}$$

This symmetric product (or singlet state) is justified as long as the spin wave function $\chi(\sigma_1, \sigma_2)$ is antisymmetric.

With (4.48), the TDHF equations (4.32) simplify to a single equation for the orbital $\phi(x, t)$. In particular, it is easily seen that the exchange term in (4.33) becomes half the Hartree potential, thus

$$\hat{\Sigma}(x, t) = \frac{1}{2}\int \mathrm{d}\bar{x}\; |\phi(\bar{x}, t)|^2\, w(x - \bar{x}). \tag{4.49}$$

In analogy to (4.37) for the ground-state problem, the self-energy has to be modified to $\Sigma_{ij}[c] = \sum_{kl}(2w_{ij,kl} - w_{il,kj})\varrho_{kl}[c]$ with the constraint $N \to N/2$. A similar expression holds for the time-dependent case (4.46). The initial state of the system is now obtained either by direct imaginary time propagation or by solving the Roothaan–Hall equations (4.35) for $\phi(x) = \phi_1^0(x)$, for example, expanded in terms of oscillator eigenfunctions[3] $\varphi_{n+1}(x) = [2^n n!\sqrt{\pi}]^{-1/2}\mathrm{e}^{-x^2/2}H_n(x)$ with $n = 0, 1, 2, \ldots$. For the specific case of $\lambda = 2$ and $\kappa = 0.1$, Fig. 4.5 shows the result of both methods.

The ITP starts from the energetically lowest oscillator eigenfunction $\varphi_1(x)$, which is the ideal reference state for $\lambda \equiv 0$ (see the thin black curve for $t = 0$ in Fig. 4.5a). Then, as time increases, $|\phi(x, -\mathrm{i}t)|^2$ evolves getting more and more broadened due to the Coulomb-like interaction (see the gray curves in Fig. 4.5a) and, finally, it converges to a stationary solution $\phi_{t\to\infty}(x)$, denoted by the thick black curve. At the same time, the effective one-particle potential V^{eff}, defined as

$$V^{\mathrm{eff}}(x, -\mathrm{i}t) = V(x) + \Sigma[\phi](x, -\mathrm{i}t), \tag{4.50}$$

[3] $H_n(x)$ denotes the Hermite polynomial of order n.

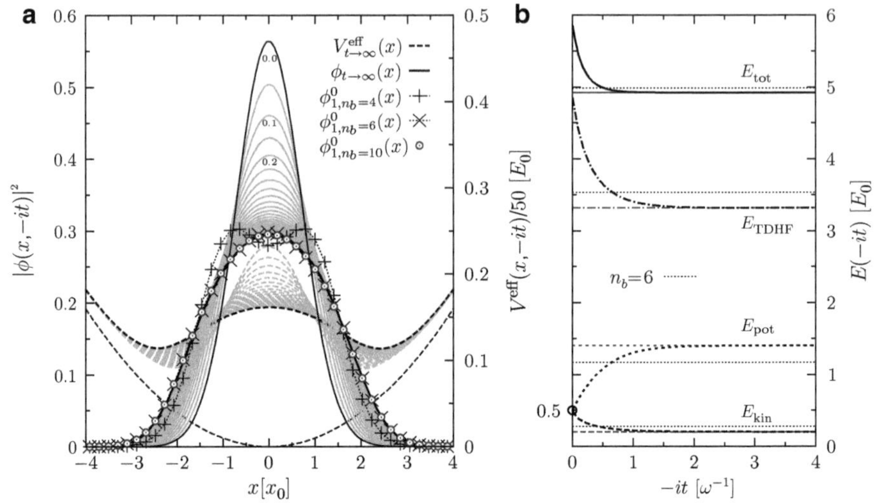

Fig. 4.5 Computation of the ground-state function $\phi(x)$ ($\lambda = 2$ and $\kappa = 0.1$) via imaginary time propagation starting from the lowest oscillator eigenstate $\varphi_1(x)$. (**a**) Orbital $\phi(x,-it)$ for different times t and the corresponding effective one-particle potential $V^{\rm eff}(x,-it)$; discretization $\Delta t = 0.01$ and $\Delta x = 0.04$. The *thick black (dashed) line* shows the converged result for $t \to \infty$. (**b**) Convergence of the different energy contributions. The *thin dotted lines* denote the (not yet converged) energies obtained from the self-consistent field method with $n_b = 6$, compare with $\phi^0_{1,n_b}(x)$

changes from a sharply peaked function in space into a smoother stationary equilibrium potential $V^{\rm eff}_{t\to\infty}(x)$; see the sequence of dashed curves. Figure 4.5b shows the (rapid) convergence of the different energies obtained from the total wave function $\Psi(x_1, x_2, t)$. Particularly, note that $E_{\rm pot}$ is computed from the single-particle potential $V(x)$, whereas $E_{\rm TDHF}$ denotes the expectation value of $\Sigma[\phi](x,-it)$. According to the initial state $\varphi_1(x)$, the kinetic and potential energy at $t = 0$ take the value of *two* independent particles in the 1D harmonic confinement, $E_{\rm kin} = E_{\rm pot} = 1/2$, while the interaction energy $E_{\rm TDHF}$ is initially much larger than the converged value. For comparison, Fig. 4.5 also shows the self-consistent field method. It reaches the same ground state in the limit $n_b \gtrsim 10$, compare with the orbitals $\phi^0_{1,n_b}(x)$ with $n_b = 4$, 6, and 10.

As an application to nonequilibrium, we consider the response of the two-fermion system to a short turnoff of the trap potential $V(x)$, where the switch-off time has been chosen to be $\delta t \approx 0.1$ and, hence, $V(x,t) = \theta(t - \delta t)V(x)$. After releasing the confinement, the initial product state of the two fermions is no longer an eigenstate of the actual system. Consequently, $\phi(x,t)$ undergoes broadening and starts to oscillate harmonically when the confinement is reactivated for $t \geq \delta t$. In conjunction with this, the potential energy $E_{\rm pot}(t)$, as computed from the total wave function, also begins to oscillate with a frequency $\omega_{\rm br}$ which we call the *breathing frequency*. It is found that this frequency depends strongly on λ as well as on the

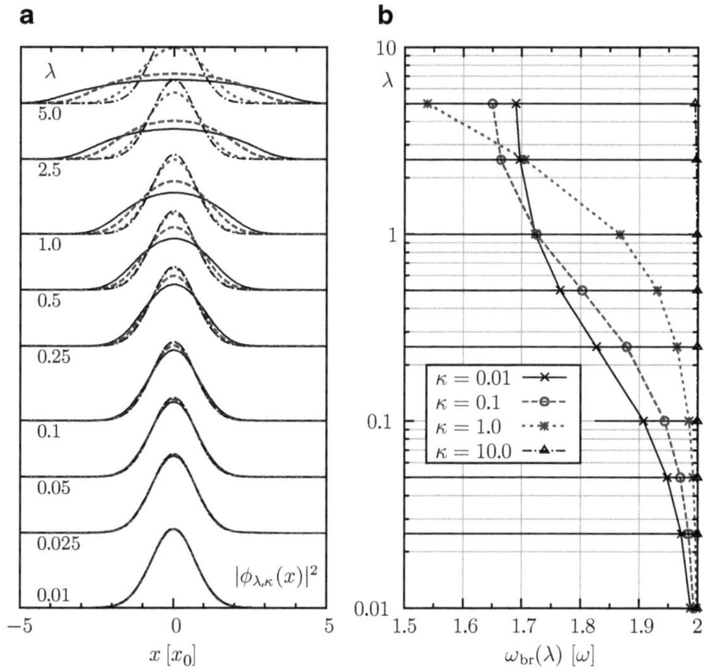

Fig. 4.6 Nonequilibrium response of the two-fermion system, Hamiltonian (4.47), after a short turnoff of the confinement. (**a**) Initial ground state $\left|\phi_{\lambda,\kappa}(x)\right|^2$ as a function on the coupling parameter λ and κ. For the assignment of the four different curves with $\kappa = 0.01$, 0.1, 1.0, and 10.0, see (**b**). For $\lambda \ll 1$, the ground state is practically independent of κ. (**b**) λ-dependence of the potential energy oscillation (breathing) frequency ω_{br} for different κ-values in units of the confinement frequency ω. The breathing frequency is obtained from a fit, $a\cos(\omega_{\mathrm{br}}t + b) + c$, applied to the time-dependent potential energy

regularization parameter κ (cf. Fig. 4.6a, b). For $\lambda \to 0$, the breathing frequency approaches the value $\omega_{\mathrm{br}} = 2\omega$ which is the well-known result for the noninteracting (ideal) system [38, 39]. With increasing λ, the frequency ω_{br} generally decreases in the considered λ-regime, and further exhibits a nontrivial behavior at moderate coupling, $\lambda \gtrsim 1$. A more detailed analysis of the breathing motion of quantum particles in traps can be found in [40].

4.4 Quantum Monte Carlo Methods

The phenomenon of physical structure formation is closely related to the exact treatment of many-body correlations. To rigorously take into account the mutual interplay between a large number of individual particles, random-number-based Metropolis Monte Carlo (MC) methods can be applied to efficiently sample the high-dimensional *configuration space*. Unlike molecular dynamics (see Chap. 10), the Monte Carlo method is stochastic rather than dynamical and thus, following the general concept, only statistical averages of *equilibrium properties* can be computed. In this section, we start the introduction with the basic Monte Carlo algorithm, which can also be used in advanced quantum simulations, as shown in the second part of this section. Here, the imaginary time path-integral representation is derived, which allows for a (quasiclassical) high-temperature approximation of the N particle density matrix and its numerical solution with efficient Monte Carlo methods.

4.4.1 Metropolis Monte Carlo Method

The original idea of this stochastic simulation method was coined by E. Fermi, J. von Neumann, S. Ulam, and N. Metropolis, who proposed in 1953 a stochastic algorithm to generate microstates according to the Boltzmann distribution, so that thermal averages could be computed easily [41]. This famous Metropolis sampling scheme has been rated as being among the top 10 algorithms having the "greatest influence on the development and practice of science and engineering in the 20th century" [42].

To describe the considered model system (4.1) by means of MC methods, the dynamical physical process has to be transformed into a stochastic one. A key element in the Metropolis Monte Carlo procedure is thus the concept of the Markov chain. This means that the immediate sequencing of a state depends only on the present state, regardless of the preceding development of the system. The Markov process generates a path in the configuration space and all quantities of interest are averaged along this trajectory, which is the probabilistic analogue to that generated by the equations of motion in molecular dynamics [43, 44] (see figure in Chap. 10).

In mathematical terms, the Markov chain is defined as a sequence of sample points i in the configuration space Ω:

$$\boldsymbol{r}_0^N \xrightarrow{W} \dots \xrightarrow{W} \boldsymbol{r}_i^N \xrightarrow{W} \boldsymbol{r}_{i+1}^N \xrightarrow{W} \boldsymbol{r}_{i+2}^N \xrightarrow{W} \cdots , \tag{4.51}$$

where the vector $\boldsymbol{r}_i^N = (\boldsymbol{r}_1, \boldsymbol{r}_2, \dots, \boldsymbol{r}_N)_i \in \Omega$ of dimension $3N$ comprises the coordinates of all N particles. The transition operator $W(\boldsymbol{r}_i^N \to \boldsymbol{r}_j^N)$ has to obey the *detailed balance* condition [45]

$$P(\boldsymbol{r}_i^N)W(\boldsymbol{r}_i^N \to \boldsymbol{r}_j^N) = P(\boldsymbol{r}_j^N)W(\boldsymbol{r}_j^N \to \boldsymbol{r}_i^N) \tag{4.52}$$

for each MC step from one to any other state. In thermal equilibrium at a fixed external heat bath temperature T, the probability $P(\boldsymbol{r}_i^N)$ of obtaining configuration $\boldsymbol{r}_i^N$ is weighted according to the Boltzmann probability distribution:

$$P(\boldsymbol{r}_i^N) = \frac{1}{Z}\,\mathrm{e}^{-\beta E(\boldsymbol{r}_i^N)}, \tag{4.53}$$

where $\beta = E_0/k_B T$ is the dimensionless inverse temperature,[4] k_B is the Boltzmann's constant, E is the (dimensionless) total system energy according to Hamiltonian (4.1), and $Z = \sum_{\Omega} e^{-\beta E(\boldsymbol{r}_i^N)}$ is the partition function of the canonical ensemble. Hence, the relative transition probability for the step $\boldsymbol{r}_i^N \to \boldsymbol{r}_j^N$ is a function of the total energy change $\Delta E = E(\boldsymbol{r}_j^N) - E(\boldsymbol{r}_i^N)$ only:

$$\frac{W(\boldsymbol{r}_i^N \to \boldsymbol{r}_j^N)}{W(\boldsymbol{r}_j^N \to \boldsymbol{r}_i^N)} = e^{-\beta \Delta E}. \tag{4.54}$$

This equation is satisfied by the Metropolis function [41]:

$$W(\boldsymbol{r}_i^N \to \boldsymbol{r}_j^N) = \begin{cases} \exp(-\beta \Delta E), & \Delta E > 0, \\ 1, & \Delta E \leq 0. \end{cases} \tag{4.55}$$

This means that if a trial move $\boldsymbol{r}_i^N \to \boldsymbol{r}_j^N$ lowers the energy, then the step is always accepted. However, if the energy is increased, the trial step is accomplished with a probability $W < 1$ only, and is otherwise rejected.

Starting from an arbitrary configuration $\boldsymbol{r}_0^N \in \Omega$, after an initial thermalization time of the simulation, the expectation value of the ensemble average of a generic physical quantity $A(\boldsymbol{r}^N)$ can be estimated as an arithmetic mean over the Markov chain of K consecutive MC steps:

$$\langle A \rangle = \sum_{i \in \Omega} P(\boldsymbol{r}_i^N) A(\boldsymbol{r}_i^N) \approx \frac{1}{K} \sum_{k=1}^{K} A(\boldsymbol{r}_k^N). \tag{4.56}$$

A central point in this context is the ergodicity of the Markov process, which refers to the condition that any state in the configuration space has to be accessible from any other state in a finite number of MC steps. An inherent problem with respect to the ergodicity in strongly correlated systems is, naturally, the (exponentially) growing autocorrelation time with the system size, which may easily exceed the simulation time. Especially at low temperatures, one has to take care that the statistics are not biased, since the expectation values of the observed quantities may seem to have converged although the system is trapped in local minima and has barely moved in the configuration space Ω. However, one should be aware that long simulation times do not automatically guarantee more accurate results generally, as discussed in [46].

Recommendable reviews on the subject of classical Monte Carlo simulations are to be found, for example, in [43,44,47,48].

[4] E_0 and l_0 are base units of energy and length, for example, in trapped systems the harmonic oscillator ground-state energy $E_0 = \hbar\omega_0$ and oscillator length $l_0 = \sqrt{\hbar/(m\omega_0)}$. A dimensionless system of units is obtained by applying the transformation rules $\{r \to r/r_0, E \to E/E_0\}$.

4.4.2 Path-Integral Monte Carlo

The PIMC simulation technique is founded on R.P. Feynman's path-integral formulation of quantum mechanics which, unlike E. Schrödinger's and W. Heisenberg's differential equation formalism, generalizes the formulation of classical mechanics, in particular Hamilton's principle of least action. In spite of its intuitive and theoretical sophistication, the evaluation of the path integrals is not at all trivial since one has to integrate over all possible states of the system for each moment in time [49].

In analogy to classical statistical mechanics, in which thermal equilibrium expectation values are defined as a canonical average of all microstates weighted by the Boltzmann factor (see (4.56)), the equilibrium state of a quantum system at a given inverse temperature, β, is fully characterized by the *many-body density operator*:

$$\hat{\rho}(\beta) = \frac{1}{Z} e^{-\beta \hat{H}} = \frac{1}{Z} \sum_n |\psi_n\rangle \, e^{-\beta E_n} \, \langle \psi_n| \, . \tag{4.57}$$

This statistical operator, $\hat{\rho}$, is defined as the superposition of pure N particle eigenfunctions, $|\psi_n\rangle$, which are exponentially weighted with the allowed energy eigenvalues, E_n, determined from the stationary Schrödinger equation $\hat{H} \, |\psi_n\rangle = E_n \, |\psi_n\rangle$. This means that the density operator $\hat{\rho}$ mixes the pure states, $|\psi_n|^2$, according to the thermal distribution and thus generalizes the concept of the wave function to finite temperatures, that is, mixed ensembles.

As seen in Chap. 3, the thermal average of an observable $\hat{A}$ in thermodynamic equilibrium is defined as

$$\left|\hat{A}\right| = \mathrm{Tr}[\hat{\rho}\hat{A}] = \sum_i \langle i|\hat{\rho}\hat{A}|i\rangle = \sum_{i,j} \langle i|\hat{\rho}|j\rangle \, \langle j|\hat{A}|i\rangle \, . \tag{4.58}$$

If $\hat{A}$ is diagonal in the chosen basis, the thermal average can be determined from the diagonal elements of the density matrix only, that is,

$$\langle \hat{A} \rangle = \sum_i \langle i|\hat{\rho}|i\rangle \, A_i \, . \tag{4.59}$$

However, a direct computation of $\hat{\rho}$ requires knowledge of the complete energy spectrum by solving the N particle Schrödinger equation, which, in most cases, is impossible for interacting systems. As we will see, we can avoid this problem by using a (path-)integral representation of the N particle density matrix, which can be evaluated efficiently with the help of numerical Monte Carlo methods.

To do so, we change into the *basis of position vectors* $\boldsymbol{r}^N = (\boldsymbol{r}_1, \boldsymbol{r}_2, \ldots, \boldsymbol{r}_N)$, in which the off-diagonal density matrix becomes a function of $6N$-particle coordinates, that is,

$$\hat{\rho} \to \rho(\boldsymbol{r}^N, \boldsymbol{r}^{N'}; \beta) \equiv \langle \boldsymbol{r}^N | e^{-\beta \hat{H}} | \boldsymbol{r}^{N'} \rangle \, . \tag{4.60}$$

Note that in this position basis, all particles are labeled. Moreover, the function values of the density matrix are positive[5] for all of its arguments and have the significance of a probability for the transition from an initial state $\boldsymbol{r}^N$ to the final state $\boldsymbol{r}^{N'}$. The nonnegativity of the density matrix elements is an essential prerequisite for the subsequent application of Monte Carlo methods. In coordinate representation, the (normalized) thermal average of operator $\hat{A}$ becomes

$$\langle \hat{A} \rangle = \frac{1}{Z} \int \mathrm{d}\boldsymbol{r}^N \, \langle \boldsymbol{r}^N | \rho \hat{A} | \boldsymbol{r}^N \rangle = \frac{1}{Z} \iint \mathrm{d}\boldsymbol{r}^N \mathrm{d}\boldsymbol{r}'^N \rho(\boldsymbol{r}^N, \boldsymbol{r}'^N; \beta) \, \langle \boldsymbol{r}^N | \hat{A} | \boldsymbol{r}'^N \rangle \tag{4.61}$$

and the partition function is written as

$$Z(\beta) = \int \mathrm{d}\boldsymbol{r}^N \rho(\boldsymbol{r}^N, \boldsymbol{r}^N; \beta). \tag{4.62}$$

In general, these functional integrals cannot be carried out since an explicit analytical form of $\rho(\boldsymbol{r}^N, \boldsymbol{r}^{N'}; \beta)$ is commonly unknown for nonideal quantum systems. To overcome this problem, we reduce the density matrix to the one known for free particles in the *high-temperature limit*. To this end, we employ the *product property of the density matrix*:

$$\hat{\rho}(\beta) = \mathrm{e}^{-\beta \hat{H}} = \underbrace{\mathrm{e}^{-\tau \hat{H}} \ldots \mathrm{e}^{-\tau \hat{H}}}_{M\text{-times}} = \prod_{s=1}^{\mathrm{M}} \hat{\rho}(\tau), \quad M\tau = \beta, \tag{4.63}$$

which allows us to expand a low-temperature density matrix into a series of density matrices at M-times higher temperature, τ. Insertion of $M - 1$ high-temperature factors gives us the density matrix in position basis as

$$\begin{aligned} \rho(\boldsymbol{r}^N, \boldsymbol{r}^{N'}; \beta) &= \langle \boldsymbol{r}^N | \mathrm{e}^{-\beta \hat{H}} | \boldsymbol{r}^{N'} \rangle \\ &= \left\langle \boldsymbol{r}^N \left| \prod_{s=1}^{\mathrm{M}} \mathrm{e}^{-\tau \hat{H}} \right| \boldsymbol{r}^{N'} \right\rangle \\ &= \int \cdots \int \mathrm{d}\boldsymbol{r}_1^N \mathrm{d}\boldsymbol{r}_2^N \ldots \mathrm{d}\boldsymbol{r}_{M-1}^N \prod_{s=0}^{M-1} \langle \boldsymbol{r}_s^N | \mathrm{e}^{-\tau \hat{H}} | \boldsymbol{r}_{s+1}^N \rangle \\ &= \int \cdots \int \mathrm{d}\boldsymbol{r}_1^N \mathrm{d}\boldsymbol{r}_2^N \ldots \mathrm{d}\boldsymbol{r}_{M-1}^N \prod_{s=0}^{M-1} \rho\left(\boldsymbol{r}_s^N, \boldsymbol{r}_{s+1}^N; \tau\right), \end{aligned} \tag{4.64}$$

where the ordered set $(\boldsymbol{r}_0^N, \boldsymbol{r}_1^N, \ldots, \boldsymbol{r}_M^N)$ represents a path in configuration space. Expression (4.64) is exact and comprises in the limit $M \to \infty$ an integration over all possible paths through configuration space linking the fixed initial and final points, $\boldsymbol{r}_0^N = \boldsymbol{r}^N$ and $\boldsymbol{r}_M^N = \boldsymbol{r}^{N'}$.

[5] Here, we do not yet consider the problematic issue of Fermi statistics.

Interestingly, in the position basis many observables $\hat{A}$ are diagonal, which implies that only *diagonal elements* of the full (low-temperature) density matrix , $\rho(\beta)$, are relevant (see (4.59)). As a result, the partition function (4.62) now becomes an integral that runs over *closed paths*

$$Z(\beta) = \int \cdots \int \mathrm{d}\boldsymbol{r}_1^N \mathrm{d}\boldsymbol{r}_2^N \ldots \mathrm{d}\boldsymbol{r}_{M-1}^N \prod_{s=0}^{M-1} \rho\left(\boldsymbol{r}_s^N, \boldsymbol{r}_{s+1}^N; \tau\right), \quad \boldsymbol{r}_0^N = \boldsymbol{r}_M^N, \quad (4.65)$$

and is determined by the *off-diagonal* elements of the high-temperature density matrices $\rho(\tau)$.

Considering the system in question (cf. (4.1)), the Hamiltonian consists in its general form,

$$\hat{H} = \hat{K} + \hat{V}, \quad (4.66)$$

of two noncommuting N particle operators, the kinetic $\hat{K}$ and the potential $\hat{V}$ operators with $[\hat{K}, \hat{V}] \neq 0$. Expansion yields a cumbersome expression for the exponential operator:

$$\mathrm{e}^{-\tau(\hat{K}+\hat{V})} = \mathrm{e}^{-\tau\hat{K}}\mathrm{e}^{-\tau\hat{V}}\mathrm{e}^{-(\tau^2/2)[\hat{K},\hat{V}]}\mathrm{e}^{-(\tau^3/6)[[\hat{V},\hat{K}],\hat{K}+2\hat{V}]} + \mathcal{O}(\tau^4) \quad (4.67a)$$

$$= \mathrm{e}^{-\tau\hat{K}}\mathrm{e}^{-\tau\hat{V}}\mathrm{e}^{-(\tau^2/2)[\hat{K},\hat{V}]} + \mathcal{O}(\tau^3) \quad (4.67b)$$

$$= \mathrm{e}^{-\tau\hat{K}}\mathrm{e}^{-\tau\hat{V}} + \mathcal{O}(\tau^2). \quad (4.67c)$$

However, *Trotter's product formula* states that for self-adjoint operators $\hat{K}$ and $\hat{V}$ (which are bounded from below in a Hilbert space) in the limit of a large number of high-temperature factors, $M \to \infty$, the total density matrix (4.63) can be approximated as a simple product of potential and kinetic density matrices by neglecting the commutators from the exact operator identity (4.67), that is,

$$\hat{\rho}(\beta) = \mathrm{e}^{-\beta(\hat{K}+\hat{V})} = \left[\mathrm{e}^{-\tau(\hat{K}+\hat{V})}\right]^M \overset{!}{=} \lim_{M\to\infty} \left[\mathrm{e}^{-\tau\hat{K}}\mathrm{e}^{-\tau\hat{V}}\right]^M$$
$$= \left[\mathrm{e}^{-\tau\hat{K}}\mathrm{e}^{-\tau\hat{V}}\right]^M + \mathcal{O}(M^{-1}). \quad (4.68)$$

Note that the validity of the approximation made for finite M in the last step of (4.68) is not at all obvious due to the propagation of the error terms with respect to $\tau = \beta/M$ [50, 51]. The error of the high-temperature representation is therefore strongly affected by the number of high-temperature factors, M. Hence, the issue of convergence involving finite M has to be checked carefully for each particular system under study.

The high-temperature matrix element $\rho(\boldsymbol{r}_s^N, \boldsymbol{r}_{s+1}^N; \tau)$ in (4.64) and (4.65) can be approximated as

$$\rho(\boldsymbol{r}_s^N, \boldsymbol{r}_{s+1}^N; \tau) \equiv \langle \boldsymbol{r}_s^N | \mathrm{e}^{-\tau(\hat{K}+\hat{V})} | \boldsymbol{r}_{s+1}^N \rangle \approx \langle \boldsymbol{r}_s^N | \mathrm{e}^{-\tau\hat{K}}\mathrm{e}^{-\tau\hat{V}} | \boldsymbol{r}_{s+1}^N \rangle$$
$$= \mathrm{e}^{-\tau V(\boldsymbol{r}_s^N)} \langle \boldsymbol{r}_s^N | \mathrm{e}^{-\tau\hat{K}} | \boldsymbol{r}_{s+1}^N \rangle, \quad (4.69)$$

where $\hat{V}$ is diagonal in the spatial coordinate representation. The kinetic energy density matrix elements of free particles are obtained by a momentum eigenstate expansion:

$$\begin{aligned}\langle \boldsymbol{r}_s^N | \mathrm{e}^{-\tau \hat{K}} | \boldsymbol{r}_{s+1}^N \rangle &= \int \mathrm{d}\boldsymbol{p}^N \, \langle \boldsymbol{r}_s^N | \boldsymbol{p}^N \rangle \exp\left[-\tau \sum_{i=1}^{N} \frac{\hat{\boldsymbol{p}}_i^2}{2m_i} \right] \langle \boldsymbol{p}^N | \boldsymbol{r}_{s+1}^N \rangle \\ &= \lambda_M^{-3N} \exp\left[-\frac{\pi}{\lambda_M^2} (\boldsymbol{r}_s^N - \boldsymbol{r}_{s+1}^N)^2 \right]. \end{aligned} \tag{4.70}$$

Here, we take advantage of the diagonality of the kinetic operator $\hat{K} = \sum_{i=1}^{N} \hat{\boldsymbol{p}}_i^2 2m_i$ in momentum space, and note that the Gaussian-type integral can be evaluated analytically after explicit expressions for the plane waves $\langle \boldsymbol{r}_s^N | \boldsymbol{p}^N \rangle$ and $\langle \boldsymbol{p}^N | \boldsymbol{r}_{s+1}^N \rangle$ have been substituted. The term $\lambda_M = \sqrt{2\pi\hbar^2\beta/mM}$ denotes the thermal De Broglie wavelength. Insertion of the high-temperature matrices (4.69) and (4.70) into (4.64) provides us the *discrete time path-integral representation* of the N particle density matrix:

$$\begin{aligned}\rho(\boldsymbol{r}^N, \boldsymbol{r}^{N'}; \beta) \approx & \int \cdots \int \mathrm{d}\boldsymbol{r}_1^N \mathrm{d}\boldsymbol{r}_2^N \ldots \mathrm{d}\boldsymbol{r}_{M-1}^N \\ & \times \lambda_M^{-3N} \exp\left(-\frac{\pi}{\lambda_M^2} \sum_{s=0}^{M-1} (\boldsymbol{r}_s^N - \boldsymbol{r}_{s+1}^N)^2 \right) \exp\left(-\tau \sum_{s=0}^{M-1} V(\boldsymbol{r}_s^N) \right), \end{aligned} \tag{4.71}$$

which is valid for quantum systems with the Hamiltonian (4.66) and a quadratic dispersion law for $\boldsymbol{k} = \boldsymbol{p}/\hbar$.

Following the analogy between Feynman's original idea of a time evolution operator $\hat{U}(t, t') = \mathrm{e}^{-\mathrm{i}\hat{H}t/\hbar}$ and the definition in (4.57), the inverse temperature β may be considered as imaginary time, where $t \to \beta\hbar/\mathrm{i}$ and the imaginary time step is $\tau = \beta/M$. Thus, the set of coordinates $\boldsymbol{r}_s^N$ at a specific integer number $s = 1, \ldots, M-1$ are commonly named "imaginary time slice," since only particle images within the same slice, $\boldsymbol{r}_s^N$, interact with each other via the weakened (iso-time) potential $v(\boldsymbol{r}_s^N) = V(\boldsymbol{r}_s^N)/M$ (see Fig. 4.7). The classical-like particle images in successive slices $\{\boldsymbol{r}_s^N, \boldsymbol{r}_{s+1}^N\}$ are linked by a spring-like energy term, which is due to the quadratic quantum-mechanical kinetic energy of the free particle and ensures a finite particle extension. Hence, in the imaginary time path-integral formulation, a quantum system becomes mapped onto a classical one such that each physical (quantum) particle is represented by a path through M-positions (here called *particle images*) in configuration space at different values in imaginary time. This path forms a classical ring polymer of M-links. Depending on the inverse temperature β and particle mass m, the spring coupling becomes more or less rigid and, consequently, the quantum particles become more or less delocalized.

Most of the thermodynamic quantities are determined by the trace of the density matrix (4.71), that is, closed imaginary time trajectories from $\boldsymbol{r}^N$ to $\boldsymbol{r}^{N'} = \boldsymbol{r}^N$.

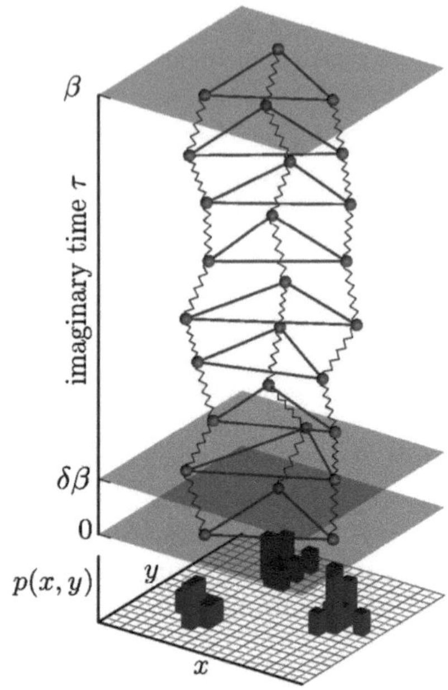

Fig. 4.7 Feynman's path-integral representation of a trapped 2D quantum system with three electrons. The probability density $p(x, y)$ is obtained by mapping of the beads along the imaginary time paths onto the 2D plane

For instance, the probability $p(\boldsymbol{r}^*)$ to observe an arbitrary particle at position $\boldsymbol{r}^*$ is given as arithmetic average over the imaginary time paths of all N particles as

$$p(\boldsymbol{r}^*) = \frac{1}{NM} \sum_{i=1}^{N} \sum_{s=0}^{M-1} \langle \delta(\boldsymbol{r}^* - \boldsymbol{r}_s^i) \rangle_{\rho_N} , \tag{4.72}$$

where $\langle \ldots \rangle_{\rho_N}$ defines the thermodynamic average according to (4.61).

So far, only quantum systems composed of distinguishable spinless particles (boltzmannons) have been considered. However, even in the case in which the Hamiltonian does not explicitly depend on particle spin, inclusion of quantum statistics requires sampling of the particle permutations in addition to the integrations in coordinate space. Specifically, the many-body density matrix (4.71) has to be properly *symmetrized* with respect to an arbitrary exchange of two indistinguishable *bosons* (e.g., bosonic atoms, molecules, or excitons), that is,

$$\rho^{\mathrm{S}}(\boldsymbol{r}^N, \boldsymbol{r}^{N'}; \beta) = \frac{1}{N!} \sum_P (+1)^P \rho(\boldsymbol{r}^N, \hat{P} \boldsymbol{r}^{N'}; \beta), \tag{4.73}$$

or *antisymmetrized* under arbitrary exchange of two indistinguishable *fermions* (such as electrons or holes with the same spin projection), that is,

$$\rho^{\mathrm{A}}(\boldsymbol{r}^N, \boldsymbol{r}^{N'}; \beta) = \frac{1}{N!} \sum_P (-1)^P \rho(\boldsymbol{r}^N, \hat{P} \boldsymbol{r}^{N'}; \beta), \tag{4.74}$$

where $\hat{P}$ is the permutation operator for particle indices and P is the parity of the permutation. In the framework of path-integral theory, the permutations can be decomposed into a sequence of two-particle exchanges along the imaginary time path. The pair exchanges are carried out by the transposition of particle positions in particular time slices, from which the paths of several particles can be merged into a single one. Such multiparticle trajectories correspond to off-diagonal elements of the density matrix, but still form closed loops.

The superposition of all $N!$-permutations of N-identical particles leads to the inherent (numerical) *fermion sign problem* since the alternating sign of the prefactor in the case of fermions (4.74) causes an essential cancelation of positive and negative contributions corresponding to even and odd permutations, respectively. Thus, an accurate calculation of such vanishing differences is strongly complicated by the increase of quantum degeneracy arising at low temperatures and high densities, where all permutations appear with equal probability.

The high-dimensional convolution integrals of the density matrix (4.71), (4.73), and (4.74) over $3N(M-1)$ degrees of freedom[6] can be numerically evaluated by a slightly modified version of the Metropolis sampling algorithm outlined for the classical systems. However, to reduce computational effort and increase the efficiency of Monte Carlo sampling, various sophisticated move strategies (e.g., the multilevel bisection sampling method or the worm algorithm [52]), approximations for the pair density matrix (e.g., using matrix-squaring technique [53, 54]), fast converging estimators with less statistical variance, and many further improvements have been developed over the last decades. For further (technical) details on this subject, we refer the interested reader to the following recommended in-depth references [45,47,55–57].

4.5 Summary

In this tutorial, we have provided an introduction to time-dependent and time-independent quantum simulations. The former part split up into an *exact* treatment of the one-particle *time-dependent* Schrödinger equation, and also an approximate investigation of many-body systems on the basis of Hartree–Fock theory (and beyond). With these two techniques, all quantum effects – for example, tunneling, quantization, and interference phenomena – can be well described and simulated to any desired accuracy. Furthermore, no approximations to external fields, such, for example, laser fields, trapping potentials, have to be introduced. The TDSE is exact for both particles of fermionic and bosonic character, whereas the Hartree–Fock method and its improvements as described here are well suited for fermionic calculations and are often used in quantum chemistry. The (exact) many-body wave function is here reduced to an (approximate) one-particle function which contains

[6] We typically use numbers of high-temperature factors M in the range $100 \leq M \leq 300$.

all pertinent information about the system under investigation. The extraction of physical properties is often a challenging task and much attention has to be paid to this point. On the other hand, the *wave function-based* TDSE methods are limited to one particle in a single-active electron approach, where the possible effects of additional particles are only included by utilizing effective external fields. Of course, such methods offer high accuracy in regard to one particle, but lack other many-body effects.

In the last part of this tutorial, we presented an introduction to the field of quantum Monte Carlo techniques. This method allows for the accurate calculation of equilibrium state properties of many-body systems with no further approximations, that is, inclusion of all correlation effects. The main advantage lies in the (efficient) sampling of the density matrix, which especially allows for a large number of *bosonic* particles. However, no phase information or corresponding wave function can be constructed in this way. Hence, this method is not suitable for investigating excitations and dynamics. Fermionic calculations are, up to now, limited to small systems due to the fermionic sign problem.

References

1. W. Press, W. Vetterling, S. Teukolsky, B. Flannery, *Numerical Recipes* (Cambridge University Press, Cambridge, 2002)
2. W. van Dijk, F.M. Toyama, preprint arXiv:physics/0701150v1 (2007)
3. K. Boucke, H. Schmitz, H.-J. Kull, Phys. Rev. A **56**, 763 (1997)
4. D. Bauer, P. Koval, Comput. Phys. Commun. **174**, 396–421 (2005)
5. D. Neuhauser, M. Baer, J. Chem. Phys. **90**, 4351–4355 (1989)
6. A. Vibok, G.G. Balint-Kurti, J. Chem. Phys. **96**, 8712–8719 (1992)
7. X.-S. Liu, X.-Y. Liu, Z.-Y. Zhou, P.-Z. Ding, S.-F. Pan, Int. J. Quant. Chem. **79**, 343–349 (2000)
8. H.-J. Kull, V.T. Tikhonchuk, Phys. Plasmas **12**, 063301 (2005)
9. S. Bauch, Diploma Thesis, University of Kiel (2008) (http://www.theo-physik.uni-kiel.de/bonitz/theses.html)
10. G.D. Mahan, *Many-Particle Physics*, 2nd edn. (Plenum, New York, 1990)
11. P.A.M. Dirac, Proc. Cambridge Philos. Soc. **26**, 376 (1930)
12. K.R. Sandhya Devi, S.E. Koonin, Phys. Rev. Lett. **47**, 27 (1981)
13. M.A. Ball, A.D. McLachlan, Mol. Phys. **7**, 501 (1963)
14. K.C. Kulander, Phys. Rev. A **36**, 2726 (1987)
15. N.E. Dahlen, R. van Leeuwen, Phys. Rev. A **64**, 023405 (2001)
16. M. Bonitz, *Quantum Kinetic Theory* (B.G. Teubner, Stuttgart, 1998)
17. W.D. Kraeft, D. Kremp, W. Ebeling, G. Röpke, *Quantum Statistics of Charged Particle Systems* (Akademie, Berlin, 1986)
18. K. Balzer, M. Bonitz, J. Phys. A: Math. Theor. **42**, 214020 (2009)
19. M. Bonitz, D. Kremp, D.C. Scott, R. Binder, W.D. Kraeft, H.S. Köhler, J. Phys.: Condens. Matter **8**, 6057 (1996)
20. H. Ehrenreich, M.H. Cohen, Phys. Rev. **115**, 786 (1959)
21. N.H. Kwong, M. Bonitz, R. Binder, S. Köhler, Phys. Stat. Sol. (b) **206**, 197 (1998)
22. J. Caillat, J. Zanghellini, M. Kitzler, O. Koch, W. Kreuzer, A. Scrinzi, Phys. Rev. A **71**, 012712 (2005)
23. M.A.L. Marques, E.K.U. Gross, Annu. Rev. Phys. Chem. **55**, 427 (2004)
24. L.P. Kadanoff, G. Baym, *Quantum Statistical Mechanics* (Benjamin, New York, 1962)

25. P. Echenique, J.L. Alonso, Mol. Phys. **105**, 3057 (2007)
26. P. Ludwig, K. Balzer, A. Filinov, H. Stolz, M. Bonitz, New J. Phys. **10**, 083031 (2008)
27. C.C.J. Roothaan, Rev. Mod. Phys. **20**, 69 (1951)
28. G.G. Hall, Proc. R. Soc. Lond. A **205**, 451 (1951)
29. N.E. Dahlen, R. van Leeuwen, J. Chem. Phys. **122**, 164102 (2005)
30. L.V. Keldysh, Zh. Eksp. Teor. Fiz. **47**, 1515 (1964) [Sov. Phys. JETP **20**, 235 (1965)]
31. K. Balzer, M. Bonitz, R. van Leeuwen, N.E. Dahlen, A. Stan, Phys. Rev. B **79**, 245306 (2009)
32. N.E. Dahlen, R. van Leeuwen, Phys. Rev. Lett. **98**, 153004 (2007)
33. K. Balzer, Diploma Thesis, Kiel University (2007) (http://www.theo-physik.uni-kiel.de/bonitz/theses.html)
34. A. Stan, N.E. Dahlen, R. van Leeuwen, Europhys. Lett. **76**, 298 (2006)
35. R. Binder, H.S. Köhler, M. Bonitz, Phys. Rev. B **55**, 5110 (1997)
36. K. Jauregui, W. Häusler, B. Kramer, Europhys. Lett. **24**, 581 (1993)
37. A.V. Filinov, M. Bonitz, Yu.E. Lozovik, Phys. Rev. Lett. **86**, 3851 (2001)
38. M.R. Geller, G. Vignale, Phys. Rev. B **53**, 6979 (1996)
39. G. Watanabe, Phys. Rev. A **73**, 013616 (2006)
40. S. Bauch, K. Balzer, D. Hochstuhl, M. Bonitz, Physica E **42**, 513 (2010)
41. N. Metropolis, A.W. Rosenbluth, M.N. Rosenbluth, A.H. Teller, E. Teller, J. Chem. Phys. **21**, 1087 (1953)
42. J. Dongarra, F. Sullivan, Comput. Sci. Eng. **2**, 22 (2000)
43. D.W. Heermann, *Computer Simulation Methods in Theoretical Physics*, 2nd edn. (Springer, New York, 1990)
44. K. Binder, *Computersimulationen*, Phys. J. **5**, 25 (2004)
45. A. Filinov, J. Böning, M. Bonitz, Lect. Notes Phys. **739**, 397 (2008)
46. J. Böning, A. Filinov, P. Ludwig, H. Baumgartner, M. Bonitz, Yu.E. Lozovik, Phys. Rev. Lett. **100**, 113401 (2008)
47. M. Bonitz, D. Semkat (eds.), *Introduction to Computational Methods for Many-Body Physics* (Rinton, Princeton, 2006)
48. W. Janke, Lect. Notes Phys. **739**, 79 (2008)
49. R.P. Feynman, A.R. Hibbs, *Quantum Physics and Path Integrals* (McGraw-Hill, New York, 1965)
50. H.F. Trotter, Pacific J. Math. **8**, 887 (1958)
51. H. De Raedt, B. De Raedt, Phys. Rev. A **28**, 3575 (1983)
52. M. Boninsegni, N. Prokof'ev, B. Svistunov, Phys. Rev. Lett. **96**, 070601 (2006)
53. R.G. Storer, J. Math. Phys. **9**, 964 (1968)
54. A.D. Klemm, R.G. Storer, Aust. J. Phys. **26**, 43 (1973)
55. D.M. Ceperley, Rev. Mod. Phys. **67**, 279 (1995)
56. B. Militzer, Ph.D. Thesis, University of Illinois at Urbana-Champaign (2000)
57. L.B. Barberà, Ph.D. Thesis, Universitat Politècnica de Catalunya (2002)

Chapter 5
Quantum Effects in Plasma Dielectric Response: Plasmons and Shielding in Normal Systems and Graphene

Norman J.M. Horing

Abstract A brief review of quantum plasma theory and phenomenology in solid-state plasmas is presented here, with attention to dynamic and nonlocal features of dielectric response. Focussing on the random-phase approximation, we discuss the RPA screening and dielectric functions in three, two, and one dimensions corresponding to bulk, quantum well, and quantum wire plasmas, respectively, taking care to distinguish quantum effects from classical ones mandated by the correspondence principle. In particular, we exhibit plasmon dispersion, damping, and static shielding in these various dimensionalities. We also review Landau-quantized magnetoplasma phenomenology, with emphasis on de Haas–van Alphen oscillatory features in intermediate strength magnetic fields and the quantum strong field limit in which only the lowest Landau eigenstate is populated.

Graphene is an exceptionally device-friendly material, with a massless relativistic Dirac energy spectrum for electrons and holes. We exhibit its RPA dynamic, nonlocal dielectric function in detail, discussing Graphene plasmons and electromagnetic modes in the THz range, self-energy, fast particle energy loss spectroscopy, atom/van der Waals interaction, and static shielding of impurity scatterers limiting dc transport in Graphene.

5.1 Introduction

5.1.1 Background

The history of quantum plasma dielectric response dates back to the early 1950s, when the effort to relieve divergencies in the calculation of correlation energy due to Coulomb interactions among huge numbers of mobile conduction electrons in solids

N.J.M. Horing (✉)
Department of Physics and Engineering Physics, Stevens Institute of Technology, Hoboken, New Jersey 07030, USA
e-mail: nhoring@stevens.edu

M. Bonitz et al. (eds.), *Introduction to Complex Plasmas*, Springer Series on Atomic, Optical, and Plasma Physics 59, DOI 10.1007/978-3-642-10592-0_5,

resulted in the recognition that it was necessary to take into account their dynamic, nonlocal screening properties.[1] In an early, illuminating quantum-mechanical formulation of the problem, the many electron Hamiltonian was employed in the form

$$H = \sum_i \frac{P_i^2}{2m} + \frac{1}{2} \sum_{i \neq j} \frac{e^2}{\left| \boldsymbol{r}_i - \boldsymbol{r}_j \right|} = H_{\text{el.}}^0 + H_{\text{e–e}}^{\text{coul.}}. \tag{5.1}$$

The first term represents free electrons, and the second term represents the long-range Coulomb interaction among the electrons. The long range of this interaction does in fact allow the electrons to behave in a collective manner, and to be specific, self-consistent collective density oscillations can exist in quantum many-electron plasmas, just as in conventional electron–ion plasmas. The significance of these plasma oscillations (plasmons) as elementary excitations of the system was brought into clear focus by the construction of a canonical transformation which put H into an approximately equivalent Hamiltonian of the form [1]

$$H \to H_{\text{el.}}^0 + H_{\text{pl.}}^0 + H_{\text{e–e}}^{\text{shield}} + H_{\text{el.–pl.}}, \tag{5.2}$$

which helped to cure the divergence problem encountered in perturbative calculations of the correlation energy. As above, $H_{\text{el.}}^0$ represents free electrons. Similarly, $H_{\text{pl.}}^0$ represents free plasmons, the self-consistent collective density oscillations which arise in consequence of the long-range character of the Coulomb electron–electron interaction. The microscopic dynamical mechanism of the plasmon will be discussed further below, and it is well known that the local plasma density oscillation (as well as its accompanying electric field oscillation which may be understood as a longitudinal photon) occurs with a natural frequency given by $\omega_{\text{p}} = (4\pi e^2 \rho_0/m)^{1/2}$, where ρ_0 is the carrier density and m is the effective mass. Associated quantum plasma energies involve multiplication by $\hbar$, so that $H_{\text{pl.}}^0$ involves a typical quantum-mechanical zero-point plasmon oscillator energy of $\hbar\omega_{\text{p}}/2$ summed over possible plasmon states contributing to the ground state of the many electron system. In addition, one has a residual short-range shielded electron–electron interaction term, $H_{\text{e–e}}^{\text{shield}}$. Finally, the fact that plasmons are not exact normal modes of the system (i.e., they are damped and have finite lifetime as elementary excitations) is reflected in the presence of an electron–plasmon interaction term, $H_{\text{el.–pl.}}$, which provides a mechanism for energy exchange between plasmons and electrons. Under equilibrium conditions the plasmon yields energy to the electrons, and for sufficiently high wave number the plasmon is so heavily damped as to be meaningless as an elementary excitation (this delimits possible plasmon states contributing to the $H_{\text{pl.}}^0$ ground state in the discussion above). However, under appropriate drifting conditions the directionality of energy flow via this mechanism can be reversed, resulting in electrons yielding energy to the plasmons, and concomitant microinstability of the plasmons.

[1] A brief introduction to the occurrence and features of quantum plasmas is given in Chap. 3. Also, see "Quantum Kinetic Theory" by M. Bonitz, Teubner, Stuttgart (1998).

5.1.2 Quantum Theory of Dielectric Response

All of these plasma features are, in fact, embodied in the structure of the longitudinal dielectric function, ε, or, equivalently, in the dynamic, nonlocal screening function, K, that is the inverse of ε. This is true both classically [2–4] and quantum mechanically. Of course, in solids it is necessary to describe free carrier dynamics quantum mechanically. The central feature of interest is the polarization of the medium in conjunction with the perturbed density, $\rho(2)$, of the plasma when subjected to an external potential, $U(2)$, at space–time point $2 = \boldsymbol{r}_2, t_2$. The perturbed density, in turn, contributes to the effective potential $V(1)$ at $1 = \boldsymbol{r}_1, t_1$ due to polarization through a nonlocal and dynamic relation as

$$V(1) = U(1) + \int \mathrm{d}^4 3\, v_\mathrm{c}(1,3)\rho(3), \tag{5.3}$$

by adding (to the impressed potential $U(1)$) the individual electron Coulomb potential contributions, $v_\mathrm{c}(1,3)$ at 1, weighted by the density, $\rho(3)$, at 3 (associated with the Poisson equation). Writing $V(1)$ of (5.3) in terms of a screening function $K(1,2)$, acting on the impressed potential $U(2)$, we have

$$V(1) = \int \mathrm{d}^4 2\, K(1,2)U(2), \tag{5.4}$$

or, equivalently, for the case of linearity, $K(1,2) = \delta V(1)/\delta U(2)$, we find that

$$\frac{\delta V(1)}{\delta U(2)} = \frac{\delta U(1)}{\delta U(2)} + \int \mathrm{d}^4 3\, v_\mathrm{c}(1,3)\frac{\delta\rho(3)}{\delta U(2)}, \tag{5.5}$$

where $\delta U(1)/\delta U(2) = \delta^4(1-2)$ by mutual independence of the variables $U(1)$ and $U(2)$ in this variational differentiation. This differentiation, which is a continuum generalization of multivariate calculus, also has the usual chain rule properties which we apply to $\delta\rho(3)/\delta U(2)$ in (5.5), leading to the random-phase approximation (RPA) integral equation, subject to the approximation for $\delta\rho(3)/\delta V(4)$ described below:

$$K(1,2) = \delta^4(1-2) + \int \mathrm{d}^4 3 \int \mathrm{d}^4 4\, v_\mathrm{c}(1,3)\frac{\delta\rho(3)}{\delta V(4)}K(4,2). \tag{5.6}$$

Of course, (5.6) remains valid beyond the RPA if one employs more accurate expressions for $\delta\rho(3)/\delta V(4)$. As $K(1,2)$ is just the inverse of the longitudinal dielectric function $\varepsilon(2,3) = \delta U(2)/\delta V(3)$ in the space–time matrix sense, that is,

$$\int \mathrm{d}^4 2\, K(1,2)\varepsilon(2,3) = \delta^4(1-3), \tag{5.7}$$

(5.6) may be rewritten directly for the dielectric function itself as

$$\varepsilon(1,2) = \delta^4(1-2) - \int d^4 3\, v_c(1,3) R(3,2), \tag{5.8}$$

where $R(3,2) = \delta\rho(3)/\delta V(2)$ is the *density perturbation response function* and the polarizability, $\alpha(1,2)$, may be identified as

$$\alpha(1,2) = -\int d^4 3\, v_c(1,3) R(3,2). \tag{5.9}$$

The role of quantum mechanics in solid-state plasmas enters in the determination of the perturbed density through the Schrödinger equation, which we initially write in terms of a retarded, unperturbed Green's function $G_0^R(\boldsymbol{r}, \boldsymbol{r}'; t-t') : (\hbar \to 1)$:

$$\left(i\frac{\partial}{\partial t_1} - h_1 \right) G_0^R(1,1') = \delta^4(1-1'), \tag{5.10}$$

where h_1 is the Hamiltonian in position-time representation in the absence of Coulombic carrier–carrier interactions and of external fields. It is readily shown that G_0^R may be written in frequency representation in terms of its energy eigenfunction expansion $[h_1\psi_n(\boldsymbol{r}) = E_n\psi_n(\boldsymbol{r})]$ as [5]

$$G_0^R(\boldsymbol{r}_1, \boldsymbol{r}_1'; \omega) = \sum_n \frac{\psi_n(\boldsymbol{r})\psi_n^*(\boldsymbol{r}')}{\omega - E_n + i0^+}. \tag{5.11}$$

The interested reader will find the derivation of (5.11) in [5, Chap. 1]. Of course, it is necessary to introduce statistical averaging in describing both classical and quantum plasmas [6], and a suitable introduction to the associated quantum-theoretic thermodynamic Green's functions [7] may be found in [6, Chap. 9], which is based on [7]. It is beyond the scope of this paper to reproduce all of this quantum many-body theory here, but a brief "road map" of the pertinent steps and relationships is provided below for interested students (others may just be satisfied with the survey of results that follows).

For a uniform, translationally invariant, normal nonrelativistic system ($h_1 = -\nabla^2/2m$), (5.10) may be Fourier transformed into $(\omega - p^2/2m)G_0^R(\boldsymbol{p}, \omega) = 1$ with $\omega \to \omega + i0^+$ (0^+ is a positive infinitesimal to enforce retardation; $\boldsymbol{p}$ is wave vector). It yields the spectral weight of the thermodynamic Green's function as $A(\boldsymbol{p}, \omega) = -2\,\mathrm{Im}\, G_0^R(\boldsymbol{p}, \omega)$, with the two parts of that many-body thermodynamic-equilibrium Green's function given by ($f_0(\omega) = [e^{\beta(\omega-\zeta)} + 1]^{-1}$ is the Fermi distribution with $\beta^{-1} = k_B T$ being the thermal energy and ζ being the chemical potential; $\Theta(T) = 1$ for $T > 0$ and $\Theta(T) = 0$ for $T < 0$ is the Heaviside unit step function)

$$G_0^{\{\lessgtr\}}(\boldsymbol{p}, \omega) = i \begin{Bmatrix} f_0(\omega) \\ -1 + f_0(\omega) \end{Bmatrix} A(\boldsymbol{p}, \omega), \tag{5.12}$$

and $G_0(1,1') = \Theta(t-t')G_0^>(1,1') + \Theta(t'-t)G_0^<(1,1')$. Introducing an external potential, $U(2)$, which polarizes the plasma and thereby produces a screened, effective potential, $V(1)$, due to Coulomb interactions in the solid-state plasma (and neglecting less important aspects of carrier–carrier interactions), we obtain a modified equation for the Green's function in place of (5.10) as

$$\left[\mathrm{i}\frac{\partial}{\partial t_1} - h_1 - V(1)\right] G(1,1') = \delta^4(1-1').$$

This may be rewritten using (5.10) as an integral equation

$$G(1,1') = G_0(1,1') + \int \mathrm{d}^4 3\, G_0(1,3)V(3)G(3,1'), \tag{5.13}$$

where G_0 is the thermodynamic Green's function which supplants G_0^{R}, and it also satisfies (5.10) (but with statistical averaging built into its structure instead of retardation). Approximating G under the integral on the right hand side of (5.13) by G_0 and taking the variation with respect to V yields the "ring diagram" as

$$\frac{\delta G(1,1')}{\delta V(2)} = G_0(1,2)G_0(2,1'). \tag{5.14}$$

The perturbed density is written in terms of G as $\rho(1) = -\mathrm{i}G^<(1,1)$, so we finally obtain R from (5.12) and (5.14) employing the eigenfunction expansion of G_0^{R} (5.11) as [6]

$$R(\boldsymbol{r}_3, \boldsymbol{r}_2; \omega + \mathrm{i}0^+) = \mathrm{i}\sum_n \sum_{n'} \frac{f_0(E_{n'}) - f_0(E_n)}{\omega + E_{n'} - E_n + \mathrm{i}0^+} \times \psi_{E_{n'}}(\boldsymbol{r}_2)\psi^*_{E_{n'}}(\boldsymbol{r}_3)\psi_{E_n}(\boldsymbol{r}_3)\psi^*_{E_n}(\boldsymbol{r}_2). \tag{5.15}$$

This is the central result of the quantum-mechanical RPA. It is the quantum generalization of the corresponding classical quantity derived from a collisionless, linearized classical Vlasov–Boltzmann equation.

5.2 Quantum Effects in Normal Solid-State Plasmas

5.2.1 Three-Dimensional Quantum Plasma

The simplest important case to examine is that of a three-dimensional (3D) uniform bulk quantum plasma, for which the foregoing considerations yield the spectral weight as

$$A(\boldsymbol{p}, \omega) = 2\pi\delta(\omega - p^2/2m).$$

Proceeding with the construction of $G_0^{\{\lessgtr\}}$ (5.12) and using (5.14) to determine $\varepsilon(\boldsymbol{p},\omega)$ yields (restore $\hbar$)

$$\varepsilon(\boldsymbol{p},\omega) = 1 - \frac{4\pi e^2}{\hbar^3 m}\int \frac{d^3k}{(2\pi)^3}\frac{f_0\left(\frac{k^2}{2m}\right)}{\left(\omega - \frac{k\cdot p}{m}\right)^2 + \hbar^2\left(\frac{p^2}{2m}\right)^2}. \tag{5.16}$$

Quantum mechanics enters this result in two distinct ways: the first is statistical in that the Pauli exclusion principle mandates that f_0 be the Fermi distribution (at finite temperature, in general), and the second is quantum dynamical, through the quantum dynamical correction term $\hbar^2(p^2/2m)^2$ in the denominator of the integrand on the right-hand side of (5.16). It would be absent in a semiclassical model based on a classical collisionless linearized Vlasov–Boltzmann equation coupled to the Poisson equation, but with initial averaging over the Fermi distribution [2–4]. In the low wave number limit $p \to 0$, the local plasma oscillation determined by the vanishing of $\varepsilon(\boldsymbol{p},\omega) \to 0$ yields $\omega^2 = \omega_\mathrm{p}^2 = 4\pi e^2\rho_0/m$, just as in gas plasma theory (ρ_0 is the equilibrium density). This may be understood in terms of the correspondence principle, since the quantum-mechanical RPA is specifically designed to take account of the long-range aspects of electron–electron Coulomb interaction: to be specific, low wave numbers correspond to long distances and for electrons interacting over such very long distances the correlation effects are controlled by essentially classical dynamics. This is to say that electron–electron interactions over long distances (i.e., low wave numbers) are in the correspondence limit in which classical features emerge from quantum mechanics. Moreover, the fully nonlocal p-dependent gas plasma dispersion function is obtained if the quantum dynamical correction is neglected and the temperature-dependent Fermi function, $f_0(\omega)$, is taken in the nondegenerate Maxwell–Boltzmann classical limit. On the other hand, in the low temperature degenerate quantum limit of a sharply cut off Fermi step function, the k-integral of (5.16) yields the well-known Lindhard dielectric function [1, 6]. With the plasmon spectrum determined by $\varepsilon(\boldsymbol{p},\omega) = 0$, alternatively, by the frequency poles of the dynamic, nonlocal screening function $K(\boldsymbol{p},\omega) = \varepsilon^{-1}(\boldsymbol{p},\omega)$, associated plasmon contributions to correlation energy are given by $\hbar\omega_\mathrm{p}$ for each such pole, and plasmon damping is described by $\mathrm{Im}\,\varepsilon(\boldsymbol{p},\omega)$, with static shielding given by the zero-frequency limit of $K(\boldsymbol{p},0) = \varepsilon^{-1}(\boldsymbol{p},0)$. Thus, all the physical features of the Hamiltonian of (5.2) are embedded in the structure of $K(\boldsymbol{p},\omega) = \varepsilon^{-1}(\boldsymbol{p},\omega)$. Furthermore, although quantum effects are small at low wave numbers (momenta), they do become significant at higher p-values, inducing new plasmon resonances and significant static shielding phenomena (such as "Friedel oscillatory" screening), as well as changing the nondegenerate classical Landau damping to a degenerate electron–hole-producing plasmon decay mechanism.

To be specific, the evaluation of $\varepsilon(\boldsymbol{p},\omega)$ for a uniform 3D bulk plasma given by (5.16) for *arbitrary* $\boldsymbol{p}$ in the degenerate limit of zero temperature yields the Lindhard dielectric function as [6] (p_F is the Fermi wave number; restore $\hbar$)

$$\begin{aligned}\varepsilon(\boldsymbol{p},\omega) =& 1 - v_{\mathrm{c}}(p)\frac{mp_{\mathrm{F}}}{2\pi^2\hbar^2}\\ &\times\left\{-1+\frac{p_{\mathrm{F}}}{2p}\left[1-\left(\frac{m\omega}{\hbar p_{\mathrm{F}}p}-\frac{p}{2p_{\mathrm{F}}}\right)^2\ln\left|\frac{1+\left(\frac{m\omega}{\hbar p_{\mathrm{F}}p}-\frac{p}{2p_{\mathrm{F}}}\right)}{1-\left(\frac{m\omega}{\hbar p_{\mathrm{F}}p}-\frac{p}{2p_{\mathrm{F}}}\right)}\right|\right]\right.\\ &\left.-\frac{p_{\mathrm{F}}}{2p}\left[1-\left(\frac{m\omega}{\hbar p_{\mathrm{F}}p}+\frac{p}{2p_{\mathrm{F}}}\right)^2\ln\left|\frac{1+\left(\frac{m\omega}{\hbar p_{\mathrm{F}}p}+\frac{p}{2p_{\mathrm{F}}}\right)}{1-\left(\frac{m\omega}{\hbar p_{\mathrm{F}}p}+\frac{p}{2p_{\mathrm{F}}}\right)}\right|\right]\right\},\end{aligned} \tag{5.17}$$

where the 3D Fourier transform of the Coulomb potential is $v_{\mathrm{c}}(p) = 4\pi e^2/p^2$. A low wave number power expansion of (5.17) yields the classical local plasmon ω_{p} described above with a nonlocal shift of order $O(p^2)$ which bears quantum corrections. However, in addition, there is a new plasmon resonance solution of $\varepsilon(\boldsymbol{p},\omega) = 0$ given by $\omega = \hbar p p_{\mathrm{F}}/m$, known as "zero sound" (which occurs at $T = 0$ and differs substantially from ordinary sound) [6]. An exact evaluation of $\mathrm{Im}\,\varepsilon(\boldsymbol{p},\omega)$ using (5.16) and the Dirac prescription $[(\omega \pm \mathrm{i}0^+)^{-1} = \wp(1/\omega) \mp \mathrm{i}\pi\delta(\omega)]$ yields ($\wp$ denotes "principal part")

$$\mathrm{Im}\ \varepsilon(\boldsymbol{p},\omega) = \frac{4e^2m^2}{\beta p^3\hbar^4}\ln\left(\frac{1+\exp\left[\zeta\beta-\left(\frac{m\omega}{p}-\frac{\hbar p}{2}\right)^2\frac{\beta}{2m}\right]}{1+\exp\left[\zeta\beta-\left(\frac{m\omega}{p}+\frac{\hbar p}{2}\right)^2\frac{\beta}{2m}\right]}\right). \tag{5.18}$$

In the nondegenerate classical limit, $\beta \to 0$, the well-known Landau damping [3] is recovered but in the degenerate limit of zero temperature, $\beta \to \infty$ (5.18) reveals strong plasmon damping under conditions of energy and momentum conservation which permit plasmon decay into electron–hole pairs by exciting an electron out of the filled Fermi sea, leaving a hole behind [6]. In the static limit, $\omega \to 0$ (5.17) provides the usual low-wave number Debye–Thomas–Fermi shielding law in accordance with the static screening function $K(\boldsymbol{p},\omega=0)=p^2/(p^2 + p_{\mathrm{DTF}}^2)$, where $p_{\mathrm{DTF}}^2 = 4\pi e^2(\partial\rho_0/\partial\zeta)_\beta$, so that the screened potential of a Coulomb center at the origin has the short-range form $V(\boldsymbol{r},t) \sim r^{-1}\exp(-p_{\mathrm{DTF}}r)$. However, the zero-frequency quantum-mechanical log singularity of (5.17) at $p = 2p_{\mathrm{F}}$ yields a long-range Friedel oscillatory contribution of the form $V(\boldsymbol{r},t) \sim r^{-3}\cos(2p_{\mathrm{F}}r)$, which does not suffer a spatial exponential decay [6].

5.2.2 Dielectric Properties of Low-Dimensional Systems

Low-dimensional systems are another important source of quantum effects in solid-state plasmas. A dip in the potential profile at a planar semiconductor–insulator junction (e.g., Si–SiO$_2$) due to band bending by an appropriate gate bias voltage can give rise to conduction electrons trapped in the dip below the Fermi energy near the interface [8]. Such electrons can move freely in the two-dimensional (2D) plane of the junction, but not out of it. This creates an "inversion layer" of plasma electrons constrained to 2D motion on a plane. So-called *quantum wells*, involving a valley in a planar heterostructure potential profile (e.g., GaAs–AlGaAs) similarly give rise to "size quantization" leaving only the lowest energy levels for motion across the valley energetically accessible, effectively confining the electrons, whereas they can move freely in the normal plane, thus constituting a 2D plasma. Similar considerations apply to potential profiles that constrain plasma electron motion to a single direction, yielding a one-dimensional (1D) "quantum wire" plasma in the single unconstrained direction. Potential profiles that constrain motion in all three directions can support a "zero-dimensional" (0D) quantum dot [9].

For a 2D plasma in the degenerate limit, at zero temperature, the dielectric function is determined by the procedures outlined above as [10] ($\rho_{2\mathrm{D}}$ is the 2D equilibrium density; $\bar{r} = (x, y)$; $\bar{p} = (p_x, p_y)$; and the 2D Fourier transform of the Coulomb potential is $v(p) = 2\pi e^2/p$) (restore $\hbar$)

$$\mathrm{Re}\ \varepsilon(\bar{p},\omega) = 1 + \frac{4\pi m e^2 \rho_{2\mathrm{D}}}{\hbar^2 p_\mathrm{F} p^2}\left\{\frac{p}{p_\mathrm{F}} - C_-\sqrt{\left(\frac{p}{2p_\mathrm{F}} - \frac{m\omega}{\hbar p_\mathrm{F} p}\right)^2 - 1} + C_+\sqrt{\left(\frac{p}{2p_\mathrm{F}} + \frac{m\omega}{\hbar p_\mathrm{F} p}\right)^2 - 1}\right\}, \tag{5.19}$$

$$\mathrm{Im}\ \varepsilon(\bar{p},\omega) = \frac{4\pi m e^2 \rho_{2\mathrm{D}}}{\hbar^2 p_\mathrm{F} p^2}\left\{D_-\sqrt{1 - \left(\frac{p}{2p_\mathrm{F}} - \frac{m\omega}{\hbar p_\mathrm{F} p}\right)^2} - D_+\sqrt{1 - \left(\frac{p}{2p_\mathrm{F}} + \frac{m\omega}{\hbar p_\mathrm{F} p}\right)^2}\right\}, \tag{5.20}$$

where

$$C_\pm = \left[\frac{p}{2p_\mathrm{F}} \pm \frac{m\omega}{\hbar p_\mathrm{F} p}\right], \qquad D_\pm = 0, \qquad \text{for } \left|\frac{p}{2p_\mathrm{F}} \pm \frac{m\omega}{\hbar p_\mathrm{F} p}\right| > 1, \tag{5.21}$$

$$C_\pm = 0, \qquad D_\pm = 1, \qquad \text{for } \left|\frac{p}{2p_\mathrm{F}} \pm \frac{m\omega}{\hbar p_\mathrm{F} p}\right| < 1. \tag{5.22}$$

At low wave numbers, the local plasmon given by $\varepsilon(\boldsymbol{p},\omega) = 0$ is

$$\omega_{2\mathrm{D}}^2 = \frac{2\pi e^2 \rho_{2\mathrm{D}} p}{m} + \mathcal{O}(p^2), \tag{5.23}$$

where the higher wave number corrections bear quantum effects, while the first term on the right of (5.23) is a classical dynamical result for the 2D plasma, as one should expect at low wave number $p \ll 2p_F$. Again, strong plasmon damping occurs when energy and momentum conservation permit plasmon decay into 2D electron–hole pairs. A low wave number analysis of 2D static shielding ($\omega = 0$) yields the classical dynamical 2D screening wave number as $p_{DTF}^{(2D)} = 2\pi e^2 \partial\rho_{2D}/\partial\zeta$, resulting in a 2D long distance shielded potential as $V(\bar{r}, t) \sim (p_{DTF}^2 r^3)^{-1}$ and $p_{DTF} = 2me^2/\hbar^2$ in the degenerate case. However, higher wave numbers involve a branch point and cut of $\sqrt{(p/2p_F)^2 - 1}$, which leads to a long-range 2D Friedel oscillatory contribution as $V(\bar{r}, t) \sim \sin(2p_F r)/(2p_F r)^2$.

In the case of a 1D quantum wire plasma, the dielectric function is given in the degenerate limit by [$\hbar \to 1$] [11, 12]:

$$\varepsilon(\boldsymbol{p}, \omega) = 1 - v_c(p) \left\{ \frac{m}{\pi p} \ln \left| \frac{\left(\omega - \frac{p^2}{2m} + \frac{p_F p}{m}\right)\left(\omega + \frac{p^2}{2m} - \frac{p_F p}{m}\right)}{\left(\omega - \frac{p^2}{2m} - \frac{p_F p}{m}\right)\left(\omega + \frac{p^2}{2m} + \frac{p_F p}{m}\right)} \right| \right.$$
$$\left. - \mathrm{i}\frac{m}{p}\left[\Theta\left(1 - \left|\frac{p}{2p_F} - \frac{m\omega}{p_F p}\right|\right) - \Theta\left(1 - \left|\frac{p}{2p_F} + \frac{m\omega}{p_F p}\right|\right)\right] \right\}. \tag{5.24}$$

In the present case, the 1D Fourier transform of the Coulomb potential, $v_c(p)$, is divergent, but can be approximated as $v_c(p) = e^2 \int_{-\infty}^{\infty} \mathrm{d}x\, \mathrm{e}^{\mathrm{i}px}/|x| \sim 2e^2 \ln(1/pa)$, where a is the radius of the quantum wire. The associated low wave number plasmon is given by (ρ_{1D} is the 1D equilibrium density)

$$\omega_{1D}^2 = \frac{2e^2 \rho_{1D} p^2}{m} \ln\left(\frac{1}{pa}\right), \tag{5.25}$$

and higher-order wave number corrections bear quantum effects, etc.

5.2.3 Dielectric Function of a Magnetized Quantum Plasma

Quantum-mechanical effects are even more prominent in the presence of a magnetic field inducing circularity of particle orbits in the plane normal to the field. As with other spatial confinement mechanisms, the magnetic field gives rise to quantization of the energy levels (Landau quantization), which produces very interesting structure in the plasmon spectrum, plasmon decay, and static shielding. The exact quantum-mechanical RPA dielectric function in the presence of an arbitrarily strong uniform, constant ambient magnetic field, $\boldsymbol{H}$, (z-direction) for a 3D bulk solid-state plasma is given by the process outlined above with a magnetic field term included in $h_1 = -(\nabla - \mathrm{i}eA)^2/2m + \mu_0 H\sigma_3$ as [13, 14] ($\boldsymbol{A} = (1/2)\boldsymbol{H} \times \boldsymbol{r}$; σ_3 is the Pauli matrix 3; $c \to 1$; μ_0 is the Bohr magneton; and ω_c is the cyclotron frequency)

$$\varepsilon(\boldsymbol{p},\omega) = 1 - \frac{4\pi e^2}{p^2}\int_0^\infty \mathrm{d}\omega'\,\frac{f_0(\omega')}{\hbar^3}\int_{c-\mathrm{i}\infty}^{c+\mathrm{i}\infty}\frac{\mathrm{d}s}{2\pi\mathrm{i}}\,\mathrm{e}^{s\omega'}\,\frac{\pi^{3/2}}{(2\pi)^3}\sqrt{\frac{2m}{s}}\,\frac{m\hbar\omega_c}{\tanh\left(\frac{\hbar\omega_c}{2}s\right)}\aleph, \tag{5.26}$$

where ($\boldsymbol{p} = \bar{\boldsymbol{p}}, p_z = p_x, p_y, p_z$; and $\bar{\boldsymbol{p}} = p_x, p_y$)

$$\begin{aligned}
\aleph = \frac{2\mathrm{i}}{\hbar}\int_0^\infty \mathrm{d}T\,\mathrm{e}^{-\mathrm{i}\omega T}\Bigg\{&\exp\left(\frac{-p_z^2}{8ms}\left([2T-\mathrm{i}\hbar s]^2+\hbar^2 s^2\right)\right)\\
&\times\exp\left(\frac{\hbar\bar{p}^2}{2m\omega_c}\,\frac{\cos[(\omega_c/2)(2T-\mathrm{i}\hbar s)]-\cosh[\hbar(\omega_c/2)s]}{\sinh[\hbar(\omega_c/2)s]}\right)\\
&-\exp\left(\frac{-p_z^2}{8ms}\left([2T+\mathrm{i}\hbar s]^2+\hbar^2 s^2\right)\right)\\
&\times\exp\left(\frac{\hbar\bar{p}^2}{2m\omega_c}\,\frac{\cos[(\omega_c/2)(2T+\mathrm{i}\hbar s)]-\cosh[\hbar(\omega_c/2)s]}{\sinh[\hbar(\omega_c/2)s]}\right)\Bigg\},
\end{aligned} \tag{5.27}$$

with m as the effective mass and $\hbar$ is restored. As indicated above the low wave number limit of this quantum RPA result yields just the plasmon dispersion relation of the Vlasov–Boltzmann equation which describes self-consistent internal plasma dynamics classically, $1 = (\omega_p^2\cos^2\phi/\omega^2) + [\omega_p^2\sin^2\phi/(\omega^2-\omega_c^2)]$ (ϕ is the angle of plasmon propagation relative to the magnetic field), which bespeaks two local principal electrostatic magnetoplasmons $\omega^2 = (1/2)(\omega_p^2+\omega_c^2)\pm(1/2)\sqrt{(\omega_p^2+\omega_c^2)^2-4\omega_p^2\omega_c^2\sin^2\phi}$ and it also yields the same low wave number Debye–Thomas–Fermi static shielding law predicted by the Vlasov–Boltzmann equation, which is devoid of magnetic field corrections as well as quantum corrections at low wave number [15]. This absence of magnetic field corrections at low wave number may be understood in terms of the correspondence principle which mandates that there be no quantum corrections as $p \to 0$: This means that the dynamics are classical, the role of the magnetic field is restricted to the classical Lorentz force which cannot do work, and is therefore unable to supply energy required to change the charge distribution associated with static shielding. Of course, quantum corrections occur as higher wave numbers are taken into consideration, and with the reorganization of the energy spectrum by Landau quantization, magnetic field effects occur in static screening at higher wave numbers. It should be noted that the occurrence of a factor $\hbar^3$ under $f_0(\omega')$ in (5.26) (as well as some other normalization factors) is actually spurious since f_0 is normalized here in accordance with

$$\rho_0 = 2\int_0^\infty \mathrm{d}\omega'\,\frac{f_0(\omega')}{\hbar^3}\int_{c-\mathrm{i}\infty}^{c+\mathrm{i}\infty}\frac{\mathrm{d}s}{2\pi\mathrm{i}}\,\mathrm{e}^{s\omega'}\,\frac{\pi^{3/2}}{(2\pi)^3}\sqrt{\frac{2m}{s}}\,\frac{m\hbar\omega_c}{\tanh\left(\frac{\hbar\omega_c}{2}s\right)}. \tag{5.28}$$

Notwithstanding the essentially classical character of the low wave number behavior of $\varepsilon(\boldsymbol{p},\omega)$, one may expect quantum corrections to arise in conjunction with consideration of the higher-order wave number terms embodied in the right-hand side of (5.26) and (5.27) for $\varepsilon(\boldsymbol{p},\omega)$. The higher-order wave number structure bearing quantum corrections affects the plasmon dispersion relation, damping [14], and static shielding as well [15]. The exact nature of these corrections depends on which statistical regime is under consideration, and what magnetic field strength is involved. Since quantum effects are most prominent and interesting in the degenerate statistical regime, we limit our discussion to the case of degeneracy, but will briefly consider quantum effects for several regimes of field strengths.

In the case of a magnetic field of intermediate strength ($\zeta \sim E_{\mathrm{F}} > \hbar\omega_{\mathrm{c}} > k_{\mathrm{B}}T$), one can expect de Haas–van Alphen (DHVA) oscillatory effects to be prominent among quantum corrections. Physically, DHVA effects arise in conjunction with the Landau quantization of energy levels which move past the Fermi energy as density varies. Analytically, DHVA oscillatory terms arise from the isolated singularities of the inverse Laplace transform s-integrand of (5.26) and (5.27) corresponding to the zeros of $\sinh[\hbar(\omega_{\mathrm{c}}/2)s]$. At zero wave number, the dispersion relation for the two local principal plasmon modes in a magnetic field stands unmodified. However, DHVA quantum corrections do indeed accompany the wave number-dependent shifts of the locations of the two principal plasmon modes. This may be seen from a wave number power expansion of the right-hand side of (5.26) and (5.27), which yields the plasmon dispersion relation with its leading wave number corrections as [14]

$$\begin{aligned}1 &= \frac{p_z^2}{p^2}\frac{4\pi e^2\rho}{m}\frac{1}{\omega^2} + \frac{\bar{p}^2}{p^2}\frac{4\pi e^2\rho}{m}\frac{1}{\omega^2-\omega_{\mathrm{c}}^2} + \frac{p_z^4}{p^2}\frac{12\pi e^2\alpha}{m^2}\frac{1}{\omega^4}\\ &\quad + \frac{\bar{p}^4}{p^2}\frac{4\pi e^2\sigma}{m^2\omega_{\mathrm{c}}^2}\left(\frac{1}{\omega^2-(2\omega_{\mathrm{c}})^2} - \frac{1}{\omega^2-\omega_{\mathrm{c}}^2}\right) - \frac{p_z^2\bar{p}^2}{p^2}\frac{4\pi e^2\sigma}{m^2\omega_{\mathrm{c}}^2}\frac{1}{\omega^2}\\ &\quad + \frac{p_z^2\bar{p}^2}{p^2}\frac{4\pi e^2\sigma}{m^2\omega_{\mathrm{c}}^2}\frac{\omega^2+\omega_{\mathrm{c}}^2}{(\omega^2-\omega_{\mathrm{c}}^2)^2} + \frac{p_z^2\bar{p}^2}{p^2}\frac{4\pi e^2\alpha}{m^2}\frac{3\omega^2+\omega_{\mathrm{c}}^2}{(\omega^2-\omega_{\mathrm{c}}^2)^3},\end{aligned} \tag{5.29}$$

where $\rho = \rho_\gamma + \sum_n \rho_{c_n}$, $\sigma = \sigma_\gamma + \sum_n \sigma_{c_n}$, $\alpha = \alpha_\gamma + \sum_n \alpha_{c_n}$ and, in the intermediate magnetic field strength regime, we have ($\Gamma(x)$ is the Γ-function; restore $\hbar$)

$$\begin{aligned}\rho_\gamma &\cong \left(\frac{m}{2\pi}\right)^{3/2}\frac{2}{\Gamma(5/2)}\frac{\zeta^{3/2}}{\hbar^3},\\ \sigma_\gamma &\cong \left(\frac{m}{2\pi}\right)^{3/2}\frac{2}{\Gamma(7/2)}\frac{\zeta^{5/2}}{\hbar^3} + \left(\frac{m}{2\pi}\right)^{3/2}\frac{1}{3\Gamma(3/2)}\frac{(\hbar\omega_{\mathrm{c}})^2\zeta^{1/2}}{\hbar^3},\\ \alpha_\gamma &\cong \left(\frac{m}{2\pi}\right)^{3/2}\frac{2}{\Gamma(7/2)}\frac{\zeta^{5/2}}{\hbar^3} + \left(\frac{m}{2\pi}\right)^{3/2}\frac{(\hbar\omega_{\mathrm{c}})^2}{6\Gamma(3/2)}\frac{\zeta^{1/2}}{\hbar^3},\end{aligned}$$

with

$$\sum_n \rho_{c_n} = \frac{(m^3\hbar\omega_{\mathrm{c}})^{1/2}}{\pi\beta\hbar^3}\sum_{n=1}^{\infty}\frac{\cos\left(\frac{2\pi n}{\hbar\omega_{\mathrm{c}}}\zeta - \frac{3\pi}{4}\right)}{n^{1/2}\sinh\left(\frac{2\pi^2 n}{\hbar\omega_{\mathrm{c}}\beta}\right)},$$

$$\sum_n \sigma_{C_n} = \frac{(m\hbar\omega_c)^{3/2}}{2\pi^2\beta\hbar^3}\sum_{n=1}^{\infty}\frac{\cos\left(\dfrac{2\pi n}{\hbar\omega_c}\zeta - \dfrac{5\pi}{4}\right)}{n^{3/2}\sinh\dfrac{2\pi^2 n}{\hbar\omega_c\beta}}\left(1-\frac{\dfrac{2\pi^2 n}{\hbar\omega_c\beta}}{\tanh\dfrac{2\pi^2 n}{\hbar\omega_c\beta}}\right)$$

$$+\frac{(m^3\hbar\omega_c)^{1/2}}{\pi\beta\hbar^3}\zeta\sum_{n=1}^{\infty}\left(\frac{\cos\left(\dfrac{2\pi n}{\hbar\omega_c}\zeta-\dfrac{3\pi}{4}\right)}{n^{1/2}\sinh\dfrac{2\pi^2 n}{\hbar\omega_c\beta}}-\frac{3\hbar\omega_c}{4\pi\zeta}\frac{\cos\left(\dfrac{2\pi n}{\hbar\omega_c}\zeta-\dfrac{5\pi}{4}\right)}{n^{3/2}\sinh\dfrac{2\pi^2 n}{\hbar\omega_c\beta}}\right),$$

$$\sum_n \alpha_{C_n} = \frac{(m\hbar\omega_c)^{3/2}}{2\pi^2\beta\hbar^3}\sum_{n=1}^{\infty}\frac{\cos\left(\dfrac{2\pi n\zeta}{\hbar\omega_c}-\dfrac{5\pi}{4}\right)}{n^{3/2}\sinh\dfrac{2\pi^2 n}{\hbar\omega_c\beta}},$$

in which DHVA oscillatory terms of the form $\cos[2\pi n\zeta/\hbar\omega_c]$ appear explicitly. In addition, "Bernstein"-type plasmon resonances, which occur undamped near each value $n\omega_c$ for $\boldsymbol{p}\perp\boldsymbol{B}$ at higher wave numbers, bear DHVA oscillatory quantum corrections. The static shielding law in the case of intermediate field strength is capable of bearing magnetic field corrections in conjunction with quantum corrections since the magnetic field modifies the electron energy spectrum through Landau quantization [16]. Indeed, the wave number power expansion of the right-hand side of (5.26) and (5.27) for zero frequency results in magnetic field quantum corrections of the DHVA type which make the Debye–Thomas–Fermi static shielding law spatially anisotropic. It is of even greater interest to consider intermediate magnetic field strength corrections to the high wave number ($p\sim 2p_F$) Friedel–Kohn "wiggle," but the analytical complexity defies brief description [17]. We shall be content to point out that the $\boldsymbol{r}$-dependent shielded potential involves the magnetic field quantum correction parameter $\hbar\omega_c/\zeta$ in the form of a "mixed" parameter $[(\hbar\omega_c/\zeta)p_F r]$, where $p_F r \gg 1$ and $\hbar\omega_c/\zeta \ll 1$. When the smallness of $\hbar\omega_c/\zeta$ is dominant, and $[(\hbar\omega_c/\zeta)p_F r] \ll 1$, then the Friedel–Kohn "wiggle" retains its long-range form. However, at larger distances when $p_F r \gg 1$ dominates, and $[(\hbar\omega_c/\zeta)p_F r] \gg 1$, then the long range of the Friedel–Kohn "wiggle" is destroyed. In essence, this happens because a large "mixed" parameter means that $\hbar\omega_c/\zeta$ is *effectively* large, in which case the static shielded potential, $V(\boldsymbol{r})$, effectively feels the internal dynamics associated with the quantum strong field limit, in which all electrons are confined to the lowest Landau level. This high field limit is discussed in greater detail below: It is characteristic of the quantum strong field limit that the log singularity responsible for the "wiggle" is replaced by its 1D counterpart $\ln|p_z - 2p_F|$. This anisotropic log singularity yields a static shielding law which is still oscillatory along the direction of the magnetic field, but which suffers an exponential decay in the transverse direction $\sim[\cos(2p_F r_z)e^{-2p_F|\bar{r}|}]$. Hence the long range of the Friedel–Kohn "wiggle" is destroyed when the "mixed" parameter is large, $[(\hbar\omega_c/\zeta)p_F r] \gg 1$. Moreover, there are no DHVA oscillatory corrections for large mixed parameter in view of the fact that $V(\boldsymbol{r})$ effectively feels the internal dynamics associated with the quantum strong field limit, with all electrons effectively confined to the lowest Landau level, so that one does not have Landau levels flicking past the Fermi energy as density varies.

The basic nature of the quantum strong field limit has already been indicated. All electrons are confined to the lowest Landau level, and $\hbar\omega_c > \zeta \sim E_F > k_B T$. Clearly one may expect maximal magnetic field quantum effects in this situation. However, the zero wave number principal plasmon mode dispersion relation again stands unmodified, as one should expect from our earlier discussion. It is the consideration of higher wave number parts of the right-hand side of (5.26) and (5.27) which reveals important magnetic field quantum effects. An exact evaluation of (5.26) and (5.27) in the quantum strong field limit yields the result [18] (Re denotes the real part; restore $\hbar$):

$$\mathrm{Re}\,\varepsilon(\boldsymbol{p},\omega) = 1 - \frac{m\omega_p^2}{2\hbar p^2}\sqrt{\frac{m}{2p_z^2\zeta}}\exp\left(-\frac{\hbar\bar{p}^2}{2m\omega_c}\right)\sum_{n=0}^{\infty}\frac{1}{n!}\left(\frac{\hbar\bar{p}^2}{2m\omega_c}\right)^n \times\left\{\ln\frac{\omega - n\omega_c - \dfrac{\hbar p_z^2}{2m} + \sqrt{\dfrac{2p_z^2\zeta}{m}}}{\omega - n\omega_c - \dfrac{\hbar p_z^2}{2m} - \sqrt{\dfrac{2p_z^2\zeta}{m}}} + (\omega \leftrightarrow -\omega)\right\}. \quad (5.30)$$

A wave number power expansion of the right-hand side of (5.30) yields wave number shifts of the locations of the two principal plasmon modes which bear magnetic field quantum corrections. It is of even greater interest to maintain the anisotropic log singularities of (5.30) intact, and investigate their effects on the plasmon spectrum [19]. One then finds that the p_z-dependent log singularities imply the existence of two plasmon resonances near each value $n\omega_c$, one of which is undamped and the other is mildly damped. These remarks hold for wave vector $\boldsymbol{p}$ not perpendicular to the magnetic field $\boldsymbol{B}$ ($p_z \neq 0$). In the limit of perpendicular propagation ($p_z = 0$) there is just one undamped resonance near each value $n\omega_c$. In regard to the static shielding law, the $\omega \to 0$ limit of the $n = 0$ log singularity takes the form $\ln|p_z - 2p_F|$. This anisotropic result gives the principal contribution to the quantum strong field counterpart of the Friedel–Kohn "wiggle," and the strong anisotropy destroys the long-range character of the "wiggle" as indicated above [16–18].

Needless to say, quantum corrections at higher wave numbers also occur in the plasmon spectrum and static shielding of 2D plasmas in a normal magnetic field [20–22].

5.3 Graphene

5.3.1 Introduction

Graphene, a single-atom-thick 2D planar layer of Carbon atoms in a hexagonal honey-combed lattice composed of two superposed triangular sublattices, has been receiving a great deal of attention, both experimental and theoretical, since the first report in 2004 of its unusual device-friendly material properties [23, 24]. These properties include:

- High mobility at elevated temperature [25], reaching 200,000 $cm^2/V\,s$, over two orders of magnitude higher than that of silicon-based materials, over 20 times that of GaAs, over twice that of InSb
- High electron density, about $10^{13}\,cm^{-2}$ in a single subband
- Long carrier mean free path, $l \sim 400\,nm$ at room temperature, opening the possibility of Graphene-based ballistic devices
- Stable to high temperatures, $\sim 3{,}000\,K$
- Quantum Hall effect occurs at room temperature
- The planar form of Graphene generally allows for highly developed top–down CMOS-type compatible process flows, a substantial advantage over Carbon nanotubes that are difficult to integrate into electronic devices and are difficult to produce in consistent sizes and electronic properties

All of these properties make Graphene an extremely promising material for future nanoelectronic devices. Such applications of Graphene are already in progress. These include:

Sensors. Schedin, Novoselov et al. [26] reported that Graphene-based chemical sensors are capable of detecting minute concentrations (1 part per billion) of various active gases and allow us to discern *individual* events when a molecule attaches to the sensor's surface. Such high sensitivity comes about because the high 2D surface/volume ratio maximizes the role of adsorbed molecules as donors/acceptors, coupled with high conductivity and low noise.

Spin valve. A simple spin valve structure has already been fabricated [27] employing Graphene to provide the spin transport medium between ferromagnetic electrodes (using advantageous properties of Graphene in regard to long spin lifetime, low spin-orbit coupling, and high conductivity).

Electromechanical resonator. Bunch et al. [28] demonstrated that Graphene in contact with a gold electrode can be used to electrostatically actuate an electromechanical resonator (employing a 2D Graphene sheet suspended over a trench in a SiO_2 substrate). The motion can be activated by an rf gate voltage superposed on a dc-voltage applied to the Graphene sheet, or by optical actuation using a laser focused on the sheet.

FET. Using Graphene, "proof-of-principle" FET transistors, loop devices, and circuitry have already been produced by Walt de Heer's group [29, 30].

Quantum interference device. A quantum interference device using a ring-shaped Graphene structure was built to manipulate electron wave interference effects.

The obvious promise of Graphene for new and improved electronic devices has resulted in a flood of research papers on its transport and optical properties, and on its technological prospects.

5.3.2 *Graphene Hamiltonian, Green's Function, and RPA Dielectric Function*

On a fundamental level the extraordinary physical features of Graphene arise from its unusual band structure, in which the conduction and valence bands touch at two nodal zero-gap ("Dirac") points (K, K') in the first Brillouin zone [31–33], with the resulting low energy electron/hole energy dispersion relation proportional to momentum (rather than its square). This single-particle energy dispersion, linear in momentum, is analogous to the energy dispersion of relativistic electrons having no mass and thus likens Graphene electron and hole carriers to massless relativistic "Dirac" Fermions. The associated Hamiltonian is written using a "pseudospin" notation that distinguishes the electron part of the spectrum from the hole part in terms of a spin-like variable. Using the usual 2D Pauli spin matrices ($\boldsymbol{\sigma} = \sigma_x, \sigma_y$) and momentum $\boldsymbol{p} = p_x, p_y$ on the Graphene plane, the Graphene Hamiltonian in pseudospin representation is given by

$$\breve{h}_1 = \gamma\, \boldsymbol{p}\cdot\boldsymbol{\sigma} = \gamma \begin{pmatrix} 0 & p_x - \operatorname{sgn}(s)\,\mathrm{i}p_y \\ p_x + \operatorname{sgn}(s)\,\mathrm{i}p_y & 0 \end{pmatrix}, \tag{5.31}$$

where

$$\operatorname{sgn}(s) = \begin{cases} +1, & s = K, \\ -1, & s = K', \end{cases} \tag{5.32}$$

and γ is given in terms of Graphene band structure parameters as $\gamma = 3\alpha a/2$ (α is the hopping parameter in the tight binding approximation and a is the lattice spacing): γ plays the role of a constant Fermi velocity independent of density. As in the study of massless relativistic neutrino fermions, *pseudohelicity*, the component of pseudospin in the momentum direction, commutes with $\breve{h}_1$ and its eigenvectors can be used as a basis in which $\hat{h}_1$ is diagonal. Introducing the transformation from pseudospin basis to pseudohelicity basis,

$$U_\mathrm{p} = \frac{1}{\gamma p}\begin{pmatrix} p_x - \operatorname{sgn}(s)\,\mathrm{i}p_y & p_x + \operatorname{sgn}(s)\,\mathrm{i}p_y \\ \gamma p & \gamma p \end{pmatrix}, \tag{5.33}$$

$\breve{h}_1$ can be diagonalized as

$$\hat{h}_1 = \left[U_\mathrm{p}\right]^+ \breve{h}_1 U_\mathrm{p} = \operatorname{diag}[\varepsilon_1(p), \varepsilon_2(p)], \tag{5.34}$$

where

$$\varepsilon_\mu = (-1)^{\mu+1}\,\gamma p. \tag{5.35}$$

As the Hamiltonian in pseudospin representation, $\breve{h}_1$, is a 2×2 matrix, the corresponding Green's functions are also 2×2 matrices. The retarded Green's function matrix satisfies the matrix counterpart of (5.10) ($\overleftrightarrow{I}$ is the unit 2 × 2 pseudospin matrix; $\hbar \to 1$)

$$\left(\mathrm{i}\overleftrightarrow{I}\frac{\partial}{\partial t} - \breve{h}_1\right)\overleftrightarrow{G}_0^{\mathrm{R}}(\boldsymbol{p},t) = \overleftrightarrow{I}\delta(t-t'). \tag{5.36}$$

In position-frequency representation, this reads as ($p_x \to X = x - x'$; $p_y \to Y = y - y'$; $T = t - t' \to \omega$; and $\gamma_\nu = \gamma\,\mathrm{sgn}(\nu) = \pm\gamma$)

$$\left[\overleftrightarrow{I}\omega - \gamma\sigma_x\frac{1}{\mathrm{i}}\frac{\partial}{\partial X} - \gamma_\nu\sigma_y\frac{1}{\mathrm{i}}\frac{\partial}{\partial Y}\right]\overleftrightarrow{G}_0^{\mathrm{R}}(\boldsymbol{R},\omega) = \overleftrightarrow{I}\delta(X)\delta(Y). \tag{5.37}$$

The individual elements satisfy

$$\omega G_{0xx}^{\mathrm{R}} - \left[\gamma\frac{1}{\mathrm{i}}\frac{\partial}{\partial X} - \gamma_\nu\frac{\partial}{\partial Y}\right]G_{0yx}^{\mathrm{R}} = \delta(X)\delta(Y), \tag{5.38}$$

$$\omega G_{0yx}^{\mathrm{R}} = \left[\gamma\frac{1}{\mathrm{i}}\frac{\partial}{\partial X} + \gamma_\nu\frac{\partial}{\partial Y}\right]G_{0xx}^{\mathrm{R}}, \tag{5.39}$$

with similar equations for G_{0yy}^{R} and G_{0xy}^{R}. The results for the retarded Green's function elements in pseudospin representation are

$$G_{xx}^{\mathrm{R}}(\boldsymbol{p},\omega) = G_{yy}^{\mathrm{R}}(\boldsymbol{p},\omega) = \frac{\omega}{\omega^2 - \gamma^2 p^2}, \tag{5.40}$$

$$G_{xy}^{\mathrm{R}}(\boldsymbol{p},\omega) = G_{yx}^{\mathrm{R}*}(\boldsymbol{p},\omega) = \gamma\frac{p_x - \mathrm{i}p_y}{\omega^2 - \gamma^2 p^2}, \tag{5.41}$$

from which the 2 × 2 spectral weight matrix, $\overleftrightarrow{A}(\boldsymbol{p},\omega)$, may be obtained using

$$\overleftrightarrow{A}(\boldsymbol{p},\omega) = -2\,\mathrm{Im}\left[\overleftrightarrow{G}_0^{\mathrm{R}}(\boldsymbol{p},\omega)\right], \tag{5.42}$$

and, correspondingly, the equilibrium thermodynamic Green's function for Graphene is given by

$$\overleftrightarrow{G}_0^{\{\lessgtr\}}(\boldsymbol{p},\omega) = \mathrm{i}\left\{\begin{matrix} f_0(\omega) \\ -1 + f_0(\omega) \end{matrix}\right\}\overleftrightarrow{A}(\boldsymbol{p},\omega). \tag{5.43}$$

On the 2D Graphene plane, the polarizability α is given in wave number–frequency representation by

$$\alpha\left(\boldsymbol{p},\omega + \mathrm{i}0^+\right) = -v_{\mathrm{c}}(\boldsymbol{p})R(\boldsymbol{p},\omega), \tag{5.44}$$

where the 2D Coulomb potential is

$$v_c(\boldsymbol{p}) = \frac{2\pi e^2}{p}, \tag{5.45}$$

and the RPA density perturbation response function, $R = \delta\rho/\delta V_{\text{eff}}$, is given by

$$R\left(\boldsymbol{p}, \omega + \mathrm{i}0^+\right) = \Im_< - \Im_>^*, \tag{5.46}$$

$$\Im_> = \int_0^{\infty} \mathrm{d}t\, \mathrm{e}^{-\mathrm{i}(\omega - \mathrm{i}0^+)t} \int \frac{\mathrm{d}^2 q}{(2\pi)^2}\, \mathrm{Tr}\left[\overleftrightarrow{G}_<(\boldsymbol{q}; -t)\overleftrightarrow{G}_>(\boldsymbol{q}-\boldsymbol{p}; t)\right], \tag{5.47}$$

$$\Im_< = \int_{-\infty}^{0} \mathrm{d}t\, \mathrm{e}^{-\mathrm{i}(\omega + \mathrm{i}0^+)t} \int \frac{\mathrm{d}^2 q}{(2\pi)^2}\, \mathrm{Tr}\left[\overleftrightarrow{G}_>(\bar{\boldsymbol{q}}; -t)\overleftrightarrow{G}_<(\boldsymbol{q}-\boldsymbol{p}; t)\right]. \tag{5.48}$$

Alternatively, one can employ (5.15) for $R(\boldsymbol{r}_3 - \boldsymbol{r}_2; \omega + \mathrm{i}0^+)$ in position-frequency representation and Fourier transform to wave number representation using the 2D Graphene eigenfunctions and eigenvalues of $\breve{h}_1$ given by (5.31) and (5.32). The resulting RPA density perturbation response function, $\delta\rho/\delta V_{\text{eff}} = R(\boldsymbol{p}, \omega) \equiv D_0 \widetilde{R}(x, \nu)$, for Graphene in the $T = 0$ degenerate limit is given by [8, 34–36] (we use the notation of Hwang, et al. [36] with dimensionless frequency and wave number variables defined by $\nu = \omega/E_{\text{F}} = \omega/\zeta$ and $x = p/p_{\text{F}}$, respectively, and $D_0 = \gamma^{-1}\sqrt{g_s g_v \rho_{2D}/\pi}$; g_s and g_v are spin and valley degeneracies, respectively; $\hbar \to 1$):

$$\widetilde{R}(x, \nu) = \widetilde{R}^+(x, \nu) + \widetilde{R}^-(x, \nu), \tag{5.49}$$

and ($\Theta(z) \equiv \eta_+(z) =$ Heaviside unit step function)

$$\widetilde{R}^+(x, \nu) = \widetilde{R}_1^+(x, \nu)\Theta(\nu - x) + \widetilde{R}_2^+(x, \nu)\Theta(x - \nu), \tag{5.50}$$

where (define $\widetilde{\Pi} = -\widetilde{R}$)

$$\begin{aligned} -\mathrm{Re}\widetilde{R}_1^+(x, \nu) = \mathrm{Re}\widetilde{\Pi}_1^+(x, \nu) = 1 - \frac{1}{8\sqrt{\nu^2 - x^2}} \Big\{ & f_1(x, \nu)\,\Theta(|2 + \nu| - x) \\ & + \mathrm{sgn}(\nu - 2 + x)\, f_1(x, -\nu)\,\Theta(|2 - \nu| - x) \\ & + f_2(x, \nu)\left[\Theta(x + 2 - \nu) + \Theta(2 - x - \nu)\right] \Big\}, \end{aligned} \tag{5.51}$$

$$\begin{aligned} -\mathrm{Re}\widetilde{R}_2^+(x, \nu) = \mathrm{Re}\widetilde{\Pi}_2^+(x, \nu) = 1 - \frac{1}{8\sqrt{x^2 - \nu^2}} \Big\{ & f_3(x, \nu)\,\Theta(x - |\nu + 2|) \\ & + f_3(x, -\nu)\,\Theta(x - |\nu - 2|) \\ & + \frac{\pi x^2}{2}\left[\theta(|\nu + 2| - x) + \theta(|\nu - 2| - x)\right] \Big\}, \end{aligned} \tag{5.52}$$

$$-\mathrm{Im}\widetilde{R}_1^+(x,\nu) = \mathrm{Im}\widetilde{\Pi}_1^+(x,\nu) = \frac{-1}{8\sqrt{\nu^2-x^2}}\left\{ f_3(x,-\nu)\,\theta(x-|\nu-2|) + \frac{\pi x^2}{2}\left[\theta(x+2-\nu)+\theta(2-x-\nu)\right]\right\}, \tag{5.53}$$

$$-\mathrm{Im}\widetilde{R}_2^+(x,\nu) = \mathrm{Im}\widetilde{\Pi}_2^+(x,\nu) = \frac{\Theta(\nu-x+2)}{8\sqrt{x^2-\nu^2}}\left\{ f_4(x,\nu) - f_4(x,-\nu)\,\Theta(2-x-\nu)\right\}, \tag{5.54}$$

and

$$-\widetilde{R}^-(x,\nu) = \frac{\pi x^2\Theta(x-\nu)}{8\sqrt{x^2-\nu^2}} + \mathrm{i}\frac{\pi x^2\Theta(\nu-x)}{8\sqrt{\nu^2-x^2}}. \tag{5.55}$$

The quantities $f_1(x,\nu)$, $f_2(x,\nu)$, $f_3(x,\nu)$, and $f_4(x,\nu)$ are defined as

$$f_1(x,\nu) = (2+\nu)\sqrt{(2+\nu)^2-x^2} - x^2\ln\frac{\sqrt{(2+\nu)^2-x^2}+(2+\nu)}{\left|\sqrt{\nu^2-x^2}+\nu\right|}, \tag{5.56}$$

$$f_2(x,\nu) = x^2\ln\frac{\nu-\sqrt{\nu^2-x^2}}{x}, \tag{5.57}$$

$$f_3(x,\nu) = (2+\nu)\sqrt{x^2-(2+\nu)^2} + x^2\sin^{-1}\frac{2+\nu}{x}, \tag{5.58}$$

$$f_4(x,\nu) = (2+\nu)\sqrt{(2+\nu)^2-x^2} - x^2\ln\frac{\sqrt{(2+\nu)^2-x^2}+(2+\nu)}{x}. \tag{5.59}$$

In the zero-frequency limit, the RPA static shielding dielectric function for Graphene given above at zero temperature reduces to [37]

$$\varepsilon(p,0) = 1+\frac{p_{\mathrm{TF}}}{p}\begin{cases} 1, & p \le 2p_{\mathrm{F}}, \\ 1+\dfrac{\pi p}{8p_{\mathrm{F}}} - \dfrac{1}{2}\sqrt{1-\left(\dfrac{2p_{\mathrm{F}}}{p}\right)^2} - \dfrac{p}{4p_{\mathrm{F}}}\sin^{-1}\left(\dfrac{2p_{\mathrm{F}}}{p}\right), & p > 2p_{\mathrm{F}}, \end{cases} \tag{5.60}$$

where (κ is the static background dielectric constant)

$$p_{\mathrm{TF}} = \frac{4e^2 p_{\mathrm{F}}}{\hbar\kappa\gamma} \tag{5.61}$$

is the 2D Thomas–Fermi shielding wave vector for Graphene. The associated shielded Coulombic impurity potential (in wave number representation),

$$v_c(p,0) = \frac{2\pi e^2}{p\,\varepsilon(p,0)}, \tag{5.62}$$

has been employed in various transport calculations for Graphene [38].

Of course, the vanishing of $\varepsilon(p,\omega)$, or

$$-\alpha(p,\omega + \mathrm{i}0^+) = v_c(p)R(p,\omega) = 1, \tag{5.63}$$

with $R(p,\omega)$ given above, identifies Graphene plasmons, and in the local limit ($p \to 0$) the fundamental plasmon mode is given by

$$\omega = bp^{1/2}, \tag{5.64}$$

where

$$b = (g_s g_v e^2 E_F / 2\kappa)^{1/2}. \tag{5.65}$$

Having the fundamental plasmon frequency (5.64) proportional to the square root of wave number is typical of local 2D plasmons. However, the coefficient b is proportional to $E_F^{1/2}$, which is proportional to the fourth root of density, $\rho_{2D}^{1/4}$, for Graphene, which is quite different from the $\rho_{2D}^{1/2}$-dependence of the fundamental plasma frequency of a normal 2D plasma. Furthermore, the leading wave number correction, given by

$$\omega = bp^{1/2}\left[1 - \frac{p_{TF}\,p}{8p_F^2}\right], \tag{5.66}$$

reduces the plasma frequency in Graphene, whereas the corresponding wave number correction would increase the fundamental plasma frequency of a normal 2D plasma.

5.3.3 Some Physical Features of Graphene

Naturally, the dielectric response properties of Graphene enter the determination of its electromagnetic normal modes. Mikhailov and Ziegler [39] recently found a new transverse electric (TE) mode in Graphene in the THz range,

$$1.667 < \frac{\hbar\omega}{E_F} < 2 \quad \text{or} \quad 15\,\text{THz} \le f \le 18\,\text{THz}, \tag{5.67}$$

and they also found a transverse magnetic (TM) plasmon–polariton mode described as

$$\frac{\hbar\omega}{E_{\mathrm{F}}} < 1.667. \tag{5.68}$$

Needless to say (5.2), the plasmons of Graphene play an important role in its correlation energy, and Das Sarma, Hwang, and Tse [40] investigated the RPA-screened Coulombic self-energy, $\sum_{\mathrm{Coul.}}$, of Graphene,

$$\sum_{\mathrm{Coul.}}(1,2) \Rightarrow -\mathrm{i}V(1-2)G(1,2), \tag{5.69}$$

where the effective screened potential is given by (analogous to (5.4))

$$V(1-2) = \int \mathrm{d}^4 3\, K(2-3) v_{\mathrm{c}}(1-3). \tag{5.70}$$

(Self-energy is a generalization of the concept of potential energy which takes account of many-body interactions that influence the dynamics of a single particle, in this case the interactions are the many screened Coulomb interactions of the multitude of charge carriers coupling with the individual single particle of interest.) Clearly, the plasma poles of the screening function, K, make distinct contributions to the Coulombic self-energy, $\sum_{\mathrm{Coul.}}$. In this study, Das Sarma et al. found that intrinsic Graphene is a marginal Fermi liquid, with the quasiparticle spectral weight vanishing near the Dirac point, but extrinsic Graphene is a well-defined Fermi liquid due to doping [41–43]. Moreover, Tse and Das Sarma [44] also examined the phonon-induced self-energy [43], $\sum_{\mathrm{phonon}}$, of Graphene,

$$\sum_{\mathrm{phonon}}(1,2) \Rightarrow -\mathrm{i}V_{\mathrm{eff}}(1,2)G(1,2), \tag{5.71}$$

where

$$V_{\mathrm{eff}} = \Gamma_{\mathrm{ep}}(\boldsymbol{x}_1)\, D^0(t_1,t_2)\, \Gamma_{\mathrm{ep}}(\boldsymbol{x}_2), \tag{5.72}$$

with Γ_{ep} measuring electron–ion interaction strength, and $D^0(t_1,t_2)$ as the free phonon Green's function [43]. They found that the phonon-mediated carrier–carrier coupling of (5.72) gives rise to a large effect on Graphene band structure renormalization.

Because of the enormous potential of Graphene, it is worthwhile to examine all possible probes of its properties. Recently, we examined Graphene energy loss spectroscopy. The dielectric response of Graphene is of central importance in this because the Coulomb potential of a fast passing charged particle polarizes this 2D plasma, and transfers energy to it by this process. In the case when the energetic (high velocity) passing particle moves parallel to the Graphene sheet, we determined its power loss to be [45] ($\lambda = (\gamma e^2 \sqrt{\rho_{2\mathrm{D}}\pi g_{\mathrm{s}} g_{\mathrm{v}}}/\kappa\hbar)^{1/2}$):

$$\frac{\mathrm{d}W}{\mathrm{d}t} = -\frac{Z^2\mathrm{e}^2\lambda^4}{2v^3}\mathrm{e}^{-\lambda^2 H/v^2}\left[K_0\left(\frac{\lambda^2 H}{v^2}\right) + K_1\left(\frac{\lambda^2 H}{v^2}\right)\right], \tag{5.73}$$

where K_0 and K_1 are modified Bessel functions and H is the height of the parallel-moving particle above the Graphene sheet, v is its speed, and Z is its charge number. In the case when the passing particle moves perpendicular to the Graphene sheet (speed v_z) and passes through it, we found the total work on it as given by [46]

$$|W| = v_z \frac{\pi D_0 p_{\mathrm{F}}}{\gamma} \left(\frac{Ze^2}{\kappa}\right)^2 \left(c\pi - 4 - \frac{2(c^2-2)}{\sqrt{c^2-4}} \cos^{-1}\left(\frac{2}{c}\right)\right), \tag{5.74}$$

where

$$c = \frac{2\pi e^2 D_0}{\kappa p_{\mathrm{F}}}. \tag{5.75}$$

The mutual polarization of nonoverlapping neutral systems (such as an atom/molecule and a nearby surface) in van der Waals (vdW) interaction also involves Graphene dielectric response in conjunction with its vdW attraction for a nearby atom/molecule. We have examined this vdW energy to second order in the Coulomb interaction between the Graphene electrons and those of the atom/molecule, with the result [47] ($\varepsilon_0 \equiv \kappa$ is the background dielectric constant here)

$$E_{\mathrm{vdW}}^{(2)} = \frac{4}{3\hbar\varepsilon_0} \sum_n{}' \int_0^\infty \frac{du}{2\pi} \frac{\omega_{n0} \left|\boldsymbol{D}_{0n}\right|^2}{u^2+\omega_{n0}^2} \int_0^\infty dp p^2 \mathrm{e}^{-2p|Z|} \frac{\alpha^{2\mathrm{D}}(p, \mathrm{i}u)}{\varepsilon_0 + \alpha^{2\mathrm{D}}(p, \mathrm{i}u)}, \tag{5.76}$$

where $|Z|$ is the distance between the atom and the 2D Graphene sheet; ω_{n0} is the energy difference of the unperturbed atomic electron levels, $\omega_{n0} = E_n^a - E_0^a$; $\boldsymbol{D}_{0n}$ is the matrix element of the atom's dipole moment operator between atomic electron levels $n, 0$; and $\alpha^{2\mathrm{D}}(p,\omega)$ is the dynamic, nonlocal polarizability (5.9) of the Graphene sheet. The prime on $\sum'$ denotes omission of the $n = 0$ term. Equation (5.76) is a useful point of departure to determine both local and nonlocal structure of the van der Waals interaction using $\alpha^{2\mathrm{D}}(p,\omega)$ of (5.44) for Graphene. For the undoped case, $\alpha^{2\mathrm{D}}(p, \mathrm{i}u)$ is given by

$$\alpha^{2\mathrm{D}}(p, \mathrm{i}u) = -\left(\frac{g_{\mathrm{s}} g_{\mathrm{v}} \pi \mathrm{e}^2}{8\varepsilon_0\hbar}\right) \frac{p}{\sqrt{u^2+\gamma^2 p^2}}. \tag{5.77}$$

It should be noted that an expansion of E_{vdw} in inverse powers of $|Z|$ is *not* available since it would involve expanding the p-integrand of (5.76) in powers of p/u, which would cause divergences in the final u-integration. However, for atomic frequencies (ω_{n0}), the polarizability is small, $\alpha^{2\mathrm{D}}(p, \mathrm{i}u) \ll 1$, and we have the p-integral of (5.76) as

$$\int_0^\infty \mathrm{d}p \ldots \cong -\left(\frac{g_{\mathrm{s}} g_{\mathrm{v}} \pi \mathrm{e}^2}{8\varepsilon_0\hbar}\right) \int_0^\infty \mathrm{d}p\, \mathrm{e}^{-2p|Z|} \frac{p^3}{\sqrt{u^2+\gamma^2 p^2}}. \tag{5.78}$$

This integral may be rewritten using

$$\int_0^\infty dx\,\frac{xe^{-\mu x}}{\sqrt{x^2+\beta^2}} = \frac{\beta\pi}{2}\left[\boldsymbol{H}_1(\beta\mu) - Y_1(\beta\mu)\right] - \beta, \tag{5.79}$$

where $\boldsymbol{H}_1(x)$ is the Struve function and $Y_1(x)$ is the Bessel function of the second kind, with the result [47]

$$\int_0^\infty dp \ldots \cong -\left(\frac{g_s g_v \pi e^2}{8\varepsilon_0 \hbar}\right)\frac{u\pi}{8\gamma^2}\frac{\partial^2}{\partial|Z|^2}\left[\boldsymbol{H}_1\left(\frac{2|Z|}{\gamma}u\right) - Y_1\left(\frac{2|Z|}{\gamma}u\right)\right]. \tag{5.80}$$

5.4 Summary

We have recounted in some detail the quantum-mechanical aspects of solid-state plasma phenomenology. Introducing the RPA, we reviewed the RPA screening and dielectric functions in three, two, and one dimensions corresponding to bulk, quantum well, and quantum wire plasmas, respectively, taking care to distinguish quantum effects from classical ones mandated by the correspondence principle. In particular, we examined plasmon dispersion, damping, and static shielding in these various dimensionalities. We also reviewed the Landau-quantized magnetoplasma in the same context.

The discussion of Graphene plasma properties in this chapter is also carried out in the RPA for its dielectric response. Its extraordinary device-friendly features and applications are discussed. Having a massless relativistic Dirac-like single-particle energy spectrum, it is also of great fundamental interest. We have exhibited its Green's function and dielectric polarizability in detail, discussing Graphene impurity shielding, plasmon dispersion, and its TE and TM electromagnetic modes. Moreover, the role of plasmons (and phonons) in Graphene self-energy is described, as is our recent work on its energy loss spectroscopy and van der Waals interaction with an atom/molecule.

It is hardly necessary to point out that there is much, much more literature pertinent to quantum plasma phenomenology that we have not discussed. For example, the vast literature on superlattices, quantum dots, heterostructure in general, low-dimensional Landau-quantized magnetoplasmas, etc. Also, we have barely scratched the surface of the huge flood of work on Graphene, omitting many very interesting studies (such as the work of Isakson et al. [48] on electronic superlattices in corrugated Graphene and the intriguing work of Matulis and Peeters [49] on Graphene quantum dots). However, this is not intended to be a comprehensive review, but as a very brief introduction to stimulate the reader's interest.

Acknowledgement This work was partially supported by DARPA grant # HR0011-09-1-0008.

References and Footnotes

I have referenced original papers and very early books that I feel are particularly informative on various topics. However, this is *not* intended as a comprehensive review, and many interesting papers, reviews, and books are omitted, with apologies to their authors. After the passage of about half of a century, many more recent, important papers and books have appeared and can be accessed by a literature search.

References

1. D. Pines, *Elementary Excitations in Solids* (W.A. Benjamin, New York, 1964) and references cited therein
2. A.I. Akhiezer, *Collective Oscillations in a Plasma* (Pergamon, London, 1967)
3. L.D. Landau, *Collected Papers* (Pergamon, Oxford, 1965), p. 445
4. J.D. Jackson, J. Nucl. Energy Part C **1**, 171 (1960)
5. E.N. Economou, *Green's Functions in Quantum Physics*, 3rd edn. (Springer, Berlin, 2006)
6. A.L. Fetter, J.D. Walecka, *Quantum Theory of Many-Particle Systems* (McGraw-Hill, New York, 1971)
7. P.C. Martin, J. Schwinger, Phys. Rev. **115**, 1342 (1959)
8. T. Ando, A.B. Fowler, F. Stern, Rev. Mod. Phys. **54**, 437 (1982)
9. P. Harrison, *Quantum Wells, Wires and Dots*, 2nd edn. (Wiley-Interscience, Chichester, 2005)
10. F. Stern, Phys. Rev. Lett. **18**, 546 (1967)
11. P.F. Williams, A.N. Bloch, Phys. Rev. B **10**, 1097 (1974)
12. M.M. Mohan, A. Griffin, Phys. Rev. B **32**, 2030 (1985)
13. N.J.M. Horing, Phys. Rev. **136**, A494 (1964)
14. N.J.M. Horing, Ann. Phys. (NY) **31**, 1 (1965)
15. N.J.M. Horing, J. Phys. Soc. Jpn. **21** (Suppl.), 704 (1966)
16. N.J.M. Horing, Ann. Phys. (NY) **54**, 405 (1969)
17. N.J.M. Horing, Ann. Phys. (NY) **68**, 337 (1971)
18. N.J.M. Horing, Phys. Rev **186**, 434 (1969) (Appendix)
19. N.J.M. Horing, in *The Many Body Problem*, ed. by L.M. Garrido, A. Cruz, T.W. Priest (Plenum, New York, 1969), p. 307
20. N.J.M. Horing, M. Yildiz, Phys. Lett. A **44**, 386 (1973)
21. N.J.M. Horing, M. Orman, M. Yildiz, Phys. Lett. A **48**, 7 (1974)
22. N.J.M. Horing, M. Yildiz, Ann. Phys. (NY) **97**, 216 (1976)
23. K.S. Novoselov, A.K. Geim et al., Science **306**, 666 (2004)
24. A.H. Castro Neto et al., arXiv:0709.1163v1[cond-mat. other], 7 September 2007
25. S.V. Morozov et al., Phys. Rev. Lett. **100**, 016602 (2008)
26. F. Schedin, K.S. Novoselov et al., arXiv.cond-mat/0610809 (unpublished data)
27. E.W. Hill, A.K. Geim, K.S. Novoselov et al., IEEE Trans. Magn. **42**, 2694 (2006)
28. J.S. Bunch et al., Science **315**, 490 (2007)
29. Georgia Tech Research News, http://gtresearchnews.gatech.edu/newsrelease/graphene.html
30. M.C. Lemme, IEEE Electron Device Lett. **28**, 282 (2007)
31. P.R. Wallace, Phys. Rev. **71**, 622 (1947)
32. G.W. Semenoff, Phys. Rev. Lett. **53**, 2449 (1984)
33. F.D.M. Haldane, Phys. Rev. Lett. **61**, 2015 (1988)
34. K.W.-K. Shung, Phys. Rev. B **34**, 979 (1986)
35. B. Wunsch et al., New J. Phys. **8**, 318 (2006)
36. E.H. Hwang, S. Das Sarma, Phys. Rev. B **75**, 205418 (2007)
37. T. Ando, J. Phys. Soc. Jpn. **75**, 074716 (2006)

38. S.Y. Liu, N.J.M. Horing, J. Appl. Phys. **104**, 043705 (2008) and references cited therein
39. S.A. Mikhailov, K. Ziegler, Phys. Rev. Lett. **99**, 016803 (2007)
40. S. Das Sarma, E.H. Hwang, W.-K. Tse, Phys. Rev. B **75**, 121406(R) (2007)
41. M. Ali Omar, *Elementary Solid State Physics* (Addison-Wesley, London, 1993) ("intrinsic" and "extrinsic" are discussed in this reference)
42. R. Enderlein, N.J.M. Horing, *Fundamentals of Semiconductor Physics and Devices* (World Scientific, Singapore, 1997). ("Intrinsic" and "extrinsic" are discussed.)
43. A.A. Abrikosov et al., *Quantum Field Theoretical Methods in Statistical Physics* (Pergamon, Oxford, 1965) (Fermi liquids and phonon self-energy are discussed here.)
44. W.-K. Tse, S. Das Sarma, Phys. Rev. Lett. **99**, 236802 (2007)
45. N.J.M. Horing, V. Fessatidis, IEEE Sensors Journal **10**, 674 (2010)
46. V. Fessatidis, N.J.M. Horing, M. Sawamura, Physics Procedia **3**, 1279 (2010)
47. N.J.M. Horing, V. Fessatidis, J.D. Mancini, in press
48. A. Isakson et al., arXiv:0709.2614v1[cond-mat.mes-hall], 17 September (2007)
49. A. Matulis, F.M. Peeters, Phys. Rev. B **77**, 115423 (2008)

Part III
Strongly Coupled and Dusty Plasmas

Chapter 6
Imaging Diagnostics in Dusty Plasmas

Dietmar Block and André Melzer

Abstract The success of dusty plasmas is based on its unique diagnostic possibilities. The ability to follow the motion of any particle in a dusty plasma allows to explore the fundamental properties of strongly coupled systems at the kinetic level (see Chap. 7). This chapter gives an introduction to imaging diagnostics in dusty plasmas. In the first part, the basics of video microscopy are reviewed and the important role of a careful image analysis is stressed. The second part reports on recent developments on three-dimensional (3D) imaging techniques, that is, diagnostic systems which are able to track simultaneously all particles in a 3D dust cloud.

6.1 Introduction

Strong coupling is a subject of intense research activity in different fields (see Chaps. 7, 8, and 10). In "normal" solid matter strongly coupled systems consist of atoms with very small interparticle distances ($b = \mathcal{O}(\text{Å})$) and their dynamical processes take place on short timescales ($\omega \gg 10^3\,\text{s}^{-1}$). Thus, spatiotemporally resolved measurements are difficult to achieve.

In dusty plasmas the situation is different. If micrometer-sized particles are immersed in a plasma environment they immediately attain high negative charges due to the higher mobility of the electrons. A rough estimate, based on the so-called *OML model* [1], for the charge $Q_\text{d} = -Z_\text{d}e$ of a spherical particle is given by

$$Z_\text{d} \leq 1{,}400\, e r_\text{p} T_\text{e}. \tag{6.1}$$

D. Block (✉)
Institut für Experimentelle und Angewandte Physik, Christian-Albrechts-Universität Kiel, Olshausenstr. 40-60, 24098 Kiel, Germany
e-mail: block@physik.uni-kiel.de

A. Melzer
Institut für Physik, Ernst-Moritz-Arndt-Universität Greifswald, Felix-Hausdorff-Str. 6, 17489 Greifswald, Germany
e-mail: melzer@physik.uni-greifswald.de

M. Bonitz et al. (eds.), *Introduction to Complex Plasmas*, Springer Series on Atomic, Optical, and Plasma Physics 59, DOI 10.1007/978-3-642-10592-0_6,

Here, the particle radius r_p has to be inserted in micrometer and the electron temperature T_e in electron volts, that is, in a typical rf-discharge with an electron temperature of a few eV and micrometer-sized particles we find $Z_d \approx 10^4$. Therefore, even at particle distances of the order of several hundred micrometer the mutual interaction energy of two particles exceeds by far their thermal energy and the system is strongly coupled (see Chap. 7). At the same time this strongly coupled system has a fairly low particle density ($n_d \approx 10\,\mathrm{mm}^{-3}$) and a favorable ratio of interparticle distance to particle radius ($b/r_p \geq 100$). This combination results in a very high optical transparency which allows to observe individual particles even at the center of clouds containing more than 10^6 particles (see, e.g., [2]).

Besides their macroscopic dimensions, dusty plasmas possess another diagnostic advantage. The charge to mass ratio of micrometer-sized dust particles is small ($Q_d/m \approx 10^{-2}\,\mathrm{C\,kg^{-1}}$). Hence any dynamic response of such a dusty plasma occurs at very low frequencies. As an example, the dust plasma frequency ω_d of a dust cloud with typical particle density and charge is of the order of tens of Hertz only. Therefore, the temporal resolution of high speed video cameras is fully sufficient to study dynamical processes in dusty plasmas.

Against this background, it is obvious why dusty plasmas offer unique diagnostic possibilities: Their transparency and slow dynamics allow to investigate liquid, solid, and gaseous phases, phase transitions, waves, and many more dynamic phenomena at a kinetic level [3, 4]. The following sections will review the prospects and limitations of diagnostics systems which are used to investigate 2D and 3D dust clouds.

6.2 Imaging 2D Systems

To study the dynamics of dust particles in a plasma two ingredients are essential: a method to record an image of micrometer-sized particles inside a luminous plasma and a precise and reliable method to extract the particle positions from the image. Although both methods have straightforward solutions for a 2D particle system, similar but more severe problems arise in 3D systems and hence it is worth to discuss the solutions for 2D systems first.

6.2.1 Imaging Particles

To image dust particles with a video camera it is necessary to discriminate them against the plasma glow. For this purpose the particles are illuminated with a thin sheet of laser light (Fig. 6.1). Such sheets with a thickness of less than a millimeter are typically produced with a cylindrical lens in front of a low power laser diode ($P \approx 20\,\mathrm{mW}$). In this setup the well-defined wavelength of the laser allows to use interference filters in front of a CCD camera to mask most of the plasma glow.

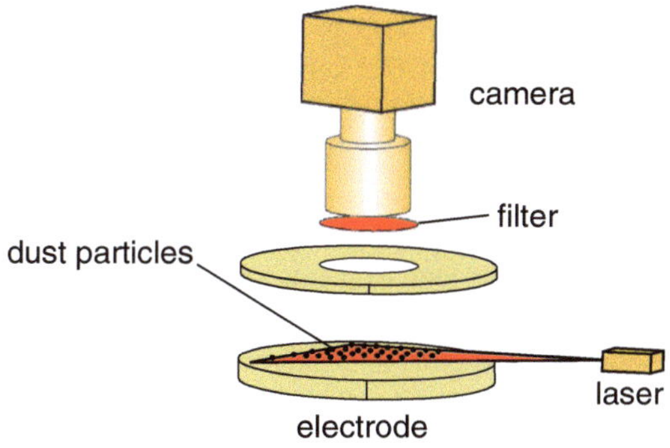

Fig. 6.1 Video microscope. The dust particles are confined in the sheath region above the electrode. A horizontal laser sheet is used to illuminate the 2D dust cloud and a camera records the scattered laser light at right angle

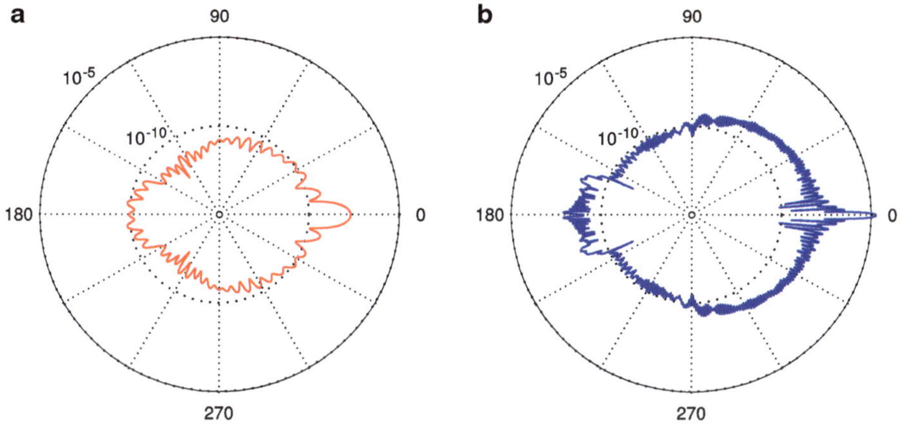

Fig. 6.2 Angular intensity distribution of light scattered by spherical glass particles with radius $r_\mathrm{p} = 2\,\mu\mathrm{m}$ (**a**) and $r_\mathrm{p} = 10\,\mu\mathrm{m}$ (**b**) according to Mie theory. The intensities are normalized to the intensity of the incoming laser light and computed for a distance of 20 cm. The angles 180° and 0° mark the direction of backward and forward scattered light, respectively. Please note the logarithmic scale

At first glance this setup seems simple and flexible but it has some noteworthy limitations. First, dusty plasmas are produced in a vacuum chamber in between an electrode arrangement and hence the distance between dust cloud and camera is typically larger than 10 cm. In addition the intensity of the light scattered by the particles is rather weak. To quantify this, for micrometer-sized spherical particles the light scattering is well described by Mie theory. For a detailed treatment of light scattering by particles we refer the reader to the book of Bohren and Huffman [5]. Here, the solution for massive, spherical glass particles is discussed. Figure 6.2 shows the angular intensity distribution of light scattering from such particles with radius $r_\mathrm{p} = 2\,\mu\mathrm{m}$ (a) and $r_\mathrm{p} = 10\,\mu\mathrm{m}$ (b) in a distance of 20 cm. The plots clearly demonstrate that only a small fraction of the light is scattered and that especially the geometrically favorable observation at right angle suffers from very low intensities. Thus, even for larger particles sensitive cameras are required to observe particles.

Although CCD cameras offer a sufficient sensitivity their usage bears a limitation: High resolution CCD sensors support only low frame rates. Presently a megapixel camera has about 25 fps (frames per second). Thus, fast processes in dusty plasmas are only accessible at the price of a reduced spatial resolution or with faster but less sensitive CMOS cameras.

Second, the field of view of the camera should cover a few hundred particles, that is, have a size of a few square centimeters. With a sensor size of up to $2.5\,\mathrm{cm}^2$ ($1''$ sensor), the magnification factor is close to one and thus macrolenses have to be used. However, macrolenses have a small depth of field if they are operated with a large aperture to collect enough light. The depth of field Δd for a typical video microscope is about $\Delta d = 1\,\mathrm{cm}$, that is, only about ten times larger than the typical interparticle distance. For a monolayer layer this is not a limitation but for 3D systems this limitation becomes important.

Finally, it is tempting to just increase the laser power to use smaller apertures or faster cameras. Unfortunately this affects the dust particles. Several experiments have shown that dusty plasmas can be manipulated even at moderate laser power. The radiation force on a particle with cross section A is given as

$$F_{\mathrm{rad}} = \gamma \frac{IA}{c}, \tag{6.2}$$

where I is the intensity of the laser (in $\mathrm{W\,m^{-2}}$) and c is the speed of light. The coefficient γ describes the type of interaction with $\gamma = 1$ for absorbed photons and $\gamma = 2$ for reflection. Although the radiation pressure radiation pressure of a laser sheet (20 mW laser diode) is just $F_{\mathrm{rad}} \approx 10^{-15}\,\mathrm{N}$, the same laser diode focused to a particle exerts a force which is comparable to other forces in the system [6]. In addition, recent experiments with high power laser beams have demonstrated that the local heating of boundaries [7] or particles [8] can considerably distort the system due to thermophoretic forces or even excite shock waves due to particle evaporation. In conclusion, stronger lasers are not suitable to improve the performance of video microscopes and thus the application of video microscopes is generally limited to nearly 2D systems.

6.2.2 *Image Analysis*

Once images have been recorded with a video microscope reliable image processing is required. A typical video microscope image is mostly black with a few bright clusters of pixels. These clusters, which are the particle images, typically have a radius of one to three pixels and an average separation of about 40 pixels. In general, the cluster size and separation depend on the camera resolution, the magnification factor of the macrolens, the degree of focussing and the brightness of the particles, that is, it strongly depends on the experimental setup and equipment. Thus, to extract precise particle positions from these images several algorithms have been proposed

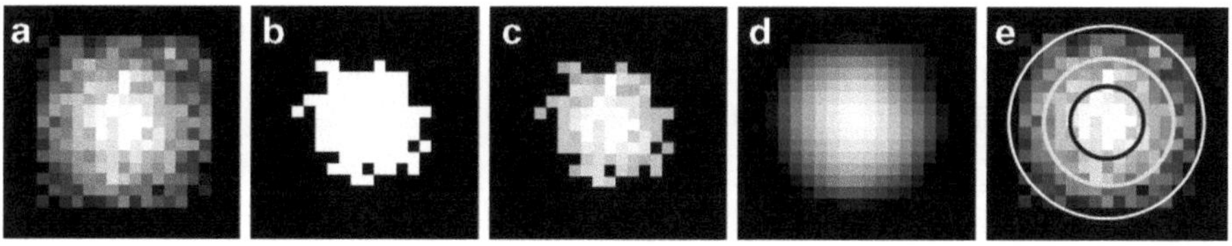

Fig. 6.3 Working principle of the particle positioning methods. (**a**) artificial particle image on the camera sensor. Note that the particle image is chosen much larger than in an experimental situation to simplify the illustration. (**b**) Particle image according to threshold method. All pixels above a chosen value are set to maximum intensity while the intensity at all other pixels is set to zero. (**c**) Particle image according to moment method. Here the intensity of all pixels below threshold is set to zero while the intensity of all other pixels remains unchanged. (**d**) same as (**c**) but after treatment with a Gaussian bandpass filter. (**e**) Fit of the particle intensity profile with local quadratic kernel method. The resulting intensity profile is indicated as a contour plot (*concentric rings*). *Darker rings* correspond to higher intensity

and were believed to give particle positions with subpixel resolution. However, when evaluating their data, Nosenko and coworkers realized that their measured velocity distributions had notable distortions. They figured out that these distortions originated from an effect called pixel locking, that is, the tendency of particle detection algorithms to favor particle positions at locations close to pixel edges [9]. Thus especially for high precision experiments like normal mode analysis [10], dust kinetic temperature measurements, wave dispersion experiments [11], etc. pixel locking can distort the result significantly.

The following paragraphs give an overview on particle detection methods of different degree of sophistication. An illustration of these methods is shown in Fig. 6.3. A detailed descriptions and error analysis of these methods have been given by Ivanov et al. [10] and Feng et al. [12]. Here only the problem of pixel locking and subpixel resolution is stressed.

6.2.2.1 Threshold Method

This method has the simplest algorithm. First, a threshold value is chosen for the entire image and all contiguous pixels with intensities above the threshold are considered as a particle (ca. Fig. 6.3a, b). In a second step the center of mass is computed for each of these pixel clusters assuming that all pixels have the same weight. Although this method is fast, it is not possible to achieve subpixel resolution [10]. The results are strongly affected by pixel locking effects.

6.2.2.2 Moment Method

This method is a close relative of the threshold method and one of the most widely applied techniques in dusty plasmas. In contrast to the threshold method, the moment method conserves the intensity of all pixels above the threshold and uses this information to improve the calculation of the center of mass (ca. Fig. 6.3a, c), that is, the particle positions are determined from

$$X_{\mathrm{C}} = \frac{\sum_k X_k (I_k - I_{\mathrm{th}})}{\sum_k I_k - I_{\mathrm{th}}}, \tag{6.3}$$

where I_k is the intensity of pixel k and X_k, Y_k its position on the camera image. Here, the threshold intensity is denoted by I_{th} and the sum runs for each cluster over all pixels which exceed the threshold. The accuracy of this method has been studied in detail by Feng et al. [12]. They showed that subpixel resolution is only achieved if the recording parameters are well adjusted and if the threshold is chosen carefully. Additionally, a detailed guide how to find the optimum setting for this method is given.

6.2.2.3 Moment Method with Gaussian Bandpass Filter

This algorithm originates from the field of colloidal suspensions [13]. Instead of applying a threshold criterion directly, a spatial bandpass filter for smoothing and background subtraction is used. The width of the Gaussian filter is chosen to match the expected particle image size. The result is shown in Fig. 6.3d. After this, contiguous pixels are identified and their weighted center of mass is computed (see moment method). This method allows to achieve subpixel resolution and minimizes pixel locking effects as well [10]. In addition the filtering process has the advantage to partly compensate imperfections during the recording process.

6.2.2.4 Least Quadratic Kernel Method

This sophisticated method was proposed by Kvarnstrom [14]. It assumes particle images with circular symmetry which is a good approximation for most experiments. Starting with a first approximation of the particle position (e.g., obtained with the moment method), a parabolic intensity profile is fitted using a Gaussian kernel:

$$K(|x - x_0|) = \frac{1}{h} \exp\left(-\frac{|x - x_0|^2}{2h^2}\right), \tag{6.4}$$

with bandwidth h and the particle center x_0. At the price of rather high computational efforts, this method has the advantage to yield very good results even if the particle shape changes with focal depth.

To summarize, particle detection seems to be a rather trivial task, but for many experiments it is not. As soon as subpixel resolution is required, pixel locking becomes a severe problem and advanced detection methods have to be used.

6.3 Imaging 3D Systems

One result of the previous section was that video microscopes are not able to resolve the depth position along the optical axis and that the optical properties of macrolenses restrict the observable volume to a thin layer. Thus, conventional video microscopes are not offhand applicable to 3D systems. This section will review four techniques which were developed to overcome these limitations: scanning video microscopes [15–17], color gradient systems [18], stereoscopic imaging techniques [19, 20], and digital holography [21].

6.3.1 Scanning Video Microscopy

Scanning video microscopes (SVM) were the first 3D diagnostics for dusty plasmas [15]. They basically consist of a standard video microscope which is mounted on a translation stage. Its construction scheme and an example are shown in Fig. 6.4. The working principle of an SVM is quite simple. The particles are illuminated with a thin laser sheet and the scattered light is detected with a CCD camera at right angle (left part of Fig. 6.4) just as for the 2D systems described above. Moving the translation stage along the y-axis with constant velocity v, a sequence of cross sections of the trapped dust cloud is recorded. The 3D particle coordinates are determined from this image sequence as follows: First, a 2D particle detection is performed for each image. If the laser sheet is significantly thinner than the interparticle distance a particle is only visible if the illuminated region almost perfectly matches its y-position. In all other frames it cannot be detected, that is, its intensity $I_k(x, z)$ is below a threshold and hence set to zero. Thus, any particle appears only in a short sequence of images and reaches its maximum brightness right at the center of the

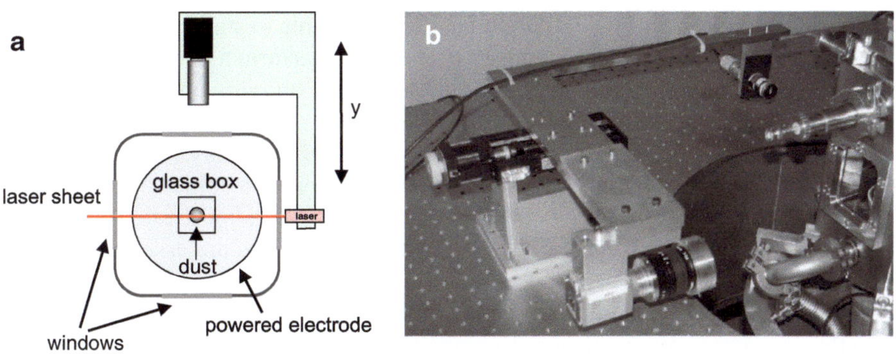

Fig. 6.4 (**a**) Scanning video microscope (SVM). The typical video microscope setup (see Fig. 6.1) is mounted on a translational stage. A scan along the y-axis allows a "tomographic" recording of a 3D dust cloud. (**b**) Picture of a SVM. The camera is located in the front while the laser is at the rear end of the construction

laser sheet. Therefore, the y-position of the ith particle can be obtained by means of an extension of the moment method to 3D, that is,

$$Y_i = v\tau \frac{\sum_k k \bar{I}_i(k)}{\sum_k \bar{I}_i(k)}, \tag{6.5}$$

where $\bar{I}_i(k)$ denotes the mean intensity of the ith particle in the kth image and $1/\tau$ is the frame rate. For slow scanning velocities ($v \ll 1$ mm/s) the 3D particle position can be recovered with an accuracy of about $\Delta x, y, z \approx 10\,\mu$m. Although this diagnostic setup is simple and allows to cover volumes of the order of cm^3, the sequential image recording limits its use to static particle configurations. Finally, this technique requires a calibration to convert the particle position into SI units. First, the magnification factor of the lens systems has to be determined and second reliable measurements of frame rate and scanning velocity are required to assure that the y-scale is computed correctly. Nevertheless, until today SVM are successfully used to investigate crystalline states of 3D dust clouds [15–17].

6.3.2 Color Gradient Method

Another possibility to encode the depth position of a particle is a color gradient [22]. This setup has been developed by the MPE[1] and is again a modification of a video microscopy setup. Here two overlapping lasers with different wavelength illuminate the particles (top of Fig. 6.5). In addition the laser beams have opposite intensity gradients which are produced with neutral density filters (bottom of Fig. 6.5). Finally, the light scattered by the particles is recorded with two cameras, where each of them is sensitive at just one wavelength. Therefore, the intensity ratio of a particle in both cameras is a measure for its depth position. Annaratone et al. [18, 23] have applied this setup successfully to investigate the properties of small dust clouds confined above an adaptive electrode. In particular, they were able to study the motion of individual particles inside these clouds [23], that is, to perform a temporally resolved 3D measurement. The trade off for the instantaneous measurement of the particle positions is a reduction of the observable volume to a few interparticle distances in depth because the aperture problem is not solved. Additionally, the sophisticated setup requires a thorough calibration to achieve good spatial resolution and a conversion of the particle positions into absolute units. Especially sophisticated calibration of the intensity gradients and adjustment of the laser beams are essential. Finally, aligned particles are only separable if an additional camera with a different perspective is used. However, including a third camera an impressive spatial resolution of less than 5 μm was reported for this system [23].

[1] Max Planck Institut fr extraterrestrische Physik, Garching, Germany

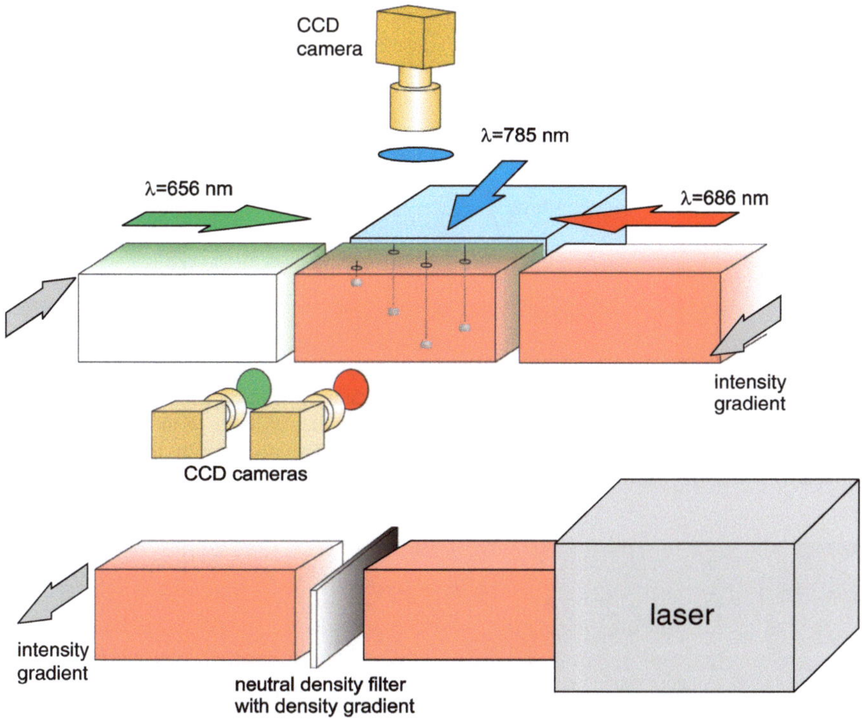

Fig. 6.5 *Top*: schematic setup of a 3D diagnostic using a color gradient to encode the depth position of a particle. The third laser beam ($\lambda = 785$ nm) and the top-view camera are used to improve the spatial resolution of the system. *Bottom*: a neutral density filter with a density gradient allows to adjust the intensity profile of the laser beam

6.3.3 Stereoscopy

The third modification of a video microscope is a stereoscopic setup which involves at least two cameras which observe a dust cloud from different perspectives [19, 20, 24]. Such a stereoscopic imaging system (SIS) is sketched in Fig. 6.6a. Three cameras with perpendicular optical axes are used to image the volume around the intersection of their optical axes. The particles are illuminated by two expanded laser beams and observed in forward scattering to maximize their brightness.

The resolution of SIS is comparable to those of SVMs and the frame rate $f > 100$ fps allows to study even fast dust dynamics, for example, even the thermal fluctuations of the particles around their equilibrium position can be followed and analyzed [10]. Further examples which nicely document the prospects of stereoscopy are given in the (Chap. 7).

To discuss the limitations of this 3D particle imaging method, contrast-enhanced camera images of a dust cloud containing $N = 102$ are shown in Fig. 6.7a–c. First, it is observed that the images of the front and side view camera are brighter than the image of the top-view camera. This is related to the particle illumination. The

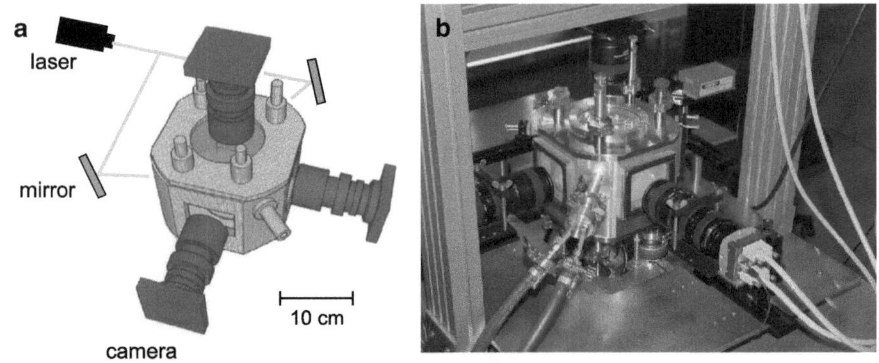

Fig. 6.6 (**a**) Stereoscopic imaging system using three cameras with perpendicular orientation of their optical axes. The particles are illuminated with an expanded laser beam. To maximize the depth of field two of the cameras record the scattered light in almost forward direction. (**b**) Technical realization at the University of Greifswald, Germany

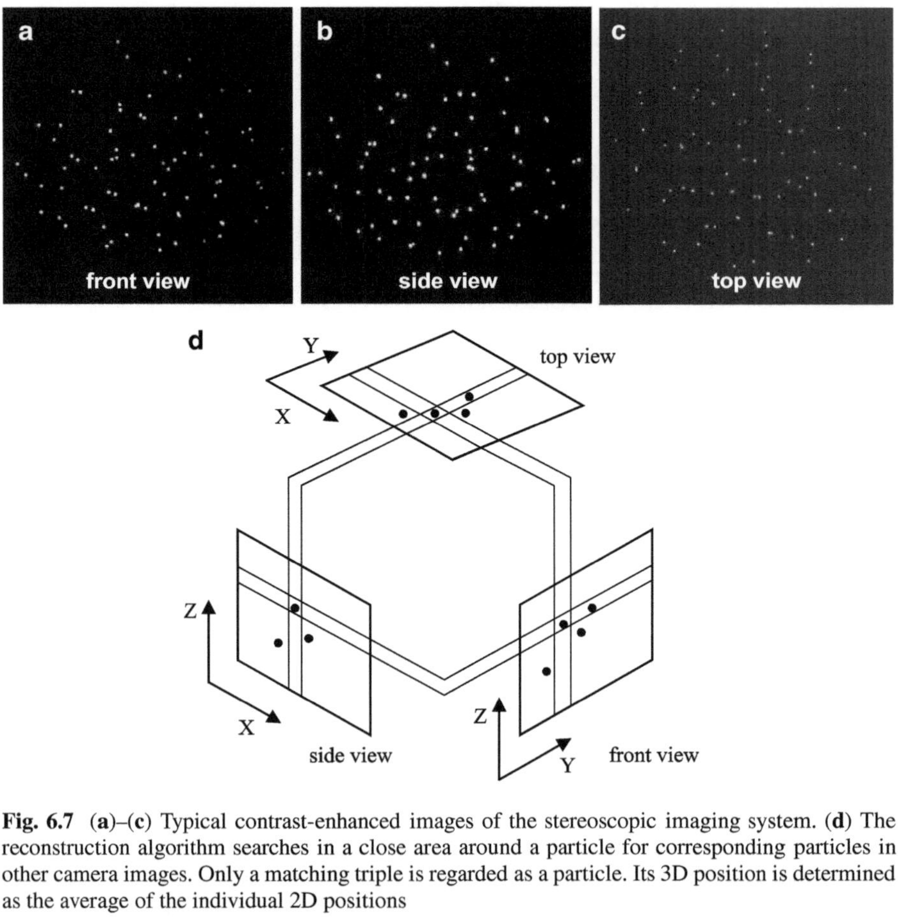

Fig. 6.7 (**a**)–(**c**) Typical contrast-enhanced images of the stereoscopic imaging system. (**d**) The reconstruction algorithm searches in a close area around a particle for corresponding particles in other camera images. Only a matching triple is regarded as a particle. Its 3D position is determined as the average of the individual 2D positions

top-view camera records the scattered light at right angle which is less favorable according to Mie theory (see Fig. 6.2). To collect enough light the top-view camera has to use a larger aperture which results in a smaller depth of field. A close inspection of Fig. 6.7c reveals that this cloud with a diameter of about 7 mm reaches the depth of field. Particles at the top and bottom of the cloud are slightly defocussed. This shows that stereoscopy is still limited by the depth of field.

Second, to compute the particle positions from images like those shown in Fig. 6.7a–c a careful calibration and data processing is essential. The processing from sequences of corresponding 2D image triples to the full 3D particle trajectories is performed in several steps. In the first step, the 2D image coordinates of all particles are determined from the two-dimensional images using standard particle identification techniques [10] as discussed in Sect. 6.2. This identification is performed for each frame of the three cameras separately. For small dust clouds it is justified to assume that in this setup the optical axes of the cameras are perpendicular and that any distortion due to the imperfections of the camera lenses and windows of the vacuum vessel can be neglected. This simplifies the data processing significantly. In a next step the 2D images of the dust cloud have to be scaled to the same size and adjusted to their respective center of mass. Finally, the search for corresponding particles is performed (Fig. 6.7d). For this purpose, the particle images are projected along the optical axes. To account for slight scaling errors and lens errors particle images should be considered as "matching" if their projections deviate by less than 20 μm. If matching particle images are found in all three camera images the particle detection is regarded successful and the 3D position is determined as the average of the individual 2D positions.

At first glance, three cameras should allow for an unambiguous determination of the 3D position of all particles even if some particle images overlap in one camera image, which is frequently observed (see Fig. 6.8a–c). However, a special situation

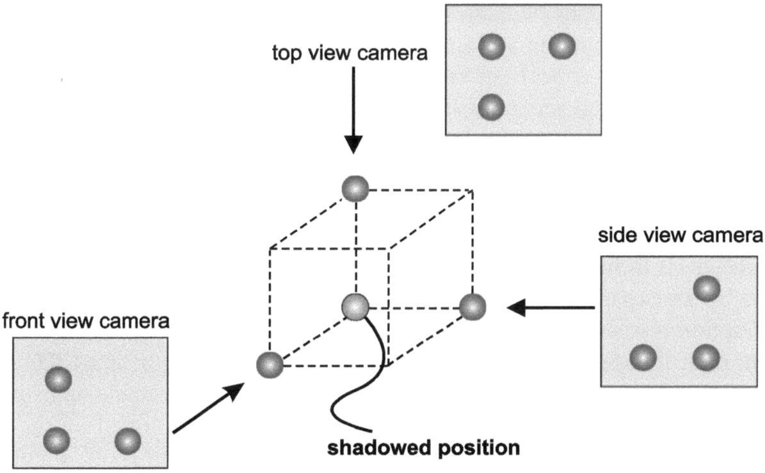

Fig. 6.8 The sophisticated reconstruction problem is additionally complicated by the finite particle size in each camera image. This results in a finite probability that a particle is shadowed in all camera images

is depicted in Fig. 6.8d. Here, one particle (shadowed) is aligned along the optical axis of each camera. Thus, the camera images do not allow one to judge whether the shadowed particle exists or not. Although this situation seems exotic, a careful inspection shows that shadowed particles have to be taken into account even for small dust clouds. Fortunately, the (thermal) motion of the particles allows to resolve this ambiguity if subsequent images are studied. However, especially for larger particle numbers or high particle densities an unambiguous position reconstruction becomes increasingly difficult.

To summarize, stereoscopy was shown to be a powerful tool. For small dust clouds, it yields good results at moderate computational efforts. It delivers the 3D trajectories of all particles with only a very few missing samples. However, its application becomes increasingly difficult for large clouds and high particle densities.

6.3.4 Digital Holography

Since most of the diagnostics discussed so far require lenses, it stands to reason to use an imaging system without lenses. A well-developed technique to encode 3D information in a single image is holography. In general, this method does not require any lenses for the imaging process and applications in other fields report promising results [25–27]. It was demonstrated that digital holography works for particle densities up to $10\,\mathrm{mm}^{-3}$ [28] and that the errors in particle position are just a few particle diameters if amplitude and phase information are used for particle detection [29]. Therefore, digital in-line holography (DIH) should be applicable to dusty plasmas as well. However, to apply holography to dusty plasmas additional problems have to be considered and solved. First, the particles are immersed in an incoherent light source, the plasma. Thus, it is questionable whether a sufficient signal to noise ratio can be achieved for a reliable hologram recording. Second, the particles are confined in a plasma chamber. This implies a number of technical challenges, that is, to cancel out vibrations and to minimize the disturbance by the plasma chamber which separates the particles and the optical setup. Third, a very short exposure time ($\tau \ll 1$ ms) is required to avoid that the thermal motion of the particles destroys the interference pattern. Finally, it has to be demonstrated that its spatial resolution can compete with other 3D diagnostics.

Lately, it was demonstrated by Kroll et al. that the mentioned obstacles can be overcome [21]. In his optical setup (Fig. 6.9), the reference wave is a plane wave realized by an expanded laser beam. It illuminates the spherical particles. Only a small fraction is scattered and forms the object wave. Since both have a fixed phase relation, their interference pattern is detectable with the sensor of a CCD camera. The typical hologram of a point-like object, which is a reasonable approximation for a spherical dust particle, is a Fresnel zone plate (FZP) plotted on the display sketched in Fig. 6.9a. It consists of concentric dark and bright rings. Figure 6.10a depicts a contrast-enhanced hologram of dust spheres with radius $r_{\mathrm{p}} = 10\pm2\,\mu\mathrm{m}$ inside a rf-discharge. The FZPs of individual dust particles are easily recognized with

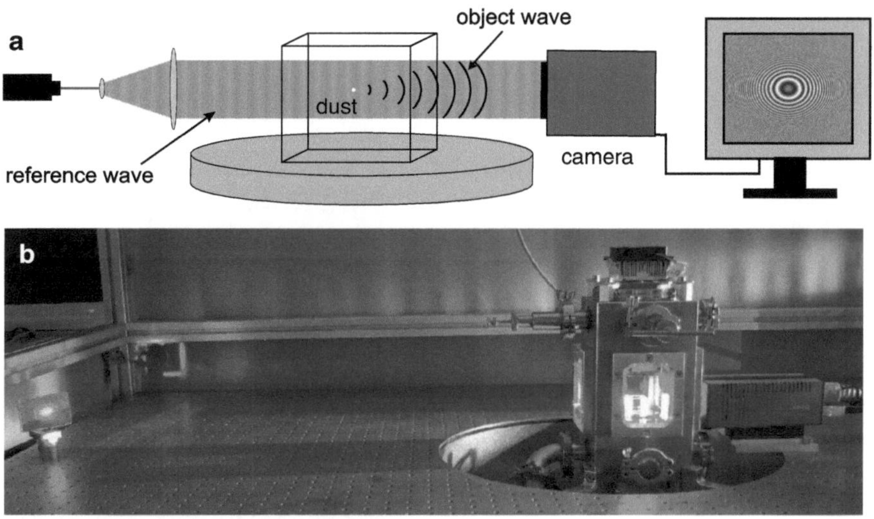

Fig. 6.9 (**a**) Digital in-line holography (DIH). A laser at 532 nm and a beam expander are used to illuminate the dust particles. The interference pattern of reference and object wave is recorded with a CCD sensor. To avoid saturation of the CCD sensor, a neutral density filter is mounted in front of the camera. (**b**) Picture of a DIH setup. The camera is located to the right directly next to the vacuum chamber. On the left, the bright spot on the mirror visualizes the diameter of the expanded laser beam. The laser and the beam expander are not shown

help of the schematic drawing below. To extract particle positions, it is necessary to reconstruct the particle cloud from the holographic recording. For a digital hologram the reconstruction is done numerically. Thus, amplitude and phase information of the reconstructed image can be used. Note that the latter is inaccessible in a classical holographic setup. The numerical reconstruction bases on Huygens principle. Every point of the hologram is regarded as the origin of an elementary wave whose amplitude is defined by the intensity of the hologram at this point. The superposition of all these elementary waves forms the reconstructed object. The mathematical formulation of this principle is the Fresnel–Kirchhoff integral [30]:

$$U(x,z) = \frac{\mathrm{i}}{\lambda} \iint_{-\infty}^{\infty} \mathrm{d}x'\mathrm{d}z'\, h(x',z') R^* \, \frac{\exp\left(-2\pi \mathrm{i}\rho/\lambda\right)}{\rho}, \tag{6.6}$$

with

$$\rho = \sqrt{(x'-x)^2 + (z'-z)^2 + d^2}. \tag{6.7}$$

Here $U(x,z)$is the reconstructed complex wave field and $h(x',z')$ is the recorded hologram. R^* is the complex conjugate of the reference wave R. We solve the Fresnel–Kirchhoff integral using a standard convolution approach [31]

$$U(x,z) = \mathcal{F}^{-1}\left\{\mathcal{F}(h \cdot R^*) \cdot \mathcal{F}(g)\right\}, \tag{6.8}$$

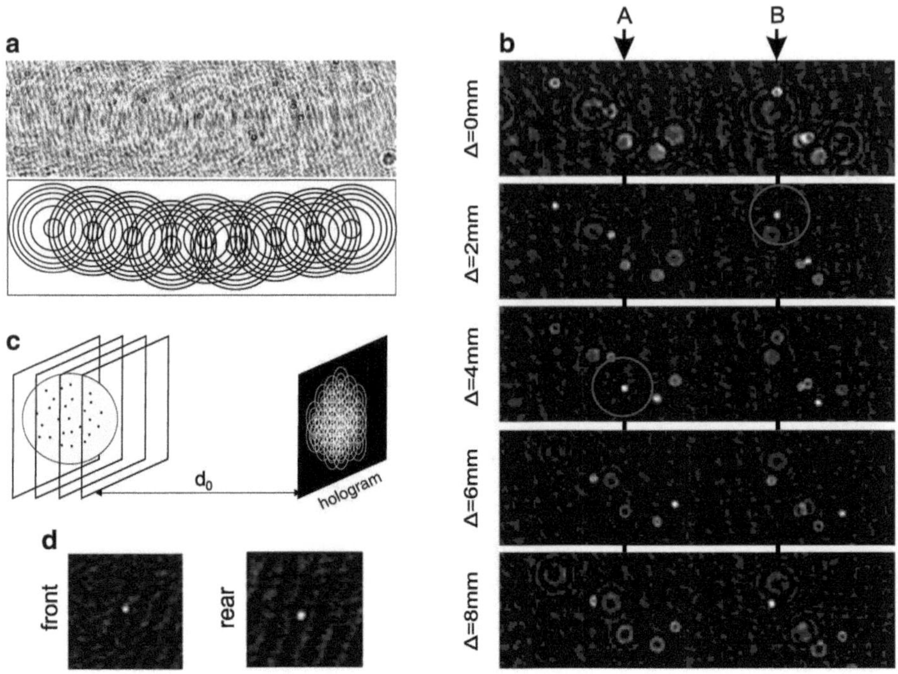

Fig. 6.10 Digital in-line holography of dusty plasmas: (**a**) Contrast-enhanced hologram of dust particles in a rf-discharge (*top*). A sketch of the hologram is plotted below. (**b**) Reconstruction of the hologram in a distance $d_0 + \Delta$. Two particles (A, B) are labeled and their focus planes are marked with circles. (**c**) Illustration of the reconstruction geometry. The reconstruction planes shown in (**b**) are indicated by *empty squares*. The dust cloud is symbolized by the *gray (yellow) sphere*. (**d**) Sharply focused particles are observed in reconstruction planes (front/rear) which are separated by 40 mm, that is, DIH overcomes the usual problem of a limited depth of field

where $\mathcal{F}$ denotes the Fourier transform. The convolution kernel is

$$g(x, z, x', z') = \frac{\mathrm{i}}{\lambda} \frac{\exp(-2\pi \mathrm{i}\rho/\lambda)}{\rho}. \tag{6.9}$$

A detailed description of this technique is given in [32]. The basic reconstruction scheme for an extended object is shown in Fig. 6.10c. The solution of (6.8) is the complex wave field at a certain distance d from the hologram plane. Thus, to reconstruct a 3D object, it is necessary to solve (6.8) for consecutive values of d. The interplane distance and hence the number of reconstruction planes has to be chosen with respect to the desired spatial resolution and the object dimensions. Typically 900–2,000 planes with an interplane distance of 10 μm are required. To get an image of the reconstructed object, the absolute value of the complex wave field has to be squared.

Images of reconstructed dust particles are shown in Fig. 6.10b. Here, $\Delta = 0$ mm denotes a reconstruction plane with $d = 18$ cm, which is close to the particle positions. Typical images of a focused particle are seen at position A ($\Delta = 4$ mm) and position B ($\Delta = 2$ mm). They appear as circular spots with a size of 3×3 pixels. Considering a pixel size of 7.4×7.4 μm and a particle size of (19 ± 2) μm, a particle image size of 3×3 pixel shows that the reconstructed image matches the original particle size. Thus, the spatial resolution within a reconstruction plane is defined by the resolution of the CCD sensor.

Looking at the particles A and B in neighboring reconstruction planes, a clear change of particle shape and brightness is observed. The defocused particles are ring shaped and their average intensity decreases. Even though the hologram (Fig. 6.10a) had a low contrast, the signal to noise ratio of the reconstructed images is fully sufficient for a reliable particle detection. The focused particles are 40 times brighter than the background. Hence, a standard particle detection can be performed to obtain the 2D positions of the particles in each reconstruction plane. Finally Fig. 6.10d shows that DIH overcomes the usual problem of a limited depth of field. Here two focused particles are plotted which are separated about 40 mm along the optical axis.

Using the shape and the intensity of the reconstructed particle images, the focal plane and thus the depth position of a particle can be extracted. However, this estimate is subject of rather large errors. An improved method, called "particle extraction using complex amplitude" (PECA), has been proposed by Pan and Meng [29]. This method uses the phase information of the complex wave field to determine the particle position. Figure 6.11 shows a typical reconstructed complex wave field around a particle position to illustrate the PECA method. Figure 6.11a–c show the real part and Fig. 6.11e–g the imaginary part of the complex wave field in direct vicinity of a particle. Figure 6.11b, f are taken at the focal plane of the particle, while Fig. 6.11a, e are taken 1 mm before, respectively, behind that focal

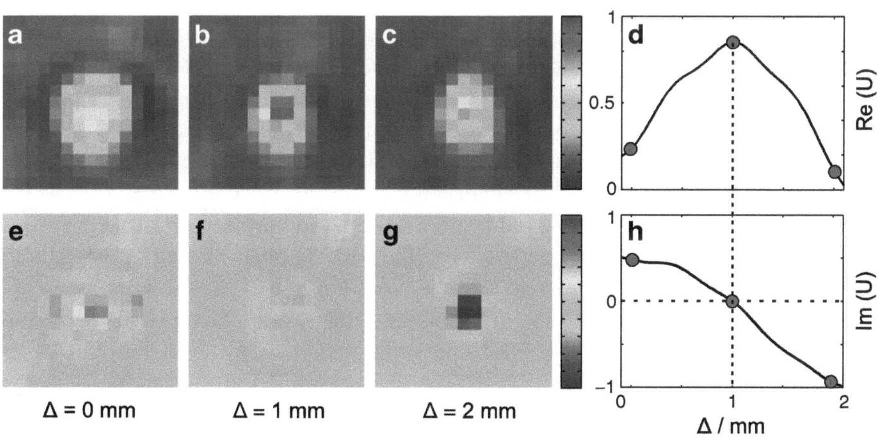

Fig. 6.11 Reconstruction of the complex amplitude with its (**a–c**) real part and (**e–g**) imaginary part at different reconstruction planes in a distance $d = d_0 + \Delta$. The real part and the intensity of a single particle of the cluster reach their maximum at the focal plane (**d**), whereas the imaginary part vanishes at the focal plane (**h**). This can be used to determine the depth position of particles

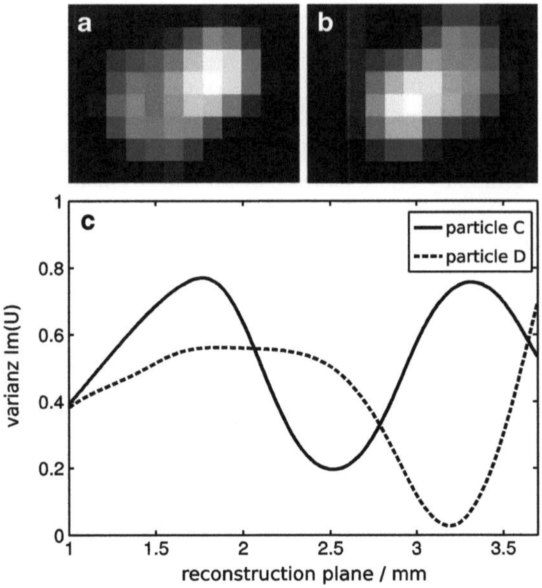

Fig. 6.12 Reconstruction of two particles which are aligned along the optical axes of the DIH setup. (**a**) Reconstruction of the hologram at the position of particle C. (**b**) Reconstruction at the position of particle D. Obviously, particles C and D are nearly perfectly aligned. (**c**) Determining the depth position of both particles using the imaginary part of the complex wave field U. The particle position is given by the minimum of the variance of $\text{Im}(U)$. Particle C is found at 2.5 mm and particle D at 3.2 mm

plane (Fig. 6.11c, g). Figure 6.11d shows the real part and Fig. 6.11h the imaginary part of the central pixel of the particle image in each plane. The dots denote the corresponding values for the planes in Fig. 6.11a–c. From Fig. 6.11, several observations can be made. First, the particle image is found smallest at the focal plane. Second, the real part has a clear maximum at the focal plane ($\Delta = 1$ mm) and the imaginary part changes sign. This zero crossing of the imaginary part is located at exactly the same plane as the maximum of the real part. The intensity curve gives only a broad maximum. As a consequence, an estimate of the position of a particle from the intensity is not favorable. The imaginary part is much better suited for an estimate of the depth position, because the determination of zero crossing provides higher accuracy than the analysis of the maximum of the real part or of the intensity. Further, this method even allows to separate particles which are aligned along the optical axis (Fig. 6.12). Thus shadowing is not an issue for DIH. Finally, the DIH setup does not require any calibration. The fixed wavelength of the laser and the lensless setup allow to reconstruct absolute particle positions with respect to the CCD sensor.

To benchmark the spatial resolution of the DIH system, Kroll et al. performed synchronous observations with DIH and video microscopy. In a first set of experiments just a few particles ($N < 10$) were confined. Top-view observations showed that the thermal motion of the particles is much smaller than the interparticle distance, that is, all particle arrangements were stable (Fig. 6.13a, d). Moreover, the particle arrangements are very regular. To determine the 3D structure of these clusters, 50 holograms with an exposure time of 100 μs were recorded and reconstructed. Figure 6.13b, e visualize the reconstruction result for clusters consisting of 4 and 7 particles. The average particle positions are indicated by circles. To visualize

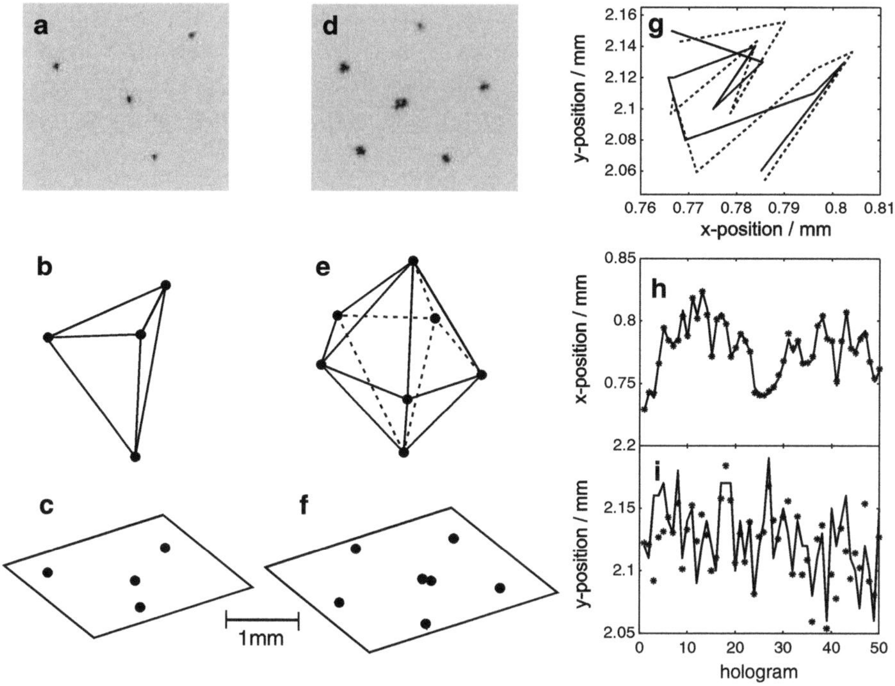

Fig. 6.13 Particle arrangement of small clusters ($N = 4$ and $N = 7$) measured simultaneously with DIH and a VM (*top view*). (**a, d**) Inverted video microscope (*top view*) images. (**b, e**) Average positions of the reconstructed clusters consisting of 4 (**b**) and 7 (**e**) particles. The marker size represents the root mean squared displacement of particles from their equilibrium position. For comparison, the top-view projections are shown below (**c, f**). (**g**) Top view of the trajectory of a particle (10 frames). The *solid line* shows the results of the DIH and the dashed line those of the video microscope (**h**) x-position of a particle in every frame. The comparison of the DIH results (—) and the video microscope (+) shows an average displacement of $\delta_x = 1.6\,\mu\text{m}$. (**i**) Same analysis for the y-position. Here, the average displacement is $\delta_y = 17\,\mu\text{m}$

the 3D arrangement, the particles are connected with lines. The projection into a horizontal plane is shown below in Fig. 6.13c, f. The marker size corresponds to the root mean squared displacement of the particles with respect to their equilibrium position due to thermal motion and due to the accuracy of our measurement. Further, a very good agreement is observed with Fig. 6.13a, c.

In a second set of experiments, Koll et al. quantified the accuracy of DIH. They traced the motion of a particle in a 2D plasma crystal with DIH and a video microscope. The perpendicular arrangement of a video microscope and a DIH system enabled the video microscope to image all particles while the DIH system imaged a side view of the dust layer. Thus, especially for the depth position the video microscope provides a reliable reference while the DIH system is operated at the most challenging observation angle. Evaluating the data of both diagnostic revealed a very good agreement. The trajectory of the particle over 10 frames is plotted in

Fig. 6.13g. Here, the solid line denotes the DIH result and the dashed line that of the video microscope. Both trajectories have striking similarities. The direction and the length of corresponding parts of the trajectory are similar. A quantitative analysis is provided by Fig. 6.13b, c, where the x-positions (perpendicular) and y-positions (along the optical axis) are compared separately. The solid lines mark the DIH result while the markers denotes the corresponding positions obtained from the video microscope data. The comparison shows an excellent agreement for the x-positions. The average displacement is just $\delta_x = 1.6\,\mu\text{m}$ which is well below the pixel size. For the y-position still an acceptable agreement is observed. On average, the particle position deviates just $\delta_y = 17\,\mu\text{m}$, which is still smaller than the particle diameter of $(19 \pm 2)\,\mu\text{m}$. Thus, the PECA method allows for a particle detection with the expected accuracy. The error along the optical axis is about one particle diameter and hence confirms the results reported by Pan and Meng [29]. Therefore, the measurements demonstrated that the spatial resolution of DIH is competitive.

However, DIH has its limits as well. First, the amplitude of the object wave is determined by the particle size. With the current setup Kroll et al. were able to record particles if their diameter is larger than $10\,\mu\text{m}$. Second, the temporal resolution of DIH is quite low. High resolution cameras with large sensors ($1''$) have frame rates below 10 Hz. And finally the hologram reconstruction and particle detection require high computational efforts.

6.4 Summary and Outlook

To conclude, imaging diagnostics in dusty plasmas has made considerable progress in the last decade. A number of diagnostics for 3D dust clouds were developed and successfully applied. This review has focused on four of them: scanning video microscopy, color gradient methods, stereoscopy, and digital holography. All of them are valuable tools to investigate 3D strongly coupled particle arrangements, although each of them has different limitations. However, many of the limitations are complementary and not yet finally tested. Thus it stands to reason that future developments in the field of imaging diagnostics will lead to diagnostic tools which are as versatile in 3D as the video microscopes are in 2D. One possibility is a combination of digital holography and stereoscopy. A combined system promises to avoid shadowing effects and sophisticated camera models and should be able to yield subpixel resolution in all directions at moderate computational efforts. First test in this direction are encouraging.

References

1. H.M. Mott-Smith, I. Langmuir, Phys. Rev. **28**, 727 (1926)
2. G.E. Morfill, H.M. Thomas, U. Konopka, H. Rothermel, M. Zuzic, A. Ivlev, J. Goree, Phys. Rev. Lett. **83**, 1598 (1999)

3. A. Bouchoule, *Dusty Plasmas: Physics, Chemistry, and Technological Impacts in Plasma Processing* (Wiley, New York, 1999)
4. G.E. Morfill, V.N. Tsytovich, H. Thomas, *Elementary Physics of Complex Plasmas (Lecture Notes in Physics)* (Springer, Berlin, 2007)
5. C.F. Bohren, D.R. Huffman, *Absorption and Scattering of Light by Small Particles* (Wiley-VCH, Berlin, 1998)
6. A. Hohmann, A. Melzer, S. Peters, A. Piel, Phys. Rev. E. **56**, 7138 (1997)
7. Y. Feng, J. Goree, B. Liu, Rev. Sci. Instrum. **78**, 053704 (2007)
8. L.-W. Teng, C.-Yu. Tsai, Y.-P. Tseng, I. Lin, in *AIP Conference Proceedings of ICPDP5*, Azores, 2008
9. V. Nosenko, J. Goree, A. Piel, Phys. Plasmas **13**, 032106 (2006)
10. Y. Ivanov, A. Melzer, Rev. Sci. Instrum. **78**, 033506 (2007)
11. S. Nunomura, J. Goree, S. Hu et al., Phys. Rev. Lett. **89**, 035001 (2002)
12. Y. Feng, J. Goree, B. Liu, Rev. Sci. Instrum. **78**, 053704 (2007)
13. J.C. Crocker, D.G. Grier, J. Colloid. Interface Sci. **179**, 298 (1996)
14. M. Kvarnstrom, Ph.D. Thesis, Chalmers University of Technology and Göteborg University, Göteborg, Sweden (2005)
15. J.B. Pieper, J. Goree, R.A. Quinn, Phys. Rev. E **54**, 5636 (1996)
16. M. Zuzic et al., Phys. Rev. Lett. **85**, 4064 (2000)
17. O. Arp, D. Block, A. Piel, A. Melzer, Phys. Rev. Lett. **93**, 165004 (2004)
18. B.M. Annaratone et al., Plasma Phys. Control. Fusion **46**, B495 (2004)
19. S. Käding, A. Melzer, Phys. Plasmas **13**, 090701 (2006)
20. E. Thomas, J. Williams, J. Silver, Phys. Plasmas **11**, L37 (2004)
21. M. Kroll, S. Harms, D. Block, O. Arp, A. Piel, Phys. Plasmas **15**, 063707 (2008)
22. D.D. Goldbeck, Ph.D. Thesis, Ludwig Maximilian University of München (2003)
23. T. Antonova, B.M. Annaratone, D.D. Goldbeck, V. Yaroshenko, H.M. Thomas, G.E. Morfill, Phys. Rev. Lett. **96**, 115001 (2006)
24. S. Käding, A. Melzer, O. Arp, D. Block, A. Piel, AIP Conf. Proc. **799**, 335 (2005)
25. K.D. Hinsch, Meas. Sci. Technol. **13**, R61 (2002)
26. F. Dubois, L. Joannes, J. Legrosm, Appl. Opt. **38**, 7085 (1999)
27. G. Indebetouw, W. Zhong, Opt. Soc. Am. A **23**, 1699 (2006)
28. Y. Pu, H. Meng, Appl. Opt. **44**, 7697 (2005)
29. G. Pan, H. Meng, Appl. Opt. **42**, 827 (2003)
30. M.V. Klein, T.E. Furtak, *Optics*, 2nd edn. (Wiley, New York, 1986)
31. T.H. Demetrakopoulos, R. Mittra, Appl. Opt. **13**, 665 (1974)
32. U. Schnars, W. Jueptner, *Digital Holography* (Springer, Berlin, 2005)

Chapter 7
Structure and Dynamics of Finite Dust Clusters

André Melzer and Dietmar Block

Abstract The structure of matter is widely determined by the action of electrostatic (Coulomb) forces. Systems with only a small number of trapped charge carriers, so-called *Coulomb clusters*, have a particular charm since their structural and dynamic properties strongly depend on the exact particle number. Dusty plasmas provide a unique possibility to trap, manipulate, and observe dust Coulomb clusters in different confinement geometries with high spatial and temporal resolution.

7.1 Introduction

Coulomb clusters are systems of a small number of charged particles trapped in an external potential. As early as 1904, such systems have been analyzed by J.J. Thomson as a model for the structure of the atom [1]. Coulomb clusters are particularly appealing because of the interplay between the Coulomb repulsion among the particles and the confinement due to an external potential. This interplay determines the structure of a cluster and also its dynamics. Structure and dynamics dramatically depend on the exact particle number N, and the formation of highly symmetric configurations with "magic" particle numbers is observed in a wide variety of situations in the classical and quantum world.

Examples of Coulomb clusters of present-day interest are found as ions in Paul and Penning traps, ions in storage rings of accelerators, electrons on the surface of

A. Melzer (✉)
Institut für Physik, Ernst-Moritz-Arndt-Universität Greifswald, Felix-Hausdorff-Str. 6, 17489 Greifswald, Germany
e-mail: melzer@physik.uni-greifswald.de

D. Block
Institut für Experimentelle und Angewandte Physik, Christian-Albrechts-Universität Kiel, Olshausenstr. 40-60, 24098 Kiel, Germany
e-mail: block@physik.uni-kiel.de

M. Bonitz et al. (eds.), *Introduction to Complex Plasmas*, Springer Series on Atomic, Optical, and Plasma Physics 59, DOI 10.1007/978-3-642-10592-0_7,

liquid helium, electrons in quantum dots, brown dwarf starts, charged particles in colloidal suspensions [2], or, as will be discussed here, charged dust particles in dusty plasmas [3–5].

The governing parameter in charged-particle systems is the Coulomb coupling parameter:

$$\Gamma = \frac{Z^2 e^2}{4\pi\epsilon_0 b k_B T}, \tag{7.1}$$

where Z is the particle charge number, b the nearest-neighbor distance, and T the particle temperature. Γ describes the ratio of Coulomb (electrostatic) and thermal (kinetic) energy. When the Coulomb energy dominates over the thermal energy the system is said to be strongly coupled. Crystallization of the particles into an ordered arrangement is found when the coupling parameter exceeds a critical value Γ_c (which is found to be $\Gamma_c \approx 170$ for 3D infinite systems).

Hence, laser-cooled ions in traps are found to crystallize at temperatures T of the order of millikelvin. In the great planets or in dwarf stars the charges are compressed to small interparticle distances b (then, however, also quantum effects become relevant). Most effectively, strong coupling is realized by a high particle charge Z. This is exploited in colloidal suspensions and, as we will show in the following, in dusty plasmas.

Dusty plasmas, that is, particles ("dust") immersed in a gaseous plasma environment, are found in a wide variety of systems. Dusty plasmas are ubiquitous in astrophysical situations, like planetary rings, comets, or interstellar clouds [6, 7]. Dust in plasmas plays also a crucial role in industrial processes such as thin film deposition or plasma etching (see Chap. 16). There, typically, the removal of the dust particles from the process chamber is a critical issue. Here, however, dusty plasmas provide an ideal tool to study fundamental properties of the structure and dynamics of crystallized dust systems.

The dust particles in the plasma attain high negative charges due to the continuous inflow of plasma electrons and ions. Typical charges are of the order of 10,000 elementary charges for a 10 μm dust particle. Due to their high charge the dust particles crystallize into ordered arrangements even at room temperature forming so-called *plasma crystals* [8–10]. The spatial and time scales of the particle motion (particle distances of the order of hundreds of microns and wave frequencies of the order of 10 Hz) allow easy observation by video microscopy. The only weak frictional damping due to the ambient neutral gas ensures that the dynamics and kinetics of individual particles become observable. Thus, dusty plasmas enable the investigation of crystal structure, solid and liquid plasmas, phase transitions, waves, and many more phenomena on the kinetic level of individual particles.

7.2 Trapping of Dust Clouds

In the plasma, the dust particles are subject to a number of forces (see, e.g., [11,12]). For the micrometer-sized particles typically used in studies on fundamental properties of dusty plasmas gravity and electric field forces are dominant. The gravitational force is simply given by

$$\boldsymbol{F}_{\mathrm{g}} = m_{\mathrm{d}}\boldsymbol{g} = \frac{4}{3}\pi a^3 \rho_{\mathrm{d}}\boldsymbol{g}, \tag{7.2}$$

where $\boldsymbol{g}$ is the gravitational acceleration, m_{d} the dust mass, a the dust radius, and ρ_{d} the mass density of the dust. The electric field force is

$$\boldsymbol{F}_{\mathrm{E}} = -Ze\boldsymbol{E}, \tag{7.3}$$

where $\boldsymbol{E}$ is the electric field and e is the elementary charge (it is assumed here that the dust is negatively charged). Furthermore, the thermophoretic force is exploited in our experiments. Thermophoresis is due to a temperature gradient ∇T_{n} in the neutral gas background and drives the particles toward colder gas regions. It is given by

$$\boldsymbol{F}_{\mathrm{th}} = -\frac{16}{15}\sqrt{\pi}\frac{a^2 k_{\mathrm{n}}}{v_{\mathrm{th,n}}}\nabla T_{\mathrm{n}}, \tag{7.4}$$

where $k_{\mathrm{n}}, v_{\mathrm{th,n}}$ is the thermal conductivity and the thermal velocity of the neutral gas, respectively. Another important force usually is the ion drag force F_{ion} which is a drag force due to a directed ion flow. This force, however, is not explicitly exploited in the formation of dust Coulomb clusters. By tailoring the other forces, finite and extended dust structures can be formed.

Extended flat (2D) dust crystals are formed in the space charge sheath above the flat electrode of parallel plate RF discharges [9, 10, 13, 14]. In these discharges, usually the sheath electric field is strong enough to balance the weight of the particles. Since the mean electric field in the sheath generally increases toward the electrode a force balance is achieved at a unique vertical position in the sheath. Due to the strong vertical forces the dust is confined to a flat, but horizontally extended structure.

These 2D systems preferably arrange in a hexagonal structure [9, 10, 13–15] which is the ground-state configuration of 2D crystals. In multilayer systems also other crystal structures like fcc or bcc have been identified [13, 16, 17]. Moreover, vertically aligned structures due to the directed ion streaming motion in the sheath are frequently observed [15, 18]. Phase transitions [19, 20] and various types of waves have been studied in 2D extended crystals (see, e.g., [21] and references therein).

Extended three-dimensional (3D) dust structures are found when dusty plasmas are investigated under microgravity conditions aboard parabolic flights [22–26] or on the International Space Station [27–31]. There, the particles are confined to the plasma bulk by the much weaker ambipolar electric field. However, the central plasma region is found to be dust-free. The dust is expelled from this so-called *void* and a sharp boundary between the void and the dust cloud is formed (see, e.g., [22]). The void boundary is generally considered to be due to the force balance of the inward electric field force and the outward ion drag force [32–35]. In the 3D dust cloud, for example, the structure [36], charging processes [37, 38], forces [26], and self-excited waves have been investigated [24, 29].

7.3 Formation of Finite Dust Clusters

Our experiments on fundamental properties of dust clusters are performed in a capacitively coupled RF discharge (13.56 MHz) between parallel plates. The discharge is operated in argon at gas pressures between 1 and 100 Pa with RF powers of 1–10 W. Monodisperse plastic microspheres of micrometer diameter are dropped into the discharge. The trapped particles are illuminated by laser beams. The particle motion is recorded with high-speed video cameras. From the video images the positions of all particles are determined and the particles are tracked through the video sequence (see Chap. 6) resulting in detailed information on the particle structure and dynamics.

Finite dust Coulomb clusters are formed by trapping only a small number of dust particles into a confinement potential provided by a combination of the earlier described forces. Such a finite system is then described by the energy

$$E = \sum_{i=1}^{N} V(x_i, y_i, z_i) + \frac{Z^2 e^2}{4\pi\epsilon_0} \sum_{i>j}^{N} \frac{\exp(-r_{ij}/\lambda_D)}{r_{ij}}, \tag{7.5}$$

where $\boldsymbol{r}_i = (x_i, y_i, z_i)$ are the particle coordinates and $r_{ij} = |\boldsymbol{r}_i - \boldsymbol{r}_j|$ is the distance between particle i and j. The first term is the potential energy of the particles in the confinement V. The second term denotes the electrostatic repulsive energy. Here, a shielded Coulomb (Debye–Hückel or Yukawa) interaction with a screening length λ_D is considered. The screening strength $\kappa = b/\lambda_D$, the interparticle distance in units of the shielding length, is used to describe shielding effects.

The confinement potential energy V is often considered as harmonic, that is,

$$\begin{aligned} V(x_i, y_i, z_i) &= \frac{1}{2} m_d \omega_x^2 x_i^2 + \frac{1}{2} m_d \omega_y^2 y_i^2 + \frac{1}{2} m_d \omega_z^2 z_i^2 \\ &= \frac{1}{2} m_d \omega_0^2 \left(x_i^2 + \beta_y y_i^2 + \beta_z z_i^2 \right). \end{aligned} \tag{7.6}$$

Here, $\omega_{x,y,z}$ is the confinement strength in x, y, z-direction and $\beta_{y,z} = \omega_{y,z}^2/\omega_x^2$ is the relative strength of the confinement with respect to the confinement in x-direction ($\omega_0 = \omega_x$). By changing the relative confinement strengths $\beta_{x,y}$ various confinement geometries from 1D to 2D and 3D can be realized. How this is done in the experiment will be demonstrated below.

7.4 Structural Transitions in 1D Dust Clusters

In the experiment described here, linear (1D) dust clusters are generated by placing a rectangular metal barrier (6 mm height and $5 \times 40\,\text{mm}^2$ inner dimension) onto the lower electrode (see Fig. 7.1). A small number of particles ($N = 1$–20) of 10.2 μm diameter (and a mass of $m_d = 8.41 \times 10^{-13}$ kg) are dropped into the barrier (see also [39]).

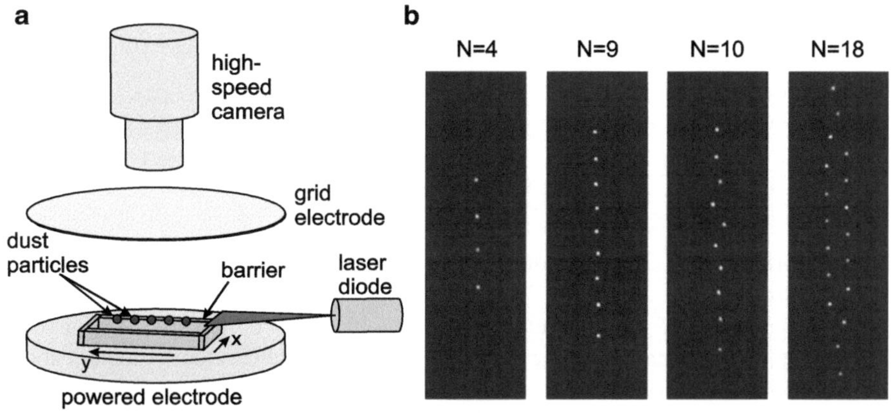

Fig. 7.1 (**a**) Experimental setup for the confinement in 1D a dust cluster. A rectangular barrier is placed onto the lower electrode. The dust particles are illuminated by a laser sheet and the particle motion is recorded by video cameras. (**b**) Snap shots of the 1D dust cluster for $N = 4$, 9, 10, and 18. A structural transition in the cluster is seen by increasing the particle number. From 9 to 10 particles a zigzag transition occurs. The size of each image corresponds to $2.8 \times 9.1\,\text{mm}^2$ (from [39])

Vertically, the particles are strongly confined due to the balance of electric field force and gravity (i.e., $\beta_z \gg 1$). Horizontally, the particles are confined due to the barrier on the electrode: The barrier distorts the electrostatic equipotential lines in the plasma sheath above the electrode and thus forms an anisotropic confinement. Since the elongation of the barrier is much larger in y-direction than in x-direction the particles are only weakly confined along y. Thus, $\beta_y \ll 1$ and the particles arrange along y in a linear 1D dust cluster.

Figure 7.1b shows the arrangement of the particles in the confinement provided by the barrier. There $N = 4$ to $N = 18$ particles are trapped above the barrier. It is seen that for $N = 4$ to $N = 9$ the particles strictly arrange in a linear arrangement. When, however, the particle number is increased from 9 to 10 a zigzag transition in the center of the chain occurs. For 18 particles a zigzag structure is seen nearly throughout the entire chain.

The reason for that is easily understandable: the confinement in the y-direction compresses the chain along its extension. The interparticle distance is smallest in the central part of the chain. When more and more particles are inserted into the chain the compression increases until it is easier for the central particles to make a transverse excursion (in the x-direction). Then, the force along y due to compression exceeds the force in x from the confinement.

The onset of the zigzag transition allows to determine the particle and plasma conditions: the particle charge is found as $Z = 11{,}000$, the screening length as $\lambda_D = 1{,}000\,\mu\text{m}$, which is twice the interparticle distance of $b = 500\,\mu\text{m}$. The ratio of longitudinal to transverse confinement is determined as $\beta_y = \omega_y^2/\omega_x^2 = 4\times10^{-3}$. Exactly then, a zigzag transition is expected to occur between 9 and 10 particles in the trap [39].

7.5 Structure of 2D Dust Clusters

For the formation of 2D finite dust clusters an electrode with a shallow circular parabolic trough is used (see Fig. 7.2). A few ($N = 3$–150) melamine-formaldehyde (MF) particles of 9.55 μm diameter and a mass of $m_{\mathrm{d}} = 6.90 \times 10^{-13}$ kg are immersed into the plasma where the particles form 2D finite Coulomb clusters above the electrode. In the experiment, it was confirmed from the side view camera that the microspheres are indeed trapped into only a single layer. With this confinement a 2D isotropic situation with $\omega_y = \omega_x$ (i.e., $\beta_y = 1$) is realized. The vertical confinement is still due to the balance of electric field force and gravity with $\beta_z \gg 1$.

These 2D clusters have a well-defined structure: the particles arrange on circular shells [3, 4]. For example, with $N = 3$ only a single shell is found, whereas $N = 7$ has a single shell with a central particle, $N = 19\,(1, 6, 12)$ has two shells with a central particle and $N = 34$ has three shells with a configuration $(1, 6, 12, 15)$. The cluster configuration is usually given as $(N_1, N_2, \ldots)$ where N_1 is the number of particles in the inner shell, N_2 that of the second shell, etc. Due to the obvious analogy the 2D clusters are said to form a "periodic table" of clusters.

The cluster structure very well reflects the interplay of confinement and Coulomb interaction. The confinement introduces a circular order whereas the Coulomb force favors an hexagonal order (in 2D). Larger clusters show in their central region already the dominance of the 2D hexagonal order which in the outer parts has to adjust to the circular boundary (see also [41]).

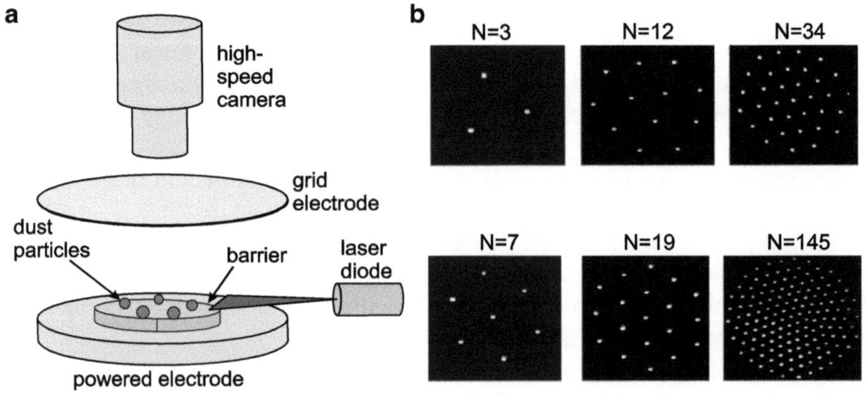

Fig. 7.2 (**a**) Experimental setup using an electrode with a parabolic trough for the formation of 2D clusters. (**b**) Snap shots of the 2D dust clusters with $N = 3$, 7, 12, 19, 34, and 145 particles (from [40])

7.6 Normal Mode Dynamics of Dust Clusters

The particle dynamics in the cluster can be studied in very detail on the kinetic level of individual particles [40]. The dynamical properties of strongly coupled finite systems are described by their normal modes. The normal modes are derived from the eigenvalues and eigenvectors of the dynamical matrix [42]:

$$A_{ij} = \frac{\partial^2 E}{\partial r_{\alpha,i} \partial r_{\alpha',j}}, \tag{7.7}$$

where $r_{\alpha,i}, r_{\alpha',i}$ is the x, y (x, y, z for 3D systems) of the ith particle. This yields a $2N \times 2N$ matrix ($3N \times 3N$ for 3D) resulting in $2N$ ($3N$) normal modes. The eigenvalues determine the mode frequencies and the eigenvectors describe the mode oscillation pattern.

In finite systems, the normal modes take the role of the dispersion relation of infinite systems. Thus, they are the central instrument to describe the dynamics of finite systems. Due to the boundary, normal modes are not purely compressional or shear modes, but can be described as compression-like or shear-like depending on their respective contribution to the mode pattern (see, e.g., [42]).

An alternative method to extract dynamical information is the singular value decomposition (SVD). There, the information is derived purely from the particle trajectories without the need for a model equation (e.g., (7.5)). Since no model is associated with the trajectories also no physical quantities can be derived from SVD. Nevertheless, it has been proven a valuable technique for the description of particle dynamics in dusty plasmas (see, e.g., [43, 44]). In the following, however, we will concentrate on the normal modes only.

Figure 7.3 shows the normal mode spectrum of the 2D cluster with $N = 19$ particles (see Fig. 7.2b) and its 38 modes. A few modes are highlighted: first, the rotation of the entire cluster at a frequency $\omega = 0$, since there are no restoring forces for a rotation in an isotropic confinement. Second, the center-of-mass mode (sloshing mode) where all particles oscillate together around the trap center is twofold degenerate (in a 2D system) and is found at the frequency of the confinement $\omega = \omega_0$. A third mode is the so-called *breathing mode* where the particles oscillate coherently in radial direction. With pure Coulomb interaction the frequency of this breathing mode is found at $\omega = \sqrt{3}\omega_0$, independent of N. The frequency of the breathing mode increases with screening and becomes (weakly) dependent on N (also the purely radial motion is affected upon screening [45]).

Finally, the intershell rotation is mentioned, here. There, two shells of the cluster rotate with respect to each other. This mode usually is the mode of lowest frequency (besides the global rotation) for medium-sized 2D clusters. The modes with the lowest frequency are those with the smallest restoring forces. Thus, such modes determine the stability of a cluster and are therefore of particular importance. This intershell rotation clearly is of shear-like nature.

Experimentally, the dynamic properties of the cluster can be derived from the thermal Brownian motion of the particles around their equilibrium positions [40]. Therefore, the thermal fluctuations with the respective velocity v_i of the ith particle are projected onto each of the normal modes ($\boldsymbol{v}_i(t) \cdot \boldsymbol{e}_{i,\ell}$, where $\boldsymbol{e}_{i,\ell}$ is the

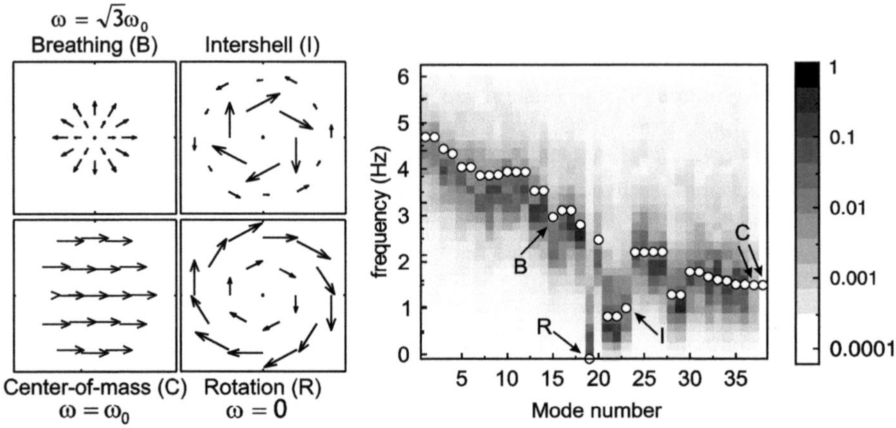

Fig. 7.3 Spectral power density (in *gray scale*) of the 38 modes of a 2D cluster with $N = 19$ particles. The *open circles* indicate the theoretically expected mode frequencies. *Left*: the oscillation pattern of four modes of special interest are indicated (from [40])

eigenvector of particle i in mode number ℓ). From that the spectral power density of the fluctuations "along" the normal modes are determined from

$$S_\ell(\omega) = \frac{2}{T}\left|\int_0^T \mathrm{d}t \sum_{i=1}^{N} \left(\boldsymbol{v}_i(t)\cdot \boldsymbol{e}_{i,\ell}\right) \mathrm{e}^{\mathrm{i}\omega t}\right|^2 . \tag{7.8}$$

The spectral power density $S_\ell(\omega)$ of all modes ℓ gives the mode spectrum of a cluster as shown in Fig. 7.3 for the 19-particle cluster. A clear mode spectrum for the cluster is observed. The spectral power density is concentrated only in a narrow stripe of frequencies.

As discussed earlier, the theoretical frequencies that follow from the eigenvalues of the dynamical matrix contain as the single unknown parameter for all modes only the strength of the confinement ω_0. By fitting this parameter, all theoretical modes are found to very well agree with the experimentally derived power spectrum: Of the four described normal modes, the breathing mode has the highest frequency, followed by the center-of-mass mode. The intershell rotation is found at decisively lower frequencies, it almost has the lowest frequency of all modes (besides rotation). This shows that shear-like modes, like in real solid matter, determine the stability of the cluster. Having determined ω_0 the particle charge and screening strength can be extracted. Here, the charge is found to be $Z = 10{,}000$ for $\kappa = 0, \ldots, 2$ [40]. These values are in excellent agreement with those obtained from active excitation of normal modes [46].

The influence of the structure on the dynamical properties becomes obvious when the breathing mode and the intershell rotation are compared for three similar clusters, namely the $N = 19$ cluster in ground-state configuration $(1, 6, 12)$ and in metastable state with $(1, 7, 11)$ as well as the $N = 20$ cluster with ground

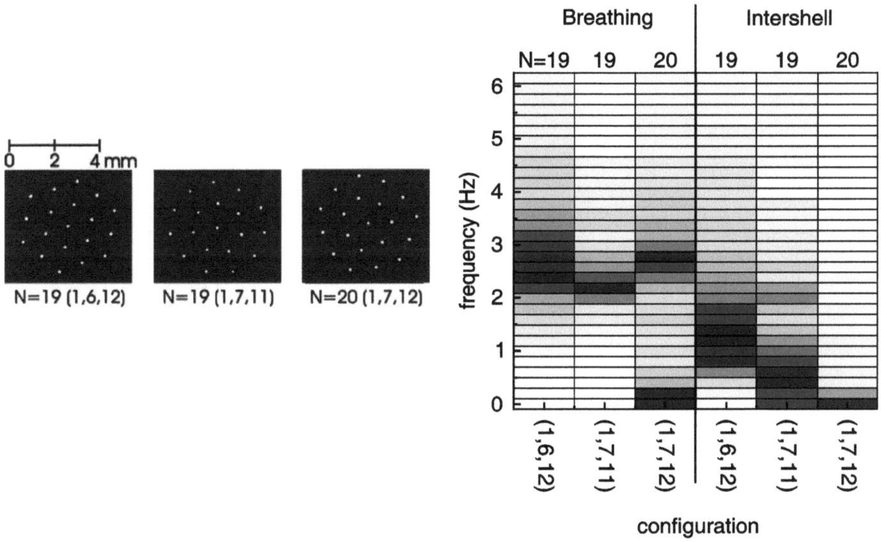

Fig. 7.4 Breathing mode and intershell rotation mode for the clusters $N = 19$ in ground state $(1, 6, 12)$ and metastable $(1, 7, 11)$ configuration as well as for the cluster $N = 20$ in ground-state configuration $(1, 7, 12)$ (from [40])

state $(1, 7, 12)$ (see Fig. 7.4). The breathing mode is found to be nearly the same for all clusters at a frequency of about 2.5–3 Hz. The intershell rotation frequency, however, strongly depends on the structure. In the $(1, 6, 12)$ configuration the intershell rotation frequency is highest while for the other two configurations it is decisively less at much lower frequencies. This shows that the $N = 19\,(1, 7, 11)$ and $N = 20\,(1, 7, 12)$ configurations are relatively unstable against this mode. The reason is easy to understand: the $(1, 6, 12)$ configuration has a commensurate number of particles in inner and outer shell. The shells are locked with respect to each other and a differential rotation is difficult to excite. The other two configurations, however, have incommensurate numbers in inner and outer ring and a locking of the shells does not occur. An intershell rotation can be easily driven. Thus, configurations with "magic" numbers in the shells also show a pronounced dynamic stability.

7.7 Formation of 3D Dust Clusters

Three-dimensional dust clusters have been produced in a novel setup that has been developed recently [5]. With this, it has become possible to investigate these interesting systems. Figure 7.5 shows the setup for the confinement of the 3D dust clusters. There, the gravitational force is to a large extent balanced by the thermophoretic force. The thermophoretic force is provided by heating the lower electrode to a temperature of 50–80°C. With this, the particle cloud is pushed more

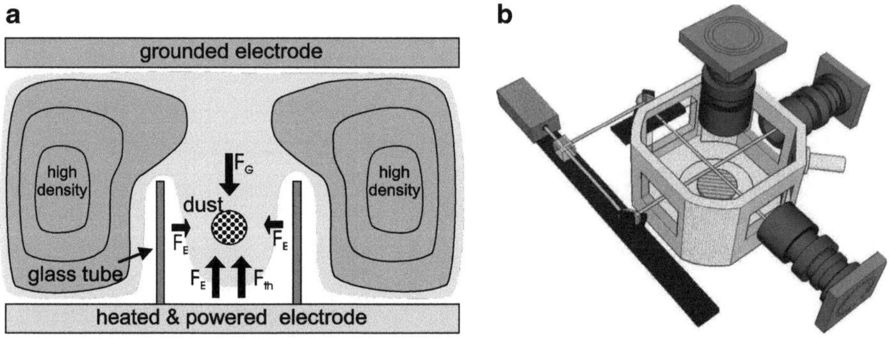

Fig. 7.5 (**a**) Experimental setup for the formation of 3D clusters. (**b**) Stereoscopic imaging system with three orthogonal cameras (after [47, 48])

toward the bulk plasma. There the weaker electric field together with thermophoresis balances gravity. Hence, the vertical confinement is much weaker than in the previous cases. The horizontal confinement is provided by an open glass box. The dielectric glass walls provide an inward electric field force on the dust. By tuning these forces a 3D isotropic potential well for the dust particle is achieved [47], where $\omega_x = \omega_y = \omega_z$ or $\beta_y = \beta_z = 1$.

Since the observation of 3D dust clusters is much more difficult more effort is required to obtain the full 3D particle trajectories (see Chap. 6). Static configurations can be observed by scanning video microscopy where successive 2D cross sections of the dust clouds are recorded and the 3D information is reconstructed from these slices [47]. For a simultaneous observation of the particles a stereoscopic imaging system with three orthogonal cameras is used [48–50] (see also Fig. 7.5b). An alternative method is the digital holography that has recently been applied to finite dust clusters [51]. Moreover, color-gradient techniques have been applied to study 3D dust clusters [52, 53].

In this confinement, 3D finite dust systems are confined by inserting plastic microspheres of 3.46 μm in diameter (mass $m_d = 3.28 \times 10^{-14}$ kg). The plasma is operated in argon at a gas pressure of 90 Pa and a discharge power between 2 and 7.5 W.

Figure 7.6 shows typical video images from the three-camera stereoscopic setup. The cameras are oriented pairwise perpendicular to each other to keep the observation geometry as simple as possible. At the same time, the spatial resolution in the three dimensions is assured to be equal. Using three cameras also reduces the problem of overlapping particle images that often leads to ambiguities if only two cameras are used: each particle should be visible in at least two cameras which helps to reconstruct the 3D positions of all particles [48].

In the first step, the 2D image coordinates of all particles are determined from the two-dimensional images using standard particle identification techniques (see, e.g., [54, 55] and Chap. 6). This identification is performed for each frame of the three cameras separately. To reconstruct the 3D particle positions and

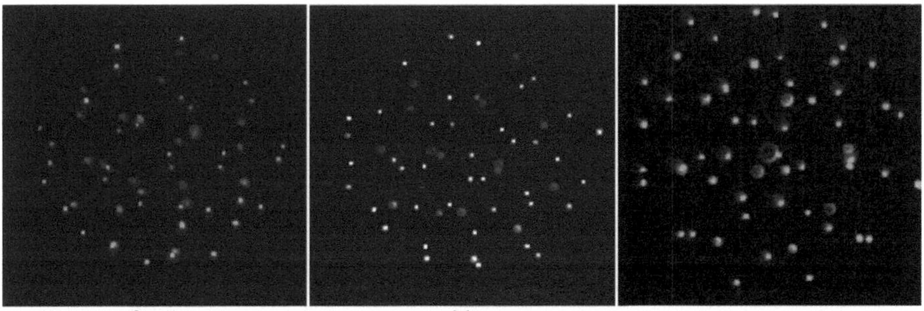

Fig. 7.6 Raw video still images of a cluster with $N = 91$ recorded by the three cameras. All particles are seen in at least two cameras (from [48])

trajectories, respectively, we rely on the perpendicular optical axes and neglect any distortion due to the imperfections of the camera lenses and windows of the vacuum vessel. For identifying corresponding particles in the different camera images, we exploit that due to the orthogonal camera arrangement each pair of cameras shares a common coordinate. Having determined corresponding particles it is easy to derive its 3D position.

7.8 Structure of 3D Dust Clusters

Figure 7.7 shows a 3D dust cluster with $N = 190$ particles. There, the particle coordinates are shown in cylindrical coordinates ρ and z, where $\rho = (x^2 + y^2)^{1/2}$. From that it is easily seen that the cluster consists of nested concentric shells. For the $N = 190$ cluster, four shells with a configuration $(2, 21, 60, 107)$ are observed. Similarly to the 2D clusters the external confinement requires the spherical arrangement of the shells. Unlike for the 2D systems, a volume order, like bcc or fcc, is not observed in the central part of the cluster.

On the shells, the particles are highly ordered. For that purpose, a Voronoi analysis of the structure on the outermost shell ($M = 4$) and on the next inner shell ($M = 3$) has been performed (see Fig. 7.7b, c). It is found that the preferred hexagonal order on the shell is interrupted by inserting pentagons, like on a soccer ball or a fullerene molecule. These pentagons are required to bend a hexagonal structure onto a sphere.

A selection of smaller clusters ($N < 100$) is shown in Fig. 7.8. There, a 3D reconstruction of the structure is shown. Also these clusters arrange in nested spherical shells ranging from one shell with central particle ($N = 17$) to three shells ($N = 91$). From this 3D bond structure also the highly ordered arrangement is readily seen. Analogously to the 2D case, also here a "periodic table"-like construction of the clusters is observed. Typically, a new shell opens up, when the inner shell has 12 particles [56, 57].

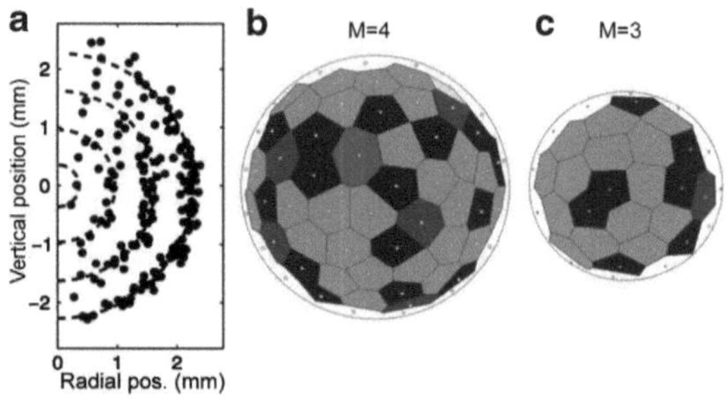

Fig. 7.7 (**a**) Particle positions of a cluster with $N = 190$ particles in cylindrical projection. The cluster consists of four concentric shells. (**b, c**) Voronoi analysis of the outermost ($M = 4$) and next inner ($M = 3$) shell of the cluster: the shell consists of pentagons (*dark gray*), hexagons (*light gray*), and defects (*medium gray*) (from [5])

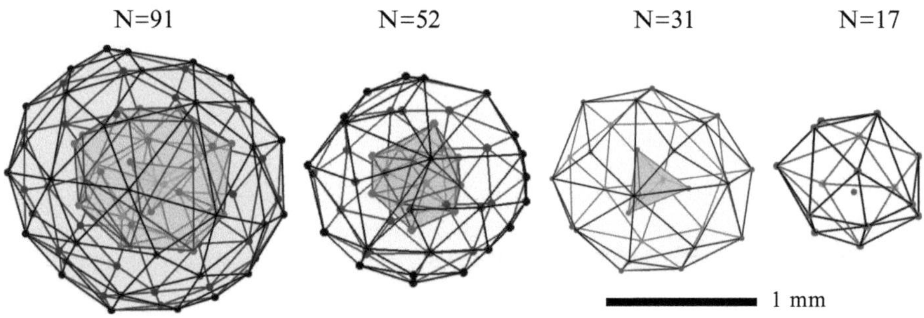

Fig. 7.8 Cluster configurations reconstructed from single video snap shots with $N = 91, 52, 31$, and 17. The Yukawa balls consist of concentric shells with the configurations $(4, 25, 62)$ for the $N = 91$, $(11, 41)$ for the $N = 52$, $(5, 26)$ for the $N = 31$, and $(1, 16)$ for the $N = 17$ cluster (from [48])

It is now interesting to analyze the occupation number N_s on the different shells as a function of total particle number. Figure 7.9 shows the occupation numbers of various experimentally observed clusters together with occupation numbers determined from MD simulations [58] for pure Coulomb interaction and screened interaction ($\kappa = 0.6$) (see also Chaps. 8 and 10). It is seen that in the experiment more particles are found on the inner shells on the expense of the outer shells as compared to a pure Coulomb interaction. The experiments clearly show a much better agreement with occupation numbers with a screened interaction. The particle interaction is therefore more adequately described by a shielded Yukawa-type interaction. Therefore, these 3D dust clusters are also termed "Yukawa balls."

The reason for the higher occupation of the inner shells as compared to pure Coulomb interaction is the following: for pure Coulomb interaction the force on a particle inside a shell of charges vanishes. For screened interaction, where near-

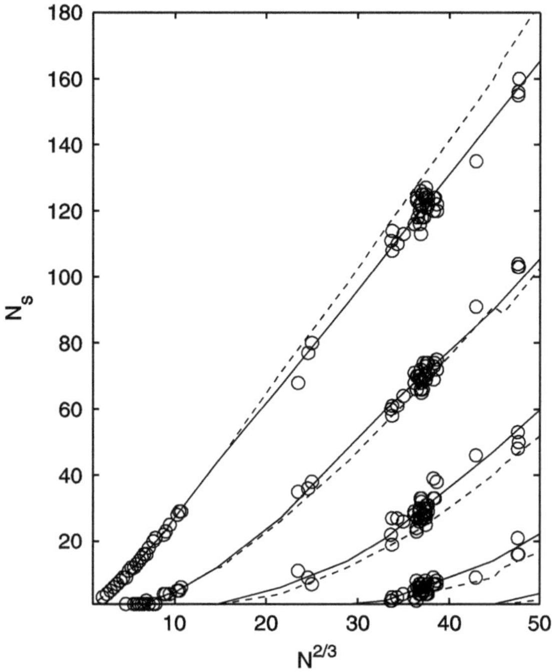

Fig. 7.9 Occupation number N_s as a function of total particle number N. The *symbols* indicate experimentally observed configurations. For $N > 100$ ($N^{2/3} > 20$) the configurations have been determined by scanning video microscopy, for $N < 100$ by stereoscopic imaging. The lines are results from MD simulations: pure Coulomb interaction (*dashed lines*), with screening using $\lambda_D = 1.7\,b$ (*solid lines*) (from [58])

est neighbors play a more dominant role, this is not true any more. For a particle inside a charged sphere, a force toward the center remains. This inward force has to be compensated by a change of radial density (i.e., occupation number), where more particles in the center balance the inward force from the outer shells (see also [59, 60]).

7.9 Metastable Configurations of Yukawa Balls

The analysis of the occupation number of the Yukawa balls certainly indicates that the particle interaction is affected by screening. When looking in more detail, especially for clusters with $N < 100$, it is often seen that the observed configurations are no ground-state configurations, even if screening is taken into account.

For example, the Yukawa ball with $N = 31$ in Fig. 7.8 shows a configuration (5, 26). By comparison to MD simulations (Chaps. 8 and 10) it is found that (5, 26) is the ground-state configuration only in an extremely tiny region of screening around $\kappa \approx 1.6$. Similarly, the observed configuration (11, 41) of the Yukawa ball with $N = 52$ never is a ground-state configuration. With increased screening, ground-state configurations are (10, 42), (12, 40) and (1, 12, 39), respectively [61]. Obviously, metastable configurations are observed quite frequently in our discharges.

To address this question in more detail we have performed dedicated experiments on the formation of metastable states. Therefore, we have repeatedly produced Yukawa balls with the same number of particles under identical plasma conditions. Since the trapping position of the Yukawa ball sensitively depends on the confinement, the cluster structure can be easily destroyed and afterwards reestablished by controlled changes of the plasma parameters. During a brief and well-defined interruption of the discharge power the particles start to fall downwards. Resetting the discharge power, the confinement is restored before the particles reach the lower electrode and the Yukawa ball is newly formed with exactly the same particles. However during the plasma interruption the initial particle arrangement is destroyed. The cluster has lost any memory on its previous structure. After the trap is reestablished the particles are allowed to relax into their new equilibrium positions for about 60 s. Afterwards the particle configuration is recorded for 30 s. This procedure can be repeatedly applied to obtain independent realizations of a cluster under identical conditions [48, 62].

The results of the repeated trapping experiment for a $N = 31$ cluster are shown in Fig. 7.10. A number of 37 realizations of the cluster have been generated at exactly the same plasma and confinement conditions. All clusters are found to consist of two shells, but the shell population differs among these clusters. Clusters with $N_1 = 4, 5$ and 6 particles on the inner shell are observed. The respective particle arrangement on the inner shell is shown in Fig. 7.10b–d. The observed structures are in perfect agreement with those expected from geometric considerations, namely a tetrahedron for $N_1 = 4$, a double tetrahedron ($N_1 = 5$), and a bipyramid ($N_1 = 6$). From

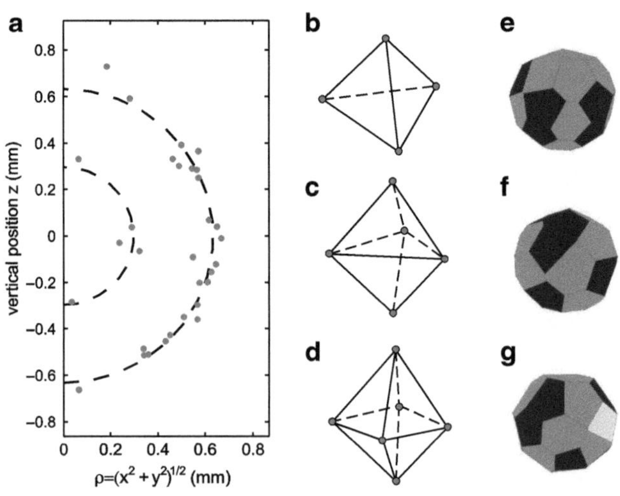

Fig. 7.10 Structure of metastable configurations of the $N = 31$ cluster. In (**a**) the (5, 26) configuration is shown in cylindrical coordinates. (**b**–**d**) Average structure of the inner shell. The particle arrangement is shown for clusters with $N_1 = 4$ (**b**), $N_1 = 5$ (**c**), and $N_1 = 6$ (**d**) particles. (**e**–**g**) Voronoi analysis (*dark gray*: pentagons, *light gray*: hexagons, *white*: defect) of the corresponding outer shell for $N_2 = 27$ (**e**), $N_2 = 26$ (**f**), and $N_2 = 25$ (**e**) particles (from [62])

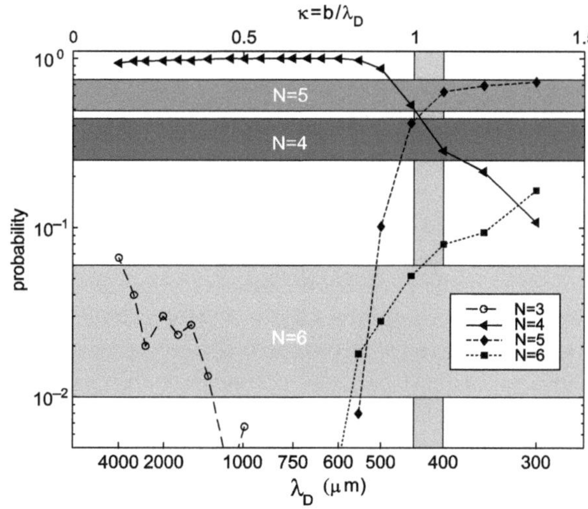

Fig. 7.11 Probability of realization of configurations with $N_1 = 3$, 4, 5 and 6 particles on the inner shell for a $N = 31$ cluster as a function of screening length λ_D. The experimental probabilities are indicated by the *horizontal stripes*. The matching value of κ is indicated by the *vertical bar* (from [62])

the Voronoi analysis in Fig. 7.10e–g, the particles on the outer shell arrange in an organized pattern of hexagons and pentagons as required for a hexagonal lattice bent onto a sphere.

Hence, well-defined crystalline clusters of different (metastable) states $(4, 27)$, $(5, 26)$, and $(6, 25)$ have been reliably produced in the experiment. The configuration $(5, 26)$ occurred in about two thirds of the cases (23 of 37 realizations), $(4, 27)$ in about one third (13 of 37). The configuration $(6, 25)$ was seen only once. Note that the metastable configuration $(5, 26)$ appears more frequently than the ground state $(4, 27)$.

This behavior is clarified by comparing these experimental observations to MD simulations. The simulation is started from random particle positions to mimic the experimental loss and recovery of the confinement. For each value of the screening length the probability of the different configuration are derived from 200 simulation runs. From the variation of the screening length λ_D in the simulation the influence of screening on the probability to find a certain metastable configuration is obtained as seen in Fig. 7.11. The configuration $(3, 28)$, which has not been observed in the experiment, is found to have a non-vanishing probability only for $\lambda_D > 1{,}000\,\mu$m ($\kappa < 0.5$). In contrast, the experimentally observed configuration $(6, 25)$ only appears for $\lambda_D < 600\,\mu$m ($\kappa > 0.78$). The configuration $(5, 26)$ also is found only for $\lambda_D < 550\,\mu$m ($\kappa > 0.85$). The ground state $(4, 27)$ shows a decreasing probability with increased screening. Nevertheless, in this entire shielding range ($\kappa < 1.5$) the ground state of lowest energy remains to be $(4, 27)$. Near $\lambda_D = 400\,\mu$m ($\kappa = 1.1$) the simulated probabilities nicely arrive at the experimental ones. This value of the screening length matches that determined from plasma simulations under similar conditions [47].

Here, we have seen that although the *energetically favored* ground-state configuration does not change with $\kappa < 1.5$ (namely the $(4, 27)$ configuration of the $N = 31$ cluster) the appearance *probability* does. That means that one has a higher probability to find a metastable cluster with higher occupation number on the inner shell. Thus, *on average* the trend of a higher population of the inner shells is also observed in finite systems with $N < 100$ [48, 62].

To form a qualitative picture, it is instructive to compare the observed abundance of the ground and metastable states with results expected from thermodynamic considerations: In a canonical ensemble with temperature T the probabilities of a state with energy E_j is proportional to $p_j = g_j \exp(-E_j/k_B T)$ where g_j is the degeneracy factor, that is, the number of states with that energy. In our case of classical distinguishable particles the number of possible realizations g of a shell configuration (N_1, N_2) simply is $(N_1 + N_2)!/(N_1!N_2!)$. Thus, the degeneracy factor of the metastable state $(5, 26)$ exceeds that of the ground state $(4, 27)$ by a factor of $27/5$. This can at least partially compensate the exponential weight. For a quantitative consideration, however, a detailed analysis of the various isomeric configurations, their eigenfrequency spectra and the nonequilibrium nature of the Yukawa balls have to be taken into account [63]. Nevertheless, a significant population of metastable states should be possible from thermodynamic considerations.

7.10 Shell Transitions in Yukawa Balls

Since metastable states are frequently observed it is tempting to search for transitions between different states. Such spontaneous shell transitions are observed by recording the Yukawa balls for long times of the order of minutes to detect whether spontaneous changes in the configurations occur.

Here, shell transitions in a somewhat larger cluster, namely for $N = 52$, are described. While the $N = 31$ cluster kept a constant number of shells for its different metastable states, the Yukawa ball with $N = 52$ particles changes its structure between two and three shells. In a long-run experiment, this cluster was observed for ten minutes. Figure 7.12a shows a fraction of the observed trajectories. Three different configurations were found during that period. The Yukawa ball remained for

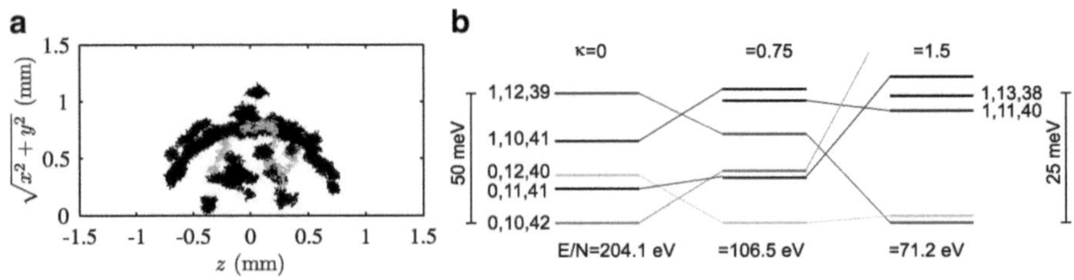

Fig. 7.12 (**a**) Trajectories of particles in a cluster with $N = 52$ over 200 s in the ρ–z plot. *Light gray trajectories* indicate transitions between shells, that is, a change in the configuration occurs. (**b**) Energies of a cluster with $N = 52$ particles for different selected configurations and three different values of the screening strength κ. The energies of the observed structures $(11, 41)$ and $(1, 11, 40)$ are close to the ground state $(12, 40)$ for $\kappa = 0.75$. The left energy scale is for $\kappa = 0$, the right for $\kappa = 0.75$ and 1.5 (from [48])

approximately 400 s in the $(11, 41)$ configuration. Then the configuration changes for a short period to $(12, 40)$ and returns to $(11, 41)$. After 440 s another transition occurred where one particle moves from the outer shell to the inner shell while another particle from the inner shell starts to travel to the center of the cluster and the configuration $(1, 11, 40)$ is observed.

The energies of a cluster with $N = 52$ particles are shown in Fig. 7.12b for different selected configurations and screening strengths. The ground-state configuration is $(10, 42)$ for $\kappa < 0.6$ [61]. For screening strengths between 0.6 and 1.4 the ground-state configuration is $(12, 40)$. The configurations with a center particle $(1, 12, 39)$ represents the ground state for $\kappa > 1.4$. Only the experimentally observed configuration $(12, 40)$ matches a simulated ground-state configuration. The other experimental configurations $(11, 41)$ and $(1, 11, 40)$ are close to simulated ground states, but differ in detail.

Again, also this cluster shows in the experiment a definite trend to higher occupation numbers of the inner shell as compared to the pure Coulomb case. This can be used to deduce the screening strength κ. While in the simulation the configuration $(10, 42)$ has a low-lying energy only for small κ and is not detected for $\kappa > 1.5$, the metastable configuration $(1, 11, 40)$ is found only for $\kappa > 0.75$. Only in a range near $\kappa = 0.75$ all three experimental configurations are found as low-energy states. This again supports the previous analysis on the screening strength and radial density distribution in Sect. 7.9.

7.11 Dynamical Properties of Yukawa Balls

Finally, similar to the 2D case, the normal modes of the Yukawa ball are determined from the thermal Brownian motion. For the 3D case, this is much more difficult since a much higher error for the reconstruction of the 3D position has to be taken into account as compared to the 2D case. For 2D the positions are relatively easily determined with subpixel resolution. In the 3D case the information of three cameras has to be reconciled (see Chap. 6).

For the experimental results presented here, special attention has been paid that *all* particle positions through *all* camera frames have been reconstructed. In some cases, visual inspection of the camera frames was necessary. Each particle then has been tracked through the entire sequence. Thus, we were able to determine the motion of a complete cluster with $N = 31$ particles through more than $F = 1{,}300$ frames. The mean reconstruction error was about 2 pixels whereas the mean Brownian particle excursions from the equilibrium were about 27 pixel. Thus, it was possible to extract the dynamics of the $N = 31$ Yukawa ball from this video sequence [64].

The resulting spectral power density is shown in Fig. 7.13 for all modes of the $N = 31$ Yukawa ball. There the modes have been ordered with increasing frequency. First, it is seen that we were indeed able to derive a reliable spectrogram. This is due to the fact that the mean reconstruction error was smaller than the mean displacement from the thermal motion. However, the spectral power density appears

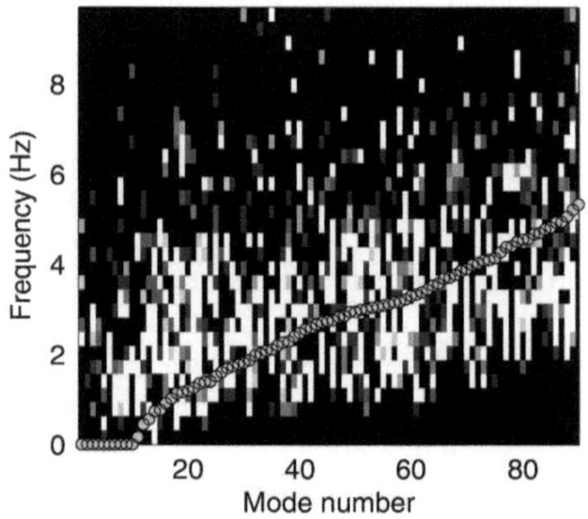

Fig. 7.13 Spectral power density (*gray scale*) as a function of frequency for the modes of the Yukawa ball. The calculated spectral power density increases from black to white. The *gray circles* are a best fit of the calculated eigenfrequencies to the spectral power density (after [64])

to be much broader in frequency than compared to corresponding analyses of 2D dust clusters (see Sect. 7.6).

The spectral power density has its maximum in a stripe increasing from low frequencies at low wave numbers to about 5 Hz at the highest wave numbers. This is very well in the expected frequency range.

Also for the 3D normal modes, we have calculated their respective shear and compressional contribution. It is again found that the low-frequency modes are dominated by a shear contribution whereas the high-frequency modes are mainly compressional. This finding is very similar to that of 2D dust clusters. Thus, also here the stability of the Yukawa balls is determined by the shear modes.

From the power spectral density information of the confinement strength and the charge are derived. By fitting the expected eigenfrequencies to the observed spectral power density a nice agreement is obtained and the confinement frequency is obtained as $\omega_0/(2\pi) = 1.6\,\text{Hz}$. The fitted frequencies are shown as gray circles in Fig. 7.13. From ω_0 the dust charge is determined as $Z \approx 900$ for our $3.47\,\mu\text{m}$ particles which is similar to the values obtained from structural considerations [47,48,62]. The particle charge is much smaller than for the 2D clusters since much smaller particles are used and the confinement is at decisively different discharge conditions.

7.12 Summary

Here, a brief review of the structure and the dynamics of finite charged-particle systems in dusty plasmas was given with an emphasis on the experimental observations. The main results can be summarized as

- Dusty plasmas are ideally suited to study finite systems of charged particles. The time and spatial scales of particle motion as well as low damping allow to measure particle trajectories with high accuracy on the level of individual particles.

- The forces in dusty plasmas can be tailored in such a way that various confinement geometries can be realized. Here, dust clusters in 1D, isotropic 2D and isotropic 3D confinement have been presented.
- Structural transitions in 1D geometry upon the change of particle number has been discussed.
- Ring and shell like structures have been observed in 2D and 3D geometries due to the interplay of shielded Coulomb repulsion and confinement boundaries. In 3D clusters, the shielded interaction leads to an observable deviation in occupation numbers and radial densities.
- Metastable configurations are often found in 3D Yukawa balls. In 3D, the metastable states have only a marginally higher energy than the ground state and are thermodynamically accessible in our discharges.
- Dynamical properties have been discussed in 2D and 3D using normal mode analysis. Shear-like modes are found to determine the stability of these systems. From the normal mode analysis governing parameters of the confinement, the particle charge and screening are obtained.

References

1. J.J. Thomson, Philos. Mag. **39**, 237 (1904)
2. G.J. Kalman, J.M. Rommel, K. Blagoev (eds.) *Strongly Coupled Coulomb Systems* (Plenum, New York, 1998)
3. W.-T. Juan et al., Phys. Rev. E **58**, 6947 (1998)
4. M. Klindworth, A. Melzer, A. Piel, V. Schweigert, Phys. Rev. B **61**, 8404 (2000)
5. O. Arp, D. Block, A. Piel, A. Melzer, Phys. Rev. Lett. **93**, 165004 (2004)
6. F. Verheest, *Waves in Dusty Space Plasmas* (Kluver, Dordrecht, 2000)
7. P.K. Shukla, A.A. Mamun, *Introduction to Dusty Plasma Physics* (Institute of Physics Publishing, Bristol, 2002)
8. J.H. Chu, J.-B. Du, I. Lin, J. Phys. D: Appl. Phys. **27**, 296 (1994)
9. H. Thomas et al., Phys. Rev. Lett. **73**, 652 (1994)
10. Y. Hayashi, K. Tachibana, Jpn. J. Appl. Phys. **33**, L804 (1994)
11. A. Melzer, in *Plasma Physics, Lecture Notes in Physics 670*, ed. by A. Dinklage, T. Klinger, G. Marx, L. Schweikhard (Springer, Berlin, 2005), p. 297
12. A. Melzer, J. Goree, in *Low Temperature Plasma Physics: Fundamentals, Technologies, and Techniques*, 2nd edn., ed. by R. Hippler, H. Kersten, M. Schmidt, K.H. Schoenbach (Wiley-VCH, Weinheim, 2008), p. 129
13. J.H. Chu, I. Lin, Phys. Rev. Lett. **72**, 4009 (1994)
14. A. Melzer, T. Trottenberg, A. Piel, Phys. Lett. A **191**, 301 (1994)
15. T. Trottenberg, A. Melzer, A. Piel, Plasma Sour. Sci. Technol. **4**, 450 (1995)
16. J. Pieper, J. Goree, R. Quinn, Phys. Rev. E **54**, 5636 (1996)
17. M. Zuzic et al., Phys. Rev. Lett. **85**, 4064 (2000)
18. A. Melzer et al., Phys. Rev. E **54**, 46 (1996)
19. A. Melzer, A. Homann, A. Piel, Phys. Rev. E **53**, 2757 (1996)
20. H. Thomas, G.E. Morfill, Nature **379**, 806 (1996)
21. A. Piel, A. Melzer, Plasma Phys. Control. Fusion **44**, R1 (2002)
22. G.E. Morfill, A. Ivlev, J. Jokipii, Phys. Rev. Lett. **83**, 971 (1999)
23. M. Klindworth, A. Piel, A. Melzer, Phys. Rev. Lett. **93**, 195002 (2004)
24. A. Piel et al., Phys. Rev. Lett. **97**, 205009 (2006)

25. W. Suzukawa, R. Ikada, Y. Tanaka, S. Iizuka, Appl. Phys. Lett. **88**, 121503 (2006)
26. M. Wolter et al., IEEE Trans. Plasma Sci. **71**, 036414 (2007)
27. A. Nefedov et al., New J. Phys. **5**, 33 (2003)
28. V.E. Fortov et al., JETP **96**, 704 (2003)
29. V.V. Yaroshenko et al., Phys. Rev. E **69**, 066401 (2004)
30. M. Mikikian, L. Boufendi, Phys. Plasmas **11**, 3733 (2004)
31. M. Kretschmer et al., Phys. Rev. E **71**, 056401 (2005)
32. J. Goree, G. Morfill, V. Tsytovich, S.V. Vladimirov, Phys. Rev. E **59**, 7055 (1999)
33. V. Tsytovich, S.V. Vladimirov, G. Morfill, J. Goree, Phys. Rev. E **63**, 056609 (2001)
34. V. Tsytovich, Phys. Scr. **T89**, 89 (2001)
35. S. Vladimirov, V.N. Tsytovich, G.E. Morfill, Phys. Plasmas **12**, 052117 (2005)
36. M. Rubin-Zuzic et al., Nat. Phys. **2**, 181 (2006)
37. A. Ivlev, G. Morfill, U. Konopka, Phys. Rev. Lett. **89**, 195502 (2002)
38. A. Ivlev et al., Phys. Rev. Lett. **90**, 055003 (2003)
39. A. Melzer, Phys. Rev. E **73**, 056404 (2006)
40. A. Melzer, Phys. Rev. E **67**, 016411 (2003)
41. M. Kong, B. Partoens, F.M. Peeters, Phys. Rev. E **67**, 021608 (2003)
42. V.A. Schweigert, F. Peeters, Phys. Rev. B **51**, 7700 (1995)
43. Y. Ivanov, A. Melzer, Phys. Plasmas **12**, 072110 (2005)
44. T. Trottenberg, D. Block, A. Piel, Phys. Plasmas **13**, 042105 (2006)
45. C. Henning et al., Phys. Rev. Lett. **101**, 045002 (2008)
46. A. Melzer, M. Klindworth, A. Piel, Phys. Rev. Lett. **87**, 115002 (2001)
47. O. Arp, D. Block, M. Klindworth, A. Piel, Phys. Plasmas **12**, 122102 (2005)
48. S. Käding et al., Phys. Plasmas **15**, 073710 (2008)
49. S. Käding, A. Melzer, Phys. Plasmas **13**, 090701 (2006)
50. S. Käding, Y. Ivanov, A. Melzer, IEEE Trans. Plasma Sci. **35**, 328 (2007)
51. M. Kroll, D. Block, A. Piel, Phys. Plasmas **15**, 063703 (2008)
52. B.M. Annaratone et al., Plasma Phys. Control. Fusion **46**, B495 (2004)
53. T. Antonova et al., Phys. Rev. Lett. **96**, 115001 (2006)
54. Y. Ivanov, A. Melzer, Rev. Sci. Instrum. **78**, 033506 (2007)
55. Y. Feng, J. Goree, B. Liu, Phys. Rev. Lett. **100**, 205007 (2008)
56. P. Ludwig, S. Kosse, M. Bonitz, Phys. Rev. E **71**, 046403 (2005)
57. S. Apolinario, B. Partoens, F. Peters, New J. Phys. **9**, 283 (2007)
58. M. Bonitz et al., Phys. Rev. Lett. **96**, 075001 (2006)
59. C. Henning et al., Phys. Rev. E **74**, 056403 (2006)
60. C. Henning et al., Phys. Rev. E **76**, 036404 (2007)
61. H. Baumgartner et al., New J. Phys. **10**, 093019 (2008)
62. D. Block et al., Phys. Plasmas **15**, 040701 (2008)
63. H. Kählert et al., Phys. Rev. E **78**, 036408 (2008)
64. Y. Ivanov, A. Melzer, Phys. Rev. E **78**, 036402 (2009)

Chapter 8
Statistical Theory of Spherically Confined Dust Crystals

Christian Henning and Michael Bonitz

Abstract Statistical methods are well established within theoretical plasma physics, which is concerned with the equilibrium and nonequilibrium properties of charged particle systems. This chapter adopts the statistical theory to spherically confined dust crystals allowing for the derivation of their ground-state density profile. First, using the canonical ensemble, the energy functional is derived whose variational problem is intimately connected with the density profile. Thereon two approximation schemes, the mean-field and the local density approximation, are introduced and the density profile is derived analytically. Comparisons with results from first-principles simulations of spherically confined dust crystals demonstrate that the two approximations complement each other, and together describe the average density profile accurately.

8.1 Introduction

The experimental realization of non-neutral plasmas in external trapping potentials has been attracting lots of interest over the last years in many fields, including trapped ions, and electrons, and positrons in Penning traps (see, e.g., [1] for an overview). One of the main advantages of such charged particle systems is the possibility to realize strong correlation effects more or less easily. Probably the most striking forms of appearance of these effects are the liquid states and the crystal formation which were predicted and observed in various geometries. In particular, the ion crystals and the recently observed spherically confined dust crystals (Yukawa balls) [2] have actuated intensive new experimental and theoretical work (see Chaps. 6, 7, and 10). As a result, the shell structure of these Yukawa balls is well explained by computer simulations of charged particles in an external parabolic

C. Henning (✉) and M. Bonitz
Institut für Theoretische Physik und Astrophysik, Christian-Albrechts Universität zu Kiel, 24098 Kiel, Germany
e-mail: henning@theo-physik.uni-kiel.de; bonitz@physik.uni-kiel.de

M. Bonitz et al. (eds.), *Introduction to Complex Plasmas*, Springer Series on Atomic, Optical, and Plasma Physics 59, DOI 10.1007/978-3-642-10592-0_8,

confinement interacting via a screened Coulomb potential, that is, a Yukawa or Debye–Hückel potential [3]. On the theoretical side, there are successful models for the description of the average particle distribution, the so-called *shell models* [4–9], in which the shell-like structure is employed a priori. The objective of these models, which are presented in Sect. 8.6, is to achieve an accurate prediction of the shell populations and of the shell radii. However, these models are somewhat artificial due to their immanent shell structure and a completely analytical theory is required.

The theoretical foundation for the determination of the average particle distribution, that is, the density profile, of spatially confined dust crystals is the thermal equilibrium statistical mechanics. It allows for this determination by using a simple variational principle. The idea behind is very basic – the equilibrium density profile minimizes the corresponding Helmholtz free energy [10]. For the case of vanishing temperature, as it is studied within this chapter, the free energy equals the energy; thus, the ground-state density profile can be obtained by minimizing the latter one.

Therefore, within the next section, an expression for the energy in dependence on the density is derived. In principle, its variation provides an equation for the determination of the ground-state density profile. However, due to incomplete knowledge of particle correlations, this is not possible in full generality. For this reason, within Sects. 8.3 and 8.5, the two most essential approximations, the mean-field and the local density approximation (LDA), are introduced, which yield analytical solutions for the ground-state density. To check the quality of these solutions, simulation results for the ground-state density are presented in Sect. 8.4.

8.2 Variational Problem of the Energy Functional

The dust crystals are characterized by a three-dimensional, classical system of N-identical particles harmonically confined by the potential

$$\phi(\boldsymbol{r}) = \frac{m\omega_0^2}{2}\boldsymbol{r}^2, \tag{8.1a}$$

and interacting by an isotropic Yukawa-type pair potential

$$v(r) = q^2\frac{\exp(-\kappa r)}{r}. \tag{8.1b}$$

The Hamiltonian is then given by[1]

$$H(\mathbf{r},\mathbf{p}) = \underbrace{\sum_{i=1}^{N}\frac{\boldsymbol{p}_i^2}{2m}}_{K(\mathbf{p})} + \underbrace{\sum_{i=1}^{N}\phi(\boldsymbol{r}_i) + \frac{1}{2}\sum_{i\neq j}^{N} v(|\boldsymbol{r}_i - \boldsymbol{r}_j|)}_{U(\mathbf{r})}. \tag{8.1}$$

[1] In the following, $3N$-dimensional vectors are written upright, $\mathbf{r} = (\boldsymbol{r}_1, \boldsymbol{r}_2, \ldots, \boldsymbol{r}_N)$.

To consider statistical quantities, of course, not only one such system but an ensemble of these has to be used. Due to the fixed particle number, the appropriate ensemble is the canonical one, which also depends on the temperature T or rather its inverse $\beta = (k_\mathrm{B}T)^{-1}$. The proper N particle distribution of the equilibrium is then given by

$$f(\mathbf{r},\mathbf{p}) = \frac{1}{h^{3N}N!}\frac{\mathrm{e}^{-\beta H(\mathbf{r},\mathbf{p})}}{Z}. \tag{8.2}$$

Thus, the quantity $f(\mathbf{r},\mathbf{p})\mathrm{d}\mathbf{r}\mathrm{d}\mathbf{p}$ represents the probability that the phase point describing the state of the system is included in the infinitesimal phase-space volume $\mathrm{d}\mathbf{r}\mathrm{d}\mathbf{p}$ at $(\mathbf{r},\mathbf{p})$. The factor $N!$ makes allowance for the indistinguishability of the particles, while the power of Planck's constant, h^{3N}, ensures the correct correspondence to quantum statistics [11]. The partition function

$$Z = \frac{1}{h^{3N}N!}\mathrm{Tr}_\mathrm{cl}\,\mathrm{e}^{-\beta H(\mathbf{r},\mathbf{p})} \tag{8.3}$$

then normalizes this probability density such that its "classical" trace, that is,

$$\mathrm{Tr}_\mathrm{cl} \equiv \int_{\mathcal{V}^N}\mathrm{d}\mathbf{r}\int \mathrm{d}\mathbf{p}, \tag{8.4}$$

yields unity. Within this trace, there is the constraint that only those states are considered, in which all particles are situated within the spatial region $\mathcal{V}$. While for unconfined, homogeneous systems only its volume is of importance, here indeed the actual region is decisive. Thus, all statistical quantities like the partition function or statistical averages are depending on $T, N, \mathcal{V}$.

The ensemble average of a physical quantity $O(\mathbf{r},\mathbf{p})$ can be calculated using the N particle distribution:

$$\langle O\rangle = \mathrm{Tr}_\mathrm{cl}\,[f(\mathbf{r},\mathbf{p})O(\mathbf{r},\mathbf{p})]\,. \tag{8.5}$$

For only spatially dependent quantities $O(\mathbf{r})$ or only momentum-dependent quantities $O(\mathbf{p})$, the latter equation can be simplified due to the factorization

$$f(\mathbf{r},\mathbf{p}) = f^\mathrm{s}(\mathbf{r})f^\mathrm{m}(\mathbf{p}) \tag{8.6}$$

into spatial and momentum distributions with

$$f^\mathrm{s}(\mathbf{r}) = \frac{\mathrm{e}^{-\beta U(\mathbf{r})}}{\int_{\mathcal{V}^N}\mathrm{d}\mathbf{r}\,\mathrm{e}^{-\beta U(\mathbf{r})}} \quad \text{and} \quad f^\mathrm{m}(\mathbf{p}) = \frac{\mathrm{e}^{-\beta K(\mathbf{p})}}{\int \mathrm{d}\mathbf{p}\,\mathrm{e}^{-\beta K(\mathbf{p})}}. \tag{8.7}$$

Hence, in those cases the ensemble averages can be written as

$$\langle O\rangle = \int_{\mathcal{V}^N}\mathrm{d}\mathbf{r}\,f^\mathrm{s}(\mathbf{r})O(\mathbf{r}) \quad \text{and} \quad \langle O\rangle = \int \mathrm{d}\mathbf{p}\,f^\mathrm{m}(\mathbf{p})O(\mathbf{p}), \tag{8.8}$$

respectively. This is applicable in the derivation of the ensemble average of the energy $E = \langle H \rangle$, which itself can be separated into momentum-dependent kinetic energy, $K(\mathbf{p})$, and into spatial-dependent potential energy, $U(\mathbf{r})$. It is given by

$$\begin{aligned} E &= \langle H(\mathbf{r},\mathbf{p}) \rangle = \langle K(\mathbf{p}) \rangle + \langle U(\mathbf{r}) \rangle \\ &= \frac{\int \mathrm{d}\mathbf{p}\, K(\mathbf{p}) \mathrm{e}^{-\beta K(\mathbf{p})}}{\int \mathrm{d}\mathbf{p}\, \mathrm{e}^{-\beta K(\mathbf{p})}} + \int_{\mathcal{V}^N} \mathrm{d}\mathbf{r}\, f^{\mathrm{s}}(\mathbf{r}) U(\mathbf{r}). \end{aligned} \tag{8.9}$$

The first expression can be easily calculated by rewriting it as a derivate of a logarithm and then using the known integral of a Maxwellian distribution, which finally yields the well-known result $3Nk_{\mathrm{B}}T/2$. To calculate the second expression, it is useful to introduce the reduced spatial distribution functions

$$f_k^{\mathrm{s}}(\boldsymbol{r}_1, \ldots, \boldsymbol{r}_k) = \int_{\mathcal{V}^{N-k}} \mathrm{d}\boldsymbol{r}_{k+1}, \ldots, \mathrm{d}\boldsymbol{r}_N\, f^{\mathrm{s}}(\mathbf{r}), \tag{8.10}$$

so that $f_k^{\mathrm{s}}(\boldsymbol{r}_1, \ldots, \boldsymbol{r}_k)\mathrm{d}\boldsymbol{r}_1, \ldots, \mathrm{d}\boldsymbol{r}_k$ represents the joint probability of finding one particle within a volume $\mathrm{d}\boldsymbol{r}_1$ at $\boldsymbol{r}_1$, *and* another particle within a volume $\mathrm{d}\boldsymbol{r}_2$ at $\boldsymbol{r}_2$, *and* so on, irrespective of the position of all the other $N-k$ particles. The most important reduced spatial distribution functions are f_1^{s} and f_2^{s}. While the former is related to the ensemble averaged density $n(\boldsymbol{r}) = \langle \sum_{i=1}^{N} \delta(\boldsymbol{r} - \boldsymbol{r}_i) \rangle$ by

$$n(\boldsymbol{r}) = N f_1^{\mathrm{s}}(\boldsymbol{r}), \tag{8.11}$$

f_2^{s} in turn is related to f_1^{s} by

$$f_2^{\mathrm{s}}(\boldsymbol{r}_1, \boldsymbol{r}_2) = f_1^{\mathrm{s}}(\boldsymbol{r}_1) f_1^{\mathrm{s}}(\boldsymbol{r}_2) \left[1 + h(\boldsymbol{r}_1, \boldsymbol{r}_2)\right]. \tag{8.12}$$

At heart, the latter is given by the pair correlation function h, which measures deviations of this probability density from the case of a statistically independent (mean-field) distribution of $\boldsymbol{r}_1$ and $\boldsymbol{r}_2$.

For binary interactions, all of the thermodynamic functions can be evaluated from knowledge of $n(\boldsymbol{r})$ and $h(\boldsymbol{r}_1, \boldsymbol{r}_2)$. Accordingly, by using (8.9)–(8.12), the ensemble averaged energy can be written as

$$\begin{aligned} E &= \frac{3}{2} N k_{\mathrm{B}} T + \sum_{i=1}^{N} \int_{\mathcal{V}^N} \mathrm{d}\mathbf{r}\, f^{\mathrm{s}}(\mathbf{r}) \phi(\boldsymbol{r}_i) + \frac{1}{2} \sum_{i \neq j}^{N} \int_{\mathcal{V}^N} \mathrm{d}\mathbf{r}\, f^{\mathrm{s}}(\mathbf{r}) v(|\boldsymbol{r}_i - \boldsymbol{r}_j|) \\ &= \frac{3}{2} N k_{\mathrm{B}} T + \int_{\mathcal{V}} \mathrm{d}\boldsymbol{r}\, n(\boldsymbol{r}) \phi(\boldsymbol{r}) \\ &\quad + \frac{N-1}{2N} \int_{\mathcal{V}^2} \mathrm{d}\boldsymbol{r}\, \mathrm{d}\boldsymbol{r}'\, n(\boldsymbol{r}) n(\boldsymbol{r}') v(|\boldsymbol{r} - \boldsymbol{r}'|) \left[1 + h(\boldsymbol{r}, \boldsymbol{r}')\right], \end{aligned} \tag{8.13}$$

wherein the integral of interaction contains both the mean-field and the correlation contribution. For the ground-state energy ($T = 0$), which subsequently plays a central role, this expression reduces to

$$E = \int_{\mathcal{V}} \mathrm{d}\boldsymbol{r}\, \left(u_{\mathrm{trap}}(\boldsymbol{r}) + u_{\mathrm{mf}}(\boldsymbol{r}) + u_{\mathrm{corr}}(\boldsymbol{r})\right) \tag{8.14}$$

with the energy densities of confinement, of mean-field interaction, and of the correlations

$$u_{\mathrm{trap}}(\boldsymbol{r}) = n(\boldsymbol{r})\phi(\boldsymbol{r}), \tag{8.15a}$$

$$u_{\mathrm{mf}}(\boldsymbol{r}) = \frac{N-1}{2N}\, n(\boldsymbol{r}) \int_{\mathcal{V}} \mathrm{d}\boldsymbol{r}'\, n(\boldsymbol{r}')v(|\boldsymbol{r}-\boldsymbol{r}'|), \tag{8.15b}$$

$$u_{\mathrm{corr}}(\boldsymbol{r}) = \frac{N-1}{2N}\, n(\boldsymbol{r}) \int_{\mathcal{V}} \mathrm{d}\boldsymbol{r}'\, n(\boldsymbol{r}')v(|\boldsymbol{r}-\boldsymbol{r}'|)h(\boldsymbol{r},\boldsymbol{r}'), \tag{8.15c}$$

respectively. This expression for the ground-state energy shows the dependence on n and h, which on their part are fixed by the equilibrium N particle distribution f. Because this distribution yields the lowest value for the energy, n and h in turn have to minimize it. This fact provides the possibility to actually calculate the density profile. For this purpose, the generally unknown pair correlation function is approximated and the minimum of the energy with respect to the density has to be determined, which can be done by a variational principle. However, within this minimization, care has to be used regarding the restrictions that the density is non-negative everywhere and that it reproduces the total particle number, that is,

$$\int_{\mathcal{V}} \mathrm{d}\boldsymbol{r}\, n(\boldsymbol{r}) = N. \tag{8.16}$$

While the former constraint requires restriction of the allowed variations, the latter constraint can be included by introducing a pertinent Lagrange multiplier μ.

Thus, in the following not the ground-state energy (8.14), but a corresponding energy functional is considered, which depends not only on N, but also on a density function. In contrast, the dependence on $\mathcal{V}$ is lapsed and the boundless space

$$\mathcal{V} = \mathbb{R}^3 \tag{8.17}$$

is used instead. This is because within this chapter not a volume restrictive density profile of confined dust crystals is considered, but a density profile which is restricted only by its confinement. Thus, the energy functional yields

$$E[n] = \int \mathrm{d}\boldsymbol{r}\, \left(u_{\mathrm{trap}}(\boldsymbol{r}) + u_{\mathrm{mf}}(\boldsymbol{r}) + u_{\mathrm{corr}}(\boldsymbol{r})\right) + \mu\left(N - \int \mathrm{d}\boldsymbol{r}\, n(\boldsymbol{r})\right). \tag{8.18}$$

8.3 Ground-State Density Profile Within Mean-Field Approximation

One of the most important approximations for the ensemble averaged energy of interaction is the mean-field approximation. This approximation utilizes the full nonlocal mean-field energy density, completely neglects the correlation contributions, that is, $u_{\text{corr}} \equiv 0$, and is associated with a structure-spanning averaging. The energy functional is then given by

$$E_{\text{mf}}[n] = N\mu_{\text{mf}} + \int \mathrm{d}\boldsymbol{r}\, n(\boldsymbol{r})(\phi(\boldsymbol{r}) - \mu_{\text{mf}}) + \frac{N-1}{2N}q^2 \int \mathrm{d}\boldsymbol{r}\mathrm{d}\boldsymbol{r}'\, n(\boldsymbol{r})n(\boldsymbol{r}')\frac{\exp(-\kappa\,|\boldsymbol{r}-\boldsymbol{r}'|)}{|\boldsymbol{r}-\boldsymbol{r}'|}. \tag{8.19}$$

This expression is formally equivalent to the electrostatics expression of a charge distribution with fixed total charge in an external confinement [13]. The factor $(N-1)/N$ leads to an effective charge per particle of $q_{\text{eff}} = q\sqrt{(N-1)/N}$. However, in contrast to electrostatics with its Coulomb interaction, the interaction in (8.19) is screened. Nevertheless, the density profile in mean-field approximation can be seen as the electrostatic charge distribution in the case of a Yukawa interaction. Especially in the limiting case of vanishing screening, this view allows for the solution of the density profile.

8.3.1 The Coulomb Limit and Electrostatics

To obtain the density profile of the harmonically confined system in the Coulomb limit, one can use the well-known textbook result of the electrostatic field of a homogeneously charged ball, which is shown in Fig. 8.1. If the ball is of radius R_C, the density is given by

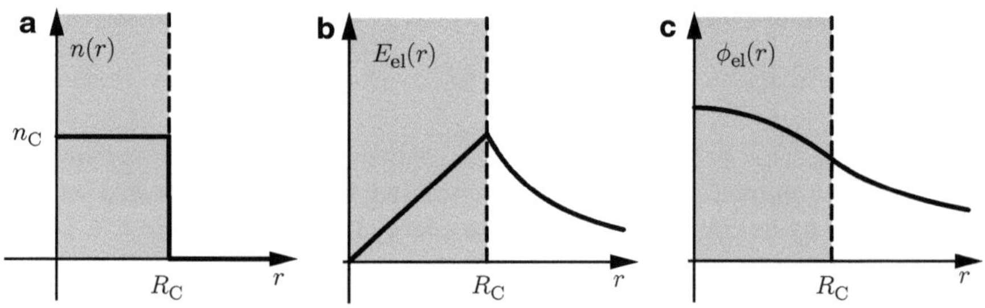

Density Electric field Electric potential

Fig. 8.1 A homogeneously charged ball with radius R_C is related to a linear electric field (**b**) and a parabolic electrostatic potential (**c**) within the ball

$$n(\boldsymbol{r}) = \begin{cases} n_{\mathrm{C}}, & |\boldsymbol{r}| \leq R_{\mathrm{C}}, \\ 0, & |\boldsymbol{r}| > R_{\mathrm{C}}, \end{cases} \tag{8.20}$$

and the charge density is $q_{\mathrm{eff}} \cdot n(\boldsymbol{r})$. The electric potential caused by this charge density can be obtained by using Gauss's law or by directly solving Poisson's equation. It yields

$$\phi_{\mathrm{el}}(\boldsymbol{r}) = \frac{4\pi}{3} R_{\mathrm{C}}^3 \, q_{\mathrm{eff}} n_{\mathrm{C}} \begin{cases} \frac{1}{R_{\mathrm{C}}} \left(\frac{3}{2} - \frac{r^2}{2R_{\mathrm{C}}^2} \right), & |\boldsymbol{r}| \leq R_{\mathrm{C}}, \\ \frac{1}{|\boldsymbol{r}|}, & |\boldsymbol{r}| > R_{\mathrm{C}}. \end{cases} \tag{8.21}$$

Hence, the electric potential ϕ_{el} is parabolic within the ball and consequently can compensate the external parabolic potential ϕ. This is the case if the density takes a specific value which can be calculated from the equilibrium condition. Namely, equilibrium is attained if the overall potential inside the ball is constant, that is,

$$q_{\mathrm{eff}} \phi_{\mathrm{el}}(\boldsymbol{r}) + \phi(\boldsymbol{r}) = \text{constant} \quad \forall \boldsymbol{r} : |\boldsymbol{r}| \leq R_{\mathrm{C}}. \tag{8.22}$$

By using (8.1a) and (8.21) the equilibrium density results in

$$n_{\mathrm{C}} = \frac{3m\omega_0^2}{4\pi q_{\mathrm{eff}}^2}. \tag{8.23}$$

Additionally, normalization allows the determination of the ball's radius and yields

$$R_{\mathrm{C}} = \sqrt[3]{\frac{q^2(N-1)}{m\omega_0^2}}. \tag{8.24}$$

In summary, from the electrostatic analogy it follows that the mean-field density profile of the harmonically confined system is homogeneous in the Coulomb limit. However, in general the dust crystals are not described by Coulomb interacting particles, but by Yukawa interacting ones. Thus, what is the effect of screening on the density profile?

8.3.2 General Solution

As outlined earlier, the problem of ascertaining the density profile n_{mf} in the general case of screening is given by determining the minimum of the energy functional (8.19). The vanishing of its linear approximation at the minimum gives then rise to the variational problem:

$$0 = \left. \frac{\delta E_{\mathrm{mf}}[n]}{\delta n(\boldsymbol{r})} \right|_{n=n_{\mathrm{mf}}}. \tag{8.25}$$

The explicit variation of the energy functional E_{mf} yields an inhomogeneous integral equation for the density:

$$0 = \phi(\mathbf{r}) - \mu_{\mathrm{mf}} + q_{\mathrm{eff}}^2 \int \mathrm{d}\boldsymbol{r}'\, n_{\mathrm{mf}}(\boldsymbol{r}') \frac{\exp(-\kappa\,|\boldsymbol{r}-\boldsymbol{r}'|)}{|\boldsymbol{r}-\boldsymbol{r}'|} \qquad \forall \boldsymbol{r} \in \mathcal{V}_{\mathrm{mf}}, \tag{8.26}$$

which is valid for all space points within the supporting region:

$$\mathcal{V}_{\mathrm{mf}} = \left\{\boldsymbol{r} \in \mathbb{R}^3 | n_{\mathrm{mf}}(\boldsymbol{r}) > 0\right\}. \tag{8.27}$$

This is due to the restriction to nonnegative densities. The region $\mathcal{V}_{\mathrm{mf}}$ is unrelated to the thermodynamic region $\mathcal{V}$ which is already set by (8.17). The density outside of $\mathcal{V}_{\mathrm{mf}}$ vanishes, that is,

$$n_{\mathrm{mf}}(\boldsymbol{r}) = 0 \quad \forall \boldsymbol{r} \notin \mathcal{V}_{\mathrm{mf}}, \tag{8.28}$$

so that there is an explicit space separation of the density. This is outlined in Fig. 8.2a. In the case of isotropic systems as is given by the Hamiltonian (8.1), the density profile is isotropic as well. Thus, $\mathcal{V}_{\mathrm{mf}}$ has to be spherically symmetric as shown by Fig. 8.2b. In this case, the integral equation can be solved for the density by successive integration (cf. [12] for details).

However, an explicit solution for the density can be obtained more smartly as well, because the kernel of the integral equation (8.26), that is, the Yukawa potential, is the Green's function of the Helmholtz operator. This means

$$(\Delta - \kappa^2)\frac{\exp(-\kappa\,|\boldsymbol{r}-\boldsymbol{r}'|)}{|\boldsymbol{r}-\boldsymbol{r}'|} = -4\pi\delta(\boldsymbol{r}-\boldsymbol{r}'). \tag{8.29}$$

Hence, application of the Helmholtz operator on (8.26) yields the explicit solution

$$n_{\mathrm{mf}}(\boldsymbol{r}) = \frac{1}{4\pi q_{\mathrm{eff}}^2}\left(\Delta\phi(\boldsymbol{r}) - \kappa^2\phi(\boldsymbol{r}) + \kappa^2\mu_{\mathrm{mf}}\right)\forall \boldsymbol{r} \in \mathcal{V}_{\mathrm{mf}} \tag{8.30}$$

for the ground-state density profile in mean-field approximation in case of an arbitrary confinement potential. However, within this expression the Lagrange multiplier μ_{mf} as well as the supporting region $\mathcal{V}_{\mathrm{mf}}$ are not yet determined.

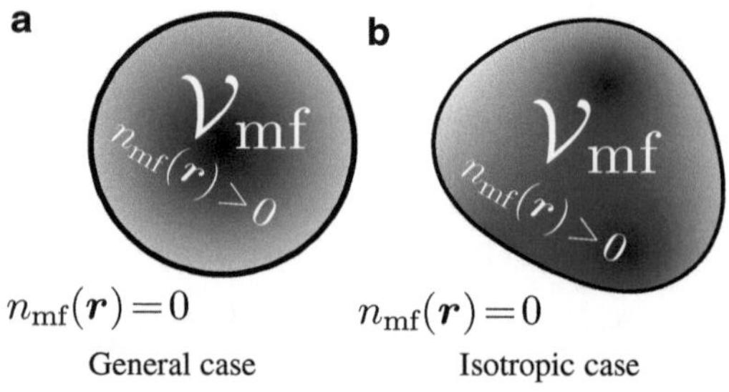

Fig. 8.2 Spatial confinement of the ground-state density

The calculation of μ_{mf} can be performed by using (8.16), that is, the constraint of normalization. Taking the results (8.17), (8.28), and (8.30) into account, this yields[2]

$$\kappa^2 \mu_{\mathrm{mf}} = \frac{4\pi q_{\mathrm{eff}}^2 N - \int_{\mathcal{V}_{\mathrm{mf}}} \mathrm{d}\boldsymbol{r}\, (\Delta - \kappa^2)\phi(\boldsymbol{r})}{|\mathcal{V}_{\mathrm{mf}}|}, \tag{8.31}$$

where $|\mathcal{V}_{\mathrm{mf}}|$ denotes the volume of the spatial region $\mathcal{V}_{\mathrm{mf}}$. Therefore, only the determination of this region $\mathcal{V}_{\mathrm{mf}}$ has to be accomplished. This can be done by inserting the solution (8.30) into its integral equation (8.26), because the latter one contains additional boundary conditions, which are disregarded in (8.30) due to the application of the differential operator. However, the extraction of the supporting region out of the resulting equation

$$4\pi\,(\mu_{\mathrm{mf}} - \phi(\mathbf{r})) = \int_{\mathcal{V}_{\mathrm{mf}}} \mathrm{d}\boldsymbol{r}' \left(\Delta\phi(\boldsymbol{r}') - \kappa^2\phi(\boldsymbol{r}') + \kappa^2\mu_{\mathrm{mf}}\right) \frac{\exp(-\kappa\,|\boldsymbol{r} - \boldsymbol{r}'|)}{|\boldsymbol{r} - \boldsymbol{r}'|} \quad \forall \boldsymbol{r} \in \partial\mathcal{V}_{\mathrm{mf}} \tag{8.32}$$

is not at all simple.

For isotropically confined dust crystals, that is, $\phi(\mathbf{r}) = \phi(r)$, the supporting region has to be spherically symmetric (cf. Fig. 8.2b) and is specified by a ball:

$$\mathcal{V}_{\mathrm{mf}} = \mathcal{B}(R_{\mathrm{mf}}), \tag{8.33}$$

which is centered at $\boldsymbol{r} = 0$, that is, at the minimum of the trap, and which has a still unknown radius R_{mf}. This circumstance significantly simplifies the issue of determining $\mathcal{V}_{\mathrm{mf}}$ from (8.32), because only one parameter, the mean-field radius R_{mf}, has to be found. Indeed, using (8.33) within (8.32) yields, after some algebra,

$$\mu_{\mathrm{mf}} = \phi(R_{\mathrm{mf}}) + \frac{R_{\mathrm{mf}}\,\phi'(R_{\mathrm{mf}})}{1 + \kappa R_{\mathrm{mf}}}, \tag{8.34}$$

which is an implicit equation for the radius for given μ_{mf}.

8.3.3 *Density Profile for Harmonic Confinement*

For the special case of a harmonic confinement (8.1a), which is considered within this chapter, the density profile (8.30) is only radially dependent and reduces to

---

[2] It should be noted that in the Coulomb case, (8.31) results

$$\int_{\mathcal{V}_{\mathrm{mf}}} \mathrm{d}\boldsymbol{r}\; \Delta\phi(\boldsymbol{r}) = 4\pi q^2 (N - 1).$$

$$n_{\mathrm{mf}}(r) = n_{\mathrm{C}} \left(\frac{R_{\mathrm{C}}^3}{R_{\mathrm{mf}}^3} + \frac{\kappa^2 R_{\mathrm{mf}}^2}{10} - \frac{\kappa^2 r^2}{6} \right) \Theta(R_{\mathrm{mf}} - r), \tag{8.35}$$

where the definitions (8.23) and (8.24) of n_{C} and R_{C} have been used. The mean-field radius R_{mf} can be obtained from (8.34) together with the result of μ_{mf}, which yields the implicit equation

$$\kappa^3 R_{\mathrm{mf}}^6 + 6\kappa^2 R_{\mathrm{mf}}^5 = 15 \left(R_{\mathrm{mf}}^3 + \kappa R_{\mathrm{mf}}^4\right) \left(\frac{R_{\mathrm{C}}^3}{R_{\mathrm{mf}}^3} - 1 \right) \tag{8.36}$$

for the radius. It has an unique positive solution for $R_{\mathrm{mf}}/R_{\mathrm{C}}$, which can be specified as a function of κR_{C}. The result is graphically shown in Fig. 8.3. While in the unscreened case ($\kappa = 0$) the mean-field radius equals the Coulomb radius R_{C}, for finite screening the mean-field radius is decreased. Thus, in comparison with Coulomb systems, the considered Yukawa systems are compressed. This is comprehensible, because the exponential weakening of the interaction results in a reduced total force acting on the outer particles, which therefore move somewhat toward the center.

Often the screening parameter κ is given in units of d_{C}^{-1}, where d_{C} is the stable distance between two charged particles in the absence of screening [3]. One obtains

$$\kappa R_{\mathrm{C}} = \kappa d_{\mathrm{C}} \sqrt[3]{\frac{N-1}{2}}, \tag{8.37}$$

and therefore Fig. 8.3 shows that not only an increase of κ but also an increase of the particle number at small κ accounts for a stronger compression with respect to the Coulomb case.

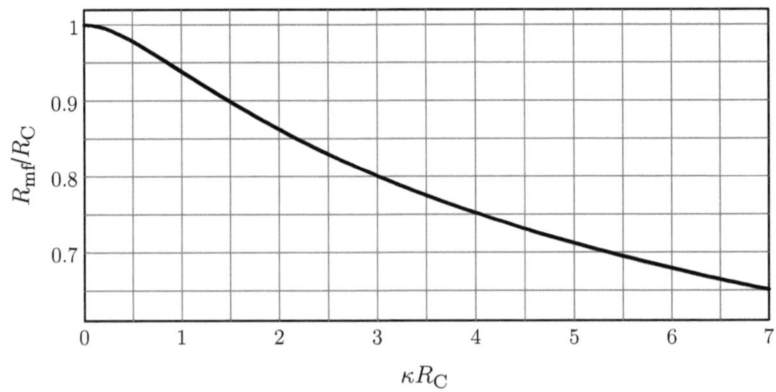

Fig. 8.3 The mean-field radius R_{mf} in units of R_{C} as a function of the normalized screening parameter κR_{C}

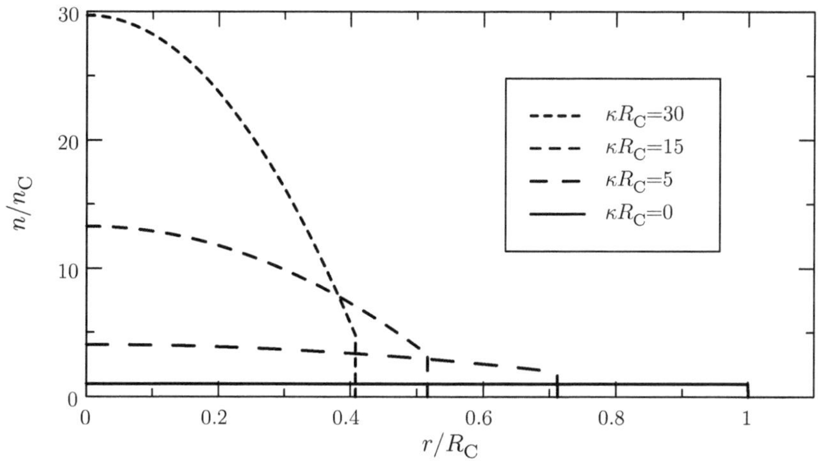

Fig. 8.4 Density profile of harmonically confined dust crystals in mean-field approximation for four screening values (*lines*), *from bottom to top*: $\kappa R_C = 0$, $\kappa R_C = 5$, $\kappa R_C = 15$, and $\kappa R_C = 30$

With the determined radius, the density profile can be calculated from (8.35). Corresponding results for different screening parameters are shown in Fig. 8.4.

On the one hand, in the Coulomb limit ($\kappa R_C = 0$), the constant density profile obtained in Sect. 8.3.1 is recovered. On the other hand, for finite screening inhomogeneous profiles emerge which are accompanied by the aforementioned compression. With increasing κR_C, the density values increase continuously, but most significantly in the center. As a result, the density profiles (8.35) are described by an inverted parabola which terminates in a discontinuity at $r = R_{mf}$ with a finite density value.

In summary, the mean-field density profile changes radically from a flat profile, in case of a long-range Coulomb interaction, to a profile rapidly decaying away from the trap center in the case of a screened Yukawa potential.

8.3.4 Force Equilibrium Within Yukawa Electrostatics

As mentioned earlier, the density profile in mean-field approximation can be regarded as the electrostatic charge distribution in case of a Yukawa interaction. How can the parabolically decaying density profile be understood from that point of view?

The determination of the electrostatic charge distribution can not only be seen as a minimizing problem of the electrostatic energy, but equivalently as the question of a local force equilibrium for all points where the density is nonzero. The forces in case of a screened Coulomb interaction can be obtained from (8.26), which represents the total potential at the point $\boldsymbol{r} \in \mathcal{V}_>$. Thus, taking the gradient results in the

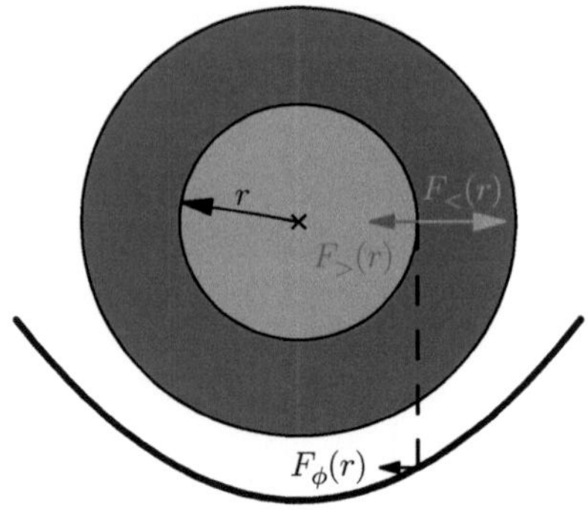

Fig. 8.5 Forces within a dust crystal for a spherical layer at distance r: external confining force $F_\phi(r)$, Yukawa repulsion $F_<(r)$ of all inner particles, and Yukawa repulsion $F_>(r)$ of all outer particles

corresponding force equation. For the harmonic confinement this equation is only radially dependent and yields

$$m\omega_0^2 r = F_<(r) + F_>(r), \tag{8.38}$$

which means that for any spherical layer at a distance r from the center the external force of the confinement $F_\phi(r) = -m\omega_0^2 r$, which acts toward the center, is balanced by the internal force due to the Yukawa repulsion between the particles. The internal force contains two parts, which are outlined in Fig. 8.5. The force

$$F_<(r) = 4\pi q_{\text{eff}}^2 \frac{\mathrm{e}^{-\kappa r}}{r}\left(1 + \frac{1}{\kappa r}\right)\int_0^r \mathrm{d}r'\, r' n(r') \sinh(\kappa r') \tag{8.39a}$$

arises from the action of all particles inside the given layer, $r' \leq r$, and acts outwards, whereas

$$F_>(r) = 4\pi q_{\text{eff}}^2 \frac{1}{r}\left(-\cosh(\kappa r) + \frac{\sinh(\kappa r)}{\kappa r}\right)\int_r^{R_{\text{mf}}} \mathrm{d}r'\, r' n(r') \mathrm{e}^{-\kappa r'} \tag{8.39b}$$

results from the action of all particles located outside, $r' \geq r$, and acts inwards. The density within these equations has to guarantee the balance of the forces so that (8.38) is fulfilled.

In the Coulomb case, the forces (8.39) simplify to

$$F_{<,\mathrm{C}}(r) = \frac{q_{\text{eff}}^2 N_<(r)}{r^2}, \tag{8.40a}$$

$$F_{>,\mathrm{C}}(r) = 0, \tag{8.40b}$$

with

$$N_<(r) = 4\pi \int_0^r \mathrm{d}r'\, r'^2 n(r')$$

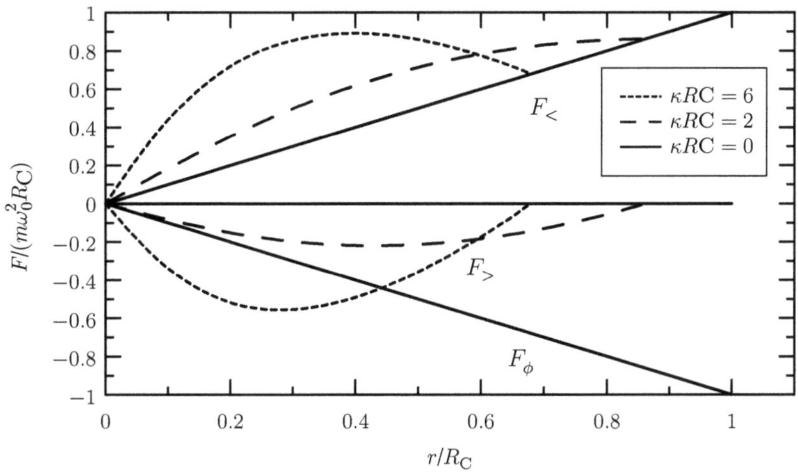

Fig. 8.6 Local force equilibrium within a dust crystal. For each $r \leq R_{\mathrm{mf}}$, the external force $F_\phi(r)$, the Yukawa repulsion $F_<(r)$ of all inner particles, and the Yukawa repulsion $F_>(r)$ of all outer particles result in a zero net force. In the Coulomb case ($\kappa R_C = 0$) the force $F_>(r)$ vanishes identically

being the particle number within the sphere of radius r. These two equations just represent two well-known results of Coulomb electrostatics: A spherically symmetric charge distribution produces the same field in the outer region as a point charge at the center of the sphere and the inside of a hollow charged sphere is free of forces. From these equations and the force equilibrium (8.38), it follows that the equilibrated Coulomb density is the constant density $n(r) = n_C$.

In the general case of finite screening the Coulomb results are not valid anymore, but the principle of equilibrium changes drastically. Now, a hollow charged sphere produces a force toward the center, and in case of a constant density this is not balanced by the outward going force $F_<$. To balance the force an additional central charge, that is, a higher central density, is required. Thus the equilibrium density is not constant, but has to increase toward the center. The resulting equilibrated forces are displayed within Fig. 8.6 for some screening parameters showing the general differences between the Coulomb and the Yukawa principles of forces.

8.4 Simulation Results of Spatially Confined Dust Crystals

The density profiles obtained in the previous section utilize the mean-field approximation. To check the quality of this approximation, the density profiles can be compared with results of numerical simulations. One of the best methods for the simulation of the canonical ensemble is the Monte Carlo method with the Metropolis algorithm (see the chapter of Bauch et al.), which can accurately calculate ensemble

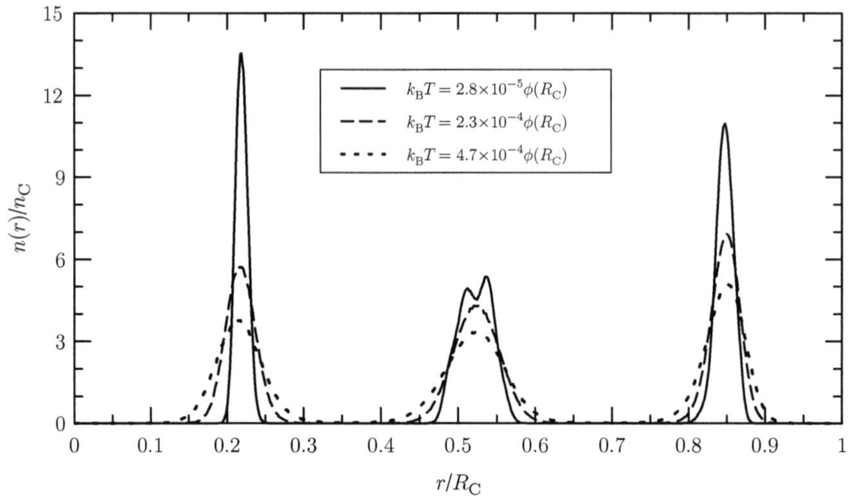

Fig. 8.7 Density profile of a Coulomb cluster with $N = 100$ particles for three low temperatures obtained from Monte Carlo simulations

averages of physical quantities for finite temperatures. As an example of such a Monte Carlo simulation, Fig. 8.7 shows the ensemble averaged density of a Coulomb cluster with $N = 100$ particles for different small temperatures. One clearly sees the shell structure, which becomes more pronounced when the temperature is reduced. For very low temperatures, even a subshell structure emerges.

8.4.1 Ground-State Simulations

In case of vanishing temperature, the Metropolis algorithm requires more and more computing time to calculate the right ensemble averages or rather to accurately determine the spatial distribution function f^s. Thus, this method is not appropriate for ground-state density profiles.

On closer inspection, it becomes apparent that the spatial distribution function at zero temperature is easy to achieve. In fact, it is pinpointed by the global minima, that is, the ground states, of the Hamiltonian (8.1). This issue is depicted by Fig. 8.8. There, the configuration-dependent energy of some system is sketched and corresponding distribution functions for three different temperatures T are drawn. For high temperatures, all low energy configurations are nearly equally probable, while for low temperatures only the configurations with lowest energy have a finite probability. Therefore, only the global minima of the Hamiltonian have to be determined to obtain the spatial distribution function and the ensemble averaged density profile for $T = 0$, respectively.

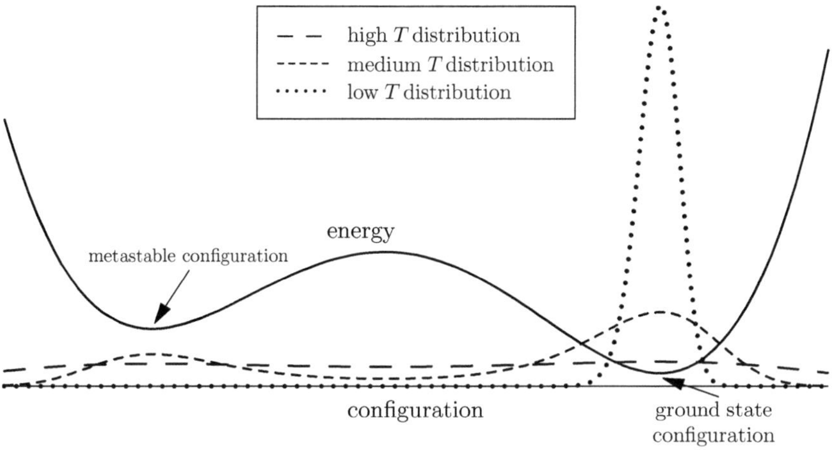

Fig. 8.8 Sketch of the configuration-dependent energy of some system and the corresponding probability distribution functions for different temperatures. The low-temperature distribution function is peaked at the global minimum, that is, the ground-state configuration

The global minima of the Hamiltonian can be found by global optimization techniques like Basin-Hopping [14] or simulated annealing (see the chapter Ott et al.). In doing so, the isotropy of the system entails that each global minimum represents a two-dimensional manifold of global minima, which merge by rotation.[3] Let $\mathbf{r}_{\min}$ denote one of these minima. Then the resulting ground-state density, which is spherically symmetric, is given by

$$n(r) = \sum_{i=1}^{N} \frac{\delta(r - |\boldsymbol{r}_{\min,i}|)}{4\pi r^2}. \tag{8.41}$$

This equation represents the exact (nonapproximated) ensemble averaged ground-state density. It shows a δ-peaked shell structure, which is displayed for the Coulomb system of $N = 100$ particles in Fig. 8.9a.

However, naturally not the structure of all the δ-peaks is called the shell structure of the cluster, but the widespread structure of grouped δ-peaks. A clearer graphical representation of this widespread structure is possible by substituting for the point particles of the ground states spherical objects of small size with either solid or cloudy consistence. By this mollifying, not only the shell structure but also the densities of the shells are revealed. For the aforementioned Coulomb system, three mollified results corresponding to three different radii of the substituted objects

[3] The rotational merging defines an equivalence relation on the set of the global minima. Within the following only one corresponding equivalence class is assumed, that is, barring the rotational symmetry there is only one global minimum. An extension to the general case of multiple equivalence classes is straightforward.

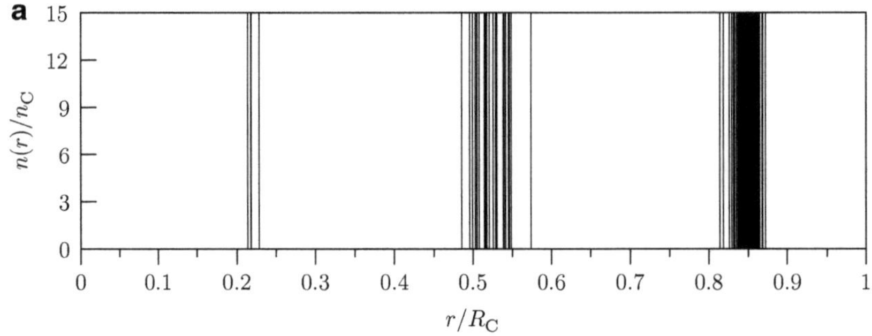

Delta-peaked structure. Each *thin line* represents a delta peak.

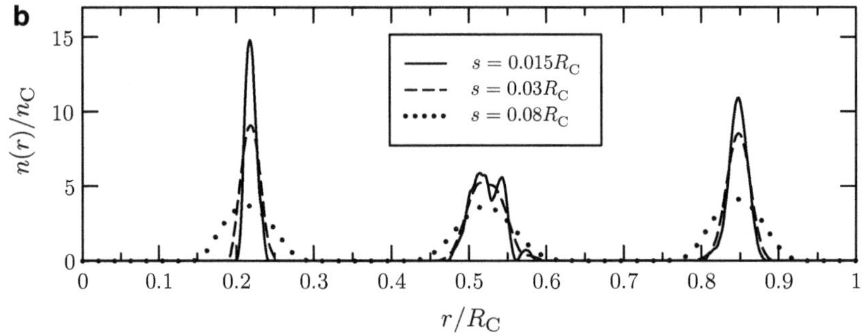

Mollified structure. For the point particles expanded spherical objects with radius s were substituted. The three different lines correspond to three different radii of these objects.

Fig. 8.9 Ground-state density profile of a Coulomb cluster with $N = 100$ particles obtained from the global minima of the Hamiltonian

are shown in Fig. 8.9b. There the shell structure including finite density values within the shells are evident and bear a resemblance to the Monte Carlo results of finite temperature (cf. Fig. 8.7). Nevertheless, finite temperature simulations with the Metropolis algorithm use distribution functions which have finite values in the vicinity of ground-state and metastable configurations (cf. Fig. 8.8). The mollified ensemble instead only uses configurations in the vicinity of the ground state, and hence gives an accurate description of the ground state.

8.4.2 Comparison of Simulation and Mean-Field Results

The mollified ground-state density profiles give a possibility of comparison with analytical results like the ones from the mean-field approximation. The direct comparison of Figs. 8.4 and 8.9b then shows that the mean-field result does not show any

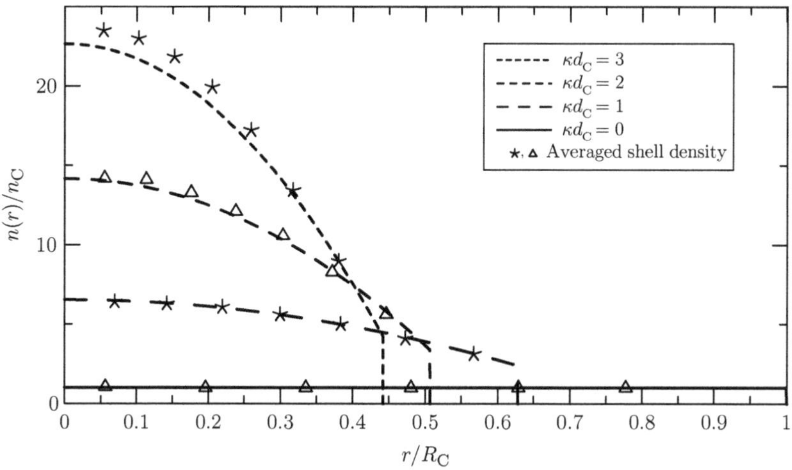

Fig. 8.10 Mean-field density profile of a harmonically confined dust crystal for four screening values (*lines*), *from bottom to top*: $\kappa d_C = 0, \kappa d_C = 1, \kappa d_C = 2$, and $\kappa d_C = 3$. The *symbols* denote the average shell densities, which are obtained from numerical simulations of a corresponding system with $N = 1{,}000$ particles

shell structure, what is caused by the neglect of correlations. However, the mean-field results should reflect the average behavior of the density profile, which itself is expressible by the average densities of the shells.

As an example, the mean-field density of a harmonically confined system with $N = 1{,}000$ particles, which is large enough to exhibit macroscopic behavior [1], is shown in Fig. 8.10. The symbols denote the average particle densities of the shells and are obtained from the mollified results of ground-state simulations. The figure shows that the mean-field results not only reflect the average behavior of the density, but quantitatively reproduces the radially decreasing average density very well. However, there are also discrepancies in case of strong screening (cf. $\kappa d_C \gtrsim 2$). These are caused by disregarding the correlation contributions in the mean-field energy functional (8.19), which become important with increasing density. Hence, to remove these deviations, the energy functional has to be extended by correlations.

8.5 Inclusion of Correlations by Using the Local Density Approximation

One way to include correlations in a simple but very successful way is to use the LDA [16], which is well known within the context of the density functional theory. This approximation is based upon the idea of replacing complicated nonlocal terms within the energy density by simple local expressions using the known energy density of the corresponding homogeneous system. Therefore, this method works fine in case of nearly homogeneous systems, but it is suitable even in case of rapidly varying densities.

8.5.1 LDA Without Correlations

To familiarize with LDA and its characteristics, it is advisable first to apply this method only to the energy density without correlations, which is given by $u_{\text{trap}} + u_{\text{mf}}$ (cf. (8.15)). To substitute this sum by the local density approximated energy density, one has to know the corresponding expression of the homogeneous system. This can be easily obtained by substituting the spatial-dependent density $n(\boldsymbol{r})$ within the proper energy density by a homogeneous density n_0. This yields

$$u_{0,\text{trap}} = n_0 \phi(\boldsymbol{r}), \tag{8.42}$$

$$u_{0,\text{mf}} = \frac{q_{\text{eff}}^2}{2} n_0 \int \mathrm{d}\boldsymbol{r}'\, n_0 \frac{\exp(-\kappa\,|\boldsymbol{r}-\boldsymbol{r}'|)}{|\boldsymbol{r}-\boldsymbol{r}'|} = q_{\text{eff}}^2 n_0^2 \frac{2\pi}{\kappa^2}, \tag{8.43}$$

wherein the infinite homogeneous system is considered.[4] Once the energy density of the homogeneous system is available, the energy density of LDA follows by the substitution $n_0 \to n(\boldsymbol{r})$:

$$u_{\text{LDA,trap}}(\boldsymbol{r}) = n(\boldsymbol{r})\phi(\boldsymbol{r}), \tag{8.44}$$

$$u_{\text{LDA,mf}}(\boldsymbol{r}) = q_{\text{eff}}^2 n(\boldsymbol{r})^2 \frac{2\pi}{\kappa^2}. \tag{8.45}$$

While the expression for $u_{\text{LDA,trap}}$ is identical to the nonapproximated expression (8.15a), this is not the case for the density of the mean-field energy. The LDA of the latter one is much simple than its proper expression (8.15b) and additionally diverges in the long-ranged Coulomb limit.

General Solution

By using the local approximated energy density, the LDA energy functional without correlations is given by

$$E_{\text{LDA}}[n] = N\mu_{\text{LDA}} + \int \mathrm{d}\boldsymbol{r}\; n(\boldsymbol{r})\Big(\phi(\boldsymbol{r}) - \mu_{\text{LDA}}\Big) + \frac{2\pi q_{\text{eff}}^2}{\kappa^2} \int \mathrm{d}\boldsymbol{r}\; n(\boldsymbol{r})^2. \tag{8.46}$$

Its minimum, the LDA ground-state density $n_{\text{LDA}}(\boldsymbol{r})$, is obtained in the same manner as the proper mean-field solution (cf. Sect. 8.3.2), and thus given by

$$n_{\text{LDA}}(\boldsymbol{r}) = \frac{1}{4\pi q_{\text{eff}}^2}\Big(\kappa^2 \mu_{\text{LDA}} - \kappa^2 \phi(\boldsymbol{r})\Big) \qquad \forall \boldsymbol{r} \in \mathcal{V}_{\text{LDA}} \tag{8.47}$$

[4] Regarding a finite homogeneous system results in finite-size effects, which are discussed in [15].

for all space points within the supporting region

$$\mathcal{V}_{\mathrm{LDA}} = \left\{ \boldsymbol{r} \in \mathbb{R}^3 | n_{\mathrm{LDA}}(\boldsymbol{r}) > 0 \right\}. \tag{8.48}$$

The Lagrange multiplier within is obtained from normalization and given by

$$\kappa^2 \mu_{\mathrm{LDA}} = \frac{4\pi q_{\mathrm{eff}}^2 N + \kappa^2 \int_{\mathcal{V}_{\mathrm{LDA}}} \mathrm{d}\boldsymbol{r}\, \phi(\boldsymbol{r})}{|\mathcal{V}_{\mathrm{LDA}}|}. \tag{8.49}$$

Both the density equation and the equation for the Lagrange multiplier are very similar to the equations of the nonlocal mean-field solution (8.30) and (8.31), but show one important difference. The Laplacian of the potential $\Delta\phi(\boldsymbol{r})$ is missing. That is a reflection of the fact that this Laplacian contains derivatives and thus information about contiguous values of the potential, which are generally suppressed within LDA.

However, the determination the supporting region $\mathcal{V}_{\mathrm{LDA}}$ cannot be accomplished as in Sect. 8.3.2. This is because the LDA energy density is local, so that minimization does only give a local condition for the density in contrast to the nonlocal integral equation (8.26). The determination can be realized instead by inserting the solution (8.47) directly into the LDA energy functional (8.46) and minimizing the resulting expression with respect to $\mathcal{V}_{\mathrm{LDA}}$, which yields after some algebra in

$$\mu_{\mathrm{LDA}} = \phi(\boldsymbol{r}) \qquad \forall \boldsymbol{r} \in \partial\mathcal{V}_{\mathrm{LDA}}. \tag{8.50}$$

Equally to (8.33), due to the symmetry of the isotropically confined dust crystals, the LDA supporting region is given by a ball:

$$\mathcal{V}_{\mathrm{LDA}} = \mathcal{B}(R_{\mathrm{LDA}}). \tag{8.51}$$

The boundary of this ball is described solely by one parameter, the LDA radius R_{LDA}, so that (8.50) results in

$$\mu_{\mathrm{LDA}} = \phi(R_{\mathrm{LDA}}), \tag{8.52}$$

which is an implicit equation for this radius.

Harmonic Confinement

For the harmonic form of the confinement, this implicit radius equation can be explicitly solved and yields

$$R_{\mathrm{LDA}} = \sqrt[5]{15 \frac{R_{\mathrm{C}}^3}{\kappa^2}}, \tag{8.53}$$

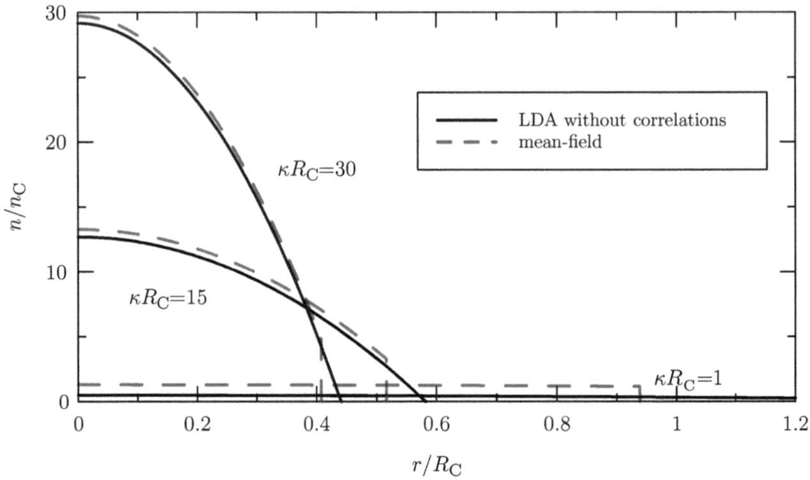

Fig. 8.11 Density profile of harmonically confined dust crystals in local density approximation without correlations for three screening parameters (*lines*), *from bottom to top*: $\kappa R_C = 1$, $\kappa R_C = 15$, and $\kappa R_C = 30$. For comparison, the corresponding nonlocal mean-field results are shown by the *dashed lines*

and the density profile is only radially dependent and reduces to

$$n_{\text{LDA}}(r) = n_C \frac{\kappa^2}{6}\left(R_{\text{LDA}}^2 - r^2\right)\Theta(R_{\text{LDA}} - r). \tag{8.54}$$

Corresponding results are shown in Fig. 8.11 for three screening parameters. The density profiles clearly possess a parabolic decrease away from the trap center until they vanish in a continuous manner. An increase of the screening parameter κR_C leads to a compression with respect to the Coulomb case. At the same time, the density values increase continuously, most significantly in the center. Thus, in the case of the harmonic potential, the LDA density profile without correlations bears qualitative resemblance to the nonlocal mean-field density profile.

However, quantitatively in two points both approximations differ from one another as can also be seen in Fig. 8.11. Firstly, the density in the LDA does not show a discontinuity at $r = R_{\text{LDA}}$, in contrast to the mean-field result. This is due to the neglect of finite-size effects in the LDA derivation. Secondly, LDA yields too small values for the density – most evidently in the diverging Coulomb limit. However, this underrating of the density is reduced with increasing values of the parameter κR_C. The reason for this improved behavior with increasing κR_C is due to the fact that an increase of κ contracts the effective area of integration within (8.43). This contraction is in favor of the accuracy of LDA, because the decreased integration volume contains a more homogeneous density. Additionally, an increase of the particle number N and consequently of R_C flattens the density profile, and will similarly improve LDA.

In summary, the LDA without correlations has shown that LDA results will of course not be as accurate as their nonlocal counterparts, but give a good approximation, especially in the case of strong screening.

8.5.2 LDA with Correlations

The LDA energy functional $E_{\mathrm{LDA}}[n]$ considered up to now contains only the energy densities of the confinement and of the mean-field interaction. The inclusion of particle correlations via LDA can be accomplished by using the energy density of correlations of the homogeneous system. An accurate approximation of this energy density is given for $n_0^{1/3} \geq 3\kappa/(20\pi)$ by

$$u_{0,\mathrm{corr}} = -\gamma_1\, q^2 n_0^{4/3} \exp\left(-\gamma_2\, \kappa\, n_0^{-1/3} + \gamma_3\left(\kappa\, n_0^{-1/3}\right)^4\right) \tag{8.55}$$

with

$$\gamma_1 = 1.444, \quad \gamma_2 = 0.375, \quad \gamma_3 = 7.4 \times 10^{-5}, \tag{8.56}$$

and was calculated in [17] from the Madelung energy of the corresponding Yukawa lattice. The local approximated energy density $u_{\mathrm{LDA,corr}}(\boldsymbol{r})$ then follows by substituting the density of the homogeneous system n_0 by the local density $n(\boldsymbol{r})$ of the inhomogeneous system. Consequently, the complete ground-state energy functional in LDA reads

$$E_{\mathrm{LDA}}[n] = N\,\mu_{\mathrm{LDA}} + \int \mathrm{d}\boldsymbol{r}\; u(\boldsymbol{r}) \tag{8.57}$$

with the energy density

$$\begin{aligned} u(\mathbf{r}) = n(\boldsymbol{r})\Big(\phi(\boldsymbol{r}) - \mu_{\mathrm{LDA}}\Big) + q_{\mathrm{eff}}^2\, n(\boldsymbol{r})^2 \tfrac{2\pi}{\kappa^2} - \gamma_1 q^2 n(\boldsymbol{r})^{4/3} \\ \times \exp\left(-\gamma_2 \kappa n(\boldsymbol{r})^{-1/3} + \gamma_3\left(\kappa n(\boldsymbol{r})^{-1/3}\right)^4\right). \end{aligned} \tag{8.58}$$

To assess the importance of the correlations within (8.58), the ratio of energy densities of correlations and of the mean-field interaction is plotted in Fig. 8.12. This ratio only depends on one parameter, $z(\boldsymbol{r}) = \sqrt[3]{n_{\mathrm{LDA}}(\boldsymbol{r})}/\kappa$, and shows three different regions. For $z(\boldsymbol{r}) \gtrsim 1$ the absolute of the ratio is very small and hence correlations are negligible. In contrast, for $z(\boldsymbol{r}) \lesssim 1$ the absolute of the ratio has higher values implying the importance of the correlations.

As before, variation of the energy functional yields the LDA ground-state density $n_{\mathrm{LDA}}(\boldsymbol{r})$, but now with correlations included. Due to the correlations, the energy

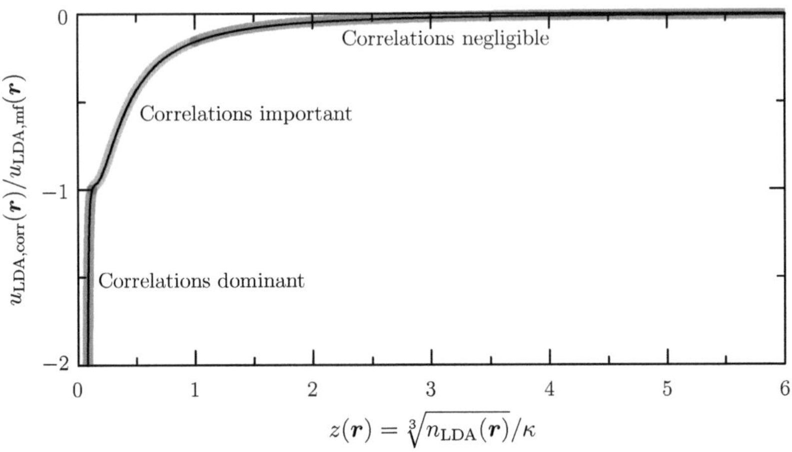

Fig. 8.12 The ratio of the energy densities of correlations and of the mean-field interaction in dependence on its local parameter $z(\boldsymbol{r}) = \sqrt[3]{n_{\mathrm{LDA}}(\boldsymbol{r})}/\kappa$ is shown. Three regions are highlighted which display the different kinds of importance of the correlations

density is strongly nonlinear and thus does not allow for an explicit solution. However, an implicit solution is possible and is given as a function of $z(\boldsymbol{r})$ by

$$0 = z(\boldsymbol{r})^3 + \frac{\phi(\boldsymbol{r}) - \mu_{\mathrm{LDA}}}{4\pi\kappa q_{\mathrm{eff}}^2} - \frac{N}{3\pi(N-1)}\Big(\gamma_1 z(\boldsymbol{r}) + \frac{\gamma_1\gamma_2}{4} - \gamma_1\gamma_3 z(\boldsymbol{r})^{-3}\Big)$$
$$\times \exp\Big(-\gamma_2 z(\boldsymbol{r})^{-1} + \gamma_3 z(\boldsymbol{r})^{-4}\Big) \quad \forall \boldsymbol{r} \in \mathcal{V}_{\mathrm{LDA}}. \tag{8.59}$$

The solution of this density equation as well as the Lagrange parameter μ_{LDA}, the supporting region, and the LDA radius R_{LDA}, respectively, have to be determined numerically. For the case of a harmonic confinement results are given in Fig. 8.13. There, the LDA ground-state density profiles with correlations are shown for three different screening parameters. For comparison the LDA results without correlations are shown too. In case of low screening, both density profiles are identical so that there is no effect of the particle correlations, in agreement with Fig. 8.12. But with increasing screening the correlation contributions within LDA alter the curvature of the profile, which rises more steeply toward the center. Hence, the particle correlations tend to increase the central density of the dust crystals.

On grounds of the limitation of (8.55), the resulting equation for the density (8.59) is not valid for densities smaller than the limiting density $n^* = 9n_{\mathrm{C}}(\kappa d_{\mathrm{C}})^3/(4{,}000\pi^2)$. However, this limitation is irrelevant as this limiting density is in all cases much smaller than the average density. For example, the values of n^* corresponding to the density profiles within Fig. 8.13 are $2.8 \times 10^{-5} n_{\mathrm{C}}$, $7.7 \times 10^{-4} n_{\mathrm{C}}$, and $6.2 \times 10^{-3} n_{\mathrm{C}}$, respectively.

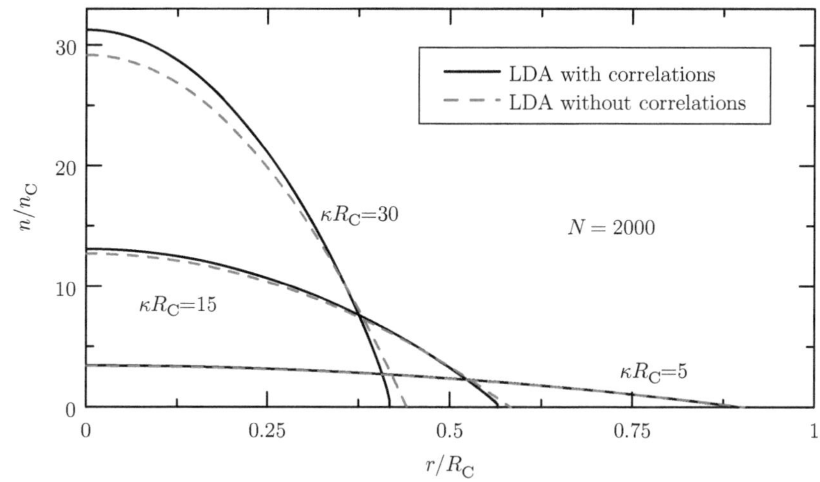

Fig. 8.13 Density profiles of harmonically confined dust crystals with $N = 2{,}000$ particles in local density approximation with correlations for three screening parameters (*lines*), *from bottom to top*: $\kappa R_C = 5$, $\kappa R_C = 15$, and $\kappa R_C = 30$. For comparison, the corresponding LDA results without correlations are shown by the *dashed lines*

8.5.3 Comparison of Simulation and LDA Results

The comparison of simulation and mean-field results in Sect. 8.4.2 revealed a very good agreement for weak screening, but some discrepancies for strong screening, which were attributed to the missing correlations within the mean-field approximation. The LDA, however, allows for the inclusion of such correlations in a simple manner and works accurately just in case of strong screening as was shown in Sect. 8.5.1. Consequently, in this case LDA results should be in good accordance with simulation results. Actually, this is the case as is shown in Fig. 8.14.

There, LDA ground-state density profiles with correlations are shown for four different screening parameters together with the average particle densities of the shells, which are obtained from mollified results of ground-state simulations. For comparison the mean-field density profiles are shown too. The figure shows that LDA allows for removing the discrepancies of the mean-field approximation, which arise in case of strong screening, but it does not feature the accuracy of the latter in case of weak screening. Therefore, both approximations complement one another in the description of the average density of harmonically confined dust crystals. To describe not only the average behavior of the density but also its shell structure the LDA is not appropriate, because it is only useful within the study of long-range correlations and does not treat the shell-causing short-range correlations correctly [11].

For a systematical treatment of correlations (including short-range correlations) the pair correlation function has to be included in the energy expression (cf. (8.14)). Therefore, different approximation schemes are available. However, this is far beyond the scope of this chapter and the interested reader is referred to [11, 18].

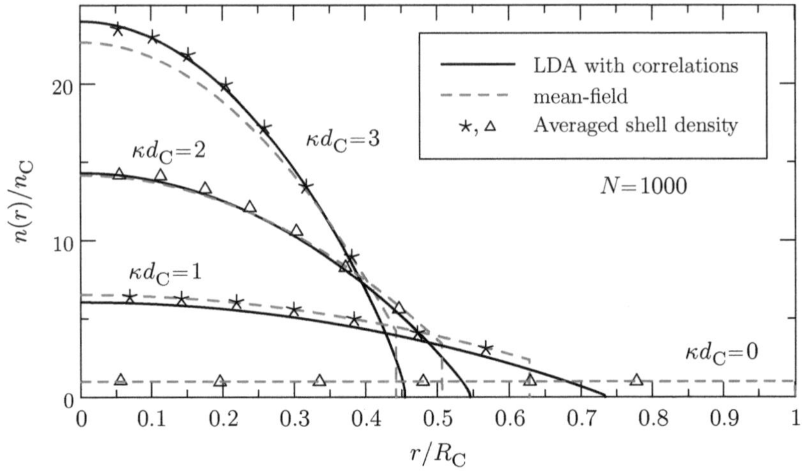

Fig. 8.14 LDA density profiles with correlations of harmonically confined dust crystals with $N = 1{,}000$ particles for four screening values (*lines*), *from bottom to top*: $\kappa d_C = 0$, $\kappa d_C = 1$, $\kappa d_C = 2$, and $\kappa d_C = 3$. The *symbols* denote the average shell densities, which are obtained from corresponding numerical simulations. For comparison, the mean-field results are shown by *dashed lines*

8.6 Shell Models of Spherical Dust Crystals

To describe the shell structure of harmonically confined dust crystals the so-called *shell models* in spite of their simplicity proved to be very successful. These models possess an immanent shell structure by making the ansatz

$$n_{\mathrm{sm}}(\boldsymbol{r}) = \sum_{\nu=1}^{L} N_\nu \frac{\delta(|\boldsymbol{r}| - R_\nu)}{4\pi R_\nu^2} \tag{8.60}$$

for the ensemble averaged density. Thus, there are L infinitely thin origin centered shells with radii R_ν and "occupation numbers" N_ν. The occupation numbers count the number of particles on each shell and hence fulfill $\sum_{\nu=1}^{L} N_\nu = N$. A sketch of the structure is shown in Fig. 8.15 for $L = 3$. In spite of the similarities to the exact density (8.41), shell models attempt to reproduce only the global shell structure of the cluster. For this purpose, the parameters L, $\{R_\nu\}$, and $\{N_\nu\}$ have to be determined, what is carried out by (numerically) minimizing a corresponding energy function $E_{\mathrm{sm}}\big(L, \{R_\nu\}, \{N_\nu\}\big)$ with respect to these parameters. This energy function then specifies the individual shell model.

The simplest shell model is the mean-field shell model, which neglects the contributions of the correlations. By analogy to electrostatics, it can be seen as a series of interlaced homogeneously charged capacitors. This model (or rather its energy function) can be easily obtained by evaluating the mean-field energy functional

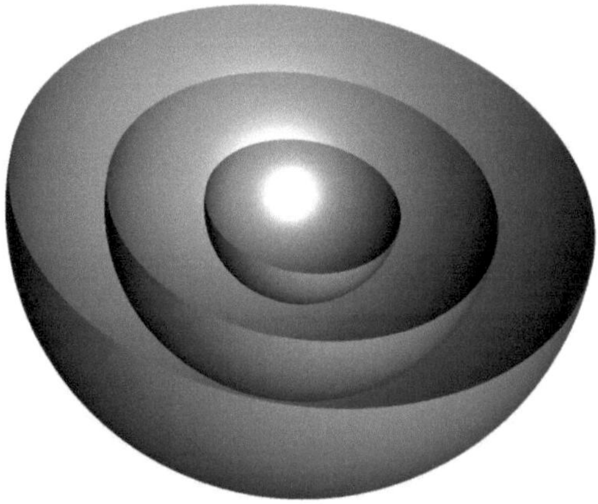

Fig. 8.15 Sectorial view of a shell model structure with $L = 3$ shells

(8.19) at the density of the ansatz (8.60), that is, $E_{\mathrm{mf}}[n_{\mathrm{sm}}]$. It yields, after some algebra [12],

$$E_{\mathrm{sm,mf}}\big(L,\{R_\nu\},\{N_\nu\}\big) = \sum_{\nu=1}^{L} N_\nu \Bigg\{ \phi(R_\nu) + q_{\mathrm{eff}}^2 \frac{\mathrm{e}^{-\kappa R_\nu}}{R_\nu} \times \left(\frac{\sinh(\kappa R_\nu)}{\kappa R_\nu}\frac{N_\nu}{2} + \sum_{\mu<\nu}^{L} \frac{\sinh(\kappa R_\mu)}{\kappa R_\mu} N_\mu \right) \Bigg\}, \tag{8.61}$$

and consists of three parts, which are common to all shell models. Beside the energy of confinement, there are the intrashell energy and the intershell energy, which take into account the interaction energy of the particles on the same shell and on different shells, respectively. The mean-field model can slightly be improved by allowing the possibility of a single central particle without self-interaction [5] and by a counting correction $N_\nu \to N_\nu - 1$ in the intrashell energy [12]. In the Coulomb limit ($\kappa \to 0$) it reduces to

$$E_{\mathrm{sm,HA}}\big(L,\{R_\nu\},\{N_\nu\}\big) = \sum_{\nu=1}^{L} N_\nu \left\{ \phi(R_\nu) + \frac{q^2}{R_\nu}\left(\frac{N_\nu}{2} + \sum_{\mu<\nu}^{L} N_\mu \right) \right\}, \tag{8.62}$$

that is, the shell model of Hasse and Avilov [4], which was designed to describe spherical ion crystals. Both mean-field models tend to the structure $L \to \infty$ with $N_\nu \to 0$ for all ν, what reflects the results of Sect. 8.3, showing that n_{mf} yields the lowest mean-field energy. Therefore, to get improved results, correlations have to be included.

In the *Coulomb* case different extended models were proposed, which differ in the expression of the intrashell energy E_{intra}. They are summarized in Table 8.1. The first extension of the mean-field model was given by Tsuruta and Ichimaru [5],

Table 8.1 Coulomb shell models of various authors and their expression for the intrashell energy

Coulomb shell model	E_{intra}/q^2
Hasse and Avilov [4]	$\sum_\nu N_\nu N_\nu/(2R_\nu)$
Tsuruta and Ichimaru [5]	$\sum_\nu N_\nu(N_\nu - \sqrt{N_\nu})/(2R_\nu)$
Kraeft and Bonitz [6]	$\sum_\nu N_\nu(N_\nu - \varepsilon\sqrt{N_\nu})/(2R_\nu)$
Cioslowski and Grzebielucha [7]	$\sum_\nu E_{\text{Th}}(N_\nu)/R_\nu$
	$\approx \sum_\nu N_\nu(N_\nu - \varepsilon_1\sqrt{N_\nu} + \varepsilon_2/\sqrt{N_\nu})/(2R_\nu)$

who took into account intrashell correlations by allocating a fixed domain to a given particle on a shell ν and distributing the $N_\nu - 1$ other particles uniformly on the shell except for that domain. The surrounding area of each particle not occupied by others was assumed to be equal to the total area of the shell divided by N_ν. This simple partitioning allows to reproduce the ground-state energy within a relative error less than 5%, but produces systematical errors for the radii and occupation numbers [6].

An improved shell model was given by Kraeft and Bonitz [6], who generalized the simple partitioning: Because there is no obvious reason why the domain of a particle is given by the total area of its shell ν divided by N_ν, they introduced a parameter ε, which accounts for a different size of the domain. The parameter ε was obtained by fitting either the energy E_{sm} or the occupation numbers to results from numerical simulations. Both methods allow a reproduction of energies, radii, and occupation numbers within very good agreement. Interestingly, for large N the parameter converges to $\varepsilon \approx 1.104$. The reason for this convergence was identified by Cioslowski and Grzebielucha [7]. They showed that the exact intrashell energy is strongly related to the energy solution $E_{\text{Th}}(N)$ of the Thomson problem with N particles, and used conjectured asymptotics of the Thomson problem to reason the convergence.

In case of *Yukawa* interaction the situation is more complex. To extend the mean-field shell model (8.61) by correlations one ansatz was made by Totsuji et al. [8], who approximated the correlation energy by that of a plane, two-dimensional Yukawa lattice obtained from numerical simulations. While the corresponding results are quite good for finite screening, the Coulomb limit is missed.

Another ansatz was made by Baumgartner et al. [9]. Analogous to the Coulomb shell model of Kraeft and Bonitz, they substituted $N_\nu \to N_\nu - \varepsilon\sqrt{N_\nu}$ within the intrashell energy E_{intra} of the Yukawa mean-field shell model. This model is able to reproduce energies, radii, and occupation numbers with good accuracy, but only for small screening values. A generally suitable shell model as it is available for the Coulomb case remains in demand for finite screening.

8.7 Summary and Discussion

Within this chapter, an introduction to statistical theories suitable to describe spherically confined dust crystals was presented. By means of a variational principle, the ground-state density could be derived from the ensemble averaged energy obtained in Sect. 8.2. Therefore, two different approximations were used, the mean-field and the local density approximation, which complement one another in the description of the ground-state density. Beyond this completely analytical theory, artificial shell models were introduced in order to describe the shell structure of the ground-state density.

While this chapter focussed on the ground-state properties, an extension to finite temperatures is possible. First results along this way have been obtained by Wrighton et al. [19], but this is beyond the scope of this chapter. For further reading see the review [20].

References

1. D.H.E. Dubin, T.M. O'Neil, Rev. Mod. Phys. **71**, 87 (1999)
2. O. Arp, D. Block, A. Piel, A. Melzer, Phys. Rev. Lett. **93**, 165004 (2004)
3. M. Bonitz, D. Block, O. Arp, V. Golubnychiy, H. Baumgartner, P. Ludwig, A. Piel, A. Filinov, Phys. Rev. Lett. **96**, 075001 (2006)
4. R.W. Hasse, V.V. Avilov, Phys. Rev. A **44**, 4506 (1991)
5. K. Tsuruta, S. Ichimaru, Phys. Rev. A **48**, 1339 (1993)
6. W.D. Kraeft, M. Bonitz, J. Phys.: Conf. Ser. **35**, 94 (2006)
7. J. Cioslowski, E. Grzebielucha, Phys. Rev. E **78**, 026416 (2008)
8. H. Totsuji, T. Ogawa, C. Totsuji, K. Tsuruta, Phys. Rev. E **72**, 036406 (2005)
9. H. Baumgartner, H. Kählert, V. Golubnychiy, C. Henning, S. Käding, A. Melzer, M. Bonitz, Contrib. Plasma Phys. **47**, 281 (2007)
10. R. Evans, Adv. Phys. **28**, 143 (1979)
11. J.P. Hansen, I.R. McDonald, *Theory of Simple Liquids* (Academic, London, 1991)
12. C. Henning, H. Baumgartner, A. Piel, P. Ludwig, V. Golubnichiy, M. Bonitz, D. Block, Phys. Rev. E **74**, 056403 (2006)
13. J.D. Jackson, *Classical Electrodynamics* (Wiley, New York, 1999)
14. D.J. Wales, J.P.K. Doye, J. Phys. Chem. A **101**, 5111 (1997)
15. C. Henning, P. Ludwig, A. Filinov, A. Piel, M. Bonitz, Phys. Rev. E **76**, 036404 (2007)
16. H. Totsuji, C. Totsuji, K. Tsuruta, Phys. Rev. E **64**, 066402 (2001)
17. H. Totsuji, J. Phys. A: Math. Gen. **39**, 4565 (2006)
18. S. Ichimaru, *Statistical Plasma Physics, Vol. II: Condensed Plasmas* (Addison-Wesley, Redwood City, CA, 1994), p. 33
19. J. Wrighton, J. Dufty, C. Henning, M. Bonitz, J. Phys. A: Math. Theor. **42**, 214052 (2009); J. Wrighton, J.W. Dufty, H. Kählert, M. Bonitz, Phys. Rev. E **80**, 066405 (2009); Contrib. Plasma Phys. **80**, 26 (2010)
20. M. Bonitz, C. Henning, D. Block, Reports on Prog. Phys. **73**, 066501 (2010)

Chapter 9
PIC–MCC Simulations of Capacitive High-Frequency Discharge Dynamics with Nanoparticles

Irina V. Schweigert

Abstract A new combined particle-in-cell Monte Carlo collision (PIC–MCC) approach is discussed for accurate and fast simulation of a radio-frequency (rf) discharge at a low gas pressure and high plasma density. Test calculations of the heating mode transition in a capacitively coupled rf-discharge in helium and argon show a good agreement with experimental data. The combined PIC–MCC algorithm is very efficient, especially for the collisionless regime of electron heating. Using this algorithm, the properties of a capacitively coupled 13.56 MHz discharge are studied in a mixture of Ar/C_2H_2 with nanoparticles of different sizes. For the description of plasma–dust interaction a kinetic model is introduced. The dust surface potential and discharge parameters are calculated self-consistently. In this model, any assumption about electron and ion energy distribution functions is not used. This allows us to calculate accurately the dust particle charging. The transport of dust is calculated by using the fluid approach. It is shown that at the initial stage of the growth, nanoparticles are accumulated near the sheath–plasma boundaries, where the ionization rate has maximum value. The nanoparticles suppress the ionization due to the absorbing of fast electrons and stimulate a quick change of the plasma parameters followed by the transition between different modes of discharge operation.

9.1 Introduction

Capacitively coupled 13.56 MHz discharges in a mixture of argon and acetylene were intensively studied in recent experiments [1, 2]. The formation and growth of hydrocarbon nanoparticles due to gas phase reactions in the discharge volume were observed. The nanoparticles grow to some critical radius, leave the central part of the discharge, and eventually drop to the bottom electrode due to gravity. After that

I.V. Schweigert (✉)
Institute of Theoretical and Applied Mechanics, Russian Academy of Sciences, Novosibirsk 630090, Russia
e-mail: ischweig@itam.nsc.ru

M. Bonitz et al. (eds.), *Introduction to Complex Plasmas*, Springer Series on Atomic, Optical, and Plasma Physics 59, DOI 10.1007/978-3-642-10592-0_9,

a new particle growth cycle starts. This periodic behavior is accompanied by strong periodic variations of all plasma characteristics [1, 2]. One of the most spectacular experimental observations is that only at the initial stage of nanoparticle growth, the discharge parameters change very rapidly. With further particle growth only a slight variation of the plasma parameters was found.

In this chapter, we concentrate on a theoretical study on the phenomena of dust particle transport and the change of plasma parameters at the initial stage of particle growth [3]. Note that the influence of movable dust on the discharge properties and the influence of the dust distribution on the initial stage of the nanoparticle growth is not completely understood. Previously, a numerical and experimental study of the capacitively coupled radio-frequency (ccrf) discharge in argon with 50 nm dust particles in the volume was carried out in [4]. In pure silane, the transition between capacitive and resistive modes of ccrf-discharge operation was observed with silicon particles in the experiment [5] and in numerical calculations [6]. Unfortunately, in simulations in [4, 6] the dust profile in the discharge volume was set by a model function. Usually the capacitive mode is characterized by much higher plasma density. The electron energy profile exhibits two maxima in the sheaths adjacent to the electrodes and a deep minimum in the center of the discharge. In the resistive mode of glow, the discharge has lower plasma density, higher electron energy, and higher power input. The mean electron energy profile has the usual shape of a single plateau. In Denysenko et al. [7], the power absorption and discharge glow were analyzed within the framework of an analytical discharge model with uniformly distributed dust for the experimental conditions of Kovačević et al. [1] and Stefanović et al. [2]. The dust transport in the fluid approximation was studied in [8] for a two-dimensional geometry of the discharge chamber. The charging of nanoparticles was treated using the orbital motion limited (O.M.L.) theory and a Maxwellian velocity distribution of electrons and ions. However, in the dusty plasma it generally is not a good approximation, because electron and ion distributions can be far from Maxwellian ones. In Chutov and Goedheer [9], the charging of unmovable dust was studied in a particle-in-cell Monte Carlo collision (PIC–MCC) simulations. Here, the authors assumed that the nanoparticles are uniformly distributed over discharge volume including the electrode sheaths. The assumption is very rough and diminishes the reliability of the conclusions. The characteristic time of dust motion is much larger than the discharge rf-time and the nanoparticles move in a time – averaged electrical field. Therefore, the negatively charged dust is trapped in the bulk plasma by the electrical potential.

Here, the combined PIC–MCC method developed in [10] is applied to calculate the properties of discharge with nanoparticles. The following new aspects in our study will be discussed (a) dust particle charging by using the electron and ion energy distribution functions taken from the kinetic PIC–MCC simulation, (b) the motion of dust is calculated self-consistently with the discharge parameters, and (c) the role of movable dust on the transition process between the different discharge regimes.

This chapter is organized as the following. In Sect. 9.2, the description of the combined PIC–MCC algorithm is presented. In Sect. 9.3, the influence of movable dust on the discharge glow is considered.

9.2 Combined PIC–MCC Approach for Fast Simulation of a Radio-Frequency Discharge at Low Gas Pressure

Modern trends of plasma technologies are directed to a reduction of the gas pressure and an increase of plasma density. Furthermore, the development of efficient methods is required for simulation of collisionless regimes in a capacitively coupled and especially in an inductively coupled discharges as the collisionless heating of electrons plays a key role in dynamics of thin skin layers. The PIC–MCC method [11] has become a standard simulation technique for a gas discharge in plasma reactors of etching or deposition. Unlike the fluid approach, the PIC–MCC algorithm requires large computer resources, but it provides a detail kinetic picture of processes in a gas discharge. However, a problem of statistical fluctuations of an electric field appears at low gas pressures, in particular for gases with a deep Ramsauer minimum in the elastic scattering cross section. At low gas pressure in argon, the ccrf-discharge operates in the capacitive regime and the electrons gain the energy in electrode sheaths. In the quasineutral plasma, the electric field is small and the electrical field numerical fluctuations can be comparably large due to the discrete representation of electrons and ions in Monte Carlo method. The electron heating on these numerical fluctuations of the electrical field essentially distorts the electron energy distribution function. Although the numerical smoothing of the charge density [12] helps to diminish the statistical noise, it is necessary to develop a novel way for reduction of the influence of statistical fluctuations. As the discharge simulation lasts more than 1,000 of discharge cycles, in [13] the averaging of the charge density over several cycles was suggested to suppress the statistical noise. However, the direct averaging can lead to the development of the numerical instability. To eliminate this problem, the electric field was calculated in [13] from the current continuity equation. Note that this approach requires an explicit distinction of electrode sheaths that is difficult for implementation in the two-dimensional case. Besides, it does not take into account inertia of electrons, which is very important at low gas pressures.

Recently, the hybrid models combining the kinetic and fluid descriptions are widely used for simulation of glow discharges (see, e.g., [14–16]). In V.A. Shveigert and I.V. Shveigert [14], the cathode sheath (including the negative glow and Faraday dark regions) of the stationary self-sustained glow discharge was calculated self-consistently. There, the motion of higher energy electrons was treated with a Monte Carlo method (see Chap. 4) to find the electron impact rate coefficients, whereas the lower energy electrons and ions were described within the drift–diffusion model. Belenguer and Boeuf [15] applied the two electron group model for the detail study the of $\alpha - \gamma$ transition in a rf-discharge in helium. In the Sommerer and Kushner hybrid model [16] N_1-rf-discharge cycles were simulated with the kinetic model and then N_2-rf-cycles with the fluid approach. In the kinetic simulations, the kinetic coefficients were updated and passed to the fluid model where the electrical field and the electron and ion densities were renewed. The iterations of N_1-kinetic and N_2-fluid rf-cycles were proceeded until the convergence of plasma parameters is established.

Below another way of noise reduction in the combined PIC–MCC approach, developed in [10], is presented. In this approach, both the kinetic and fluid models are used over a rf-discharge cycle that provides high accuracy of results, in particular for Ramsauer gases and for low gas pressure $P = 0.001$–0.1 Torr.

9.2.1 Combined PIC–MCC Approach

Our combined PIC–MCC model includes the electron and ion kinetic equations (three dimensional over velocity and one dimensional in space), the auxiliary equations for electron and ion densities and fluxes and the Poisson equation. The velocity distribution functions of electrons $f_e(t, x, \boldsymbol{v})$ and ions $f_i(t, x, \boldsymbol{v})$ are obtained from Boltzmann kinetic equations:

$$\frac{\partial f_e}{\partial t} + \boldsymbol{v}_e \frac{\partial f_e}{\partial x} - \frac{e\boldsymbol{E}}{m}\frac{\partial f_e}{\partial \boldsymbol{v}_e} = J_e, \quad n_e = \int f_e \mathrm{d}\boldsymbol{v}_e, \tag{9.1}$$

$$\frac{\partial f_i}{\partial t} + \boldsymbol{v}_i \frac{\partial f_i}{\partial x} + \frac{e\boldsymbol{E}}{M}\frac{\partial f_i}{\partial \boldsymbol{v}_i} = J_i, \quad n_i = \int f_i \mathrm{d}\boldsymbol{v}_i, \tag{9.2}$$

where v_e, v_i, n_e, n_i, and m, M are the electron and ion velocities, densities, and masses, respectively. J_e and J_i are the collisional integrals for electrons and ions. The density and the flux of electrons and ions are calculated from the auxiliary equations based on the momentums of the kinetic equations (9.1) and (9.2):

$$\frac{\partial n'_e}{\partial t} + \frac{\partial j'_e}{\partial x} = Q, \tag{9.3}$$

$$\frac{\partial n'_i}{\partial t} + \frac{\partial j'_i}{\partial x} = Q, \tag{9.4}$$

$$\frac{\partial j'_e}{\partial t} = -\frac{\partial T'_e n'_e}{\partial x} - \frac{eE}{m} n'_e - \nu_e j'_e - Q_e, \tag{9.5}$$

$$\frac{\partial j'_i}{\partial t} = -\frac{\partial T'_i n'_i}{\partial x} + \frac{eE}{M} n'_i - \nu_i j'_i - Q_i, \tag{9.6}$$

where

$$Q = N_g \int v_{ex} \sigma_i f_e \mathrm{d}\boldsymbol{v}_e \tag{9.7}$$

is the ionization rate, σ_i is the ionization cross sections, N_g is the gas density, and

$$T'_e = \frac{\int v_{ex}^2 f_e \mathrm{d}\boldsymbol{v}_e}{\int f_e \mathrm{d}\boldsymbol{v}_e}, \quad T'_i = \frac{\int v_{ix}^2 f_i \mathrm{d}\boldsymbol{v}_i}{\int f_i \mathrm{d}\boldsymbol{v}_i} \tag{9.8}$$

are the effective electron and ion temperatures, respectively.

$$Q_e = N_g \int \nu_{ex} |\boldsymbol{v}_e| \sigma_t f_e \mathrm{d}\boldsymbol{v}_e - \nu_e \int \nu_{ex} f_e \mathrm{d}\boldsymbol{v}_e, \tag{9.9}$$

$$Q_i = N_g \int \nu_{ix} |\boldsymbol{v}_i| \sigma_r f_i \mathrm{d}\boldsymbol{v}_i - \nu_i \int \nu_{ix} f_i \mathrm{d}\boldsymbol{v}_i \tag{9.10}$$

describe the friction for electrons and ions, the efficient frequencies

$$\nu_e = \frac{N_g \int |\boldsymbol{v}_e| \sigma_t f_e \mathrm{d}\boldsymbol{v}_e}{\int f_e \mathrm{d}\boldsymbol{v}_e}, \quad \nu_i = \frac{N_g \int |\boldsymbol{v}_i| \sigma_r f_i \mathrm{d}\boldsymbol{v}_i}{\int f_i \mathrm{d}\boldsymbol{v}_i}, \tag{9.11}$$

where σ_t is the electron transport cross section and σ_r is the ion resonance charge exchange cross section. After calculating the auxiliary values of electron n'_e and ion n'_i densities, the electric field is found from the Poisson equation

$$\Delta\phi = 4\pi e(n'_e - n'_i), \quad E = -\frac{\partial \phi}{\partial x}. \tag{9.12}$$

Notice that in the usual fluid approach, the terms Q_e and Q_i are supposed to be zero, which is correct only for the constant scattering frequencies.

9.2.2 Description of the Algorithm

The presented combined PIC–MCC algorithm cannot be separated in "fluid" and "kinetic" stages. In this algorithm "fluid" and "kinetic" calculations are performed during one rf-cycle and, therefore, the approximation of kinetic coefficients by analytical expressions is not needed. The scheme of algorithm is shown in Fig. 9.1. Over one rf-cycle there is three characteristic time steps: δt_i for ions and δt_e for electrons in Monte Carlo calculations and δt_f for solving the auxiliary equations and the Poisson equation. Note that $\delta t_i \gg \delta t_e \gg \delta t_f$. First suppose the plasma parameters have some initial distribution obtained from the analytical estimations. The auxiliary equations and the Poisson equation (APE) are solved with δt_f step over a rf-cycle. As a result we obtain the electrical field distribution in which the electron kinetic equation (EKE) is solved every $10\delta t_f$ steps. The electron and ion weights are fitted with the densities n'_e, n'_i calculated from the APE model. Using Monte Carlo method we allow all electrons to move during one time step $\delta t_e = 10\delta t_f$. The resulting electron distribution function is used for calculation of the kinetic coefficients Q, T'_e, T'_i, Q_e, and Q_i with (9.7)–(9.11). Every $10\delta t_f$ steps the APE model is executed with updated kinetic coefficients. For every time interval $\delta t_i = 10\delta t_e = 100\delta t_f$, we solve the ion kinetic equation (IKE) with the Monte Carlo method. All ions make one step with δt_i, and the ion density distribution and the electron–ion secondary emission from electrodes are calculated. The ion time step $\delta t_i = 1/(\omega \times i_0)$, where $i_0 = 60$–100 and $\delta t_i \approx 10^{-9}\,\mathrm{s}^{-1}$. The kinetic coefficient Q, T'_e, T'_i, Q_e, and Q_i

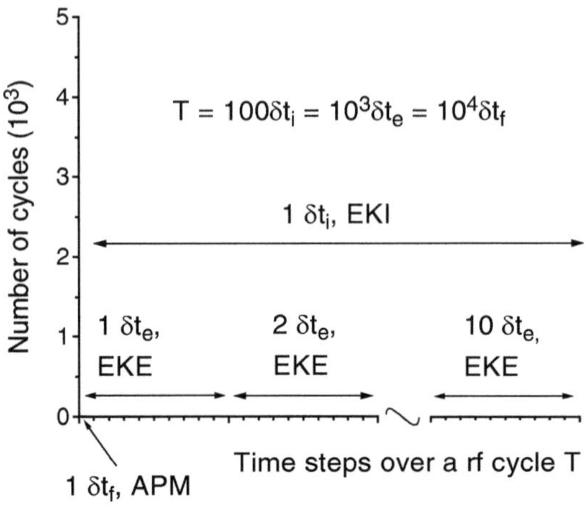

Fig. 9.1 Combined PIC–MCC algorithm. T is the rf-cycle time, δt_{i} and δt_{e} are the time steps of the ion (EKI) and electron (EKE) kinetic equations, respectively, δt_{f} is the time step of the transport and Poisson equations (APE)

are the functions of position x_k and time moment t_{i} over a rf-cycle (where x_k is the nodes of simulation grid and i varies from 1 up to i_0). The reduction of statistical noise is reached with averaging the kinetic coefficients over many rf-cycles and with smoothing over the spatial coordinate. For averaging a function $F(x_k, t_{\mathrm{i}})$ over preceding rf-cycles we use the following algorithm

$$F(x_k, t_{\mathrm{i}})^n = \alpha F(x_k, t_{\mathrm{i}})'^n + (1-\alpha) F(x_k, t_{\mathrm{i}})^{n-1}, \tag{9.13}$$

where $F(x_k, t_{\mathrm{i}})'^n$ is the value on the n – the rf-cycle and $\alpha = 0.01$–0.1 depending on plasma parameters. The spatial smoothing is chosen as in [12]

$$F(x_k, t_{\mathrm{i}}) = \frac{F(x_{k+1}, t_{\mathrm{i}}) + 2F(x_k, t_{\mathrm{i}}) + F(x_{k-1}, t_{\mathrm{i}})}{4}. \tag{9.14}$$

The spatial smoothing is very important for resolving the space charge in the quasineutral part of a discharge, where the charge is a small difference of two large and almost equal values (ion and electron densities).

It can be easily seen that (9.3)–(9.6) are direct consequences of the kinetic equations (9.1) and (9.2). As we calculate the kinetic coefficients Q, T'_{e}, T'_{i}, Q_{e}, and Q_{i} by solving the kinetic equations, the obtained densities n'_{e} and n'_{i} have to coincide with a good accuracy with values from the kinetic equations (9.1) and (9.2). Equations (9.3) and (9.5) are approximated with implicit finite-difference schemes:

$$\frac{n^i_{\mathrm{e},k} - n^{i-1}_{\mathrm{e},k}}{\Delta t} + \frac{j^i_{\mathrm{e},k+1/2} - j^i_{\mathrm{e},k-1/2}}{x_{k+1/2} - x_{k-1/2}} = Q^i_k, \tag{9.15}$$

$$\frac{j^i_{\mathrm{e},k+1/2} - j^{i-1}_{\mathrm{e},k+1/2}}{\Delta t} + \nu^i_{\mathrm{e}} j^i_{k+1/2} = -\frac{a_{k+1/2} n^i_{\mathrm{e},k+1} - b_{k+1/2} n^i_{\mathrm{e},k}}{x_{k+1} - x_k} - Q^i_{\mathrm{e},k+1/2}, \tag{9.16}$$

where $x_{k+1/2} = (x_{k+1} + x_k)/2$, $j_{e,k+1/2} = j'_e(x_{k+1/2})$, and $n_{e,k} = n'_e(x_k)$; the index i denotes number of time step; the coefficients a_k and b_k are calculated with the Scharfetter and Gummel scheme [17]

$$a_{k+1} = c_{k+1/2}\frac{T'_{e,k+1}}{\exp(c_{k+1/2}) - 1}, \quad b_{k+1} = c_{k+1/2}\frac{T'_{e,k}\exp(c_{k+1/2})}{\exp(c_{k+1/2}) - 1}, \tag{9.17}$$

and $c_{k+1/2} = 2e(\phi^{i-1}_{k+1} - \phi^{i-1}_k)/(T'_{e,k+1} + T'_{e,k})$. A similar scheme was used for ions transport calculations (9.4) and (9.6). The boundary conditions for the auxiliary equations includes the secondary emission as in [18]. For small grid spacing $T_e \gg e|\phi_{k+1} - \phi_k|$ this finite-difference scheme has a second order accuracy in Δx and gives a correct result on rough grids for the Boltzmann electron distribution. Like for the explicit PIC–MCC method, there exists a restriction on time step $\omega_p \Delta t < 1$, where ω_p is the plasma frequency. For the solution of (9.3)–(9.6) with the Poisson equation (9.12), therefore, the time step is $\delta t_f < 10^{-11}\,\text{s}^{-1}$.

The restriction on a time step is usually set by the gas pressure and the plasma density. For example, in argon the electron collision frequency $\nu \sim 10^7$–$10^{10}\,\text{s}^{-1}$ at $P = 0.001$–1 Torr and $\omega_p \sim 5 \times 10^8$–$2 \times 10^{10}\,\text{s}^{-1}$ at $n_e = 10^8$–$10^{11}\,\text{cm}^{-3}$. The minimum step of the spatial grid must be much less than the electrode sheaths thickness, which decreases with pressure increasing. For the grid values in the range $\delta x = 0.01$–0.1 cm and the electron energy $U_e = 10$ eV the time of electron flight through a grid step $\delta x/v_e$ is 5×10^{-10}–5×10^{-11} s. Note that in some schemes used for the charge distribution and electric field calculation in the PIC–MCC method, there is a restriction for the maximal grid step $\lambda_e / \Delta x > 0.3$, where λ_e is the electron Debye radius [19]. Here, we have chosen a scheme, which conserves the energy [20], where the last restriction is not important.

In the kinetic model, we use 5,000 simulation particles for each charged species and the cloud-in-cell charge assignment scheme. In this scheme, the particles are considered as having finite size, with a diameter roughly that of the cell. The null collision method is applied to calculate the real electron scattering rates. In this method we apply the following numerical steps: First, for all of the collisions the cross sections, which are a function of the electron total kinetic energy E, have to be included. Cross sections used for electrons are specified by σ_{max}. Knowing all of the collision frequencies between the charged particles and neutrals and the total collision frequency $\nu = N_g \sigma_t E v$ (where N_g is the gas density, σ_t is the total cross section, and v is the electron velocity) the probability of collision of an electron (or ion) in a time step δ_t is $P = 1 - \exp[-N_g \sigma_t E v \delta_t]$.

The next step is to compare P with R, where R is a uniform random number. For $P > R$, the electron is to be scattered in the step δ_t. Then we determine which scattering process occurs. This is done by assuming that the probability of process P_m is proportional to $\sigma_m(E)/\sigma_t(\text{E})$. The energy conserving scheme with a second order of accuracy to solve the equations of motion [12, 20].

Since the cross section of the electron Coulomb scattering is $\sigma_{ee} \sim n_e/U_e^2$ (where U_e is the mean electron energy), a correct discharge simulation of some regimes requires accounting for electron Coulomb collisions. For description of these collisions we apply the method [21], where the Langevin force and friction of electrons are introduced and defined from their distribution function. Note also that the Coulomb collisions do not change the total electron momentum of motion. In our model the kinetic equations, the auxiliary equations and the Poisson equation are solved self-consistently. The calculation typically proceeding during a few thousands rf-cycles leads to the convergence of kinetic coefficients and all discharge characteristics. The electron energy and density from the kinetic and fluid models are in good agreement. The algorithm is numerically stable and provides a significant acceleration of the PIC–MCC method due to two factors. At first, the time step for solving the kinetic equations (9.1) and (9.2) with the implicit scheme does not depend on the plasma frequency [12, 19]. Second, averaging over many rf-cycles allows one to reduce greatly (by factor of 5–20) the total number of simulation particles in the PIC–MCC method without an increase of the statistical noise.

9.2.3 How Many Simulation Particles We Need?

In the following, we are interested in the study of a ccrf-discharge in argon and helium with the combined PIC–MCC approach for the experimental conditions of Godyak et al. [22]. The ccrf-discharge operates with the frequency $\omega = 13.56\,\mathrm{MHz}$ and with the sinusoidal shape of the discharge current j. One electrode is grounded and the voltage on another electrode is calculated self-consistently to sustain the desired amplitude of the discharge current. The spatial grid has typically 81 nodes for the interelectrode distance $d = 2\,\mathrm{cm}$ and 181 nodes for $d = 6.7\,\mathrm{cm}$, condensing in electrode sheaths. The minimum grid spacing is decreased with the gas pressure rise, thus the sheath contains approximately the constant number of nodes. The cross sections of electron scattering in helium are taken from [23], and for argon from [13, 23]. The ion–electron emission from electrodes is taken into account with coefficient 0.2 in helium and 0.1 in argon.

It is well known that the statistical error of Monte Carlo methods decreases as $1/N^2$. The statistical noise leads to the systematical error in the electron cooling or heating. Therefore, we have studied the influence of the number of simulation particles on an accuracy of results with three different methods: the standard PIC–MCC [11], the PIC–MCC with the spatial smoothing (PIC–MCC SS) [12] and with our combined PIC–MCC. The mean electron energy in the discharge center calculated with three methods and measured in [22] is shown in Fig. 9.2. The simulations are performed for two values of argon pressures $P = 0.1,\ 0.3\,\mathrm{Torr}$, the interelectrode distance $d = 2\,\mathrm{cm}$, and the discharge current $j = 2.65\,\mathrm{mA/cm^2}$. The results of the standard PIC–MCC method with different N show the significant role of electric field fluctuations under lower gas pressure (squares in Fig. 9.2a). The standard PIC–MCC considerably overestimates the value of U_e for $N = 4{,}000$–$256{,}000$. In contrast, the second method, PIC–MCC SS, gives much better results (circles in Fig. 9.2a). Here, the spatial smoothing indeed decreases the statistical noise, but the feasibility of this technique is restricted, since it distorts

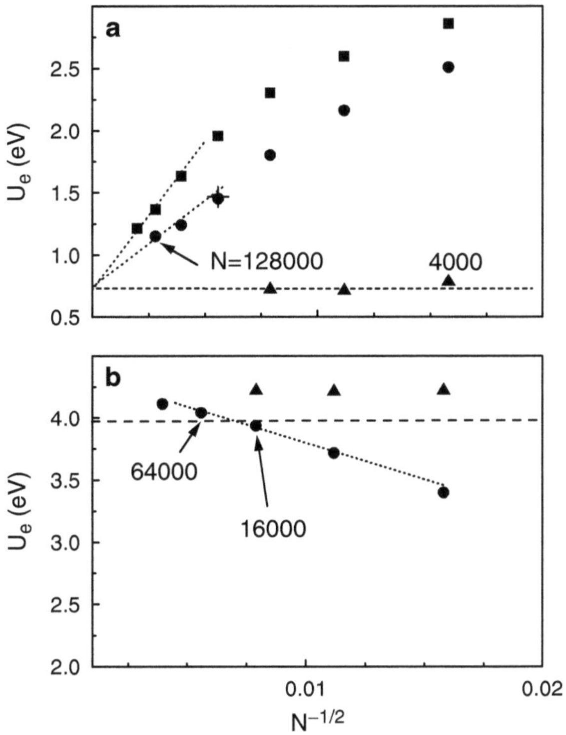

Fig. 9.2 Mean electron energy, U_e, at the discharge center as a function of the total number of simulation particles for $P = 0.1$ Torr (**a**) and $P = 0.3$ Torr (**b**) calculated using the standard PIC–MCC method (*squares*), with the PIC–MCC method with spatial smoothing of the space charge and electrical field distributions (PIC–MCC SS) (*dots*), and with our combined algorithm (*triangles*). The *cross* indicates a calculation from [12] with $N = 32{,}000$, $d = 2$ cm, and $j = 2.65$ mA/cm^2. The *dashed lines* show the measured U_e [22]

the space charge in the electrode sheath. The PIC–MCC SS is not able to guarantee convergence to the measured value of U_e (dashed line in Fig. 9.2a) even with $N = 256{,}000$. At higher gas pressure $P = 0.3$ Torr, the computed U_e shown in Fig. 9.2b increases with N, and the PIC–MCC SS method gives accurate results (within 10%) with comparably small number of simulation particles $N = 10{,}000$. Nevertheless, the fact that at low gas pressure the enormous number of simulation particles is needed for accurate simulation makes the standard PIC–MCC methods inapplicable in this range of the gas pressure. As seen in Fig. 9.2 the electron energy obtained with the combined PIC–MCC method is very close to the experimental one already with 4,000 simulation particles.

9.2.4 Simulation Results of a CCRF-Discharge in Helium and Argon

The transition between different modes of a rf-discharge glow is well studied experimentally [22, 24] and numerically (see, e.g., [11, 15, 25, 26]). In our simulation, we considered the heating mode transition in the ccrf-discharge with changing gas pressure. The electron density and the mean electron energy in the midplane are

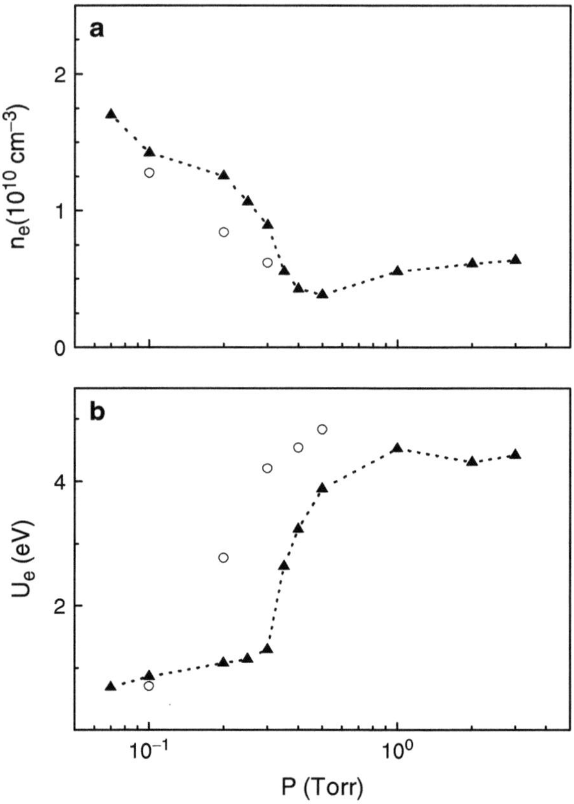

Fig. 9.3 Electron density (**a**) and mean electron energy (**b**) in the argon discharge center ($x = 1$ cm) computed (*circles*) and measured in [22] (*triangles*) for $d = 2$ cm, $j = 2.65$ mA/cm^2, and $N = 5{,}000$

shown in Fig. 9.3 as a function of argon gas pressure for $j = 2.65$ mA/cm^2 and $d = 2$ cm. The computed n_e and U_e demonstrate a jump at some critical gas pressure that agrees with the experimental data [22]. The collision electron heating takes place due to elastic scattering of electrons on atoms, when the directed velocity transfers into the thermal one. At high gas pressures the collisional (or ohmic) heating determines the electron energy in the quasineutral part of the discharge. At low gas pressure the electrons are heated due to interaction with moving sheaths boundaries and the ohmic heating in bulk is very small. We studied the discharge operation in non-Ramsauer (helium) and Ramsauer (argon) gases. The noticeable difference in discharge behavior was found with gas pressure variation. The results for $P = 0.03$ Torr and $P = 0.3$ Torr, $j = 1$ mA/cm^2 and $d = 6.7$ cm are shown in Figs. 9.4 and 9.5 for helium and argon, respectively. To illustrate the difference in the electron kinetics in non-Ramsauer and Ramsauer gases, we calculated the *electron energy probability functions* (EEPF) in helium and argon. The EEPF, which are averaged over discharge rf-cycles, are shown in Fig. 9.6 for several gas pressures. In the experiment [22] the electron energy probability function was measured with Langmuir probes at the midplane of a radially confined, symmetrically driven rf-discharge in argon and in helium. In the calculations as in the experiment [22],

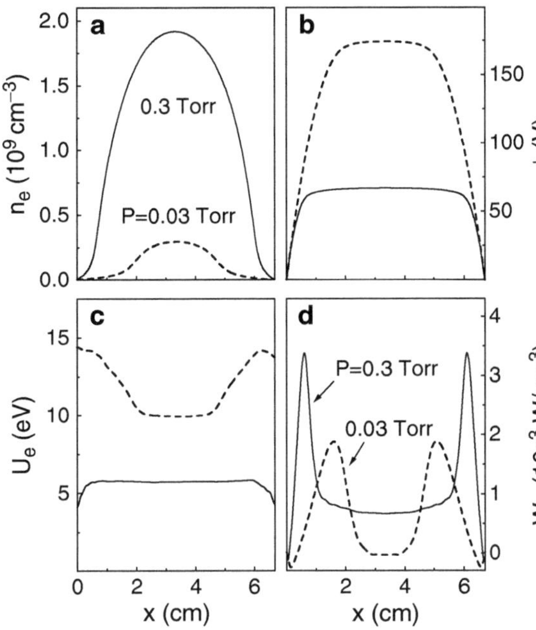

Fig. 9.4 Spatial distributions of the over several rf-cycles averaged electron density (**a**), potential of electric field (**b**), mean electron energy (**c**), and electron heating rate (**d**) in helium for gas pressures $P = 0.03$ (*dashed lines*) and 0.3 Torr (*solid lines*), $d = 6.7$ cm, $j = 1$ mA/cm^2, and $N = 5{,}000$

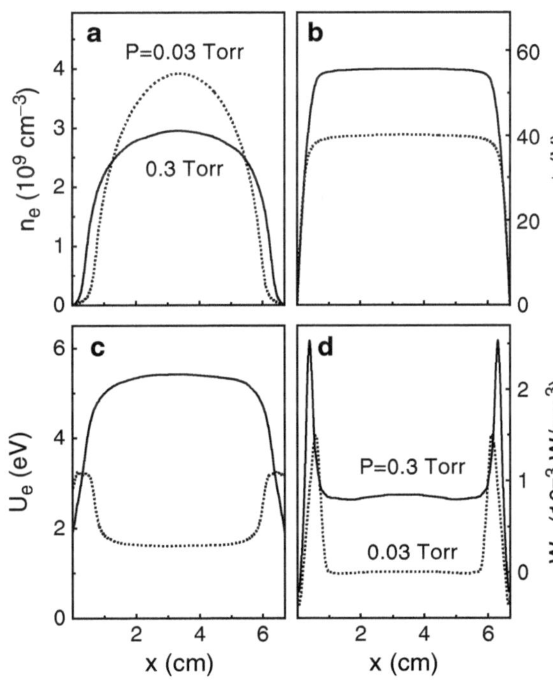

Fig. 9.5 Spatial distributions of the over several rf-cycles averaged electron density (**a**), potential of electric field (**b**), mean electron energy (**c**), and electron heating rate (**d**) for the same parameters as in Fig. 9.4, but here for argon gas

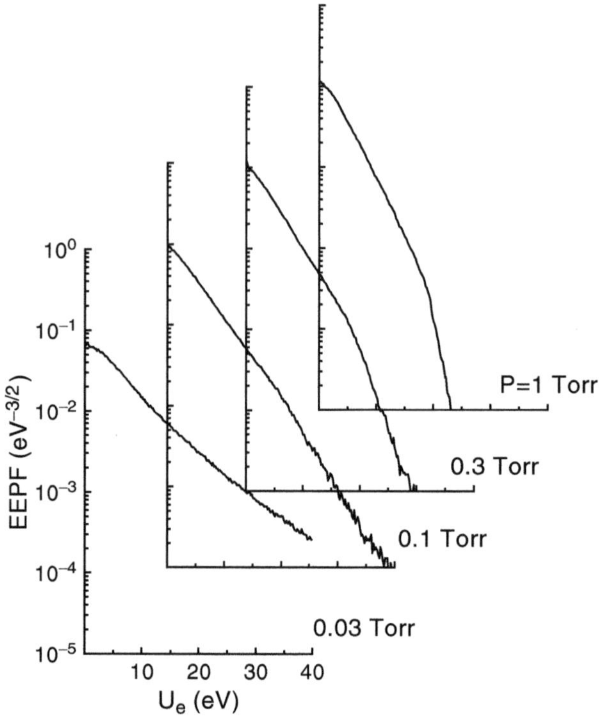

Fig. 9.6 Electron energy probability function in helium in the discharge center ($x = 3.35$ cm) for different gas pressures, $d = 6.7$ cm, $j = 1$ mA/cm^2, and $N = 5{,}000$

the shape of EEPF in helium (see Fig. 9.6) retains Maxwellian for all gas pressures. For higher gas pressure of argon (see Fig. 9.7), the EEPF are Druyvesteyn-like with $\delta f/\delta\epsilon \rightarrow 0$ as $\epsilon \rightarrow 0$. For higher gas pressure, when the frequency of electron–atom collisions $\nu_{ea} > \omega$, where ω is the discharge frequency, the flattening of the EEPF as $\epsilon \rightarrow 0$ is a consequence of the Ramsauer effect. The low-energy electrons accelerate without collision, thereby escaping the low-energy region of the EEPF. For lower gas pressure, $\nu_{ea} < \omega$, the EEPF has a peak at low energy. The low-energy electrons oscillate without collisions and consequently they gain no energy, and thus remain in the low-energy group. The EEPF changes from a Druyvesteyn type to a bi-Maxwellian one with decreasing P. At low gas pressure the electrons are separated into two groups. The cold electrons are not able to reach the sheath boundary and their ohmic heating is very weak due to Ramsauer minimum in the elastic cross section (see Fig. 9.8). The fast electrons heated in the sheaths maintain the discharge operation and provide the gas ionization. Figure 9.9 presents the computed and measured [22] electron temperature ($T_e = 2U_e/3$) in the discharge center ($x = 3.35$ cm). A decrease of the gas pressure is accompanied with a drop of T_e in argon, whereas in helium T_e monotonically increases. A comparison with experimental data shows a good agreement (within 20–30%) within a pressure range

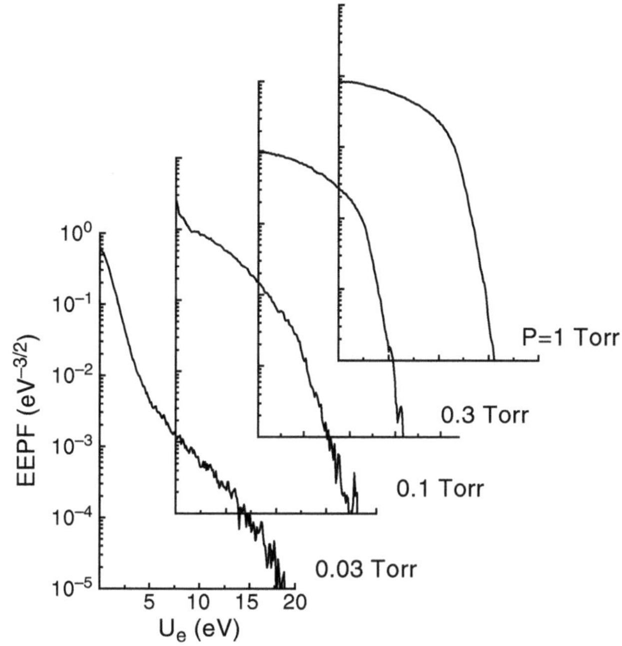

Fig. 9.7 Electron energy probability function for the same parameters as in Fig. 9.6, but here for argon

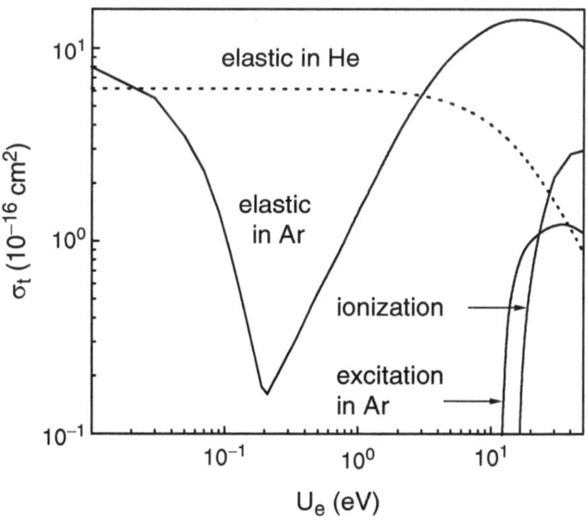

Fig. 9.8 Electron–atom elastic cross sections in argon (*solid line*) and in helium (*dotted line*) as functions of the electron energy

$P = 0.03$–0.3 Torr, but at $P > 0.3$ Torr the calculations give larger T_e. This discrepancy between computed and measured data is likely due to the contribution of metastable states in the ionization kinetics, especially in helium [25]. For higher

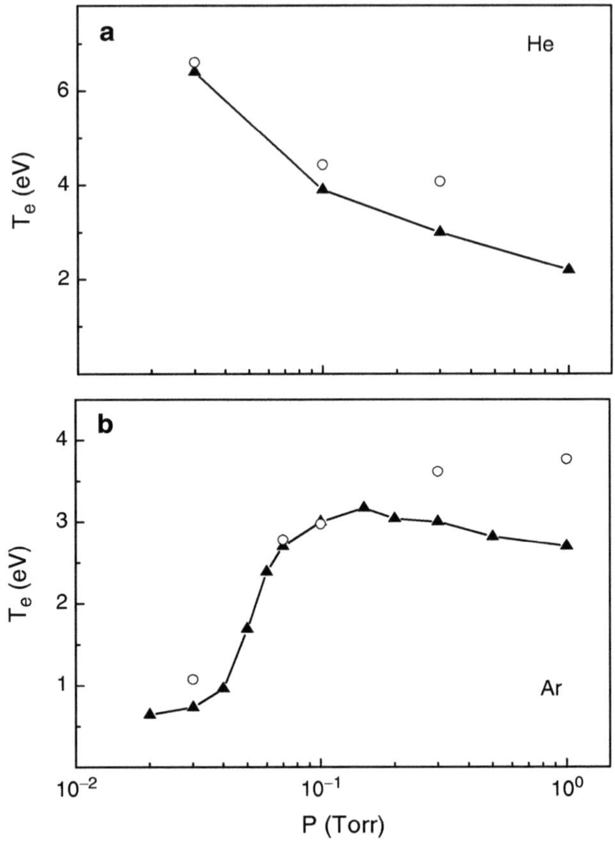

Fig. 9.9 Effective electron temperature ($T_e = 2U_e/3$) in the discharge center ($x = 3.35$ cm) in helium (**a**) and argon (**b**). Computed T_e (*circles*) and measured T_e [22] (*triangles*) for $d = 6.7$ cm, $j = 1$ A/cm^2, and $N = 5{,}000$

gas pressure the multistep ionization (ionization from the metastable states) plays important role. In our model we do not take into account the multistep ionization, therefore the computed mean electron temperature is higher than the measured one. At low gas pressures the agreement is better because the metastable atoms are deactivated on electrodes and the influence of multistep ionization reduces. The study of ionization kinetics in noble gases is out of the scope of this introductory chapter.

In conclusion, we have presented the combined PIC–MCC model for fast simulation of rf-discharges over a wide range of gas pressures and current densities. The validity of this approach is justified by comparison with the experiment data. The advantage of our approach is the considerable reduction of the number of simulation particles N without increase of the statistical noise. We are able to reach a speed-up factor of 10 for the collision regime and even more for the collisionless regime compared with the standard PIC–MCC calculations.

9.3 Physical Model of Discharge Plasma with Movable Dust

The physical model of the gas discharge plasma with charged massive dust particles is based on the kinetic description of the electron and ion motion and their interaction with charged nanoparticles. The distribution functions for electrons $f_e(t, x, \boldsymbol{v})$ and for ions $f_i(t, x, \boldsymbol{v})$ are calculated from the Boltzmann equations:

$$\frac{\partial f_e}{\partial t} + \boldsymbol{v}_e \frac{\partial f_e}{\partial x} - \frac{e\boldsymbol{E}}{m_e} \frac{\partial f_e}{\partial \boldsymbol{v}_e} = J_e, \quad n_e = \int f_e \mathrm{d}\boldsymbol{v}_e, \tag{9.18}$$

$$\frac{\partial f_i}{\partial t} + \boldsymbol{v}_i \frac{\partial f_i}{\partial x} + \frac{e\boldsymbol{E}}{m_i} \frac{\partial f_i}{\partial \boldsymbol{v}_i} = J_i, \quad n_i = \int f_i \mathrm{d}\boldsymbol{v}_i, \tag{9.19}$$

where v_e, v_i, n_e, n_i, m_e, m_i are the velocities, densities and masses of electrons and ions, respectively, E is the electrical field, J_e and J_i are the collisional integrals for electrons and ions, which include elastic and inelastic scattering with Ar atoms and C_2H_2 molecules, as well as scattering on charged carbon nanoparticles. The absorption cross sections of the nanoparticle surface for electrons with energy $\epsilon_e > \phi_d$ is $\sigma_{ed} = \pi r_d^2(1 - \phi_d/\epsilon_e)$ and for ions is $\sigma_{id} = \pi r_d^2(1 + \phi_d/\epsilon_i)$, where r_d and ϕ_d are the dust particle radius and surface potential, ϵ_e and ϵ_i are the electron and ion energies, respectively. Additionally for accurate solution of the Poisson equation, the density and the flux of electron and ions are calculated from the auxiliary equations (9.3) and (9.6) based on the momentums of the kinetic equations (9.18) and (9.19), as it was described in Sect. 9.2.1. The charge of the dust particles $Z(x)$ as a function of the coordinate x is calculated from the balance equation of the electron and ion currents on the dust surface:

$$j_{ed}(x) = j_{id}(x). \tag{9.20}$$

The density of charged nanoparticles n_d is described with the continuity equation in the drift–diffusion approximation:

$$\frac{\partial n_d}{\partial t} = \operatorname{div} j_d, \tag{9.21}$$

where

$$j_d = \frac{\partial D_d n_d}{\partial x} - v_d n_d$$

and

$$\frac{\partial v_d}{\partial t} = \frac{F_{ele} - F_{drag} - F_{fric}}{m_d}.$$

Here, D_d, v_d, and m_d are the diffusion coefficient, the velocity, and the mass of the nanoparticle, and F_{ele}, F_{drag}, and F_{fric} are the electrostatic, ion drag, and gas friction forces acting on the dust, respectively. The internal pressure in the particle system can be neglected. For our system of nanoparticles the coupling parameter $\Gamma = V/ \mid r \mid k_B T$ (where V is the electrostatic energy of the particle interaction

Fig. 9.10 Discharge with initial nanoparticle density distribution

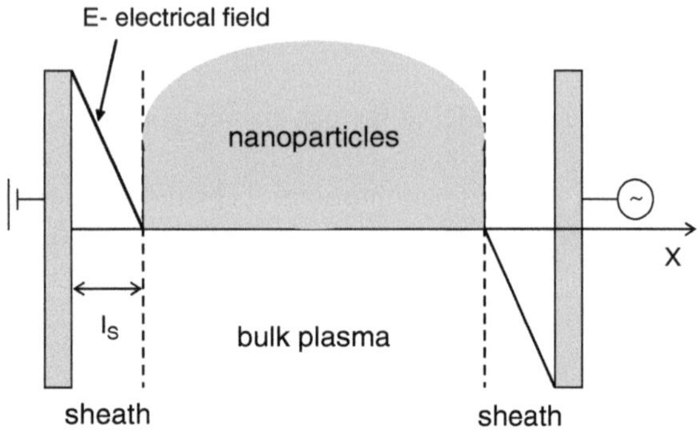

and $k_B T$ is the particle kinetic energy) varies from 1 to 10. This value is far below the critical coupling parameter for crystallization, which is 170 for a 3D Coulomb system.

The electrical potential distribution ϕ obeys the Poisson equation

$$\Delta\phi = 4\pi e(n_e - n_i + Zn_d), \quad E = -\frac{\partial\phi}{\partial x}. \tag{9.22}$$

The boundary conditions for the Poisson equation are $U(d,t) = U_0 sin(\omega t)$ and $U(0,t) = 0$, where U_0 is the applied voltage amplitude and $x = 0$, $x = d$ are the coordinates of electrodes.

As follows from the measured mass spectra in [1] several types of ions are generated in the mixture of Ar/C_2H_2 and the majority of ions are C_nH_2. To simplify the discharge model we consider only one type of ions, $C_4H_2^+$. Only elastic scattering of $C_4H_2^+$ ions in Ar background is considered as the mixture consists of 94.2% Ar and 5.8% C_2H_2. The electron scattering cross sections were taken from [27, 28] for argon and from [29] for acetylene.

In the calculations, the discharge glows between symmetrical parallel electrodes and the interelectrode distance is 7 cm. One electrode is powered and another one is grounded. The voltage waveform is sinusoidal $U(t) = U_0 \sin(\omega t)$, where U_0 is the voltage amplitude. In Fig. 9.10 the scheme of the discharge and the initial distribution of nanoparticle density are shown.

9.3.1 Algorithm of Calculation

The computational algorithm can be separated to the following main steps:

(a) The calculation starts from some initial distributions for the electrons and ions, which are taken from the analytical approximation of discharge plasma. For dust particles density

$$\begin{aligned} n_d(x) &= N_d \sin(\pi x/d), \quad l_s < x < d - l_s, \\ n_d(x) &= 0, \quad x < l_s, x > d - l_s, \end{aligned} \tag{9.23}$$

where l_s=1.5 cm is the initial width of the electrode sheath and N_d =10^7 cm^{-3}. The initial nanoparticle floating potential is 3 V. The conditions of quasineutrality of bulk plasma at $l_s < x < d - l_s$ is fulfilled and $n_e - n_i + Zn_d = 0$.

(b) The electrical field distribution is calculated from the Poisson equation (9.22). Here we take into account the space charge of dust particles, electrons, and ions.

(c) The *electron energy distribution function* (EEDF), the *ion energy distribution function* (IEDF) and all macroscopic discharge parameters are calculated. Above we analyzed the electron energy probability function f_p, which is $f_p = f_e/\sqrt{\varepsilon}$ (where f_p is EEPF and f_e is EEDF). The kinetic equations of motion for electrons and ions (1) and (2) are integrated in the electrical field of discharge taking into account collisional events with the background gas and dust particles. The probabilities of collisional events (elastic scattering, excitation of molecules, ionization, absorption on the dust surface, etc.) are set by the cross sections depending on the electron/ion energy and calculated with the Monte Carlo technique and the null-collision method [11]. The cross section of electron/ion absorption is proportional to the dust density.

(d) As a next step we calculate the dust surface charge. The PIC–MCC method gives the opportunity for a direct calculation of the electron and ion currents on the dust surface. During the Monte Carlo calculation of the electron and ion trajectories, we simply integrate the absorption events for electrons and ions on the dust particle surfaces. The electron and ion currents on the dust surface are averaged over the last 10–50 rf-cycles (depending on the dust radius) to reduce the statistical fluctuations. After every rf-cycle we adjust the floating potential to improve the electron and ion current balance. This change is always less than 10% from the previous value to avoid any numerical instability due to a local jump of the electrical field.

(e) The forces acting on the charged particles are the electrostatic force from the discharge electrical field and the ion drag force from the ion flux directed to the electrode. The electrostatic and ion drag forces are calculated at every rf-cycle using the electrical field, ion density, and ion temperature averaged over the last 100 rf-cycles, as described in next section.

(f) The dust density distribution is found from the solution of the continuity equation (9.21). The interaction of dust particles is calculated through the iterative self-consistent solution of the kinetic equations (9.18) and (9.19), the continuity equation (9.21) and the Poisson equation (9.22). The dust particles move in the electrical field created by the distribution of the dust particles, electrons, and ions.

Then we return to (b) and calculate the electrical field from the Poisson equation for the new dust charge, electron, and ion distributions. The iterative procedure is continued till the steady state solution is reached. Typically, it takes about 1,000 rf-cycles for discharge without dust and 5,000 rf-cycles with dust to reach convergence. Note that in an every run the nanoparticle size is constant. The discharge can reach the steady state with the constant nanoparticle radius, because the characteristic time of plasma relaxation depending on electron and ion motion is much less than the characteristic time of dust growth. The number of artificial particles to mimic the electrons and ions in our PIC–MCC simulations is $N_e = N_i = 10^4$–10^5 depending on the discharge parameters. The time step is 10^{-11} s for electrons and 3×10^{-10} s for ions. The interelectrode distance is subdivided by a grid with 201 nodes for the definition of the macroscopic plasma parameters and the calculation of the dust density distribution.

9.3.2 Ion Drag Force

For the definition of the ion drag force F_{dr} we applied the method described in [30]. In this method, we consider the dust particle with charge Q in the cylindrical calculation cell shown in Fig. 9.11. In our model, the flowing plasma is described with a kinetic approach for the ions and the electrons obeying a Boltzmann distribution. In most experiments [31, 32] the behavior of the charged particles is studied in a high-frequency gas discharge operating in noble gases. The main scattering processes for ions in these gases are resonance charge exchange collisions of ions with neutrals. The ion velocity distribution function (IDF) $f_i(t, \boldsymbol{v}, \boldsymbol{r})$ is described with the kinetic equation in cylindrical coordinates (two dimensional in the space and three dimensional over the velocity) [33]:

$$\frac{\partial f_i}{\partial t} + \boldsymbol{v}\frac{\partial f_i}{\partial \boldsymbol{r}} - \frac{e\nabla\phi_E}{m}\frac{\partial f_i}{\partial \boldsymbol{v}} = \int d\boldsymbol{v}'\nu(|\boldsymbol{v}-\boldsymbol{v}'|)\left(f_g(\boldsymbol{v})f_i(\boldsymbol{v}') - f_g(\boldsymbol{v}')f_i(\boldsymbol{v})\right),$$

where $\boldsymbol{v}$ and m are the ion velocity and mass, $\nu(v) = v\sigma(v)N$ is the frequency of the ion–neutral atoms resonance charge exchange collisions, σ is the cross section, N is the gas density, $\phi_E = \phi_0 + \phi$, where ϕ_0 and ϕ are the electric potentials

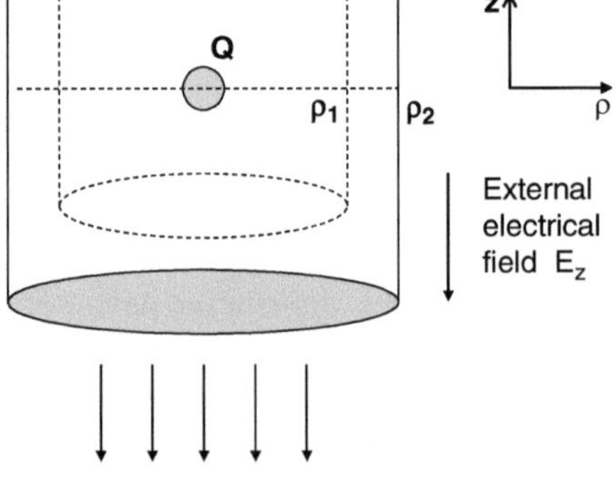

Fig. 9.11 Cylindrical calculation cell

of the undisturbed plasma and the dust particle (screened with a plasma flux), respectively. The neutral gas atoms have a Maxwellian velocity distribution $f_g(\boldsymbol{v}) = (m/2\pi T)^{3/2} \exp(-mv^2/2T)$, where T is the gas temperature. Note that the ion velocity distribution just after the collision is given by the same Maxwellian distribution. The ion density is calculated with

$$n_i(t, \boldsymbol{r}) = \int f_i(t, \boldsymbol{v}, \boldsymbol{r}) d\boldsymbol{v}. \tag{9.24}$$

The external electrical field has only a z-component, that is, $\phi_0 = -zE$, and therefore we use cylindrical coordinates $\boldsymbol{r} = (\boldsymbol{\rho}, z)$, where the z-axis is oriented against the ion flow.

The boundary condition for (9.24) far upstream and downstream from the particle corresponds to the unperturbed IDF $f_i(|z| \to \infty, \boldsymbol{v}) = f_0(\boldsymbol{v})$. This $f_0(\boldsymbol{v})$ is found from the kinetic equation

$$\frac{eE}{m}\frac{\partial f_0}{\partial v_z} = \int d\boldsymbol{v}' \nu(\boldsymbol{v} - \boldsymbol{v}') \left(f_g(\boldsymbol{v}) f_0(\boldsymbol{v}') - f_g(\boldsymbol{v}') f_0(\boldsymbol{v}) \right), \tag{9.25}$$

with the normalization $n_0 = \int f_0 d\boldsymbol{v}$, where n_0 is the unperturbed ion density. At the particle surface the boundary condition is $f_i(|\boldsymbol{r}| = R, \boldsymbol{r} \cdot \boldsymbol{v} > 0) = 0$ (where R is the particle radius) and corresponds to the absorption of ions hitting the microparticle.

Since the particle surface potential $U = eZ/R$ usually is comparable to the electron temperature T_e, the particle electric field weakly affects the electron motion. Therefore, the electric potential perturbation ϕ around the dust particle with the negative charge eZ can be found from the linearized Poisson equation:

$$\Delta\phi - \phi/\lambda_e^2 = 4\pi e \left(Z\delta(\boldsymbol{r}) - n \right), \tag{9.26}$$

where $\lambda_e = \sqrt{T_e/4\pi e^2 n_e}$ is the electron Debye length and $n = n_i - n_0$ is the ion density perturbation. The dust particle potential decays far from the particle and, thus, the boundary condition for (9.26) is $\phi(|\boldsymbol{r}| \to \infty) = 0$.

The ion density and the potential distribution around the dust particle in the plasma flux are calculated by solving numerically (9.24)–(9.26) with the PIC–MCC algorithm [11, 12]. Solving the ion kinetic equation we use in our simulation 5×10^5 particles, the cloud-in-cell charge assignment scheme, and the null-collision technique to find the time of ion free flight. To describe the ion–neutral collisions, instead of using the Maxwell model $\nu = \text{const.}$, we took the more realistic collision frequency $\nu = e/m\mu$ with the ion mobility $\mu = 3e\sqrt{\pi/mT}(1 + 9\pi^2 eE/512NT\sigma)^{-1/2}/16N\sigma$ [33], which depends on the external electric field. The ion flux is directed from $+z$ to $-z$ (see Fig. 9.11). The origin of our coordinate system is placed on the dust particle, that is, $(z, \rho) = (0, 0)$, and around it the grid is uniform within the region $|z| < z_1$, $\rho < \rho_1$. For $|z| > z_1$, $\rho > \rho_1$ the grid spacing is increased linearly up to the boundaries of the simulation region $|z| = z_2$, $\rho = \rho_2$, where the dust particle potential ϕ is set to be zero. Since the dust particle potential decays slowly with distance from the dust particle we choose $z_2, \rho_2 \gg z_1, \rho_1$. The typical size of the simulation region varies with the gas pressure in the range $z_2, \rho_2 \sim (0.3\text{–}0.5)$ cm.

The ions injected through the upper boundary $z = z_2$ are uniformly distributed over the radial coordinate. The ions trajectories are followed by the Monte Carlo method until they leave the simulation region far downstream from the particle at $z < -z_2$. The ion density, however, is calculated only inside the first region $|z| < z_1$, $\rho < \rho_1$ to avoid an artificial ion depletion nearby the boundaries z_2, ρ_2 caused by finite size effects. When solving the kinetic equation of ion motion (9.24), we use a variable time step in the vicinity of the dust particle.

To improve the accuracy of the finite-difference representation of the Poisson equation the Dirac-delta function in (9.26) is replaced by a uniformly charged sphere of radius of 30–75 μm, depending on the ion density. Thereafter, the potential distribution created by a negative point charge with a surrounding positive charged sphere is added to the solution [34]. Equations (9.24) and (9.26) are solved self-consistently with the PIC–MCC technique and the steady-state solution is reached after $(1\text{–}2) \times 10^4$ iterations. Thereafter, the solution is averaged over the next $(4\text{–}8) \times 10^4$ iterations to minimize the statistical noise.

We have calculated the ion drag force for a set of plasma parameters using the density of the particle $n_\mathrm{d} = 10^6\text{–}10^7\,\mathrm{cm}^{-3}$. The interparticle distance is 60–120 μm and the radius of dust is 10–90 nm. The ion Debye screening length λ_D is 30–60 μm for $n_\mathrm{i} = 2 \times 10^9\text{–}5 \times 10^8\,\mathrm{cm}^{-3}$, and $T_\mathrm{i} = 0.03\text{–}0.5$ eV. From self-consistent simulations we found that for these experimental conditions [3] the Barnes formula describes accurately the ion drag force for nanoscale dust. The ion drag forces calculated using the PIC–MCC method [30] and with the Barnes formula [35] coincide within 10% accuracy which is the statistical error of our Monte Carlo calculations. Therefore, for the calculation of dust motion we use the orbit part of the ion drag force that refers to the Coulomb interaction of ions with a nanoparticle $F_\mathrm{dr}(x) = 4\pi m_\mathrm{i} n_\mathrm{i} v_\mathrm{i} v_\mathrm{s} b_\mathrm{p}^2 L$, where $b_\mathrm{p} = Ze^2/2\epsilon_\mathrm{i}$ is the effective interaction radius, v_i is the ion drift velocity, $v_\mathrm{s} = (2\epsilon_\mathrm{i}/m_\mathrm{i})^{1/2}$ is the mean ion velocity, $L = 0.5\log[(\lambda_\mathrm{D}^2 + b_\mathrm{p}^2)/(b_\mathrm{p}^2 + b_\mathrm{c}^2)]$ is the Coulomb integral, and $b_\mathrm{c}^2 = r_\mathrm{d}^2(1 + \phi_\mathrm{d}/\epsilon_\mathrm{i})$ is the collection radius. The contribution to the ion drag force from captured ions is small for $r_\mathrm{d} < 100$ nm.

9.3.3 Transition Between Different Modes

The simulations and measurements were performed for the 13.56 MHz discharge operating in a mixture of C_2H_2/Ar (1:16) at $P = 70$ mTorr [3]. The discharge glows between the parallel plate electrodes with 7 cm interelectrode distance (see Fig. 9.10). The detailed description of this experimental setup and measuring technique can be found elsewhere [1, 2]. In the simulations, we assumed the voltage waveform to be sinusoidal $U(t) = U_0 \sin(\omega t)$, where U_0 is the voltage amplitude. The growth of nanoparticles from gas phase reactions is not considered

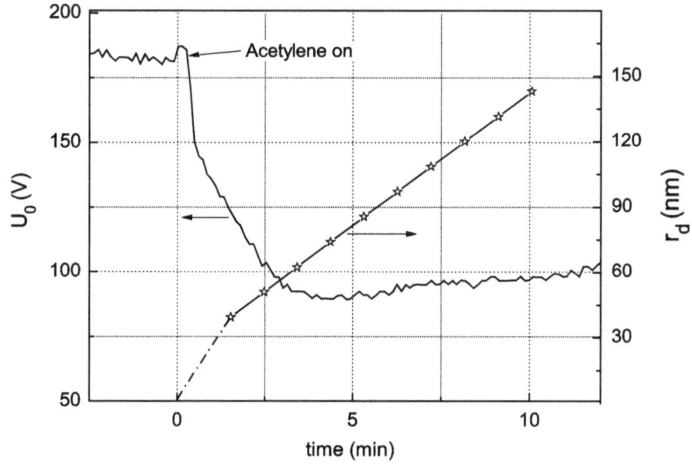

Fig. 9.12 Measured voltage amplitude (*solid line*) and measured dust radius (*symbols*) as a function of time [3]

in calculations. We took values of the voltage amplitude and dust radius from the experiment. In Fig. 9.12 the measured voltage amplitude and nanoparticle radius for $t > 1.5$ min are given as a function of time. At first the discharge glows in pure argon, and at $t = 0$ acetylene is added and the particle formation starts. Since there is no data about the dust radius on a very early stage of the growth, the dust radius is assumed to be growing linearly at $t < 1.5$ min. In our simulation as well as in the experiment the dust has a monodisperse size. In our calculations the nanoparticle radius varies from 10 to 90 nm.

Without nanoparticles the argon discharge glows in the capacitive mode, which is characterized with the large plasma density and the low electron energy in the bulk plasma. Acetylene adding initiates the particle formation and the quick change of plasma parameters. The dust acts as a electron and ion sink and therefore the plasma parameters are sensitive to an increase of nanoparticle surface area. However, only during the first 1.7 min the electron density n_e shown in Fig. 9.13 rapidly decreases. With further dust growth n_e remains constant up to the end of the dust growth cycle. Figure 9.14 shows the evolution of the mean electron energy ϵ_e and ionization rate profiles. For smaller dust radii, $r_d < 20$ nm, the mean electron energy has maxima near the sheath–plasma boundaries. For larger radii, $r_d = 30$ and 40 nm, the ϵ_e profile changes qualitatively, denoting the transition from the capacitive (C) to volume-dominated (VD) mode. The ionization rate distribution also demonstrates this transition (Fig. 9.14b). Initially when $r_d = 0$ the ionization preferably takes place near the sheath–plasma boundary and it is very low in the bulk plasma. There the electrons are trapped by the electrical potential, thermalize and have a Maxwellian energy distribution. The dust growing to 20 nm suppresses the ionization, but the discharge still glows in the capacitive regime. For 30 and 40 nm dust the new VD regime is associated with an increase of the ionization rate in the

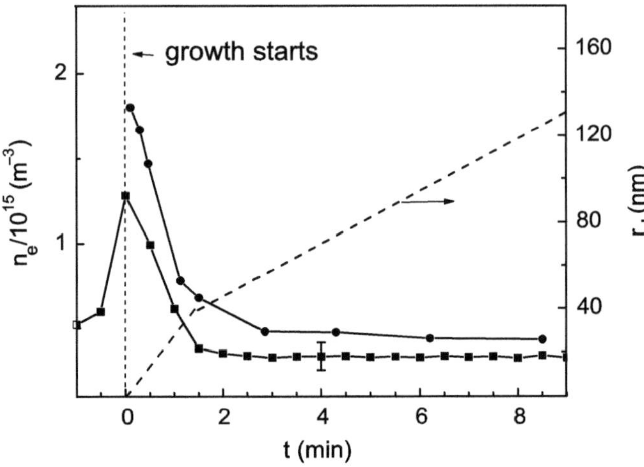

Fig. 9.13 Measured (*squares*) and calculated (*circles*) electron density in the midplane and measured nanoparticle radius (*dashed line*) as a function of time

Fig. 9.14 Distribution of the mean electron energy (**a**) and discharge power deposited to the ionization (**b**) for different dust radii: $r_d = 0$ (1), 20 (2), 30 (3), and 40 nm (4)

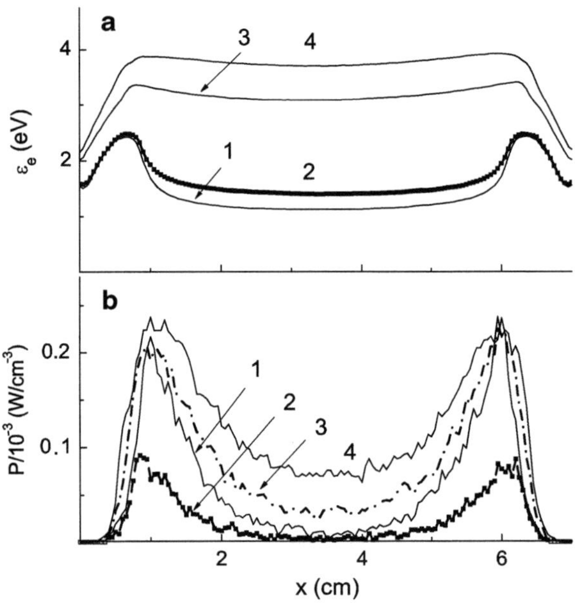

midplane (curves 3 and 4 in Fig. 9.14b). Actually the growing dust causes an increase of the electrical field in the plasma, which heats the electrons and enhances the discharge current. The electron energy distribution function (EEDF) in the midplane is shown in Fig. 9.15 for different nanoparticle radii. It is seen that the EEDFs loose the Maxwellian shape for $r_d > 30$ nm and the hot electron population becomes larger.

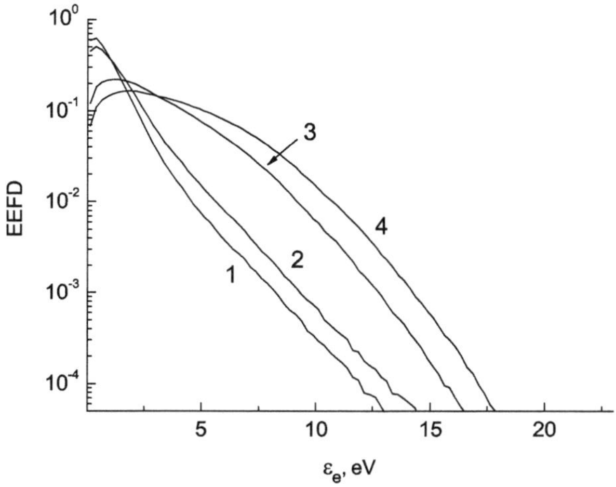

Fig. 9.15 Electron energy distribution function for different dust radii: $r_d = 0$ (1), 20 (2) , 30 (3), and 40 nm (4)

9.3.4 Dust Motion Effect

In simulations and in experiments it was found that at the initial growth stage, the nanoparticles are accumulated near the sheath–plasma boundaries. Note that in previous studies the dust was considered to be uniformly distributed over the bulk plasma. A void (area without dust) in the plasma was found only for large dust particles. In our case, we observe the void formation for 10–20 nm nanoparticles and the larger nanoparticles have a more uniform distribution. The measured distribution of laser light scattering signal from the dust particles and the calculated dust density are shown in Fig. 9.16. The nanoparticle profile changes the shape from peaked to flat during the C–VD transition, which takes place for $r_d > 20$ nm. In Fig. 9.17 the mean electron energy in the midplane and the electrode sheath width l_s are shown for different dust radii. The ϵ_e and l_s remain constant during the first minute ($r_d < 20$ nm), while the ionization is large enough to sustain the capacitive mode of discharge operation. For larger r_d the ϵ_e increases to provide the ionization in the plasma and this is accompanied with visible sheath compression. As a consequence, the dust surface potential ϕ_d and dust charge Z increase, because the calculated ϕ_d and ϵ_e coincide within 20–30% for our plasma parameters. The larger dust charge requires the larger electrical field to trap it, therefore the sheath width decreases with the dust growth. Figure 9.18 shows the ion drag F_{dr} and electrostatic F_{el} forces, which increase by a factor of 4 during the C–DV transition. Using Barnes formula [35] we obtain the ratio of the ion drag to electrostatic force in the bulk plasma (assuming $v_i \sim \mu_0 E$ for subsonic ion motion):

$$F_{dr}/F_{el} \sim Zn_i/\epsilon_i^{3/2},$$

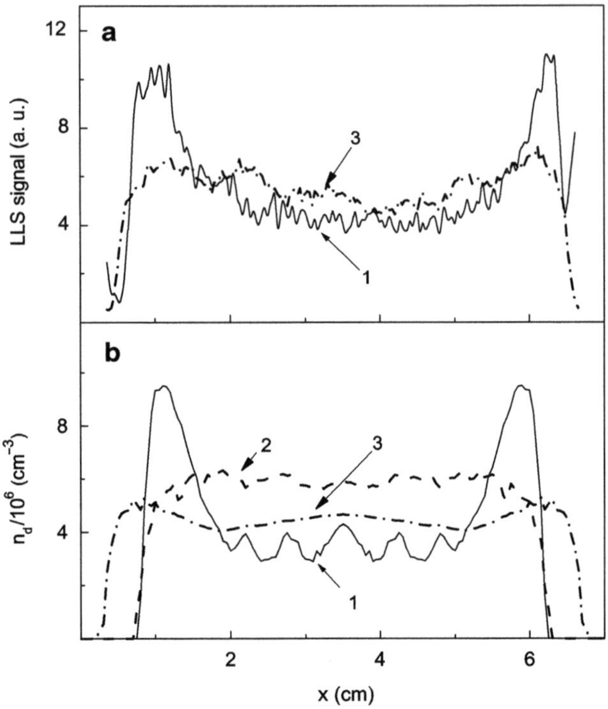

Fig. 9.16 Distribution of nanoparticles from measured laser light scattering signal (**a**) and from simulations (**b**) for different radii: $r_d = 20\,\text{nm}$ ($t = 47\,\text{s}$ in the experiment) (1), $r_d = 30\,\text{nm}$ (2), and $r_d = 90\,\text{nm}$ ($t = 5.8\,\text{min}$ in the experiment) (3)

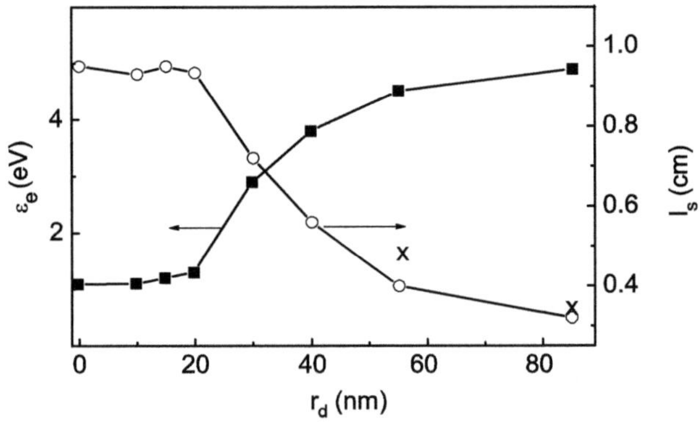

Fig. 9.17 Calculated sheath width and mean electron energy as a function of nanoparticle radius. The *two crosses* indicate experimental measurements

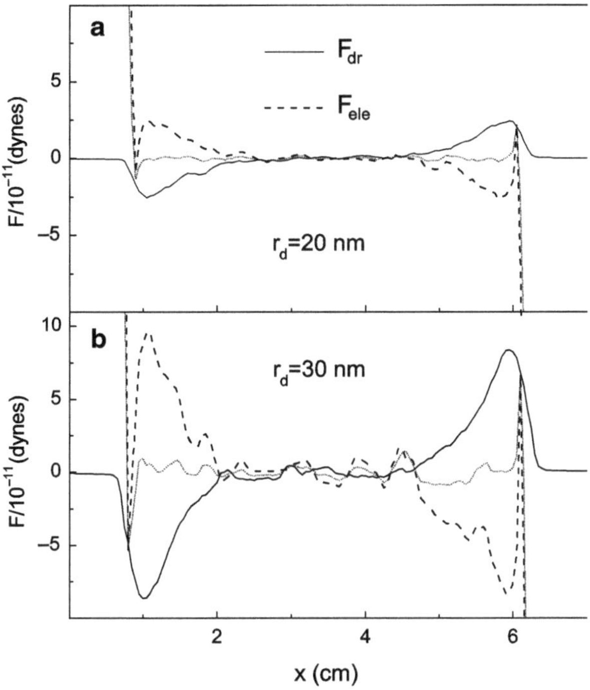

Fig. 9.18 Computed ion drag (*solid lines*) and electrostatic force (*dashed lines*) distributions for $r_\mathrm{d} = 20\,\mathrm{nm}$ (**a**) and $r_\mathrm{d} = 30\,\mathrm{nm}$ (**b**)

and near the sheath–plasma boundary (assuming $v_\mathrm{i} \sim \mu_1 E^{1/2}$ for supersonic ion motion)

$$F_\mathrm{dr}/F_\mathrm{el} \sim Zn_\mathrm{i}/E^{1/2}\epsilon_\mathrm{i}^{3/2}.$$

In the beginning of the growth, nanoparticles are shifted by the ion drag force to the sheath–plasma boundary. Growing dust induces the C–VD transition, plasma parameters change and dust distributes uniformly over the plasma. The ion density and electrical field distributions before and after the transition are shown in Fig. 9.19. For 30 nm sized dust, the ion density is smaller and the electrical field and dust charge are larger, compared to the 20 nm case. After transition the ratio $F_\mathrm{dr}/F_\mathrm{el}$ decreases and dust rearranges uniformly over the plasma. As we discussed in the previous section, the plasma density decreases due to the change of discharge operation regime and an increase of the dust surface area.

The discharge power deposited to the excitation of the background gas by fast electrons also demonstrates the transition between the capacitive and volume-dominated modes. In Fig. 9.20 the measured light distribution of the ArI 696.5 nm emission line and the calculated excitation power are shown. The calculated and experimental profiles are similar and have peaks near the sheath–plasma boundaries for smaller dust radii. This is a characteristic feature of the capacitive mode. For larger dust radii, in the volume-dominated regime the profiles become flat.

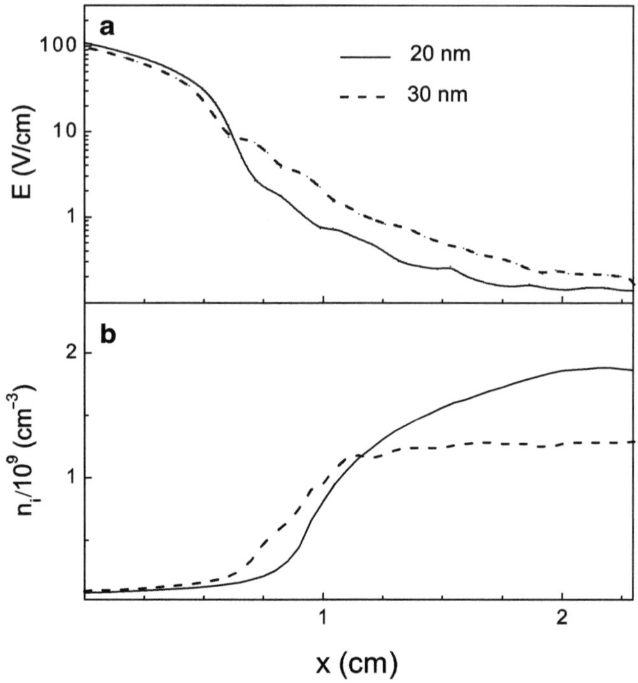

Fig. 9.19 Electrical field (**a**) and ion density (**b**) distributions near the electrode for 20 and 30 nm dust radii

We also performed simulations with the unmovable dust with fixed profile given by (9.23) for the same discharge parameters and dust radii. Under this condition the transition between different regimes was smoothed and took place at much larger nanoparticle radii. Thus, we found that the consideration of dust motion is very important for an accurate description of transient processes observed in experiments.

9.4 Conclusion

In conclusion, we have presented a kinetic model for the ccrf-discharge with mobile charged nanoparticles. The dust surface potential, dust motion, and discharge parameters are calculated self-consistently using the PIC–MCC method. In our model we do not use any assumption about EEDF and IEDF. This allows us to calculate accurately the dust particle charging. The results of our simulations and experiments are in good agreement, showing the transient processes induced by the dust presence. Calculations and experiments show the dust accumulation near the sheath–plasma boundaries at the initial stage of the nanoparticle growth. At some critical dust size the transition from the capacitive to volume-dominated regime was

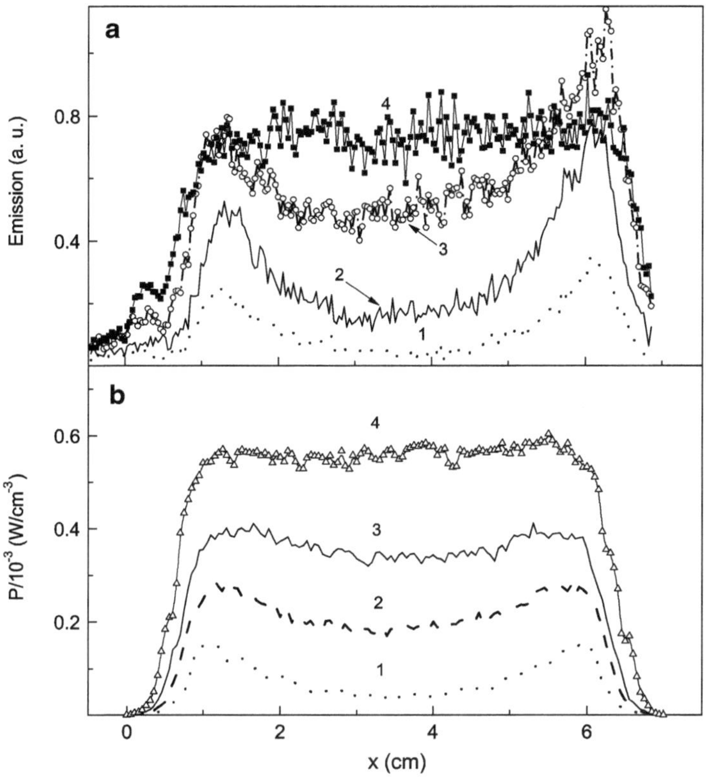

Fig. 9.20 Light distribution of the ArI 696.5 nm emission line, measured at 12 s (1), 24 s (2), 2 min (3), and 6 min (4) (**a**) and calculated power deposition to excitation of C_2H_2 molecules for $r_d = 20$ (1), 30 (2), 40 (3), and 70 nm (4) (**b**)

observed, which was accompanied by a quick change of the plasma parameters. In the capacitive mode of discharge operation the ionization generally takes place around the sheath–plasma boundaries and nanoparticles accumulated here suppress the ionization, absorbing fast electrons. For larger nanoparticles the ionization is not large enough to compensate the electron and ion losses on the dust surface. Therefore the discharge transits to more resistive VD mode with the enhanced ionization in the plasma bulk. The change of plasma parameters initiate the rearrangement of the dust profile from peaked to flat. Further dust growth does not affect the plasma density. The growing absorption of electrons and ions on the dust surface is balanced with an increasing ionization in the discharge plasma. The C–VD transition was monitored with evolution of the plasma density, mean electron energy, ionization rate, and electron energy distribution function. It was shown that the presence of movable dust is responsible for a rapid change of plasma parameters observed in the experiments.

References

1. E. Kovačević et al., J. Appl. Phys. **93**, 2924 (2003)
2. I. Stefanović et al., New J. Phys. **5**, 39.1 (2003)
3. I.V. Schweigert et al., Phys. Rev. E **78**, 026410 (2008)
4. P. Belenguer et al., Phys. Rev. A **46**, 7923 (1992)
5. C. Bohm, J. Perrin, J. Phys. D **24**, 865 (1991)
6. J.P. Boeuf, Ph. Belenguer, J. Appl. Phys. **71**, 4751 (1992)
7. I. Denysenko et al., Phys. Plasmas **13**, 073507 (2006)
8. M.R. Akdim, W.J. Goedheer, Phys. Rev. E **67**, 066407 (2003)
9. Y.I. Chutov, W.J. Goedheer, IEEE Trans. Plasma Sci. **31**, 606 (2003)
10. I.V. Schweigert, V.A. Schweigert, Plasma Sour. Sci. Technol. **13**(2), 315 (2004)
11. C.K. Birdsall, A.B. Langdon, *Plasma Physics via Computer Simulation* (McGraw-Hill, New York, 1985), p. 479
12. C.K. Birdsall, IEEE Trans. Plasma Sci. **19**, 65 (1991)
13. V.V. Ivanov, A.M. Popov, T.V. Rakhimova, Sov. Plasma Phys. **21**, 548 (1995)
14. V.A. Shveigert, I.V. Shveigert, J. Appl. Mech. Tech. Phys. **29**(4), 471 (1988)
15. Ph. Belenguer, J.-P. Boeuf, Phys. Rev. A **41**, 4447 (1990)
16. T.J. Sommerer, M.J. Kushner, J. Appl. Phys. **71**(4), 1654 (1992)
17. D.L. Scharfetter, H.K. Gummel, IEEE Trans. Electron. Dev. **31**, 1912 (1984)
18. J.-P. Boeuf, Phys. Rev. A **36**, 2782 (1987)
19. C.K. Birdsall, E. Kawamura, V. Vahedi, Rep. Inst. Fluid Sci. **10**, 39 (1997)
20. R.W. Hockney, J.W. Eastwood, *Computer Simulation Using Particles* (McGraw-Hill, New York, 1981), p. 540
21. W.M. Manheimer, M. Lampe, G. Joyce, J. Comput. Phys. **138**, 563 (1997)
22. V.A. Godyak, R.B. Piejak, B.M. Alexandrovich, Plasma Sour. Sci. Technol. **1**, 36 (1992)
23. R. Lagushenko, J. Maya, J. Appl. Phys. **55**, 3293 (1984)
24. S.M. Levitskii, Zh. Tekh. Fiz. **27**, 1001 (1957)
25. G.J. Parker et al., Phys. Fluids B **5**, 646 (1993)
26. V. Vahedi et al., Plasma Sour. Sci. Technol. **2**, 261 (1993)
27. R. Lagushenko, J. Maya, J. Appl. Phys. **55**, 3293 (1984)
28. V.V. Ivanov, A.M. Popov, T.V. Rakchimova, Plasma Phys. Rep. **21**, 548 (1995) (in Russian)
29. M. Hayashi, in *Nonequilibrium Processes in Partially Ionized Gases*, ed. by M. Capitelli, J.N. Bardsley (Plenum, New York, 1990)
30. I.V. Schweigert, V.A. Schweigert, F.M. Peeters, Phys. Plasmas **12**, 113501 (2005)
31. V.E. Fortov et al., Phys. Usp. **47**, 447 (2004)
32. V.E. Fortov et al., Phys. Rev. E **70**, 046415 (2004)
33. E.W. McDaniel, E.A. Mason, *The Mobility and Diffusion of Ions in Gases* (Wiley, New York, 1973), p. 372
34. V.A. Schweigert et al., JETP **88**, 482 (1999)
35. M.S. Barnes et al., Phys. Rev. Lett. **68**, 313 (1992)

Chapter 10
Molecular Dynamics Simulation of Strongly Correlated Dusty Plasmas

Torben Ott, Patrick Ludwig, Hanno Kählert, and Michael Bonitz

Abstract This chapter gives a tutorial introduction to the molecular dynamics (MD) technique as a first-principle description of classical many-particle dynamics. The goal is to provide practical insight into the current status of theoretical dusty plasma research as well as to present the necessary ingredients for a successful MD simulation in one place. As typical examples of the application of MD, we concentrate on two directions of current research interest: (1) the structural properties of spherical dust crystals in traps and (2) the transport properties such as diffusion of liquid unconfined, infinite dust systems.

10.1 Introduction

The field of mesoscopic and macroscopic complex (dusty) plasmas has become an important part of plasma physics in recent years. The research interest was initiated in 1994 by the experimental discovery of a new state of (soft) matter – the *plasma (Wigner) crystal* [1–4]. In a sheath of a noble gas radio-frequency discharge, highly charged dust grains of micrometer size were investigated for the first time under laboratory conditions. Due to their high charge of several thousand elementary charges, these microspheres are strongly coupled and enable the researchers to observe liquid behavior with short-range order and even macroscopic *Coulomb crystals* of *hcp*, *fcc*, and *bcc* lattice structure.

The occurrence of dusty plasma effects exceeds by far basic research interests and has practical importance in micro- and nanotechnology [5]. In the industrial plasma processing, the presence of charged dust particulates can completely change characteristic plasma parameters such as electron and ion densities, temperature, and plasma potential. This makes it difficult to run technological processes at optimum

T. Ott (✉), P. Ludwig, H. Kählert, and M. Bonitz
Institut für Theoretische Physik und Astrophysik, Christian-Albrechts-Universität zu Kiel, 24098 Kiel, Germany
e-mail: ott@theo-physik.uni-kiel.de; ludwig@theo-physik.uni-kiel.de; kaehlert@theo-physik.uni-kiel.de; bonitz@physik.uni-kiel.de

M. Bonitz et al. (eds.), *Introduction to Complex Plasmas*, Springer Series on Atomic, Optical, and Plasma Physics 59, DOI 10.1007/978-3-642-10592-0_10,

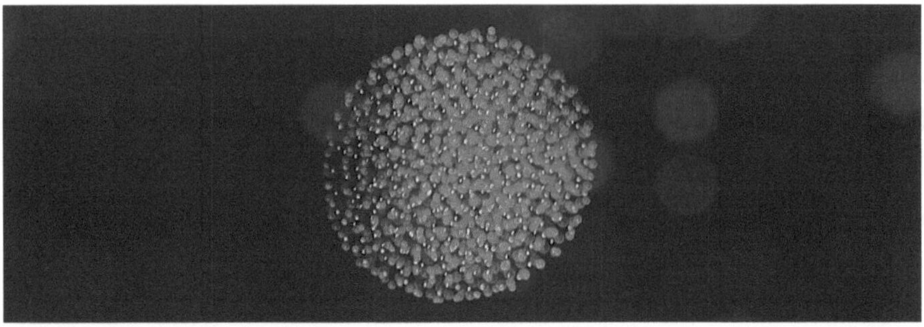

Fig. 10.1 Yukawa ball consisting of several hundred dust particles (*white dots*). The ball has a diameter of about 7 mm. The dust grains are 3.5 μm sized. One clearly recognizes the spherical shape and the nested shell structure of the cluster. Even at room temperature laboratory conditions, a dust–dust coupling parameter on the order of $\Gamma \approx 1{,}000$ and thus strongly correlated many-particle behavior can be achieved. For experimental details, see Chap. 6

settings [6]. Additionally, self-assembling of dust particles plays a crucial role in the fabrication of microchips and solar cells, where growing dust particulates can have both devastating as well as advantageous effects. On the one hand, during the manufacture of highly integrated electronic circuits, the so-called *chip-killing particles* can destroy the damageable plasma-etched nanostructures [7], while on the other hand dust grains included in polymorphous solar cells reduce the degradation of these cells [8]. Besides these technological situations, dusty plasmas are, for example, of great interest in various astrophysical phenomena. For instance, the formation and stability mechanisms of dusty plasma systems are of central interest for the understanding of protoplanetary, protostellar, and accretion disk formation, as well as planetary ring systems [9, 10].

In contrast to the mainly *weakly coupled macroscopic plasmas* in space and technology, in this chapter, we focus on the numerical simulation and analysis of *strongly coupled plasmas* such as spherical Coulomb and Yukawa balls in traps (see Fig. 10.1 for an example and Chap. 7 for details). These finite systems are subject of exceptional current interest since their recent experimental generation, for example, in dusty plasmas [11]. As a second example, we will concentrate on dynamical processes (transport quantities) in macroscopic dusty plasmas. In particular, we discuss the question about the occurrence of superdiffusion in the transition region from a purely *three-dimensional* to a quasi-*two-dimensional* system [3].

10.2 Basics of Molecular Dynamics Simulation

The molecular dynamics (MD) method was originally introduced by B.J. Alder and T.E. Wainwright in the late 1950s with the aim to calculate many-body correlations of classical hard-sphere systems exactly by means of "electronic computers [sic]" [12–14] (see Fig. 10.2). Many valuable insights concerning the collective behavior of interacting many-body systems emerged from their studies with the

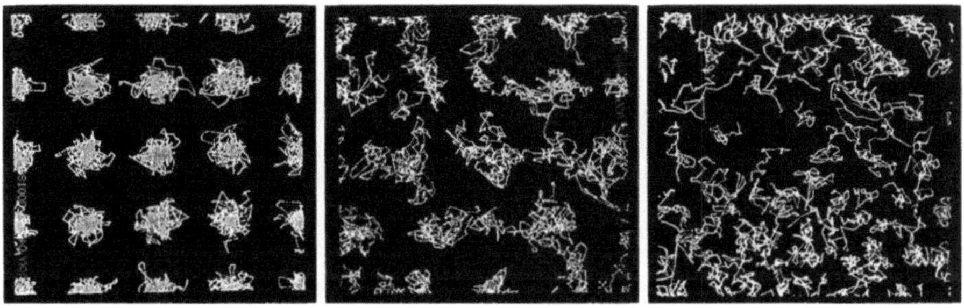

Fig. 10.2 Molecular dynamic motion simulated by means of the first mainframe computers by Alder and Wainwright [13, 14]. Shown are the oscillograph traces representing the trajectories of several ten hard-sphere particles with periodic boundary conditions in the quasi-long-range ordered solid phase (*left*), after it has transformed to the short-range ordered fluid phase (*middle*), and in the liquid–vapor region (*right*). Each picture follows the system for 3,000 collisions (reprinted with permission from [13])

first supercomputers. Prominent examples are the melting/crystallization transition of hard spheres and the long-time decay of autocorrelation functions in fluids. The term "long-time tail" describes the fact that in some systems, the correlations do not decay exponentially with time, but algebraically as $t^{-\gamma}$. Notably, Alder recently retrospected the discovery of the long-time tails for the velocity–velocity correlation function:

> It took us two years to figure out and believe this long-time tail. That's really one of the most qualitatively stunning results, and we just didn't believe it. How could a particle remember for some one hundred collisions its initial velocity [15].

Down to the present day, MD has evolved into the probably most frequently used method to study structures, thermodynamics, and time-dependent processes in many-particle physics, chemistry, biology, astrophysics, materials science, industrial engineering and development, and many other fields. The tremendously fast development of digital computers in the last decades, and in particular the growing availability of cheap computer equipment such as desktop PCs and computer clusters enables researches nowadays to solve many of the fundamental equations of physics without mathematical simplifications – *from first principles*. MD simulations are a central part in the different directions in sciences due to the fact that:

- Computer simulations do not require extensive laboratory facilities.
- Model "experiments" can be repeated arbitrarily often and easily modified.
- Comprehensive and concurrently detailed scans of large parameter sets and ranges can help to discover the conditions of exceptional physical phenomena or optimal settings for real physical experiments.
- Computer-simulated experiments can provide maximum information at the microscopic level and therefore help to give a deeper understanding of laboratory measurements.

In contrast to other particle-based *bottom-up approaches* such as random number-based *stochastic simulation models* (so-called *Monte Carlo methods*;

see Chap. 4), the strict *deterministic integration* of the classical equations of motion for many-particle systems are commonly named *molecular dynamics*. Once the positions and velocities of all individual particles are known at some initial time, their dynamical propagation through the *position–momentum phase space* can be directly computed.

10.2.1 Simulation Model of Strongly Coupled Dusty Plasmas

The existence of a plasma crystal phase was firstly reported by Ikezi [16] in 1986. By means of MD simulations he identified the plasma parameters at which a dusty plasma undergoes a phase transition into an ordered crystalline state. In fact, the theoretical description of complex plasmas turns out to be extremely difficult due to their strongly heterogeneous composition: This leads to challenging multiscale behavior due to the drastic differences in the underlying space and time scales of the plasma constituents (electrons, ions, and dust particles) [17, 18]. An accurate treatment of the multiscale problem requires to include the effect of streaming ions and collisions on the dynamical charging and screening of the dust particles self-consistently. This challenging task necessitates very large computational capabilities and is out of the scope of this review.

Instead of explicitly taking into account all interactions among the constituent particles, here we will rely on a simple model for the dust–plasma interaction which, nevertheless, allows one to reproduce many experimental observations with high accuracy. The high mass of ions and dust particles allows us to treat the particles classically. A *dynamically* screened (anisotropic) pair potential, which includes the impact of the streaming plasma environment on the dust–dust interaction, is given by

$$V(\boldsymbol{r}_i, \boldsymbol{r}_j) = \frac{1}{(2\pi)^3} \int \mathrm{d}^3\boldsymbol{k}\, \mathrm{e}^{\mathrm{i}\boldsymbol{k}\cdot(\boldsymbol{r}_i - \boldsymbol{r}_j)} \frac{V_\mathrm{C}(k)}{\epsilon(\boldsymbol{k}, \boldsymbol{k}\cdot\boldsymbol{u} + \mathrm{i}\nu_\mathrm{i})}, \tag{10.1}$$

with $V_\mathrm{C}(k) = 4\pi q^2/k^2$ being the Fourier transform of the Coulomb potential. The velocity vector $\boldsymbol{u}$ describes the constant ion flow, which is directed toward the lower electrode (see Chap. 6). The plasma response is embedded in the so-called *dielectric function*:

$$\epsilon(\boldsymbol{k}, \omega) = 1 + \frac{1}{k^2\lambda_{\mathrm{D,e}}^2} - \frac{\omega_i^2}{k^2} \frac{\int \mathrm{d}^3\boldsymbol{v}\, \frac{\boldsymbol{k}\cdot\nabla_{\boldsymbol{v}} f_{\mathrm{i}0}(\boldsymbol{v})}{\boldsymbol{k}\cdot\boldsymbol{v} - \omega}}{1 - \mathrm{i}\nu_\mathrm{i} \int \mathrm{d}^3\boldsymbol{v}\, \frac{f_{\mathrm{i}0}(\boldsymbol{v})}{\boldsymbol{k}\cdot\boldsymbol{v} - \omega}}, \tag{10.2}$$

which includes Landau damping and collisional damping (ν_i is the ion–neutral collision frequency and $f_{\mathrm{i}0}(\boldsymbol{v})$ the ion distribution function). The electron Debye length is given by $\lambda_{\mathrm{D,e}} = \sqrt{\varepsilon_0 k_\mathrm{B} T_\mathrm{e}/q_\mathrm{e}^2 \bar{n}_\mathrm{e}}$, where $\bar{n}_{\mathrm{e(i)}}$ and $T_{\mathrm{e(i)}}$ refer to the mean electron (ion) density and temperature, respectively. In the limit of $|\boldsymbol{u}| \to 0$ we recover the *static* case, where the surrounding plasma of free electrons and ions is taken into account by means of an isotropically screened Coulomb potential – the *Debye–Hückel potential* or *Yukawa potential*

$$V(\boldsymbol{r}_i, \boldsymbol{r}_j) = \frac{q^2}{4\pi\varepsilon_0} \frac{\mathrm{e}^{-\kappa|\boldsymbol{r}_i - \boldsymbol{r}_j|}}{|\boldsymbol{r}_i - \boldsymbol{r}_j|}, \tag{10.3}$$

which effectively depends only on the magnitude of the pair distance, that is, $V = V(|\boldsymbol{r}_i - \boldsymbol{r}_j|)$. Here, q denotes the charge of the dust component and ε_0 the vacuum permeability. The range of the dust pair potential V is characterized by the screening parameter $\kappa = \lambda_\mathrm{D}^{-1}$, defined as the inverse Debye screening length,

$$\lambda_\mathrm{D} = \left(\frac{q_\mathrm{e}^2 \bar{n}_\mathrm{e}}{\varepsilon_0 k_\mathrm{B} T_\mathrm{e}} + \frac{q_\mathrm{i}^2 \bar{n}_\mathrm{i}}{\varepsilon_0 k_\mathrm{B} T_\mathrm{i}} \right)^{-1/2}, \tag{10.4}$$

which incorporates the combined effect of static electrons and static ions (Fig. 10.3). Note that the Yukawa potential includes the pure Coulomb interaction as the special case $\kappa = 0$.

To accurately model the experimental conditions of Yukawa balls (see Sect. 10.6), we employ a spherical parabolic confinement potential, which is independent of the screening parameter κ [19]

$$V^\mathrm{ext}(r) = \frac{m}{2}\omega_0^2 r^2, \tag{10.5}$$

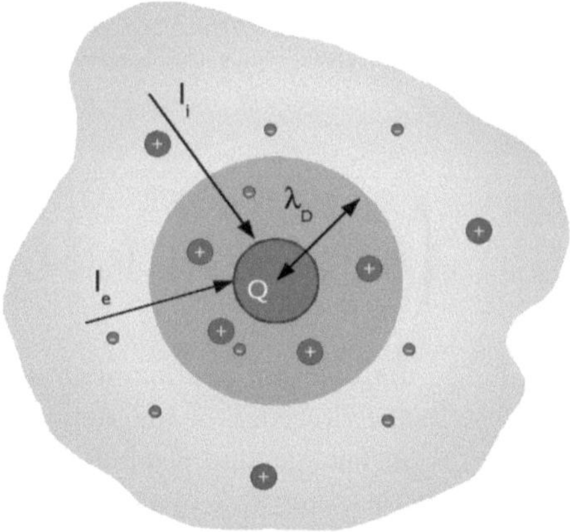

Fig. 10.3 Illustration of a dust particle immersed in a plasma. The negatively charged grain is surrounded by a cloud of positive ions and negative electrons shielding the bare Coulomb potential. The grain charge is determined by the ion and electron currents $I_{\mathrm{i,e}}$. Due to the greatly different mobility of electrons and much heavier ions, the dust grains acquire high negative charges (on the order of 10,000 elementary charges) by collecting more electrons than ions. The Debye length λ_D indicates the effective range of the dust potential

where $r = |\boldsymbol{r}|$ denotes the radial particle position (distance from the trap center). The physical properties of the reduced model of N identical dust grains of equal charge q and mass m are defined by the N particle Hamiltonian:

$$\mathcal{H}(\boldsymbol{p}_i, \boldsymbol{r}_i) = \sum_{i=1}^{N} \frac{\boldsymbol{p}_i^2}{2m_i} + \sum_{i=1}^{N-1} \sum_{j>i}^{N} V\left(|\boldsymbol{r}_i - \boldsymbol{r}_j|\right) + \sum_{i=1}^{N} V^{\text{ext}}(\boldsymbol{r}_i, t) . \qquad (10.6)$$

It is interesting to note that this classical system is characterized by only two parameters: (1) the coupling strength Γ defined by the ratio of the mean (nearest neighbor) interaction energy to the average kinetic energy and (2) the screening parameter κ [20]. Using MD, we solve the N particle problem of dust grains from first principles.

10.2.2 Equations of Motion of a One-Component Plasma

In MD, the time propagation of the N particle Hamiltonian (10.6) is achieved by high-precision numerical integration of the N coupled Newtonian differential equations $m_i (\mathrm{d}^2/\mathrm{d}t^2)\boldsymbol{r}_i = \boldsymbol{F}_i$. These equations can be split into a coupled system of first-order ordinary differential equations (ODEs):

$$\frac{\mathrm{d}}{\mathrm{d}t}\boldsymbol{r}_i = \boldsymbol{v}_i , \qquad (10.7)$$

$$\frac{\mathrm{d}}{\mathrm{d}t}\boldsymbol{v}_i = \frac{\boldsymbol{F}_i}{m_i} , \qquad (10.8)$$

for the motion of the particles $i = 1, \ldots, N$ of mass m_i, position $\boldsymbol{r}_i$, velocity $\boldsymbol{v}_i$, and the total force

$$\boldsymbol{F}_i = -\nabla_{\boldsymbol{r}_i} \left(V^{\text{ext}}(\boldsymbol{r}_i) + \sum_{j \neq i}^{N} V^{\text{int}}\left(\boldsymbol{r}_i - \boldsymbol{r}_j\right) \right) . \qquad (10.9)$$

The considered force is due to the external confinement and the mutual particle–particle interaction energies. In the presence of magnetic fields or friction forces (due to friction with the neutral gas background, see below), $\boldsymbol{F}_i$ becomes a function of the velocity $\boldsymbol{v}_i(t)$, that is, nonconservative.

If we know the solution to this system of equations at a time t, the solution at a later time $t' = t + \Delta t$ can be easily obtained:

$$\boldsymbol{r}_i(t') = \boldsymbol{r}_i(t) + \int_t^{t+\Delta t} \mathrm{d}\tau\, \boldsymbol{v}_i(\tau), \qquad (10.10)$$

$$\boldsymbol{v}_i(t') = \boldsymbol{v}_i(t) + \int_t^{t+\Delta t} \mathrm{d}\tau\, m_i^{-1} \boldsymbol{F}_i(\tau). \qquad (10.11)$$

Now making the simple approximation that the forces $\boldsymbol{F}_i$ and the velocities $\boldsymbol{v}_i$ remain constant during the finite time interval Δt, these equations reduce to

$$\boldsymbol{r}_i(t') \approx \boldsymbol{r}_i(t) + \Delta t \cdot \boldsymbol{v}_i(t), \tag{10.12}$$

$$\boldsymbol{v}_i(t') \approx \boldsymbol{v}_i(t) + \Delta t \cdot m_i^{-1} \boldsymbol{F}_i(t). \tag{10.13}$$

This so-called *Euler integration scheme* is just a Taylor expansion truncated after the first term and only the most simple time propagation approach of solving the equations of motion (10.7) and (10.8). Its (local) error is of the order $\mathcal{O}(\Delta t^2)$,[1] that is, the scheme is correct only up to the first order of Δt. The structure of a rudimentary MD code is sketched in Fig. 10.4.

It is well known that trajectories are exponentially sensitive to small perturbations of the initial conditions. Also, a finite time step always introduces numerical errors and therefore perturbations of the trajectories. A quantity that can be used to estimate the accuracy of a MD simulation is the total energy. For an explicitly time-independent Lagrangian L with $\partial L/\partial t = 0$, Noether's theorem states that the total energy is a conserved quantity. In the following, we will discuss two more advanced integration schemes which achieve in many cases a much better energy conservation (in the microcanonical ensemble, see below) than the simple Euler scheme.

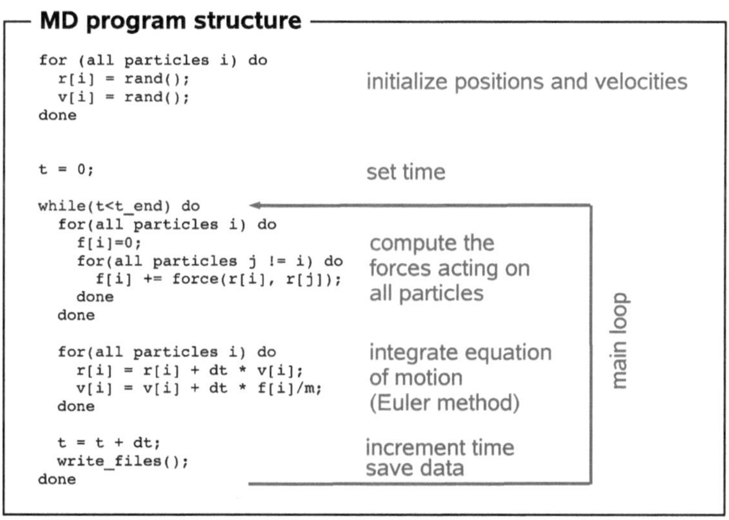

Fig. 10.4 Pseudocode showing the structural elements of a molecular dynamics program with an Euler time propagation scheme (10.12) and (10.13). The force calculation is carried out according to (10.9)

[1] The order of the integrated (global) error, which determines the accuracy of a simulation, varies in general by one factor of Δt less than the order of the local truncation error of the algorithm.

10.2.3 Velocity Verlet Integration Scheme

The choice of the MD integration scheme and step size is of crucial importance for any MD program. Integrators differ in speed, stability, memory requirements, and the number of force evaluations per MD step. A main criterion for choosing an integrator should be the degree of energy conservation and the ability to reproduce certain time- and space-dependent correlations [21]. Since energy conservation is connected with time reversibility, integrators are classified according to this feature.

A conceivably great variety of integration schemes have been proposed over the course of the last decades. One of the most popular schemes has been originally proposed by Verlet [22] and modified by Swope [23]. The so-called *velocity Verlet* or simply *Swope algorithm* reads

$$\boldsymbol{r}_i(t') \approx \boldsymbol{r}_i(t) + \Delta t \boldsymbol{v}_i(t) + \frac{\Delta t^2}{2m}\boldsymbol{F}_i(t), \tag{10.14}$$

$$\boldsymbol{v}_i(t') \approx \boldsymbol{v}_i(t) + \frac{\Delta t}{2m}\left(\boldsymbol{F}_i(t) + \boldsymbol{F}_i(t+\Delta t)\right). \tag{10.15}$$

This scheme offers exceptionally good stability and energy conservation.[2] It is implemented by the following four steps:

1. Calculation of the positions at $t + \Delta t$ using (10.14)
2. Calculation of the velocities at midstep as

$$\boldsymbol{v}\left(t + \frac{1}{2}\Delta t\right) = \boldsymbol{v}(t) + \frac{1}{2}\Delta t\, \boldsymbol{a}(t) \tag{10.16}$$

3. Calculation of the new forces at time $t + \Delta t$
4. Completion of the velocity step by

$$\boldsymbol{v}(t+\Delta t) = \boldsymbol{v}\left(t + \frac{1}{2}\Delta t\right) + \frac{1}{2}\Delta t\, \boldsymbol{a}(t+\Delta t) \tag{10.17}$$

Insight into the reasons for the high stability of the velocity Verlet algorithm can be obtained by a derivation by means of the Liouville operator [24, 25], which also gives a physical justification for the form of the velocity Verlet integration.

The phase-space coordinates $\boldsymbol{r}$ and $\boldsymbol{p}$ describing the $2\mathcal{D}N$ degrees of freedom of a $\mathcal{D}$-dimensional system evolve in time according to the action of the propagator $\exp(\mathrm{i}\mathcal{L}t)$, where $\mathrm{i}\mathcal{L} = \dot{\boldsymbol{r}}\partial_r + \dot{\boldsymbol{p}}\partial_p$ is the Liouville operator, because any function $f(\boldsymbol{r}(t), \boldsymbol{p}(t))$ evolves according to

[2] By eliminating the velocity in (10.14) using (10.15), one obtains the original Verlet scheme [22]. The two schemes thus deliver the same particle trajectory in configuration space and differ only in their estimates for the velocity.

$$\dot{f} = \dot{\boldsymbol{r}}\frac{\partial f}{\partial \boldsymbol{r}} + \dot{\boldsymbol{p}}\frac{\partial f}{\partial \boldsymbol{p}} = \mathrm{i}\,\mathcal{L}\,f, \tag{10.18}$$

with the formal solution $f(\boldsymbol{r}(t), \boldsymbol{p}(t)) = \exp(\mathrm{i}\,\mathcal{L}\,t)\,f(\boldsymbol{r}(0), \boldsymbol{p}(0))$. If the momentum (position) part of $\mathcal{L}$ is zero, $\mathcal{L}$ acts as a simple translation operator which shifts $\boldsymbol{r}$ by $\dot{\boldsymbol{r}}\,\Delta t$ ($\boldsymbol{p}$ by $\dot{\boldsymbol{p}}\Delta t$). Such a decomposition of $\mathcal{L}$ into a position part and a momentum part allows one to write

$$\mathrm{i}\,\mathcal{L} = \mathrm{i}\,\mathcal{L}_r + \mathrm{i}\,\mathcal{L}_p = \dot{\boldsymbol{r}}\cdot\frac{\partial}{\partial \boldsymbol{r}} + \dot{\boldsymbol{p}}\cdot\frac{\partial}{\partial \boldsymbol{p}}. \tag{10.19}$$

These two parts of the Liouvillian do not necessarily commute:

$$\exp(\mathrm{i}\,\mathcal{L}\,t) \neq \exp(\mathrm{i}\,\mathcal{L}_r\,t)\cdot\exp(\mathrm{i}\,\mathcal{L}_p\,t). \tag{10.20}$$

But by applying the *Trotter identity*, the following approximation can be obtained:

$$\exp(\mathrm{i}\,\mathcal{L}\,\Delta t) \approx \exp(\mathrm{i}\,\mathcal{L}_p\,\Delta t/2)\exp(\mathrm{i}\,\mathcal{L}_r\,\Delta t)\exp(\mathrm{i}\,\mathcal{L}_p\,\Delta t/2), \tag{10.21}$$

with an error of order $\mathcal{O}(\Delta t^3)$. If the right-hand side of (10.21) is applied to the phase-space coordinates, the effect is a propagation of the momentum variables by $(1/2)\dot{\boldsymbol{p}}\Delta t$, followed by a propagation of the position variables by $\dot{\boldsymbol{r}}\,\Delta t$ and another propagation of the momentum variables as before. This sequence is exactly the steps carried out in the velocity Verlet integration scheme.

The reasons for the stability of the algorithm can be seen now: Each of the three transformations according to (10.21) has a Jacobian of unity, so the total Jacobian is also unity. The algorithm is therefore phase-space conserving (symplectic). The time reversibility follows, because every step is time reversible by itself.

10.2.4 Runge–Kutta Integration Scheme

Higher-order schemes are not always superior to low-order schemes. Typically, they perform better if the time step can be chosen such that the advantage of the larger time step is not foiled by the necessity of several force evaluations per step. For regions where the forces are rapidly varying, a small time step must be chosen. On the other hand, in regions where the forces are well behaved and vary considerably only on larger scales, one can get away with a larger time step. To obtain a performance improvement for a given limit of accuracy, variable time step methods have been developed, which can significantly enhance the performance of the numerical time integration.

One of the most popular schemes for the solution of any system of ODEs is the family of Runge–Kutta integrators. An adaptive, step size controlled algorithm rests upon the *fifth-order* Runge–Kutta formula:

$$\boldsymbol{k}_1 = \boldsymbol{f}(t_n, \boldsymbol{y}_n)\Delta t, \tag{10.22}$$

$$\boldsymbol{k}_2 = \boldsymbol{f}(t_n + a_2\Delta t, \boldsymbol{y}_n + b_{21}\boldsymbol{k}_1)\Delta t, \tag{10.23}$$

$$\vdots$$

$$\boldsymbol{k}_6 = \boldsymbol{f}(t_n + a_6\Delta t, \boldsymbol{y}_n + b_{61}\boldsymbol{k}_1 + \cdots + b_{65}\boldsymbol{k}_5)\Delta t, \tag{10.24}$$

$$\boldsymbol{y}_{n+1} = \boldsymbol{y}_n + c_1\boldsymbol{k}_1 + c_2\boldsymbol{k}_2 + c_3\boldsymbol{k}_3 + c_4\boldsymbol{k}_4 + c_5\boldsymbol{k}_5 + c_6\boldsymbol{k}_6 + \mathcal{O}(\Delta t^6). \tag{10.25}$$

This scheme requires six (force) function evaluations to advance the solution through the interval Δt from t_n to $t_{n+1} \equiv t_n + \Delta t$. Another combination of the six functions yields an embedded *fourth-order* Runge–Kutta formula:

$$\boldsymbol{y}^*_{n+1} = \boldsymbol{y}_n + c^*_1\boldsymbol{k}_1 + c^*_2\boldsymbol{k}_2 + c^*_3\boldsymbol{k}_3 + c^*_4\boldsymbol{k}_4 + c^*_5\boldsymbol{k}_5 + c^*_6\boldsymbol{k}_6 + \mathcal{O}(\Delta t^5), \tag{10.26}$$

where a_i, b_{ij}, c_i, and c^*_i are the Cash–Karp coefficients [26]. The difference between the fourth- and fifth-order accurate estimates of $y(t + \Delta t)$ gives an appropriate estimate of the local numerical truncation error:

$$\delta y(\Delta t) \equiv \left|\boldsymbol{y}_{n+1} - \boldsymbol{y}^*_{n+1}\right| = \left|\sum_{i=1}^{6}(c_i - c^*_i)\boldsymbol{k}_i\right|, \tag{10.27}$$

which is employed to adapt the step size Δt in a way that the desired degree of predetermined accuracy in the trajectories is achieved with minimum computational effort. In particular in (trapped) few-particle systems with small particle numbers, where the minimum two-particle distance (and hence the force field amplitude) strongly alternates during the simulation run, the adaptive algorithm provides drastic performance gains by a factor of 10–100 compared with the standard fixed step size Runge–Kutta method of fourth order. For more details on the Runge–Kutta integrator, see [27].

10.3 Equilibrium Simulations: Thermodynamic Ensembles

Standard MD simulations conserve the total energy of the system and therefore mimic a microcanonical *NVE*-ensemble.[3] To simulate other ensembles such as the canonical (*NVT*) or the isothermal–isobaric (*NPT*) (see Table 10.1), it is necessary

Table 10.1 Comparison of different ensembles depending on the particle number N, the system volume V, the total energy E, and the pressure P

Condition	Constant properties	Ensemble
Free	N, V, E	Microcanonical
Constant temperature	N, V, T	Canonical
Constant temperature and pressure	N, P, T	Isothermal–isobaric

[3] It is sometimes cautioned that the ensemble described by MD with periodic boundary conditions (cf. Sect. 10.4) subtly differs from real microcanonical preparations due to the conservation of linear momentum. For a discussion of this aspect, see [28].

to appropriately modify the dynamics of the system. Here, we will concentrate on the canonical *NVT*-ensemble which is commonly employed to model equilibrated systems at a given finite temperature.

10.3.1 Velocity Scaling

In general, the initial conditions of the particles will not represent an equilibrium configuration. Therefore, it is necessary to equilibrate the system for some time prior to the measurement. A simple and routinely used thermostatting in the equilibration phase of a simulation is achieved by velocity rescaling (also called "isokinetic thermostat"). The instantaneous temperature $T = 2E_k/(\mathcal{D}k_B)$ is calculated from the kinetic energy E_k and each velocity is rescaled by a factor $\lambda = \sqrt{T_t/T}$ to guarantee the required temperature T_t in the system. During this premeasurement period, the total momentum (i.e., the center-of-mass momentum) should be removed after every rescaling. After the system has reached an equilibrium configuration, the measurement can be performed without a thermostat or with a more suitable thermostat.

For small nonperiodic systems such as small clusters, velocity rescaling should be avoided in finite temperature simulations altogether, because the coupling between high-frequency (e.g., local oscillations) and low-frequency degrees of freedoms (e.g., total angular momentum and total momentum) is too small to establish equipartition between all degrees of freedom (this is the so-called "flying ice-cube effect" [29]). This can result, for instance, in a quickly rotating cluster in which the relative particles positions remain largely fixed.

A scheme similar to the simple velocity rescaling is due to Berendsen [30]. Here, the scaling factor applied to the velocities to achieve the required temperature T_t is given by the expression

$$\lambda = \left[1 + \frac{\Delta t}{\tau}\left(\frac{T}{T_t} - 1\right)\right], \tag{10.28}$$

where Δt is the integration step size and the constant τ is the "rise time" of the thermostat. It is inversely related to the coupling strength of the heat bath to the system. For high values of τ, the system fluctuates less around its average values than for smaller τ.

10.3.2 Stochastic Thermostats

A second class of thermostats can be subsumed by the term *stochastic*. In the *Andersen thermostat* [31], the particles are assumed to be coupled to a heat reservoir at stochastically distributed times. The strength of the coupling between the system and the heat bath is determined by the collision frequency ν which is a parameter of the thermostat. To decide whether a particular particle is in contact with the heat bath, a random value between 0 and 1 is drawn for each particle. If the value is less than the product of the collision frequency ν and time step Δt, that particle's velocity is redrawn from a Maxwellian distribution. Note that singular kicks to the particles make the Andersen thermostat local in nature and destroy temporal and spatial correlations.

An alternative is the *dissipative particle dynamics* (DPD) thermostat (e.g., [32]); the equations of motion are modified to

$$\dot{\boldsymbol{r}}_i = \frac{\boldsymbol{p}_i}{m_i}, \tag{10.29}$$

$$\dot{\boldsymbol{p}}_i = \boldsymbol{F}_i + \boldsymbol{F}_i^{\mathrm{D}} + \boldsymbol{F}_i^{\mathrm{R}}. \tag{10.30}$$

The dissipative force $\boldsymbol{F}_i^{\mathrm{D}}$ and the random force $\boldsymbol{F}_i^{\mathrm{R}}$ are sums over pairwise forces, $\boldsymbol{F}_i^{\{\mathrm{D,R}\}} = \sum_{j\neq i}^{N} \boldsymbol{F}_{ij}^{\{\mathrm{D,R}\}}$, which in turn depend on the velocity difference and the positional distance of particles i, j (for details see [32, 33]). Here, the important point lies in the fact that the DPD fulfills Newton's third law even in the presence of stochastic forces $\boldsymbol{F}_i^{\mathrm{R}}$.

10.3.3 Nosé–Hoover Thermostat

An elegant route to thermostatting has been proposed by Nosé and further developed by Hoover [34, 35]. It is based on introducing an additional dimensionless degree of freedom into the Lagrangian formalism. The equations of motion for the thermostatted system are derived to read [36]

$$\dot{\boldsymbol{r}}_i = \frac{\boldsymbol{p}_i}{m_i}, \tag{10.31}$$

$$\dot{\boldsymbol{p}}_i = \boldsymbol{F}_i - \xi \boldsymbol{p}_i, \tag{10.32}$$

$$Q\dot{\xi} = \sum_i \frac{\boldsymbol{p}_i^2}{m_i} - \mathcal{D} N k_{\mathrm{B}} T_t, \tag{10.33}$$

where $\mathcal{D}$ is the dimensionality of the system. The mass Q of the thermostat controls the coupling of the heat bath to the particles. Because $\dot{\boldsymbol{p}}$ depends explicitly on the velocity $\boldsymbol{v} = \boldsymbol{p}/m$, these equations cannot be integrated with the velocity Verlet algorithm. Using a Trotter derivation similar to the one employed above for the velocity Verlet scheme, integrators for Nosé–Hoover systems can be devised [37].

10.3.4 Langevin Dynamics Simulation

A MD simulation can be easily extended to include friction. The fundamental equation of motion then becomes a stochastic Langevin equation:

$$m_i \ddot{\boldsymbol{r}}_i = \boldsymbol{F}_i - m_i \nu \boldsymbol{v}_i + \boldsymbol{F}_i^{\mathrm{G}}, \tag{10.34}$$

where ν is the friction coefficient and $\boldsymbol{F}_i^{\mathrm{G}}$ is a random Gaussian white noise of zero average and standard deviation:

$$\left\langle F_{\alpha,i}^{\mathrm{G}}(t_0) F_{\beta,j}^{\mathrm{G}}(t_0+t)\right\rangle = 2k_{\mathrm{B}}T\nu\delta(t)\delta_{ij}\delta_{\alpha\beta}, \qquad \alpha,\beta\in\{x,y,z\}. \tag{10.35}$$

Equation (10.35) is the *fluctuation–dissipation theorem* which relates microscopic fluctuations to macroscopic dissipation. The friction and noise term balance each other to yield a particle temperature T. The coupling to the heat bath is determined by the value of the damping parameter ν.

The solution of (10.34) is more difficult than Newtonian MD. Available integration schemes are generally less advanced in terms of reliability and stability. Two simple integration schemes are the *symplectic low-order* (SLO) scheme by Mannella [38] and the scheme proposed in [39] which will be referred to as "Li" (for "liquid").

The *SLO algorithm* was designed to be symplectic in the limit $\nu\to 0$ and reads

$$\begin{aligned}
\boldsymbol{r}\left(t+\frac{\Delta t}{2}\right) &= \boldsymbol{r}(t)+\frac{\Delta t}{2}\boldsymbol{v}(t),\\
\boldsymbol{v}(t+\Delta t) &= c_2\left[c_1\boldsymbol{v}(t)+\Delta t\frac{\boldsymbol{F}(t+\frac{\Delta t}{2})}{m}+d_1\delta\boldsymbol{r}^{\mathrm{G}}\right],\\
\boldsymbol{r}(t+\Delta t) &= \boldsymbol{r}\left(t+\frac{\Delta t}{2}\right)+\frac{\Delta t}{2}\boldsymbol{v}(t+h),
\end{aligned} \tag{10.36}$$

where $\delta\boldsymbol{r}^{\mathrm{G}}$ is a $\mathcal{D}$-dimensional vector whose components are independent Gaussian random variables with standard deviation one and average zero. The constants are

$$c_1 = 1-\frac{\nu\Delta t}{2}, \quad c_2 = \frac{1}{1+\nu\Delta t/2}, \quad d_1 = \sqrt{\frac{2k_{\mathrm{B}}T\nu\Delta t}{m}}. \tag{10.37}$$

The *Li algorithm* reduces to the velocity Verlet scheme in the limit $\nu\to 0$. It reads [39, eq. (9.24)]

$$\begin{aligned}
\boldsymbol{r}(t+\Delta t) &= \boldsymbol{r}(t)+c_1\Delta t\boldsymbol{v}(t)+c_2(\Delta t)^2\cdot\frac{\boldsymbol{F}(t)}{m}+\delta\boldsymbol{r}^{\mathrm{G}},\\
\boldsymbol{v}(t+\Delta t) &= c_0\boldsymbol{v}(t)+(c_1-c_2)\Delta t\cdot\frac{\boldsymbol{F}(t)}{m}+c_2\Delta t\cdot\frac{\boldsymbol{F}(t+\Delta t)}{m}+\delta\boldsymbol{v}^{\mathrm{G}}.
\end{aligned} \tag{10.38}$$

Each pair of components of $\delta\boldsymbol{r}^{\mathrm{G}}$ and $\delta\boldsymbol{v}^{\mathrm{G}}$ is sampled from a bivariate Gaussian distribution with average zero and the following variances and correlation coefficient c_{rv}:

$$\sigma_{\mathrm{r}}^2 = (\Delta t)^2\frac{k_{\mathrm{B}}T}{m}(\nu\Delta t)^{-1}\left[2-(\nu\Delta t)^{-1}\left(3-4\mathrm{e}^{-\nu\Delta t}+\mathrm{e}^{-2\nu\Delta t}\right)\right], \tag{10.39}$$

$$\sigma_{\mathrm{v}}^2 = \frac{k_{\mathrm{B}}T}{m}\left(1-\mathrm{e}^{-2\nu\Delta t}\right), \tag{10.40}$$

$$c_{\mathrm{rv}} = \frac{k_{\mathrm{B}}T}{m\nu}\left(1-\mathrm{e}^{-\nu\Delta t}\right)^2. \tag{10.41}$$

The constants are given by

$$
\begin{aligned}
c_0 &= \mathrm{e}^{-\nu\Delta t} \approx 1-(\nu\Delta t)+\frac{1}{2}(\nu\Delta t)^2-\frac{1}{6}(\nu\Delta t)^3+\frac{1}{24}(\nu\Delta t)^4-\cdots,\\
c_1 &= \frac{1-c_0}{\nu\Delta t} \approx 1-\frac{1}{2}(\nu\Delta t)+\frac{1}{6}(\nu\Delta t)^2-\frac{1}{24}(\nu\Delta t)^3+\cdots,\\
c_2 &= \frac{1-c_1}{\nu\Delta t} \approx \frac{1}{2}-\frac{1}{6}(\nu\Delta t)+\frac{1}{24}(\nu\Delta t)^2-\cdots.
\end{aligned}
\tag{10.42}
$$

Using the low $\nu\Delta t$ expansion given in the equations above, it is easy to see how the *Li scheme* reduces to the velocity Verlet integration. Due to computational limitations, it is often necessary to employ this expansion in the calculation of the constants and similar expansions for the variances for low values of $\nu\Delta t$.

10.3.5 Dimensionless System of Units

Due to *universal scaling laws* [40] in plasmas physics, similar physical correlation phenomena, such as plasma crystallization, are present in completely different physical regimes. By introducing dimensionless energy and length scales system specific parameters can be reduced and theoretical results become universally valid. Nevertheless, they are still easily applicable to a particular experimental measurement. In computer codes, all quantities are often reduced to a dimensionless form by choosing characteristic units. The most suitable base units for classical trapped systems are given in Table 10.2.

Here, the base length r_0 corresponds to the equilibrium distance of two identical classical particles in a harmonic confinement that interact via a Coulomb potential. Energies are expressed in terms of the pair interaction energy E_0 for this distance. The inverse trap frequency is a convenient time unit, $t_0 = \omega_0^{-1}$, and the unit of temperature is related to the unit of energy by $T_0 = E_0/k_\mathrm{B}$.

By performing the transformations $\{\mathcal{H} \to \mathcal{H}/E_0, r \to r/r_0, p \to p/p_0, \kappa \to \kappa/\kappa_0\}$, the Hamiltonian (10.6) can be recast in the dimensionless form

$$
\mathcal{H} = \sum_{i=1}^{N} \boldsymbol{p}_i^2 + \sum_{i=1}^{N-1}\sum_{i<j}^{N} \frac{\mathrm{e}^{-\kappa|\boldsymbol{r}_i-\boldsymbol{r}_j|}}{|\boldsymbol{r}_i-\boldsymbol{r}_j|} + \sum_{i=1}^{N} \boldsymbol{r}_i^2. \tag{10.43}
$$

Table 10.2 Overview on the common system of classical base units

Quantity	Relation
Distance	$r_0 = (q^2/2\pi\varepsilon_0 m\omega_0^2)^{1/3}$
Energy	$E_0 = q^2/(4\pi\varepsilon_0 r_0)$
Temperature	$T_0 = E_0/k_\mathrm{B}$
Time	$t_0 = \omega_0^{-1}$
Force	$F_0 = m\omega_0^2 r_0$
Screening parameter	$\kappa_0 = r_0^{-1}$
Momentum	$p_0 = m\omega_0 r_0$

The only remaining (free) parameter is the screening parameter κ. In a harmonically trapped system, the strength of the external confinement potential is controlled by the trap frequency ω_0. The latter defines in the classical case the characteristic scales r_0 and E_0, but does not affect the obtained dimensionless results. This fact clearly underlines the universal scaling behavior of classical many-body effects. However, quantum effects naturally limit the universal scaling to finite temperatures and low densities.

10.4 Simulation of Macroscopic Systems

The computationally most expensive part of an MD simulation is the calculation of the pairwise forces by which the particles interact, which is a $\mathcal{O}(N^2)$ task. For mesoscopic systems (up to $\sim$ 500 particles), direct calculation of all forces is possible on current workstation computers. For larger systems of particle numbers in the thousands and ten thousands, approximations have to be made.

10.4.1 Potential Truncation

The most straightforward simplification is the truncation of the interaction potential beyond a *cutoff radius* r_c:

$$V^{\text{trunc}}(r) = \begin{cases} V^{\text{int}}(r), & r < r_c, \\ 0, & r \geq r_c, \end{cases} \tag{10.44}$$

The impact of this approximation depends on the size of the cutoff radius r_c and on the quickness of the decay of the interaction potential. For the Yukawa-type grain–grain interaction in dusty plasmas, the interaction potential decays as $\exp(-r)/r$. Hence, the cutoff radius can typically be chosen large enough that the perturbation introduced into the simulation by the use of $V^{\text{trunc}}(r)$ instead of $V^{\text{int}}(r)$ is negligible.

The potential truncation is generally possible if the particles effectively interact via a *short-range pair potential*, that is, if the number of interaction partners grows slower than the potential decays. This is the case if the integral

$$\int_0^\infty \mathrm{d}r\, V^{\text{int}}(r) r^{\mathcal{D}-1} \tag{10.45}$$

converges, see Table 10.3. By use of a cutoff radius the computational effort of the force calculation reduces to $\mathcal{O}(N)$; see below on how to find neighboring particles efficiently.

Table 10.3 Decay of different types of pair potentials V in view of utilization of a cutoff radius in the case of an short-range interaction

Coulomb	r^{-1}	Long range
Yukawa	$r^{-1}\exp(-r)$	Short range
Dipole ($\mathcal{D} = 1, 2$)	r^{-3}	Short range
Dipole ($\mathcal{D} = 3$)	r^{-3}	Long range

In dimension $\mathcal{D}$, an interaction potential is called *long ranged* when the number of interaction partners grows faster than the potential decays, otherwise it is called *short ranged* (see Eq. (10.45)).

10.4.2 Electrostatic Interactions

For purely Coulombic interactions, a simple potential truncation cannot be applied. Systems with periodic boundary conditions (see below) can be dealt with by the well-known *Ewald method* [41], which however exhibits an unfavorable scaling of $\mathcal{O}(N^2)$ or $\mathcal{O}(N^{3/2})$. A class of algorithms known as *Mesh–Ewald techniques* has been developed to reduce this computational effort to $\mathcal{O}(N \log N)$ [42]. The basic idea of these algorithms is to interpolate the charge distribution of the particles onto a regular mesh which can be Fourier-transformed by FFT. The reciprocal space contributions to the electrostatic energy can then be efficiently calculated and transformed back into real space by an inverse FFT.

The three main implementations of this Mesh–Ewald idea are the *Particle–Particle–Particle–Mesh* (P3M) method by Hockney and Eastwood [43], the *Particle–Mesh–Ewald* (PME) method by Darden [44] and the *smooth particle–Mesh–Ewald* (SPME) by Essmann et al. [45]. They differ in the way the particle charges are assigned to the mesh (and vice versa), the Green's function used to solve the Poisson equation in reciprocal space and the type of differentiation of the potential to obtain the forces. Additionally, different aspects of these methods are sometimes combined. Of these methods, the P3M method by Hockney and Eastwood seems to be the most popular because of high accuracy and the availability of error estimates.

Apart from Mesh–Ewald methods, *multipole techniques* can be used to deal with electrostatic forces. Variants of this idea can scale as favorably as $\mathcal{O}(N)$, but a higher computational effort per step can put the break-even point between Mesh–Ewald and multipole techniques to astronomically large particle numbers [46].

10.4.3 Finding of Neighboring Particles

The use of a cutoff radius r_c reduces the cost of computing the forces from an $\mathcal{O}(N^2)$ to an $\mathcal{O}(N)$ effort. However, the computation of the forces then requires the knowledge of all particles inside the cutoff radius r_c around a specific particle. A straightforward implementation of this task requires $\mathcal{O}(N^2)$ calculations.

Two strategies can be employed to reduce this effort [21]. The first is the so-called *neighbor list method*, in which the list of neighbors is created with an increased radius $r_c + \Delta r$. At each time step after the neighbor list creation, the modulus of the fastest particle's velocity is added to a cumulative velocity V_m. Only when the cumulative velocity exceeds the condition

$$V_m = \sum_{\text{steps}} \max_i |v_i| > \frac{\Delta r}{2\Delta t}, \tag{10.46}$$

the neighbor list is updated. The choice of Δr influences the frequency of the updates and must be balanced for optimal performance gain.

The second strategy is called *chaining mesh technique* or cell subdivision: The primary simulation box is subdivided into a rectangular grid of mesh size h. The mesh size h is chosen in such a way that it is greater than r_c and that it evenly divides the box size L. By using the floor function,

$$k_i = \left\lfloor \frac{x_i}{h} \right\rfloor, \tag{10.47}$$

and analogous expressions for the other components of the position vector, it is possible to assign each particle to a grid cell in an $\mathcal{O}(N)$ operation. The search for neighboring particles for any one particle can then be limited to its own grid cell and the $3^{\mathcal{D}} - 1$ neighboring cells. By taking advantage of Newton's third law of motion, it is sufficient to consider only half of these neighboring cells. In three-dimensional space, the relevant 13 neighboring cells of a cell $\{j, k, l\}$ can be chosen by the following prescription:

$$\{j+1, k+s, l+t\}, \{j, k+1, l+t\}, \{j, k, l+1\} \quad \text{for} \quad s, t \in \{-1, 0, 1\}, \tag{10.48}$$

and a similar expression can be given for two dimensions. The two mentioned approaches can be combined to achieve even higher efficiency.

10.4.4 Periodic Boundary Conditions

Even with the utilization of a cutoff radius, MD can typically only be applied to particle numbers from $N = 10^2$ to $N = 10^6$. In such small systems (in a thermodynamic sense), surface effects can play a dominant role [28, 36]. For example, in a three-dimensional thermodynamic system of $N = 10^{21}$ particles, the number of particles near a wall is of order $N^{2/3}$, that is, a fraction of $w_{\text{thd}} \approx 10^{-7}$. For an MD simulation of $N = 1{,}000$ particles, this fraction is close to 10%, six orders of magnitude higher than w_{thd}. To avoid surface effects in a bulk simulation (as opposed to a cluster simulation), it is common to apply *periodic boundary conditions* to the particle system. The simulation box (which needs to be space filling, usually cubic) is repeated along all spatial directions. In each of these image cells, the same particles as in the primary cell are present, where the origin is taken relative to each cell origin (see Fig. 10.5). When a particle crosses the boundary between two cells,

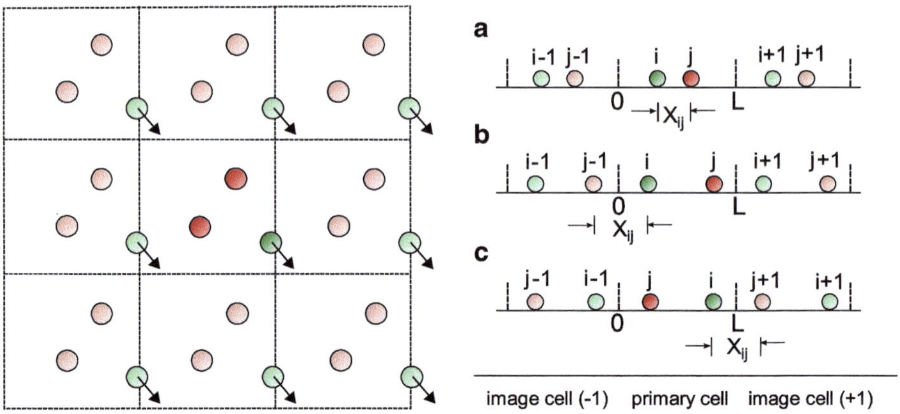

Fig. 10.5 *Left*: application of periodic boundary conditions in two dimensions. The primary cell (*center*) contains three particles (*dark red* and *dark green*) and is surrounded by eight image cells. The events in the primary cell are replicated in each of the image cells (*light red* and *light green*). As a particle crosses the boundary of the primary cell (*dark green*), one of the replicas enters the primary box from the opposite cell (*light green*). *Right*: application of the minimum image convention (MIC). The primary cell (*center*) contains two particles (*dark green* and *dark red*). The image cells to the *left* and the *right* contain the image particles (*light green* and *light red*). The sketch shows three cases of the primary cell and the two neighboring cells: (**a**) $x_{ij} < L/2$, (**b**) $x_{ij} < -L/2$, and (**c**) $x_{ij} > L/2$. Due to the MIC, in the latter two cases, the pair distance between particles i and j is calculated as $x_i - x_j + L$ and $x_i - x_j - L$, respectively (adapted from [28])

one of its images enters through the opposite boundary, which ensures conservation of the particle number and the total linear momentum. Because the newly entering particle has a different angular momentum with respect to the center of the box, MD with PBC does not conserve the total angular momentum of the system. Owing to the high number of particles leaving and entering the primary simulation cell during the course of one simulation, the total angular momentum fluctuates about zero, with the magnitude of the fluctuation being determined by the number of particles.

Each particle interacts with all particles in the primary cell and all image particles, including images of itself. When the cutoff radius can be chosen smaller than half of the box width L, the problem of calculating the interaction energy can be substantially simplified: Every particle interacts only with the nearest image of any particle (this also excludes images of itself, which are at least a distance L away).

The calculation of pair distances in systems with PBC has to be carried out according to the *minimum image convention* (see Fig. 10.5). If x_{ij} is the distance between two particles i and j, the effective distance x'_{ij} is computed as following:

$$x'_{ij} = \begin{cases} x_{ij} & : |x_{ij}| < L/2, \\ x_{ij} - L & : x_{ij} > L/2, \\ x_{ij} + L & : x_{ij} < -L/2. \end{cases} \tag{10.49}$$

The use of PBC effectively removes surface effects from the simulation. However, the limited number of particles in the simulation gives rise to finite-size effects. For instance, the reciprocal space is restricted to wave vectors of the form $\boldsymbol{k} = (2\pi/L)(n_x, n_y, n_z)$ with the box length L and integers $n_{x,y,z}$. Also, collective excitations, such as transverse (sound) waves, can propagate through the system and reenter due to the PBC. They can then cause interference effects which result in artifacts in the correlation functions such as the velocity autocorrelation function (VACF). Therefore, the sensible time for measurements is restricted to times less than the time needed for a sound wave to propagate through the box.

10.5 Input and Output Quantities

MD simulations require, in principle, only the $2\mathcal{D}N$ initial conditions $\boldsymbol{r}_{i,0}$, $\boldsymbol{v}_{i,0}$, $i = 1, \ldots, N$, and a pair interaction potential V_{ij} (which may depend on individual properties of the particles i and j such as the masses m_i, m_j or the charges q_i, q_j) as input parameters. From these input quantities, a MD simulation generates phase-space trajectories $\boldsymbol{Q}_i(t) = (\boldsymbol{r}_i, \boldsymbol{p}_i)$ at discrete times, $\boldsymbol{Q}_i(t_0), \boldsymbol{Q}_i(t_1), \boldsymbol{Q}_i(t_2), \ldots$, which are the basis for the subsequent analysis.

The availability of the microscopic phase-space information allows for the calculation of arbitrary static and dynamic properties of the system which can be defined as averages of phase-space trajectories. Instead of discussing in detail every quantity accessible by MD, here we concentrate on some of the most important and refer the reader to the standard literature for more details.

10.5.1 Pair Distribution Function and Static Structure Factor

Beside the particle density $n(\boldsymbol{r})$, the distribution of particle pairs is central in statistical physics. This is quantified by the radial pair distribution function (RPDF)[4] which gives the probability of finding a specific distance r between two particles in the system relative to the probability of finding that distance in a completely random particle distribution of the same density (Fig. 10.6). A definition in terms of an ensemble average is

$$g(r) = \frac{V}{N^2}\left\langle \sum_i \sum_{j \neq i} \delta\left(|\boldsymbol{r}_{ij}| - r\right) \right\rangle. \tag{10.50}$$

A numerical evaluation of $g(r)$ consists of constructing a histogram $H(r)$ of pair distances in the system (averaged over different time steps or configurations) with an

[4] This quantity is also often radial distribution function or simply pair distribution function.

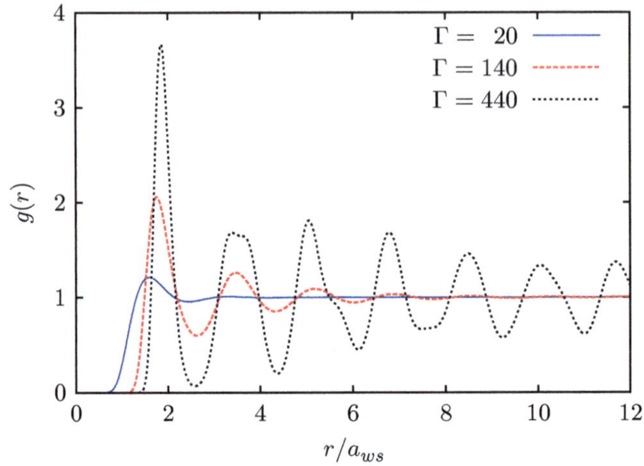

Fig. 10.6 Pair distributions of a 2D system with $\kappa = 2.0$ in the solid phase ($\Gamma = 440$), strongly coupled liquid phase ($\Gamma = 140$), and weakly coupled liquid phase ($\Gamma = 20$)

appropriate histogram width Δr. The probability of finding a pair distance between r and $r + \Delta r$ in a random configuration of N particles with density $\rho = N/V$ is $N\rho^* V_A/2$, where $\rho^* = (N-1)/V$ is the reduced density and V_A is the volume (area) of a hollow sphere (circle) with radius r and shell width Δr. Therefore, $g(r)$ can be calculated numerically as

$$g(r) = H(r)\left[\frac{N(N-1)}{2} \cdot \frac{V_A}{V}\right]^{-1}. \tag{10.51}$$

The RPDF is related to the experimentally accessible static structure factor $S(\boldsymbol{k}) = N^{-1}\langle|\rho(\boldsymbol{k})|^2\rangle$[5] by

$$S(\boldsymbol{k}) - 1 = \rho \int \mathrm{d}\boldsymbol{r}\ \exp(\mathrm{i}\boldsymbol{k}\cdot\boldsymbol{r})g(r), \tag{10.52}$$

or, after angular integration,

$$S(k) - 1 = 4\pi\rho \int_0^\infty \mathrm{d}r\ \frac{\sin(kr)}{kr} g(r). \tag{10.53}$$

[5] $\rho(\boldsymbol{k})$ is the spatial Fourier transform of the number density, that is,

$$\rho(\boldsymbol{k}) = \int \rho(\boldsymbol{r}) \exp(\mathrm{i}\boldsymbol{k}\cdot\boldsymbol{r})\mathrm{d}\boldsymbol{r} = \sum_{i=1}^{N} \exp(\mathrm{i}\boldsymbol{k}\cdot\boldsymbol{r}_i).$$

10.5.2 Transport Properties

The calculation of transport coefficients is one of the most powerful applications of MD simulations. Two different approaches of calculation of transport coefficients have to be distinguished. In *nonequilibrium methods*, a perturbation is introduced into the system and the system's response to this perturbation is used to deduce the transport coefficient. *Equilibrium methods*, on the other hand, make use of the *(Helfand-)Einstein* or the *Green–Kubo relations* to obtain information about the transport process from systems in thermodynamic equilibrium.

The Green–Kubo relations [47–49] are of the form

$$K_{\mathrm{A}} = \int_0^\infty \mathrm{d}\tau \, \langle \dot{A}(\tau)\dot{A}(0) \rangle \,, \tag{10.54}$$

where K_{A} is the transport coefficient and $\dot{A}$ the associated dynamic variable. The integrand in (10.54) is the autocorrelation function of $\dot{A}$. For example, for the self-diffusion coefficient D, $A(t) = x(t)$ and therefore

$$D = \frac{1}{\mathcal{D}} \int_0^\infty \mathrm{d}\tau \, \langle \boldsymbol{v}(\tau)\boldsymbol{v}(0) \rangle \,. \tag{10.55}$$

The Einstein relation for diffusion [50] relates the self-diffusion coefficient to the mean-squared displacement (MSD) of the particles position:

$$D = \lim_{t\to\infty} \frac{1}{2\mathcal{D}t} \left\langle |\boldsymbol{r}(t) - \boldsymbol{r}(t_0)|^2 \right\rangle . \tag{10.56}$$

For MD simulations with PBC, the unfolded (or "infinite-checkerboard") positions are used in the evaluation of (10.56). Generalized Einstein relations for other transport coefficients K_{A} were given by Helfand [51]. They are, however, not readily applicable to systems with PBC [52].

10.6 Applications I: Mesoscopic Systems in Traps

In this section, we focus on structure formation in dusty plasmas in 3D traps. From numerous experiments and simulations, finite *Coulomb systems* in a parabolic confinement potential are known to arrange themselves in nested concentric rings (in 2D) or shells (in 3D systems) (see, e.g., Fig. 10.1). These arrangements have characteristic radii R_i and occupation numbers $(N_1, N_2, \ldots)$, where N_i denotes the number of particles on the ith ring or shell (starting from the center). However, *Yukawa balls* in harmonically confined dusty plasmas show remarkable differences in their structural properties and finite-size behavior, which will be briefly presented here.

A striking feature common to trapped few-particle systems is that their structure and properties are very sensitive to the exact particle number. In recent years, different physical situations were identified, in particular those involving so-called *magic* configurations, with an unusually high cluster symmetry and stability (see e.g. [53, 54]). Even without change of density or temperature, qualitative transformations of the physical properties can be achieved just by adding or removing a single charged particle [55]. This behavior of few-particle systems reflects the basis of chemistry: even the change of the particle number by one can result in drastically different collective properties (structural, electronic, magnetic, transport, or optical).

The main parameters that determine the configuration (ground or metastable states) of a dust cluster for a given N are the screening parameter κ and the friction coefficient ν. The dependence on κ and ν is therefore discussed in detail.

10.6.1 Simulated Annealing

To obtain an understanding of the structure of finite dust crystals, it is advantageous to start with the investigation of the lowest energy states of Coulomb and Yukawa clusters, respectively. The special case of a *pure Coulomb interaction* ($\kappa = 0$) is, among others, of direct practical importance for (laser-cooled) ion crystals in traps [56,57]. Mimicking the crystallization processes that occur in nature, the classical ground-state configurations are explored by *simulated annealing*, that is, by minimizing the kinetic and potential energy in (10.6) during the MD simulation in small steps by inclusion of a velocity scaling algorithm [58, 59]. While local optimization methods can easily be trapped in a local minimum, the simulated annealing method does not suffer from this problem. For that reason, it is often used when a global minimum is hidden among many other local minima.[6]

Finding the *global minimum* on a multidimensional potential energy surface is conceptually simply, but in practice a challenging problem. To assure that the correct N particle ground state is identified, independent runs have to be repeated up to several thousand times, where each run starts with different random initial velocities and positions of all particles (see Fig. 10.7 and discussion below). The quadratic growth of the number of force computations and the exponential complexity of the problem (due to a strongly growing number of energetically very close metastable states, see Table 10.4) with respect to N makes the computation very demanding and requires various code optimizations and an optimal cooling rate. To speed up the computations, we tune the numerical truncation error $\delta y(\Delta t)$ (see Eq. 10.27) during the simulated annealings, in dependence on the particular kinetic energy per particle, continuously from 10^{-3} to 10^{-8}.[7] However, in particular for clusters with $N \geq 100$, reliable data can only be obtained within a feasible time (i.e., within a few

[6] Note that in contrast to quantum systems, in the classical case the kinetic energy vanishes at zero temperature. This means that all particle velocities become zero, which leads to infinitely strong coupling according to the definition of the coupling parameter Γ.

[7] For $E_{\text{kin}}/N > 10^{-3}$, we use a fixed minimum accuracy $\delta y(\Delta t) = 10^{-3}$ to avoid numerical instabilities.

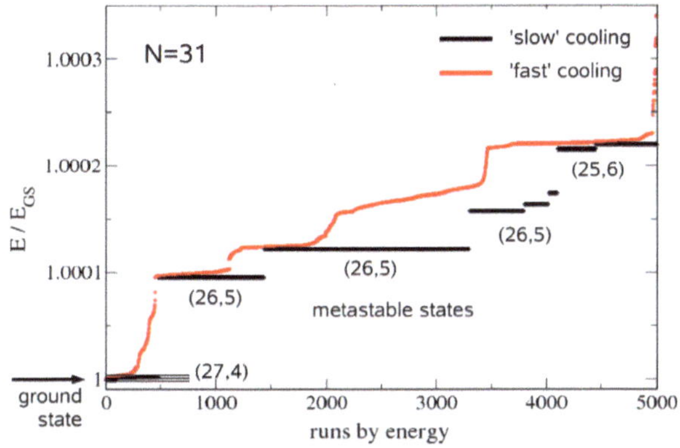

Fig. 10.7 Stationary states observed in the MD simulations for $N = 31$, $\kappa = 1.4$ for cooling to a minimum kinetic energy $E_{\rm kin}^{\rm min} = 10^{-8}$. The runs are sorted by the energy of the final state. $E_{\rm gs}$ denotes the ground-state energy. For slow cooling (*black bars*, $\nu = 0.05$), one can clearly distinguish distinct energy states. The length of the *bold lines* is proportional to the occurrence frequency of a state in a total of 5,000 runs. In the case of strong friction (*red*, *dashed line*, $\nu = 5.3$), the particles often lose their kinetic energy before they can settle into the equilibrium positions and the "fine structure" states in the energy spectrum (i.e., different states with the same shell configuration) cannot be resolved. Note the small relative energy difference between ground and metastable states (cf. Table 10.4) (from [60])

Table 10.4 Energy per particle and shell configurations for the cluster with $N = 31$ particles as seen in Fig. 10.7

$\Delta E/N$	Configuration
3.030266	(27,4)
0.000006	(27,4)
0.000009	(27,4)
0.000291	(26,5)
0.000372	(26,5)
0.000479	(26,5)
0.000499	(26,5)
0.000530	(26,5)
0.000656	(25,6)
0.000669	(25,6)

The energy difference between metastable states and the ground state is given by italic numbers. States with the same shell configuration but different energy differ only by the arrangement of the particles on the same shell (fine structure, see [58]).

weeks) by using several, typically 10–20, CPUs. Instead of parallelizing the code, we are typically running an ensemble of independent simulations on distributed processors. This strategy offers best scalability and makes the most efficient use of the available computing power.

10.6.2 *Effect of Screening*

Metastable states of finite dust clusters have been experimentally observed in recent experiments (see Chap. 7). Here, it was found that Debye shielding has an effect on the structure formation in dusty plasmas. In particular, metastable states often occurred with a higher probability than the ground state. For a theoretical explanation of these results, MD can be utilized. Therefore, we will at first investigate the effect of screening on the energetically most favorable many-particle states.

A detailed numerical analysis of the ground states of screened Coulomb clusters by simulated annealing reveals the following general trends upon increase of the screening parameter κ (see Fig. 10.8 for one representative example):

1. Screening weakens the repulsive interaction between the dust grains which consequently leads to a compression of the cluster as a whole.
2. As a consequence, the shell radii (normalized to the mean interparticle distance) are found to be independent of screening.
3. By increase of κ, the shells broaden and shell splitting as well as the emergence of subshells is observed (see Fig. 10.8 at $\kappa = 20$).
4. The number of shells is, in most cases, independent of Debye shielding.
5. In contrast, the shell occupation numbers are found to be highly sensitive to a change of screening.
6. At large values of κ, the screened Coulomb interaction becomes "hard sphere"-like. This induces a structural change from a nested shell configuration to a bulk-like close-packed symmetry.

The most remarkable finding is the effect of screening on the detailed shell population and structure of the individual shells. In general, it can be stated that the occupation numbers on the inner (outer) shells gradually increase (decrease) with κ.[8] This implies that a Yukawa system contains a smaller (or equal) number

Fig. 10.8 Ground-state configuration of a Yukawa ball composed of $N = 25$ dust grains in a parabolic trap. Upon increase of the screening length κ, the cluster size decreases and the shell configuration changes from (2,23) in the Coulomb regime, to (3,22) in the range $0.3 < \kappa < 2.2$ and finally, for large κ-values, to (4,21). *Different colors* denote particles on different (sub)shells. Note that not the plotted cluster size, but the point size is scaled with κ

[8] This statement holds with the exception of a very few special cases at very large κ that were recently detected [61].

Table 10.5 Screening dependence of the ground-state shell configurations obtained from MD simulations of the spherical Yukawa cluster $N = 190$ in comparison with experimental data [11] (last row)

κ	N_1	N_2	N_3	N_4
0.0	1	18	56	115
0.2	1	18	57	114
0.3	2	20	57	111
0.4	2	20	58	110
0.5	2	21	58	109
0.6	*2*	*21*	*60*	*107*
0.8	3	22	60	105
1.0	4	24	60	102
Experiment	2	21	60	107

Beginning in the cluster center, $N_1, \ldots, N_4$ denote the particle numbers on the ith shell. If the screening parameter κ is increased, the number of shells remains equal to 4, but the inner (outer) shell become gradually more (less) populated.

of particles on the outer and a higher (or equal) number of particles on the inner shell than a comparable unscreened Coulomb system. A representative example for this very systematic trend is shown in Table 10.5. An extensive comparison of MD simulations with experimental measurements proves that the change in the shell population numbers can only be attributed to screening [62]. Therefore, our simulations allowed us to determine the Coulomb screening parameter $\kappa_{\text{exp}} \approx 0.62/r_0$ from experimentally measured shell configurations. Further Monte Carlo and molecular dynamics simulations show that a finite temperature and the inclusion of size variations of the individual dust grains (in view of dust charge fluctuations of up to 10%) are negligible compared to the effect of screening and do not lead to significant deviations from the ground-state shell configurations [59]. Therefore, the shell occupation numbers are found to be very suitable quantities giving rise to a noninvasive diagnostics for the Debye screening parameter in dusty plasma experiments.

10.6.3 Effect of Friction

A second key quantity that determines a dusty plasma is the friction coefficient ν (see Fig. 10.7). For slow cooling ($\nu = 0.05$), the particles are not significantly hindered by friction and can freely move according to the interparticle and confinement forces. They continuously lose kinetic energy until they are finally trapped in a local minimum of the potential energy. Here, they are further damped until the simulation is stopped. In the case of strong damping ($\nu = 5.3$) the situation is different. Here, the particles are readily slowed down after the initialization process in the box. Their motion is strongly affected by friction and interrupted even before they may be

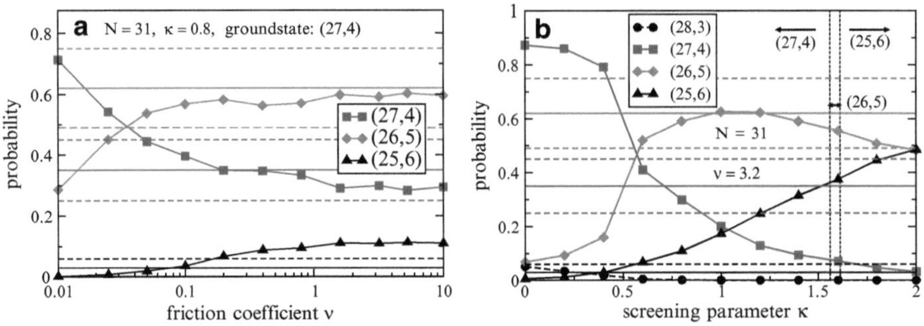

Effect of friction on the occurrence probabilities at $\kappa = 0.8$. The general trend shows that slow cooling favors the ground state (27, 4) over metastable states

Effect of screening for $\nu = 3.2$. Arrows indicate the different ground-state configurations to the left or right from the vertical line. Solid and dashed horizontal lines indicate experimental mean and standard deviation, respectively, from top to bottom: (26, 5), (27, 4), (25, 6)

Fig. 10.9 Probability of ground and metastable states for the $N = 31$ cluster (from [63])

trapped in a local minimum. Indeed, it is not ensured that the particles are in a *stable state*. The reason is that due to the rapid damping, they can be sufficiently slowed down even though they are not in a potential minimum but on a descending path and would reach the stable configuration at a later time.

Figure 10.9 shows the influence of friction on the occurrence probabilities in more detail. For a fixed screening parameter, the probability of finding the ground-state configuration increases when the friction coefficient is decreased. Here, the particles are cooled down more slowly and it is more likely that they reach the system's true ground state. During the cooling process they still have a sufficiently high kinetic energy and time to escape from a local minimum until the force on each particle vanishes.

In the case of strong friction the particles quickly fall into a nearby energy minimum. Leaving it becomes more difficult due to the rapid loss of kinetic energy since the particles are pushed straight along the gradient of the potential energy surface until they have completely lost their kinetic energy and the simulation is stopped. Note that in the case of strong damping the results strongly depend on the system's initial particle positions and temperature. For $\nu > 2$, that is, in the overdamped regime, the probabilities have practically saturated. For fast cooling, that is, large friction, metastable states can occur with a comparable or even higher probability than the ground state. Also, it is interesting to note that the dusty plasma experiments of Yukawa balls are performed in the overdamped regime, that is, ν is of the order of 3–6 [63]. Since in this limit the probabilities depend only very weakly on the damping rate, the results presented in the following for $\nu = 3.2$ should hold for any such damping coefficient [60].

For a fixed friction coefficient in the overdamped limit, the effect of screening is demonstrated in the right panel of Fig. 10.9. As in the undamped case, at some finite value of κ, a configuration with an additional particle on the inner shell becomes the ground state. Let us now consider the probability to observe the ground and metastable states. For weak screening, the ground state (27,4) is the most probable state. At the same time, the probability of the configuration with one more particle on the inner shell grows with κ, until it eventually becomes even more probable than the ground state. Note that this occurs much earlier (at a significantly smaller value of κ) than the ground-state change. Interestingly, for $N = 31$ this trend is observed twice: the probability of the configuration (26,5) first increases with κ and reaches a maximum around $\kappa \approx 1$. For $\kappa > 2$ this configuration becomes less probable than the configuration (25,6), that is, again a configuration with an additional particle on the inner shell becomes more probable with increased screening.

Figure 10.10 shows the states with the highest probability for a wide range of particle numbers. The shell filling mechanism in Fig. 10.10 is strictly monotonic upon increase of κ and N.[9] The trend that increased screening leads to a higher occupation of inner shells is confirmed. This figure may be a valuable reference for experiments, where one is interested in the configuration of a specific state and its screening dependence. The comparison with the ground states of Yukawa balls [61] yields an interesting picture: In most cases, the ground state itself is the most probable state.

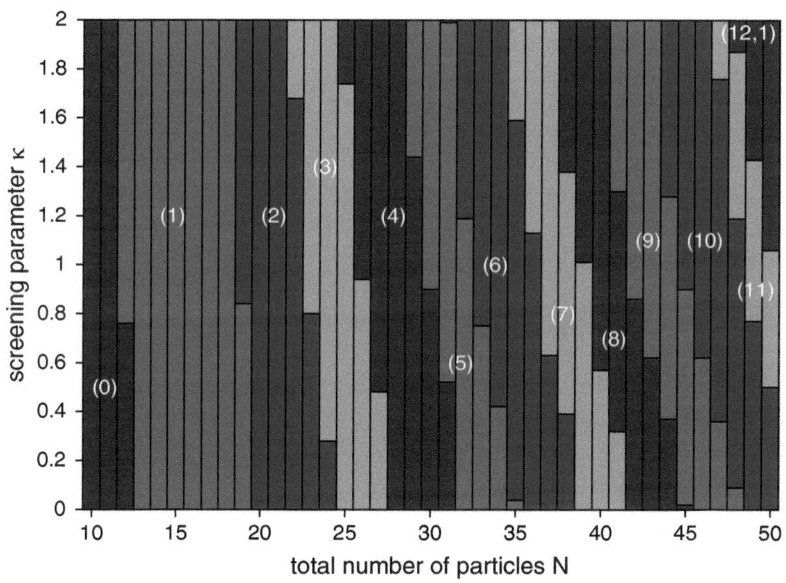

Fig. 10.10 Diagram of most probable states as found in the MD simulations. Results are obtained with a friction coefficient $\nu = 3.2$. Numbers in brackets denote the occupation of the innermost shell

[9] Anomalous behavior is only observed for $N = 47$–49 and large screening where adding one particle gives rise to a configuration change involving two particles. Here, the inner shell configuration of the most probable state changes from $(11, 0)$ to $(12, 1)$.

In some case, the ground state is identified as a configuration with one particle less on the inner shell than the most probable state. However, exceptions from this rule can be found. For 31 particles, there exists an interval of the screening parameter where the ground state has more particles on the inner shell than the most probable state. Such anomalous behavior is also found for $N = 11, 26, 30, 33, 34, 40–42, 47–50$. In the case $N = 38$, the most probable state can have *two* more particles on the inner shell than the ground state.

10.7 Applications II: Macroscopic Systems

As a second example of particular interest, we will study the effect of *superdiffusion* in macroscopic systems at the crossover from 2D to 3D system dimensionality. During the last decade, a number of experiments on diffusion in quasi-two-dimensional dusty plasmas have been performed (e.g., [64–70]). In many of these, diffusion was found to behave anomalously, that is, to not follow Einstein's law (10.56). Instead, the MSD follows the relation

$$\left\langle |\boldsymbol{r}(t) - \boldsymbol{r}(t_0)|^2 \right\rangle \propto t^\gamma, \tag{10.57}$$

where $\gamma > 1$, that is, the MSD grows faster than linearly with time (see Table 10.6). This superdiffusion is a peculiarity of two-dimensional systems and has apparently not been observed in three-dimensional systems in equilibrium to date. It is closely connected to the long-time decay of the VACF found already by Alder and Wainwright via the Green–Kubo relation (10.54). If the VACF decays too slowly for the integral to converge, this means that no diffusion coefficient can be defined for the system and it does not obey the standard laws of diffusion.

We have performed equilibrium MD simulation to examine the influence of the dimensionality on the appearance of superdiffusion [71]. To control the dimensionality in the system, we started from a thin three-dimensional system and used different confining potentials to restrict the particles' movement to a thin slab near

Table 10.6 Classification of the diffusion exponent γ

$\gamma = 1$	Normal diffusion
$\gamma < 1$	Subdiffusion
$\gamma > 1$	Superdiffusion
$\gamma = 2$	Ballistic motion

There are four cases for the diffusion exponent γ which are used to classify a diffusive process: $\gamma = 1$ corresponds to normal diffusion according to Fick's laws on a macroscopic level and by the Einstein relation (10.56) on a microscopic level; $\gamma = 2$ describes the behavior of an undisturbed motion $\boldsymbol{r}(t) \propto \boldsymbol{v} \cdot t$, and constitutes an upper limit for γ. The two anomalous cases of sub- and superdiffusion are of special research interest.

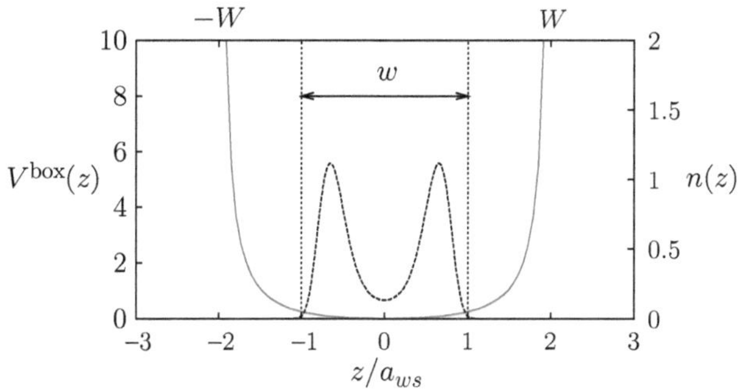

Fig. 10.11 Soft-box potential V^{box} (*solid line*) and the corresponding particle density $n(z)$. The soft-box confinement is impenetrable at $\pm W$ and the particles can typically explore the trap width w (here, $w = 2a_{\text{ws}}$)

the x, y-plane. The Hamiltonian of such a system is given by (10.6), where $V^{\text{ext}}(\boldsymbol{r})$ is a symmetric function of z. Two different confinements were used:

$$V^{\text{harm}}(z) = f \frac{Q^2}{4\pi\varepsilon_0 a_{\text{ws}}^3} \frac{z^2}{2}, \tag{10.58}$$

$$V^{\text{box}}(z) = \frac{Q^2}{4\pi\varepsilon_0} \left(\frac{\mathrm{e}^{-(z+W)/\lambda_{\text{D}}}}{z+W} + \frac{\mathrm{e}^{-(-z+W)/\lambda_{\text{D}}}}{-z+W} \right), \tag{10.59}$$

where a_{ws} is the Wigner–Seitz radius for two-dimensional systems and $W = (w/2) + a_{\text{ws}}$. The two parameters controlling the width of the system are the trap amplitude f and the width w, respectively. A harmonic trap such as (10.58) is a first approximation to any experimentally realized confinement. It is, however, not possible to maintain a constant three-dimensional density in a harmonically confined system when the trap is relaxed. Therefore, the particle number was left unchanged for different trap amplitudes in the simulations. The "soft-box" confinement (10.59), on the other hand, has steep "walls" and therefore allowed us to define a volume and increase the particle number as necessary to keep the 3D density constant. For illustration see Fig. 10.11.

By simulating dusty plasma systems with $N \geq 6{,}000$ particles for different trap amplitudes f and box widths w, the diffusion exponent γ of (10.57) can be directly obtained from the trajectories of the dust grains. The Coulomb coupling parameter in these simulations was $\Gamma = 300$.

10.7.1 Simulation Results

The time dependence of the MSD $\langle |\boldsymbol{r}(t) - \boldsymbol{r}(t_0)|^2 \rangle$ is shown in Fig. 10.12. The slope of the MSD curve amounts, in this double-logarithmic plot, to the diffusion

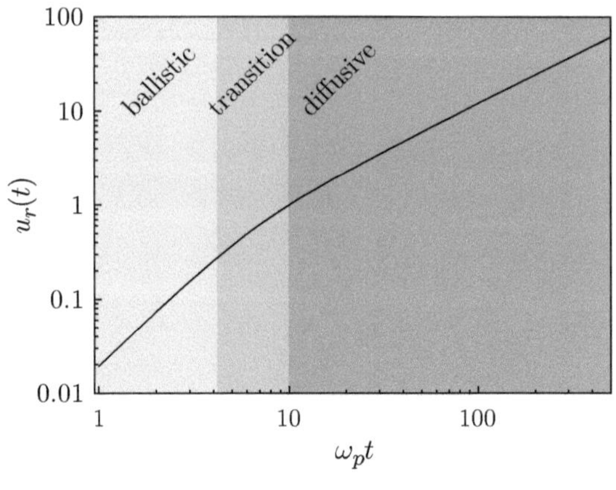

Fig. 10.12 Typical behavior of the MSD $u_r(\omega_p t)$ with three regimes: ballistic motion, transition phase, and diffusive motion. The time is given in units of the inverse plasma frequency ω_p^{-1}

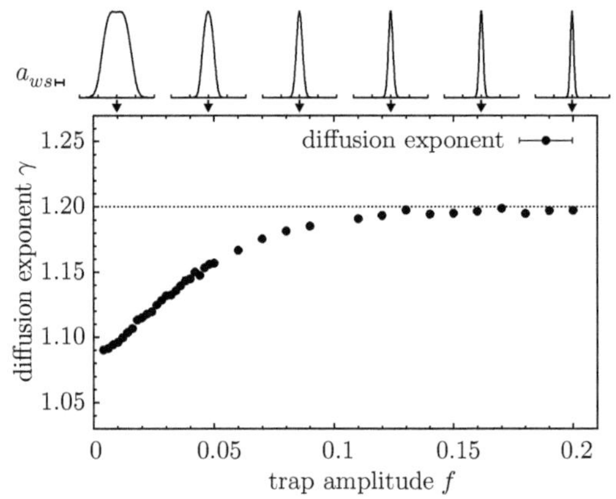

Fig. 10.13 The diffusion exponent for different trap amplitudes and $\kappa = 3.0$ in the harmonic confinement. The straight line is a guide for the eyes. The *top graphs* show the density profile in the confined direction from $z = -4a_{ws}$ to $z = 4a_{ws}$ at the trap amplitude indicated by the *arrows*. The $n(z)$ distributions are normalized here to unit amplitude

exponent γ. The initial slope of $\gamma = 2$ corresponds to scattering-free ballistic motion and is not of interest here. The diffusion exponent is extracted from the slope between $t\omega_p = 100$ and 300. The dependence of the diffusion exponent on the harmonic trap amplitude is depicted in Fig. 10.13. As expected, the degree of superdiffusivity is reduced for increasingly broader systems. Interestingly, the beginning of this decline can be connected with the density profile as shown on top of Fig. 10.13. Only when the width of the system exceeds typical interparticle distances does superdiffusion begin to vanish.

The same effect – only much more pronounced – can be seen in Fig. 10.14 which shows the same data for the soft-box confinement. Due to the constant 3D density (i.e., the particle number is increased with increasing system width), the system forms more layers than in the harmonic case. The different background colors in Fig. 10.14 mark different number of layers, and the connection between the diffusion exponent and the number of layers is apparent. These results impressively show that superdiffusion is an effect which strongly depends on the dimensionality of the system.

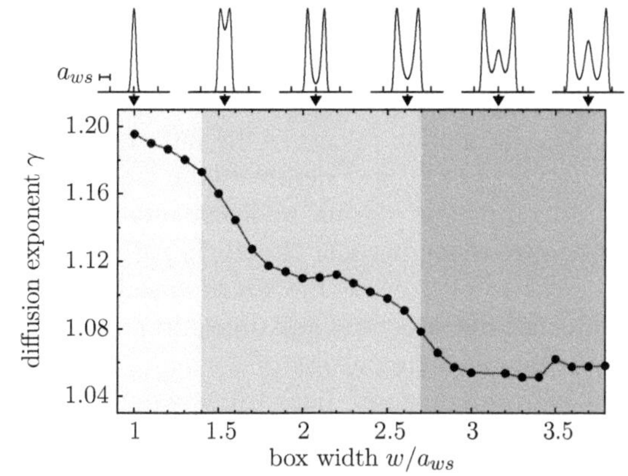

Fig. 10.14 The diffusion exponent for different trap widths and $\kappa = 3.0$ in the soft-box confinement. The *top graphs* show the density profile from $z = -4a_{\mathrm{ws}}$ to $z = 4a_{\mathrm{ws}}$ in the confined direction at the trap amplitude indicated by the *arrows*. The $n(z)$ distributions are normalized here to unit amplitude

10.8 Conclusion

Numerical methods such as molecular dynamics or Monte Carlo simulations have become increasingly important with the availability of low cost computer power in the last decade(s). They are able to treat the interaction between classical particles from first principles and are, in general, not restricted to approximate solutions such as analytical models. The relatively young field of dusty plasma physics allows one to directly study strongly correlated charged particles. The description of dust particles in an external confinement by numerical methods helps to understand recent experimental observations regarding the crystallization of the particles, especially the observed shell structure. In particular, the mechanisms that determine the stationary state probabilities of dusty plasma crystals were systematically analyzed over broad ranges of N and κ and give insightful explanations for experimental measurements. Summarizing, the MD simulations confirm the remarkable observation that in spherical Yukawa clusters the ground state is not necessarily the most probable state. Often, a metastable state with more particles on the inner shell is observed substantially (in some cases up to five times) more frequently. While slow cooling generally leads to a high probability of the ground state, strong friction is found to be responsible for the high abundances of metastable states in dusty plasma experiments. Moreover, the screening parameter, and thus the range of the interaction potential, strongly affects the shell occupation. In Coulomb systems with long-range interaction, states with only a few particles on the inner shells have a high probability. The probability of states with a high occupation number on the inner shell was shown to increase with the screening parameter.

A second application of MD simulations concerned recent dusty plasma experiments of quasi-2D setups, which have studied the diffusion process by direct optical monitoring of the dust grains' motion [64, 65]. As an intriguing result, it was found in these experiments that the diffusion process is anomalous and is described by superdiffusive motion. MD simulations allowed us to study the influence of the dimensionality on the superdiffusion and observe a gradual transition to normal Fick diffusion at the crossover from 2D to 3D systems.

The simulation results of the one-component static Yukawa model are found to be in very good agreement with the measurements (see Chap. 7). Nevertheless, the multicomponent plasma environment requires a systematic and careful analysis of collective many-particle effects. In particular in cases where the streaming velocity is close to or exceeds the sound speed, the effect of streaming ions can strongly influence the structural and transport properties of a strongly coupled dusty plasma and requires the inclusion of *dynamically screened pair potentials* which are computed from a dynamic dielectric function (see [72–75] and references therein). These anisotropic and nonmonotonic (wake) potentials take into account the presence of an electric field, as encountered in the plasma sheath region above the lower electrode, on the ion and electron distribution functions which determine the effective interaction between the dust grains. With this advanced model it is possible to additionally include effects such as plasma instabilities, non-Newtonian dynamics due to the nonreciprocal forces, Landau damping, or ion–neutral collisions, which are missing in the considered model system.

References

1. J.H. Chu, I. Lin, Phys. Rev. Lett. **72**, 4009 (1994)
2. H. Thomas, G.E. Morfill, V. Demmel, J. Goree, B. Feuerbacher, D. Möhlmann, Phys. Rev. Lett. **73**, 652 (1994)
3. Y. Hayashi, K. Tachibana, Jpn. J. Appl. Phys. **33**, L804 (1994)
4. A. Melzer, T. Trottenberg, A. Piel, Phys. Lett. A **191**, 301 (1994)
5. V.E. Fortov, A.V. Ivlev, S.A. Khrapak, A.G. Khrapak, G.E. Morfill, Phys. Rep. **421**, 1 (2005)
6. A. Bouchoule, *Dusty Plasmas: Physics, Chemistry and Technological Impacts in Plasma Processing* (Wiley, New York, 1999)
7. G.S. Selwyn, J. Singh, R.S. Bennett, J. Vac. Sci. Technol. A **7**, 2758 (1988)
8. P. Roca i Cabarrocas, J. Non-Cryst. Solids **31**, 266 (2000)
9. F. Verheest (ed.), *Waves in Dusty Space Plasmas* (Kluwer, Dordrecht, 2002)
10. K. Qiao, T.W. Hyde, Adv. Space Res. **34**, 2390 (2004)
11. O. Arp, D. Block, A. Piel, A. Melzer, Phys. Rev. Lett. **93**, 165004 (2004). For a recent overview, see M. Bonitz, C. Henning, D. Block, Reports Prog. Physics **73**, 066501 (2010)
12. B.J. Alder, T.E. Wainwright, J. Chem. Phys. **27**, 1208 (1957)
13. B.J. Alder, T.E. Wainwright, J. Chem. Phys. **31**, 459 (1959)
14. B.J. Alder, T.E. Wainwright, J. Chem. Phys. **33**, 1439 (1960)
15. D. Mac Kernan, M. Mareschal, SIMU Newsletter **15** (2002)
16. H. Ikezi, Phys. Fluids **29**, 1764 (1986)
17. D.H.E. Dubin, T.M. O'Neill, Rev. Mod. Phys. **71**, 87 (1999)
18. G. Morfill, H. Kersten, New J. Phys. **5** (2003)
19. O. Arp, D. Block, M. Klindworth, A. Piel, Phys. Plasmas **12**, 122102 (2005)
20. S. Ichimaru, Rev. Mod. Phys. **54**, 1017 (1982)
21. D.C. Rapaport, *The Art of Molecular Dynamics Simulation* (Cambridge University Press, Cambridge, 2004)
22. L. Verlet, Phys. Rev. **159**, 98 (1967)
23. W.C. Swope, H.C. Andersen, P.H. Berens, K.R. Wilson, J. Chem. Phys. **76**, 637 (1982)

24. M. Tuckerman, B.J. Berne, G.J. Martyna, J. Chem. Phys. **97**, 1990 (1992)
25. J.P. Hansen, I.R. McDonald, *Theory of Simple Liquids* (Academic, London, 2006)
26. J.R. Cash, A.H. Karp, ACM Trans. Math. Softw. **16**, 201 (1990)
27. W.H. Press, S.A. Teukolsky, W.T. Vetterling, B.P. Flannery, *Numerical Recipes* (Cambridge University Press, Cambridge, 1992)
28. J.M. Haile, *Molecular Dynamics Simulation: Elementary Methods* (Wiley, New York, 1976)
29. S.C. Harvey, R.K.Z. Tan, T.E. Cheatham III, J. Comput. Chem. **19**, 726 (1998)
30. H.J.C Berendsen, J.P.M. Postma, W.F. van Gunsteren, A. DiNola, J.R. Haak, J. Chem. Phys. **81**, 3684 (1984)
31. H.C. Andersen, J. Chem. Phys. **72**, 2384 (1980)
32. T. Soddemann, B. Dünweg, K. Kremer, Phys. Rev. E **68**, 46702 (2003)
33. P. Warren, P. Espanol, Europhys. Lett. **30**, 191196 (1995)
34. S. Nosé, J. Chem. Phys. **81**, 511 (1984)
35. W.G. Hoover, Phys. Rev. A **31**, 1695 (1985)
36. D. Frenkel, B. Smit, *Understanding Molecular Simulation* (Academic, Orlando, 2001)
37. G.J. Martyna, M.E. Tuckerman, D.J. Tobias, M.L. Klein, Mol. Phys. **87**(5), 1117 (1996)
38. R. Mannella, Phys. Rev. E **69**, 41107 (2004)
39. M. Allen, D. Tildesley, *Computer Simulation of Liquids* (Clarendon, Oxford, 1987)
40. O. Vaulina, S. Khrapak, G. Morfill, Phys. Rev. E **66**, 016404 (2002)
41. P.P. Ewald, Ann. Phys. **369**, 253 (1921)
42. A.Y. Toukmaji, J.A. Board, Comput. Phys. Commun. **95**, 73 (1996)
43. R.W. Hockney, J.W. Eastwood, *Computer Simulation Using Particles* (Institute of Physics, Bristol, 1988)
44. C. Sagui, T.A. Darden, Annu. Rev. Biophys. Biomol. Struct. **28**, 155 (1999)
45. U. Essmann, L. Perera, M.L. Berkowitz, T. Darden, H. Lee, L.G. Pedersen, J. Chem. Phys. **103**, 8577 (1995)
46. E.L. Pollock, J. Glosli, Comput. Phys. Commun. **95**, 93 (1996)
47. M.S. Green, J. Chem. Phys. **19**, 1036 (1951)
48. M.S. Green, Phys. Rev. **119**, 829 (1960)
49. R. Kubo, J. Phys. Soc. Jpn. **12**, 570 (1957)
50. A. Einstein, Ann. Phys. **322**, 549 (1905)
51. E. Helfand, Phys. Rev. **119**, 1 (1960)
52. J.J. Erpenbeck, Phys. Rev. E **51**, 4296 (1995)
53. V.A. Schweigert, F.M. Peeters, Phys. Rev. B **51**, 7700 (1995)
54. A. Filinov, M. Bonitz, Yu.E. Lozovik, Phys. Rev. Lett. **86**, 3851 (2001)
55. M. Bonitz, V. Golubnychiy, A.V. Filinov, Yu.E. Lozovik, Microelectron. Eng. **63**, 141 (2002)
56. S.L. Gilbert, J.J. Bollinger, D.J. Wineland, Phys. Rev. Lett. **60**, 2022 (1988)
57. A. Mortensen, Dissertation, University of Aarhus (2005)
58. P. Ludwig, S. Kosse, M. Bonitz, Phys. Rev. E **71**, 046403 (2005)
59. H. Baumgartner, H. Kählert, V. Golubnychiy, C. Henning, S. Käding, A. Melzer, M. Bonitz, Contrib. Plasma Phys. **47**, 281 (2007)
60. H. Kählert, P. Ludwig, H. Baumgartner, M. Bonitz, D. Block, S. Käding, A. Melzer, A. Piel, Phys. Rev. E **78**, 036408 (2008)
61. H. Baumgartner, D. Asmus, V. Golubnychiy, P. Ludwig, H. Kählert, M. Bonitz, New J. Phys. **10**, 093019 (2008)
62. M. Bonitz, D. Block, O. Arp, V. Golubnychiy, H. Baumgartner, P. Ludwig, A. Piel, A. Filinov, Phys. Rev. Lett. **96**, 075001 (2006)
63. D. Block, S. Käding, A. Melzer, A. Piel, H. Baumgartner, M. Bonitz, Phys. Plasmas **15**, 040701 (2008)
64. B. Liu, J. Goree, Phys. Rev. Lett. **100**, 055003 (2008)
65. S. Ratynskaia, K. Rypdal, C. Knapek, S. Khrapak, A.V. Milovanov, A. Ivlev, J.J. Rasmussen, G.E. Morfill, Phys. Rev. Lett. **96**, 105010 (2006)
66. R.A. Quinn, J. Goree, Phys. Rev. Lett. **88**, 195001 (2002)
67. Y.-J. Lai, I. Lin, Phys. Rev. Lett. **89**, 155002 (2002)
68. W.-T. Juan, I. Lin, Phys. Rev. Lett. **80**, 3073 (1998)

69. W.-T. Juan, M.-H. Chen, I. Lin, Phys. Rev. E **64**, 016402 (2001)
70. S. Nunomura, D. Samsonov, S. Zhdanov, G. Morfill, Phys. Rev. Lett. **96**, 015003 (2006)
71. T. Ott, M. Bonitz, Z. Donkó, P. Hartmann, Phys. Rev. E **78**, 026409 (2008). More recent results on superdiffusion can be found in the papers by T. Ott et al., Phys. Rev. Lett. **103**, 099501 (2009) and Phys. Rev. Lett. **103**, 195001 (2009)
72. M. Lampe, G. Joyce, G. Ganguli, V. Gavrishchaka, Phys. Plasmas **7**, 3851 (2000)
73. M. Lampe, V. Gavrishchaka, G. Ganguli, G. Joyce, Phys. Rev. Lett. **86**, 5278 (2001)
74. G. Joyce, M. Lampe, G. Ganguli, Phys. Rev. Lett. **88**, 095006 (2002)
75. M. Lampe, G. Joyce, G. Ganguli, IEEE Trans. Plasma Sci. **33**, 57 (2005)

Part IV
Reactive Plasmas, Plasma–Surface Interaction, and Technological Applications

Chapter 11
Nonthermal Reactive Plasmas

Jürgen Meichsner

Abstract Reactive low-temperature plasmas have found broad field of applications in the last decades. Beside the thermal plasmas used for coating of material surfaces (e.g., plasma spraying), synthesis of nano/microparticles and plasma chemical conversion of waste, the focus in this chapter is directed on nonthermal reactive plasmas or cold plasmas which are characterized by strong nonequilibrium conditions. Such nonequilibrium plasmas are implemented in many innovative technologies for surface treatment and synthesis of novel materials. Prominent examples are the plasma etching and patterning in semiconductor processing, the surface modification of polymers due to the incorporation of new functional molecular groups which determine the interactions with surrounding media, and/or the deposition of thin films with novel physical and chemical properties.

The nonthermal reactive plasma represents a multispecies system consisting of hot electrons and a mixture of several charged and neutral reactive atoms and molecules. The increasing interest in reactive plasmas containing hydrocarbons, fluorocarbons, and organosilicons needs more fundamental knowledge in both the plasma physics and plasma chemistry. Furthermore, the plasma–surface interaction has to be investigated including the plasma sheath in front of surfaces and the chemical reactions at the phase boundary in connection with the volume plasma chemistry. The aim of this chapter is (1) to give an introduction into nonthermal reactive plasmas and plasma–surface interaction, (2) to present useful diagnostics for characterization of the reactive plasmas and plasma–surface interaction, and (3) to show examples of reactive plasmas in contact with material surfaces.

11.1 Introduction

The considered nonthermal reactive plasmas represent ideal and partially ionized plasmas at nonequilibrium conditions. That means *no thermodynamic equilibrium* exists between translational energy of different plasma species (electrons, ions, and

J. Meichsner (✉)
Institute of Physics, Ernst-Moritz-Arndt-University Greifswald, 17487 Greifswald, Felix-Hausdorff-Str. 6, Germany
e-mail: meichsner@physik.uni-greifswald.de

M. Bonitz et al. (eds.), *Introduction to Complex Plasmas*, Springer Series on Atomic, Optical, and Plasma Physics 59, DOI 10.1007/978-3-642-10592-0_11,

neutrals), their internal energies, and the plasma radiation. Therefore, the quantitative description of these plasmas by only two parameters, the *temperature* and *pressure*, as in the case of plasmas in *complete thermodynamic equilibrium* (CTE) is not sufficient. Within the temperature–density plane, the nonthermal plasmas fill the area in electron temperature of between 10^4 and 10^5 K or averaged electron energy from 1 to 10 eV (1 eV corresponds 11.600 K), respectively, and in electron density of between 10^{14} and $10^{18}\,\mathrm{m}^{-3}$. In the approximation of the local *partial thermodynamic equilibrium* (PTE), the nonthermal plasma is characterized by different temperatures for:

1. Translational energy of electrons (T_e), ions (T_+), and neutrals (T_n)
2. Internal energy in excited electronic (T_exc), vibrational (T_vib), and rotational (T_rot) states for each ionization level, respectively

The electron temperature is much higher than the ion and neutral gas temperature, $T_\mathrm{e} \sim 10^4\,\mathrm{K} \gg T_+ > T_\mathrm{n}$. The neutral gas temperature remains at 300–400 K, typically.

In nonthermal plasmas, also denominated as "cold plasmas," the hot electrons represent the main energetic species. Whereas the translational energy distribution of the heavy ions and neutrals in the plasma bulk can be well described by the Maxwell distribution, the *electron energy distribution function* (EEDF) has to be investigated carefully. In many gas discharges no Maxwell distribution of electron energy is achieved. The deviations from the Maxwell distribution have their origin mainly in the electron heating in the external electric field and the inelastic collisions of electrons of the high energetic tail of the EEDF (excitation, ionization, and dissociation). Furthermore, the distribution of the excited states of heavy species may be different to the Boltzmann distribution. According to the energy input from a high-energy level by the hot electrons the population of the upper excited levels may be higher than under equilibrium conditions.

The inelastic electron–neutral collisions play the most important role in nonthermal molecular plasmas because of the production of charged and highly reactive species, simultaneously, for example,

$$\mathrm{e} + \mathrm{O_2} \rightarrow \mathrm{O^{(+)}} + \mathrm{O} + (2)\mathrm{e}$$
$$\mathrm{e} + \mathrm{CF_4} \rightarrow \mathrm{CF_3^{(+)}} + \mathrm{F} + (2)\mathrm{e}$$
$$\mathrm{e} + \mathrm{C_6H_{18}Si_2O} \rightarrow \mathrm{C_5H_{15}Si_2O^{(+)}} + \mathrm{CH_3} + (2)\mathrm{e}$$

The generated reactive species due to the dissociation of precursor molecules combined with excitation and/or ionization of atoms and molecules initiate chemical reactions in the gas volume and at phase boundaries. New gaseous compounds are produced and a thin surface layer is generated with modified physical and chemical properties. Therefore, the knowledge in the EEDF, the density of reactive species in ground and excited states, and the plasma chemical reaction products is of fundamental interest in nonthermal plasma physics and chemistry. The main task is the determination of the EEDF experimentally by means of plasma diagnostics and/or by numerical calculation using kinetic equations.

The classical *Langmuir probe technique* and determination of the EEDF by the *Druyvesteyn method* [1] fails in most cases due to reactive processes at the probe tip. Therefore, the noninvasive optical emission spectroscopy or microwave interferometry represents appropriate diagnostic tools to analyze the plasma electrons in reactive plasmas. Generally, the methods based on optical spectroscopy (emission, absorption, and laser techniques) have become amongst the most important ones because they provide a means of analyzing the shape of emission/absorption lines or line ratios allowing for the determination of temperatures and/or the population densities of species in both ground and excited states. Furthermore, specific mass spectrometric methods are applied to analyze the plasma ions and their translational energy distribution in the plasma sheath in front of surfaces (walls and discharge electrodes) or to determine densities of neutral transient species and plasma chemical reaction products. In both cases, the careful extraction of the plasma species out of the plasma vessel is necessary by use of a small orifice and a differentially pumped mass spectrometer. The surface and thin film characterization involves a wide spectrum of diagnostic tools. Here, the specific interaction of particle beams or electromagnetic waves provides information about the elementary composition, molecular or crystal structure, and the macroscopic properties such as surface topography, thin film thickness, and refractive index. Figure 11.1 gives an overview about the useful experimental techniques for the diagnosis of reactive plasmas, surfaces, and thin films compared with applied methods of modeling and simulation.

Nonthermal reactive plasmas are of increasing interest in both, in fundamental research to study complex systems, and in applications to establish innovative plasma technologies.

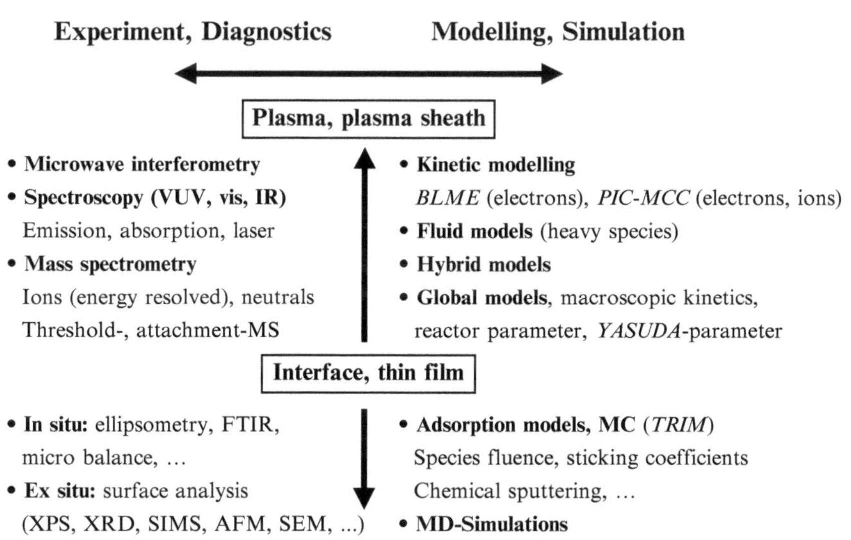

Fig. 11.1 Overview about the diagnostic tools of reactive plasmas and surface/thin film analysis compared with adequate modeling and simulation methods

The nonthermal plasma sources are working either at low-pressure condition from 10^{-1} to 10^3 Pa (e.g., radio-frequency plasmas, microwave plasmas, and magnetron plasmas) or at total pressure around 10^5 Pa (e.g., corona and dielectric barrier discharges [2]). In the last case, no expensive vacuum equipment is necessary and the plasma processing can be easily implemented into conventional surface technologies. Nevertheless, plasma sources at low and atmospheric pressure have their specific advantages concerning the elementary processes in plasma generation and balancing between surface (electrode) and volume processes, the transport processes (diffusion, convection, and field drift), as well as the diagnostics and up-scaling.

Nonthermal reactive plasmas are found in a broad field of applications such as surface treatment in semiconductor processing, surface activation and cleaning, thin film deposition, synthesis of new materials, ozone generation, and treatment of exhaust gases. The advantage of using such nonequilibrium plasmas are the:

- Production of nanostructures due to plasma etching, for example, in semiconductor processing
- Treatment of sensitive materials, for example, polymers, living human, and animal tissues
- Synthesis of novel materials which is not possible in the conventional thermochemistry, for example, thin amorphous hydrocarbon or fluorocarbon films, composite films with embedded nanoparticles
- Good environmental compatibility due to low-energy consumption and material turnover

In particular, the plasma–surface modification including deposition of thin (composite) films provides new functionalities of material surfaces and thin films (see Fig. 11.2).

In many cases, the developments in the field of reactive nonthermal plasmas are driven by empirical and technological progress through trial and error. The situation

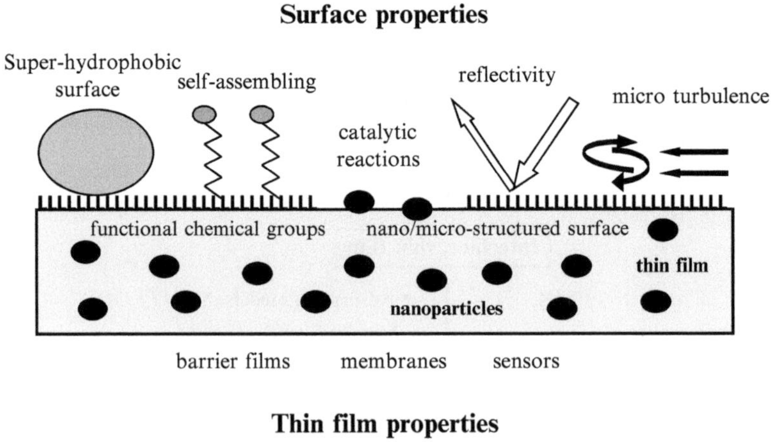

Fig. 11.2 Novel physical and chemical properties of surfaces and thin films due to reactive plasma–surface interaction

is further complicated due to the fact that reactive plasmas represent a multidisciplinary field which requires knowledge in physics, chemistry, and materials science. Nevertheless, nonthermal reactive plasmas represent a hot research field with a fascinating variety of well-established processes and novel potential applications such as the involvement of new plasma sources using microplasmas [3] or the treatment of living human tissues by means of an atmospheric pressure plasma jet in medicine.

11.2 Nonthermal Plasma Conditions

The existence of the nonthermal plasmas in gas discharges is based on the following physical reasons [4, 5].

Firstly, the electrons are strongly accelerated in electric fields and they gain much more kinetic energy per time unit in comparison with the heavy ions due to their low inertia. In collision-dominated and weakly ionized plasmas, the power absorption of electrons per volume unit in the electric field is given by *Joule's heating*, determined by the electron–neutral collision frequency (ν_e) and the circular frequency (ω) of the external electric field strength ($\vec{E}$):

$$P_{\text{abs}} = Re\{\vec{j}_e \cdot \vec{E}\} = \sigma_{\text{eff}} \cdot E_{\text{eff}}^2 = \frac{n_e \cdot e^2 \cdot \nu_e}{m_e \cdot (\omega^2 + \nu_e^2)} \cdot E_{\text{eff}}^2 = \frac{n_e \cdot e^2}{m_e \cdot \nu_e} \cdot \frac{(\nu_e/\omega)^2}{1 + (\nu_e/\omega)^2} \cdot E_{\text{eff}}^2. \tag{11.1}$$

The absorbed electric power, P_{abs}, increases with E_{eff}^2. The optimum electric power absorption is achieved if $\nu_e = \omega = \nu_{\text{max}}$. In the case of low collision frequency ($\nu_e < \nu_{\text{max}}$) P_{abs} increases proportional to ν_e, and at high collision frequency ($\nu_e > \nu_{\text{max}}$) P_{abs} decreases reciprocal with ν_e. At a very high electric field frequency, the power absorption becomes inefficient.

Taking into consideration the low-pressure radio-frequency (rf) plasma at 13.56 MHz with electron temperature $T_e = 10^4$ K, the optimum electric power absorption in the plasma bulk is achieved at $\nu_{\text{max}} = \omega = 85.2$ MHz which requires a total pressure of about 10 Pa.

In the DC case with constant electric field strength ($\omega = 0$, $E_{\text{eff}}^2 = E_{\text{DC}}^2$) the electron current density and power absorption corresponds to the *Drude model* of quasi-free electrons.

Secondly, an efficient kinetic energy transfer is necessary into the heavy species system (ions and neutrals) to achieve the thermal equilibrium state with uniform temperature in translational and internal energy distribution. According to the conservation of kinetic energy and momentum in elastic collisions, the kinetic energy transfer depends on the mass ratio of the interacting species. In elastic electron–neutral collision, the momentum gain of the neutral species is expressed by $\Delta p_n = m_e \cdot \Delta v_e$ which results in an increase of kinetic energy of the neutral species with the mass (M):

$$\frac{\Delta p_n^2}{2M} = \frac{(m_e \cdot \Delta v_e)^2}{2M} = \frac{m_e}{M} \cdot \left(\frac{m_e}{2} \cdot \Delta v_e^2\right) = \frac{m_e}{M} \cdot \Delta \varepsilon_T^e. \tag{11.2}$$

An efficient energy transfer is achieved when the collision partners have comparable masses. In the case of electron–neutral collision the mass ratio amounts to about 10^{-5} which leads to a low kinetic energy transfer. Taking into account the total rate of electron kinetic energy variation, the maximum difference between the electron and neutral gas temperature in the two temperature model can be approximated as follows [5]:

$$T_e - T_n = \frac{\sigma_{\text{eff}} \cdot M \cdot E_{\text{eff}}^2}{3 \cdot k_B \cdot m_e \cdot n_e \cdot \nu_e} = \frac{e^2 \cdot M \cdot E_{\text{eff}}^2}{3 \cdot k_B \cdot m_e^2 \cdot (\omega^2 + \nu_e^2)}. \tag{11.3}$$

With decreasing elastic electron–neutral collision frequency and electric field frequency, the temperature difference increases and the nonequilibrium state is enforced.

Thirdly, the existence of gradients in particle concentration and temperature, as well as the presence of external electric and magnetic forces cause particle fluxes out of the active plasma region to electrodes and walls. This particle flux together with charge carrier recombination determines essentially the energy loss processes of the plasma. Furthermore, the plasma boundaries are not in thermal equilibrium with the plasma radiation as in the case of black body radiation. The effective energy loss at the boundaries by absorption and transmission of plasma radiation contributes additionally to the formation of the nonthermal conditions.

Summarized, the energy confinement time τ_E in nonthermal plasmas is lower than the effective time which is necessary to transfer the kinetic energy of electrons into the heavy species system by elastic collisions, τ_{en}

$$\tau_E = W / P_{\text{loss}} < \tau_{en}, \tag{11.4}$$

where W is the energy input and P_{loss} the power loss processes. This situation is observed either in stationary and weakly ionized low-pressure plasmas ($<10^4$ Pa) because of the low elastic electron–neutral collision frequency and electron concentration ($<10^{18}\,\text{m}^{-3}$), or in electric gas discharges of short duration (<100 ns), for example, single discharge filaments in the dielectric barrier discharge and streamers in the corona discharge, in spite of the higher local electron concentration ($>10^{20}\,\text{m}^{-3}$) and elastic electron–neutral collision frequency at atmospheric gas pressure.

11.3 Plasma Kinetics and Plasma Chemical Reactions

11.3.1 Boltzmann Equation

The knowledge of the energy distribution function of plasma species is of fundamental interest to describe the elementary processes in nonthermal plasmas including chemical reactions. The most important kinetic equation to calculate energy distribution functions of plasma species at low-pressure conditions is the *Boltzmann equation* (BLME) which describes the evolution of the distribution function f_i of a single species i in the phase space [6, 7]

$$\frac{\partial f_i}{\partial t} + \vec{v}_i \cdot \frac{\partial f_i}{\partial \vec{r}} + \frac{\vec{F}}{m_i} \cdot \frac{\partial f_i}{\partial \vec{v}} = \left(\frac{\partial f_i}{\partial t}\right)_{\text{coll}}. \tag{11.5}$$

The BLME consists of a *differential operator* on the left involving terms to describe the explicit time dependence and the particle flow along trajectories in the phase space. The term on the right is the collision term which describes the change of the velocity due to collisions by means of an *integral operator*. The force ($\vec{F}$) in the differential operator represents averaged macroscopic forces, for example, external electric and magnetic fields as well as macroscopic space charge fields, whereas the integral operator includes the binary elastic and inelastic collisions controlled by microfields and the quantum physics of atomic or molecular interactions at short distances.

Considering the single species i any binary collision represents one of the channels to change the velocity and translational energy, respectively, of the species i

$$\begin{aligned} &\frac{\partial f_i}{\partial t} + \vec{v}_i \cdot \frac{\partial f_i}{\partial \vec{r}} + \frac{\vec{F}}{m_i} \cdot \frac{\partial f_i}{\partial \vec{v}} = \left(\frac{\partial f_i}{\partial t}\right)_{\text{coll}} \\ &= \sum_{\substack{klmn \\ i \ni klmn}} \left[\iint \mathrm{d}\Omega\, \mathrm{d}\vec{v}_l \cdot |\vec{v}_l - \vec{v}_k| \cdot \sigma_{klmn}(v, \vartheta) \cdot (f_m f_n - f_l f_k)\right]^{(i)} \end{aligned} \tag{11.6}$$

$$\mathrm{A}_k + \mathrm{B}_l \xrightarrow[\Delta\varepsilon_{klmn}]{\sigma_{klmn}} \mathrm{C}_m + \mathrm{D}_n$$

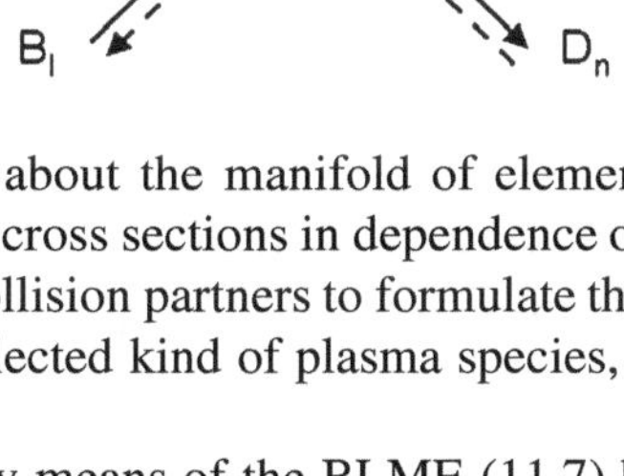

A lot of knowledge is necessary about the manifold of elementary processes in plasmas concerning their collision cross sections in dependence on the translational energy and quantum state of the collision partners to formulate the integral operator and to solve the BLME for the selected kind of plasma species, such as electrons, ions, and neutrals.

The calculation of the EEDF by means of the BLME (11.7) has to consider all the important elastic and inelastic collisions of electrons including their production and loss processes [8]:

$$\frac{\partial f_{\mathrm{e}}}{\partial t} + \vec{v} \cdot \frac{\partial f_{\mathrm{e}}}{\partial \vec{r}} - |e|\frac{\vec{E}}{m_{\mathrm{e}}} \cdot \frac{\partial f_{\mathrm{e}}}{\partial \vec{v}} = \left(\frac{\partial f_{\mathrm{e}}}{\partial t}\right)_{\text{coll}} = C^{\text{el}} + \sum_n C_n^{\text{inel}}. \tag{11.7}$$

In particular, the electrons in the high-energy tail of the EEDF have enough kinetic energy for ionization, dissociation, and excitation of atoms and molecules, for example, in oxygen plasma:

$$
\begin{aligned}
\mathrm{e} + \mathrm{O_2} &\rightarrow \mathrm{O_2^+} + 2\mathrm{e} && \text{electron impact ionization}\\
&\rightarrow \mathrm{O^+} + \mathrm{O} + 2\mathrm{e} && \text{dissociative ionization}\\
&\rightarrow \mathrm{O^*} + \mathrm{O} + \mathrm{e} && \text{dissociative excitation}\\
&\rightarrow \mathrm{O^-} + \mathrm{O} && \text{dissociative electron attachment}
\end{aligned}
$$

Furthermore, the excitation of atomic and molecular oxygen species in different quantum states (electronic, vibrational, and rotational states) have to be involved, which requires energy-dependent cross sections of the most important elementary collision processes.

The heating of electrons by the external electric field and the inelastic electron collisions causes deviations from the Maxwell distribution. Considering the anisotropy of the EEDF, the calculation by means of BLME may be performed by different approximations depending on the electric field strength, from a two-term approximation with an isotropic (Maxwellian) and weakly anisotropic part at low electric field strengths, over the multiterm approximation using Legendre polynomials up to the highly anisotropic electron beam approximation (runaway electrons) at very high electric field strength.

In the same way, the kinetic description of the collision processes between heavy species (ions and neutrals) has to be taken into account. These collisions produce new chemical compounds, for example, fluorocarbon plasmas containing hydrogen:

$$
\begin{aligned}
\mathrm{CF} + \mathrm{F_2} &\rightarrow \mathrm{CF_2} + \mathrm{F}\\
\mathrm{CF_3} + \mathrm{H} &\rightarrow \mathrm{CF_2} + \mathrm{HF}\\
\mathrm{CF_2} + \mathrm{H} &\rightarrow \mathrm{CF} + \mathrm{HF}
\end{aligned}
$$

11.3.2 Reaction Rate Coefficient

At known energy distribution functions of the plasma species (electrons, ions, and neutrals), the macroscopic transport equations in space are obtained by multiplying the Boltzmann equation with a conserved quantity (moment $M \cdot \vec{v}^{\,n}$) and the integration over all velocities. Here, the moment of zero order ($M = 1, n = 0$) is of special interest. It provides the particle balance equation with the macroscopic sources and sinks of the selected species (i), respectively

$$\frac{\partial n_i}{\partial t} + \operatorname{div} \vec{j}_i = S_i - L_i . \tag{11.8}$$

For example, the source term S_i may describe the production of single positive charged $CF_3{}^+$ ions by dissociative electron impact ionization of neutral CF_4 molecules in the ground state (11.9):

$$e + CF_4 \xrightarrow{k_{CF_3{}^+}} CF_3{}^+ + F + 2e \tag{11.9}$$

The corresponding ionization rate can be calculated by means of the rate coefficient $k_{CF_3{}^+}$, taking into calculation the electron energy distribution function (f_e), the partial ionization cross section ($\sigma_{CF_3{}^+}$), and the relative velocity (v) of the collision partners according to

$$\left[\frac{dn_e}{dt}\right]_+ = \left[\frac{dn_{CF_3{}^+}}{dt}\right]_+ = S_+^{CF_3} = n_{CF_4} \cdot n_e \cdot k_+^{CF_3{}^+}$$
$$= n_{CF_4} \cdot n_e \cdot \left\langle v \cdot \sigma_{CF_3{}^+} \right\rangle = n_{CF_4} \cdot n_e \cdot \int_{\text{thres}}^{\infty} f_e(v) \cdot v \cdot \sigma_{CF_3{}^+}(v) \cdot dv, \tag{11.10}$$

where n_e is the electron density, n_{CF_4} the CF_4 density in the ground state, and $k_{CF_3{}^+}$ the rate coefficient for dissociative electron impact ionization.

In the case of chemical reactions in nonthermal plasmas the rate coefficient (k) represents the *reaction rate coefficient*. Contrary to plasmas in thermodynamic equilibrium state, in *nonequilibrium plasmas* the reaction velocities for the forward and backward reaction in any elementary process are quite different. That means the law of mass action, usually applied for chemical reactions in thermodynamic equilibrium, is not valid.

Taking into consideration the binary collision of different species in various states of internal energy (p, q, r, s), for example,

$$A(p) + B(q) \underset{\leftarrow k_{CD}}{\overset{k_{AB}\rightarrow}{\longleftrightarrow}} C(r) + D(s) \tag{11.11}$$

the reaction rate of the product concentration [C] is proportional to the concentrations of the reagents [A] and [B] where the quantity k_{AB} describes the reaction rate coefficient of this chemical reaction [7].

$$\frac{d[C(p, qIr, s)]}{dt} = k_{AB}[C, D; r, sIA, B; p, q] \cdot [A(p)] \cdot [B(q)]. \tag{11.12}$$

By means of averaging $\langle \sigma_{AB} \cdot v_{AB} \rangle$ over the distribution function f_{AB}, and summation over all internal states of the reaction partners, the *total reaction rate* k of the product concentration [C] is obtained:

$$k_{AB} = \langle \sigma_{AB} \cdot v_{AB} \rangle = \int \sigma_{AB} \cdot v_{AB} \cdot f(v_{AB}) \cdot dv_{AB}, \tag{11.13}$$
$$k(C, DIA, B) = \sum_{pqrs} x_A(p) \cdot x_B(q) \cdot k_{AB}(C, D; r, sIA, B; p, q), \tag{11.14}$$

where

$$x_A(p) = \frac{[A(p)]}{\sum_p [A(p)]}, \quad x_B(q) = \frac{[B(q)]}{\sum_q [B(q)]}. \tag{11.15}$$

11.4 Plasma–Surface Interaction

The plasma–surface interaction describes all phenomena which are connected with the interaction of charged and neutral plasma species as well as photons with a solid or liquid phase boundary and their response and influence on the bulk plasma. The particle flux to the surface and the plasma radiation may initiate manifold elementary processes in the interface of the condensed matter due to exchange of energy, momentum, mass and charge, for example, surface recombination, displacement of atoms, dissociation, excitation, chemical reactions, and secondary species emission. Therefore, the complex plasma–surface interaction includes collective and synergistic processes involving charged species, fast neutrals and metastable excited species, energetic photons, and chemically reactive atoms and molecules.

In front of any plasma boundary a transition region, the *plasma sheath*, is observed. The plasma sheath properties are determined by the internal plasma parameters, external biasing, and the surface material. Therefore, the transport of charged species and their kinetic energy depends on the electric space charge field and collision processes inside the plasma sheath. For example, positive ions which receive sufficient kinetic energy in the plasma sheath in front of negatively charged surfaces can significantly contribute in reactive ion etching, chemical sputtering, or ion-assisted thin film deposition.

11.4.1 Plasma Sheath

Taking into consideration a stationary, uniform and ideal *nonthermal plasma* consisting of electrons and the same kind of single charged positive ions with Maxwellian energy distribution, $T_e \sim 10^4\,\mathrm{K} \gg T_+$, an immersed chemically inert and totally absorbing material surface will be negatively charged in relation to the plasma potential, φ_{pl}. The reason is the higher mobility of electrons compared with the heavy ions $[\sim (m_+ \cdot T_e/m_e \cdot T_+)^{1/2}]$. The negatively charged surface at the potential $-\varphi_s$ is shielded by a positive space charge sheath in front of the surface (see Fig. 11.3). The resulting self-consistent electric field of the space charges determines the charge carrier transport to the surface.

Under the assumption of no collisions in the sheath and cold plasma ions, the sheath can be described analytically by use of the energy conservation of positive ions and the Boltzmann distribution of electron density in the space charge sheath. Furthermore, the solution of the Poisson equation in the space charge sheath demands the continuous transition of the electric potential from the space charge

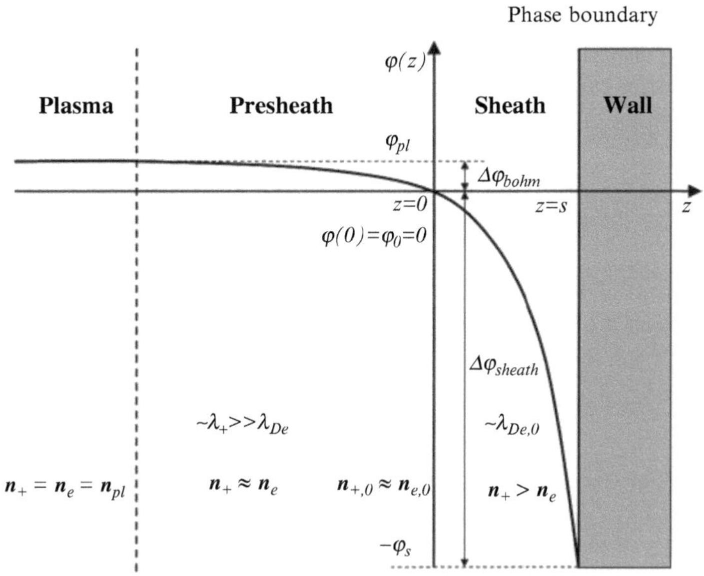

Fig. 11.3 Transition region between bulk plasma and surface – the plasma sheath

sheath to the quasineutral plasma. This requires a minimum of ion velocity at the sheath edge which means the acceleration of cold positive ions on the *Bohm velocity* in a presheath. This requirement is well known as the *Bohm criterion* [9]:

$$\left[\frac{1}{2 \cdot k_{\mathrm{B}} \cdot T_{\mathrm{e}}} - \frac{1}{2 \cdot m_{+} \cdot v_{+}^{2}(0)}\right] > 0. \tag{11.16}$$

The presheath region (see Fig. 11.3) is defined by the necessary potential drop, $\Delta\varphi_{\mathrm{bohm}}$, from the plasma potential in the bulk, φ_{pl}, to the potential at the sheath edge, φ_0:

$$\frac{m_{+}}{2} \cdot v_{\mathrm{bohm}}^{2} = \frac{1}{2} \cdot k_{\mathrm{B}} \cdot T_{\mathrm{e}} = |e| \cdot \Delta\varphi_{\mathrm{bohm}} = |e| \cdot (\varphi_{\mathrm{pl}} - \varphi_0). \tag{11.17}$$

11.4.2 Surface on Floating Potential

Taking into consideration an electric insulating substrate immersed in a quasineutral plasma consisting of cold ions and Maxwellian electrons (plasma density $n_{\mathrm{pl}} = n_{\mathrm{e}} = n_{+}$), the net charge flow to the substrate surface will be zero under steady-state conditions:

$$j_{+,\mathrm{sf}} + j_{\mathrm{e,sf}} = 0|_{\mathrm{surface}}. \tag{11.18}$$

For planar geometry, these current densities at the surface on floating potential are expressed by use of the *Bohm current density* at the sheath edge for positive ions and the electron current density by means of the Boltzmann factor with retarding surface potential $-\varphi_{\rm sf}$:

$$|e|\cdot n_{\rm pl}\cdot \exp(-1/2)\cdot\sqrt{\frac{k_{\rm B}\cdot T_{\rm e}}{m_+}} - |e|\cdot n_{\rm pl}\cdot \exp(-1/2)\cdot\sqrt{\frac{k_{\rm B}\cdot T_{\rm e}}{2\cdot\pi\cdot m_{\rm e}}}\cdot \exp\left(\frac{|e|\cdot\varphi_{\rm sf}}{k_{\rm B}\cdot T_{\rm e}}\right) = 0. \tag{11.19}$$

The total potential drop between the plasma potential, φ_{pl}, and the floating surface potential results in

$$\varphi_{\rm pl} - \varphi_{\rm sf} = \Delta\varphi_{\rm bohm} + \Delta\varphi_{\rm sh} = \frac{1}{2}\cdot\frac{k_{\rm B}\cdot T_{\rm e}}{|e|}\cdot\ln\left(0.43\cdot\frac{m_+}{m_{\rm e}}\right). \tag{11.20}$$

Using characteristic plasma parameters for low-pressure nonthermal plasmas, $n_{\rm pl} = 10^{16}\,{\rm m}^{-3}$ and $k_{\rm B}T_{\rm e} = 2\,{\rm eV}$, as well as single charged argon ions, the maximum kinetic energy of ions at the surface amounts to about 10 eV, and the corresponding ion current density $1.3\times10^{15}\,{\rm cm}^{-2}{\rm s}^{-1}$, respectively.

11.4.3 High-Voltage Plasma Sheath, Radio-Frequency Plasma Sheath

In the case of high sheath voltages $|e\cdot\Delta\varphi_{\rm sh}| \gg k_{\rm B}\cdot T_{\rm e}$ two approximations are usually applied to describe the space charge sheath without collisions ($\lambda_+ \ll s$), the *matrix sheath approximation* and the *Child–Langmuir sheath model*. In the matrix sheath model a constant positive ion density is assumed, and the electrons are completely neglected because of the strong retarding electric field. On the other hand, the conservation of the ion flux and energy is taken into account in the Child–Langmuir sheath model which results in the well known $\varphi^{3/2}$ law for the space charge limited ion current density:

$$j_{+,\rm child} = \frac{4}{9}\cdot\varepsilon_0\cdot\left(\frac{2\cdot|e|}{m_+}\right)^{1/2}\cdot\frac{\varphi_{\rm s}^{3/2}}{s^2}. \tag{11.21}$$

By use of the Bohm current density at the sheath edge, the sheath thickness is expressed in (11.22), scaled with the electron *Debye length* $\lambda_{\rm De}$:

$$s_{\rm child} = \frac{2^{1/2}}{3}\cdot\lambda_{\rm De}\cdot\left(\frac{2\cdot|e|\cdot\varphi_{\rm s}}{k_{\rm B}\cdot T_{\rm e}}\right)^{3/4}, \quad \text{where} \quad \lambda_{\rm De} = \sqrt{\frac{e^2\cdot k_{\rm B}\cdot T_{\rm e}}{\varepsilon_0\cdot n_{\rm e}}}. \tag{11.22}$$

In asymmetric, capacitively coupled rf-plasmas (CCP) with different effective electrode areas between powered and grounded electrode ($A_{\rm powered} < A_{\rm grounded}$), the

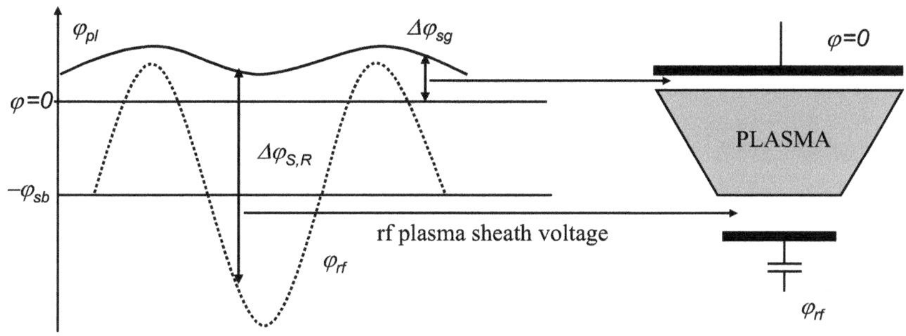

Fig. 11.4 Electric potentials (idealized) of an asymmetric, capacitively coupled rf-plasma (CCP). The difference between the plasma potential (φ_{pl}) and the shifted rf-potential (φ_{rf}), as well as the ground potential ($\varphi = 0$), represents the sheath voltage at the powered and grounded electrode, respectively

formation of a negative DC self-bias voltage is observed at the powered electrode. The electric potentials over the rf-cycle are shown in Fig. 11.4 for an idealized asymmetric, capacitively coupled rf-discharge. The maximum negative self-bias voltage -φ_{sb} is limited by the half of the peak-to-peak rf-voltage because of the condition for the plasma potential as the most positive potential. At the powered electrode the strong modulation of the sheath voltage can be clearly seen, from the full sheath expansion phase, with nearly the peak-to-peak voltage, to the very low sheath voltage during the sheath collapse. This sheath dynamics contributes essentially to the electron heating during the sheath expansion phase, as seen in phase-resolved optical emission spectroscopy in Sect. 11.5.1. On the other hand, the plasma sheath in front of the grounded electrode is weakly modulated at low mean sheath voltage. The different plasma sheath conditions at the powered and grounded electrode in strong asymmetric, capacitively coupled rf-plasmas result in a different kinetic energy of positive ions at these electrodes. At low pressure the maximum ion kinetic energy at the powered electrode is strongly coupled to the self-bias voltage and can reach some hundreds electron volts, whereas the weakly modulated sheath voltage at the grounded electrodes can be described by a DC-like sheath with single ion peak of 10–25 eV in the ion energy distribution function.

Assuming a time averaged sheath potential and current density in rf-plasma, the corresponding formula similar to the Child–Langmuir law is obtained [9]:

$$\bar{j}_+ = |e| \cdot n_0 \cdot v_{\text{bohm}} = C_+ \cdot \varepsilon_0 \cdot \left(\frac{2 \cdot |e|}{m_+}\right)^{1/2} \cdot \frac{\bar{\varphi}_{\text{s}}^{3/2}}{s_{\text{max}}^2}, \tag{11.23}$$

with $C_+ = 200/243$ for the completely modulated rf-sheath voltage, compared with $C_+ = 4/9$ in the DC case (Child–Langmuir). The comparison of the sheath thickness with the DC case results in

$$s_{\text{max}} = \sqrt{50/27} \cdot s_{\text{child}}. \tag{11.24}$$

Taking into consideration the ion transport with collisions inside the sheath an appropriate formula (11.25) for the transition regime ($\lambda_+ \geq s$) is derived in [9] by use of the mean free path length for positive ions λ_+:

$$\bar{j}_+ = |e| \cdot n_0 \cdot v_{\text{bohm}} \approx 1.68 \cdot \varepsilon_0 \cdot \left(\frac{2 \cdot |e|}{m_+}\right)^{1/2} \cdot \frac{\bar{\varphi}_s^{3/2} \cdot \lambda_+^{1/2}}{s_{\max}^{5/2}}. \tag{11.25}$$

11.5 Low-Pressure Oxygen rf-Plasma

The low-pressure oxygen plasma represents an example of complex multispecies plasmas involving electrons, positive/negative and molecular/atomic ions, reactive neutrals, as well as (metastable) excited species. Beside many applications of plasmas containing oxygen due to their high reactivity in the interaction with material surfaces, for example, oxidation, formation of reactive functional groups on polymer surfaces and plasma chemical etching, or the production of ozone in the bulk plasma, they represent an appropriate model system for studying complex molecular and electronegative plasmas, their stability and reactive interaction with surroundings.

In the following, the capacitively coupled low-pressure rf-plasma at 13.56 MHz in oxygen is taken into consideration more in detail. The investigations were performed by use of a stainless steel vacuum chamber, equipped with vacuum pumps (base pressure 10^{-5} Pa), and plasma process controlling for total gas pressure (5–100 Pa) as well as process gas flow (1–10 sccm). The discharge arrangement consists of the rf-powered circular stainless steel electrode with a typical diameter of 100 mm and a grounded electrode, which is either the parallel plate electrode of same diameter in distance to the powered electrode of between 2 and 4 cm, or the chamber wall. The rf-electrode was powered by the rf-generator (13.56 MHz) and a fully tunable π-type matching network. The power input of 5–100 W in the pressure range from 5 to 100 Pa is connected with the appearance of negative self-bias voltage of −70 to −600 V at the powered electrode.

The low-pressure oxygen rf-plasma was characterized by means of different diagnostic techniques, such as:

- Electric probe measurement for determination of plasma parameters (n_{pl}, T_{e}) [10]
- Microwave interferometry (n_{e}) [11]
- Energy-resolved mass spectrometry (O_2^+, O^+, O_2^-, O^-) [10, 12, 13]
- Spatiotemporally resolved optical emission spectroscopy (excited atomic oxygen) [14, 15]
- Two-photon laser-induced fluorescence (TALIF; ground-state atomic oxygen) [16, 17]

Furthermore, the interaction of oxygen plasma with polymer surfaces is studied by means of in situ *Fourier transform infrared* (FTIR) spectroscopy [18] and spectroscopic ellipsometry [19] of plasma-treated thin polymer films, as well as the mass spectrometry of charged and neutral reaction products in gas phase.

11.5.1 Plasma Characterization

11.5.1.1 Electric Probe Measurement, Positive Ion Density

The experimental determination of the internal plasma parameters in nonthermal plasmas (the plasma density and electron energy distribution function, respectively, the electron temperature) is the most important task in nonthermal plasma physics.

Invasive electric probe techniques can be used in plasmas without deposition processes only. For correct interpretation of the current–voltage characteristics and calculation of plasma parameters appropriate plasma sheath models for electron and positive ion sampling by the probe tip have to be involved. Additional problems have to be considered, for example, the plasma potential modulation in rf-plasmas, the external magnetic field, several kinds of positive ions in molecular plasmas, as well as the presence of negative ions.

Figure 11.5 represents the spatial distribution of the plasma density (positive ion density) nearby the powered electrode in axial and radial direction for a strong asymmetric, capacitively coupled oxygen plasma at 13.56 MHz and low pressure ($p = 5$ Pa). The plasma density was measured by means of a passively rf-compensated electric probe. The analysis of the ion saturation current of the probe characteristics provides typically plasma densities of 10^{15}–10^{16}m^{-3} [10]. The axial

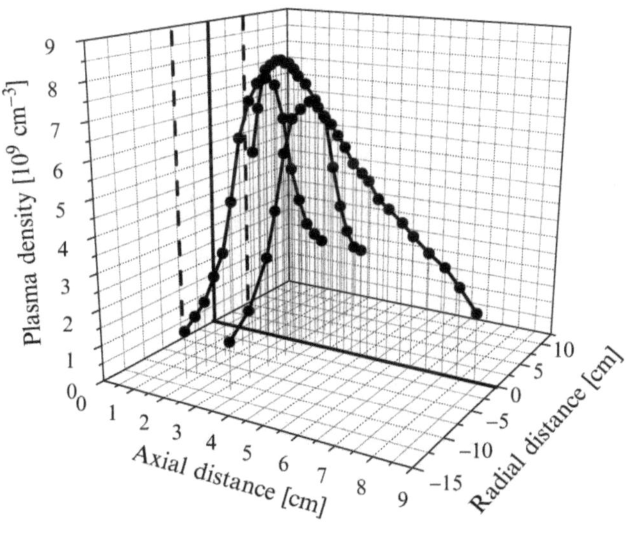

Fig. 11.5 Spatial plasma density distribution in an asymmetric capacitively coupled oxygen rf-plasma at low pressure. Axial distance measured from the powered electrode with diameter of 10 cm. Langmuir probe (passively compensated) measurement and calculation of the plasma density from positive ion saturation current by use of $O_2{}^+$ ions as dominant positive oxygen ion, $p = 5$ Pa, $U_{pp} = 800$ V

density distribution of the positive ions shows a significant density maximum at the plasma sheath edge, whereas in radial direction over the electrode diameter a reduction of the ion density is observed resembling a *Bessel function*, which is caused mainly by ambipolar diffusion.

11.5.1.2 Microwave Interferometry, Electron Density

The microwave interferometry compares the microwave propagation through the plasma with a microwave reference path. The measured phase shift between both beams provides the line-integrated electron density in the plasma without any model assumption. By use of the quasioptical model and a *Gaussian beam approximation* the detection volume in the active plasma zone can be reduced by appropriate design of the beam waist. The realized microwave interferometry at 160 GHz ($\lambda = 1.87$ mm) with Gaussian beam propagation is shown schematically in Fig. 11.6 [20]. The optical axis of the microwave beam crosses the asymmetric, capacitively coupled rf-plasma in radial direction at 20 mm distance to the powered electrode.

The effective diameter of the microwave beam is about 10 mm in the discharge axis. Figure 11.5 shows the measured line-integrated electron density in dependence on the rf-power with the total oxygen pressure as parameter. In dependence on the selected pressure, the electron density reveals a characteristic behavior by comparison of the range of low rf-power with that of higher rf-power. At low power the electron density remains at low level until a characteristic rf-power, where a

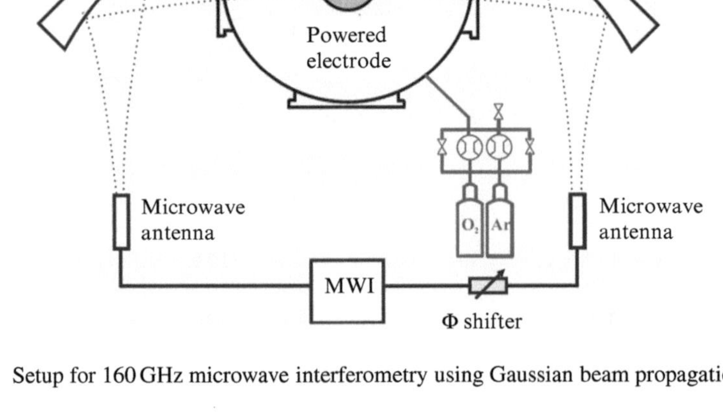

Fig. 11.6 Setup for 160 GHz microwave interferometry using Gaussian beam propagation

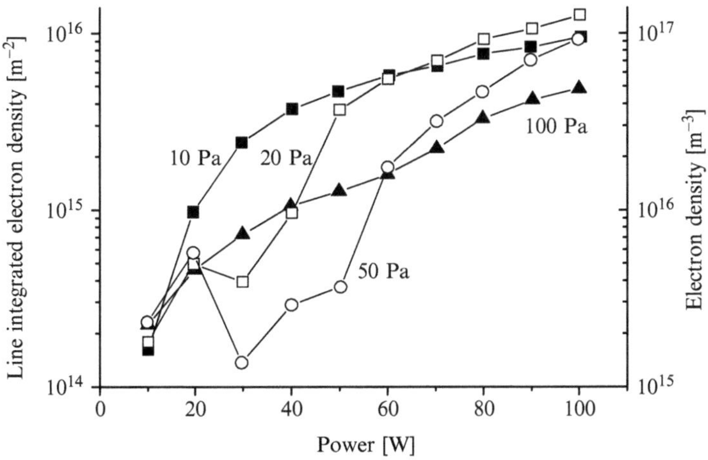

Fig. 11.7 Line-integrated electron density in oxygen rf-plasma (*left*) and electron density (*right*) using a constant density over the electrode diameter (100 mm) versus rf-power at different total pressure

significant increase with the rf-power is observed. The reason may be the strengthened formation of negative ions at low power input in spite of increasing plasma density (positive ion density). Above the characteristic rf-power the more dominant negative ion detachment processes may cause the strong electron density growth.

The determination of local electron densities from line-integrated density requires knowledge about the spatial (radial) electron density distribution. Using the simple model assumption of constant electron density over the powered electrode diameter of 100 mm and neglecting the electron density outside this region, the resulting radial averaged electron density is scaled in the right ordinate in Fig. 11.7 and amounts to 10^{15}–10^{17} m^{-3}.

11.5.1.3 Ion Analysis at Discharge Electrodes (Positive and Negative Oxygen Ions)

The ion analysis was performed by means of energy-resolved mass spectrometry. The orifice of the mass spectrometer (plasma monitor *HIDEN EQP 300*) is integrated into the rf or grounded discharge electrode, respectively [12]. After ion extraction via the orifice and extractor lens the ions are guided in the drift tube to the entrance of the electric sector field energy analyzer. The transmission of ions through this energy analyzer takes place at the constant pass energy of the ions. That means ions which have not this pass energy after extraction have to be accelerated or decelerated in order to reach this kinetic energy. Tuning the necessary reference voltage, the transmission of ions is selected in respect to their kinetic energy. In the following quadruple system, the ions are separated with respect to their charge-to-mass ratio and detected in the channeltron ion detector. The result is the energy-resolved flux of ions corresponding to the extraction position on the discharge electrode or wall.

Generally, the polarity of necessary voltages and potentials of the mass spectrometer can be changed, and the ion energy distribution is measured either for positive or negative ions.

The energy-resolved mass spectrometry in rf-plasmas provides time averaged ion energy distributions. In strong asymmetric, capacitively coupled rf-plasmas the different plasma sheath potential at the powered and grounded electrode influences the ion energy distribution significantly. At the powered electrode the strong modulated sheath voltage causes saddle-shaped structures in the ion energy distribution for positive ions in the case of no collisions in the sheath. In the case of charge transfer collisions a multipeak structure is observed which has its origin in the overlapping saddle-shaped structures as the result of thermal ions produced in charge transfer collisions inside the sheath (see Fig. 11.8) [12,21]. The dominant positive ion is the molecular oxygen ion O_2^+. Comparing the mass spectrometric measured intensities for molecular and atomic oxygen ions, the fraction of the atomic positive ions amounts to about 10%, only.

On the other hand, the ion energy distribution at the grounded electrode with low sheath voltage is characterized by a single peak corresponding to the mean sheath voltage (see Fig. 11.8) [12].

At low pressure (<10 Pa) the maximum and also the mean kinetic energy are strongly coupled to the negative self-bias voltage at the powered electrode (see Fig. 11.10). With increasing total pressure the elastic ion–neutral collisions have more and more influence which results in reduced ion intensity at higher energy, disappearing multipeak structure, and continuously increasing ion intensity toward

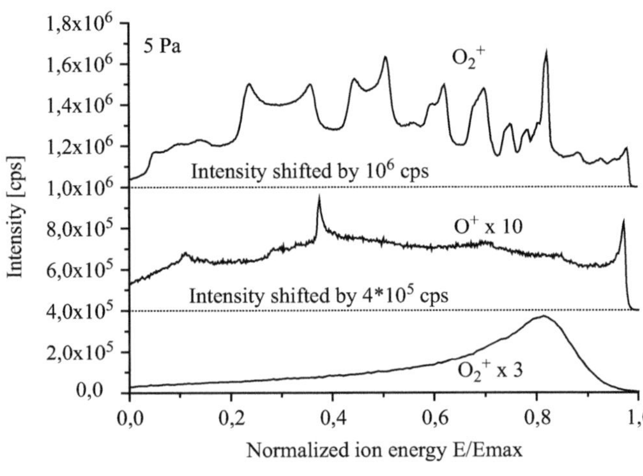

Fig. 11.8 Ion kinetic energy distribution of positive ions at powered electrode, $U_{SB} = 400\,V$, $E_{max}(O_2^+) = 430\,eV$, $E_{max}(O^+) = 580\,eV$, and at grounded electrode $E_{max}(O_2^+) = 15\,eV$

lower kinetic energy. At constant peak-to-peak rf-voltage of 1,000 V the mean ion kinetic energy (O_2^+) decreases from about 235 to 115 eV with raising pressure from 5 to 50 Pa (see Fig. 11.11).

In electronegative plasmas the extraction of negative ions fails in most cases. The negative ions at thermal energies are captured in the plasma bulk due to the negatively charged surfaces. Usually, they cannot penetrate the plasma sheath because of their low kinetic energy. In the case of negative ion generation (O^-, O_2^-) at the powered electrode surface or inside the rf-plasma sheath these negative ions are accelerated toward the bulk plasma and achieve a maximum kinetic energy of the order of some hundreds eV. At low-pressure conditions and low distance to the powered electrode, these ions can be extracted into the energy-resolved mass spectrometer (integrated in the grounded electrode) and analyzed with respect to their kinetic energy and intensity (see Fig. 11.9). The maximum and mean kinetic energy are coupled to the negative self-bias voltage at the powered electrode like those for positive ions (see Figs. 11.10 and 11.11). Note that the negative ions have to pass the plasma sheath in front of the grounded electrode before extraction which results in a small reduction of kinetic energy depending on this sheath voltage.

11.5.1.4 Optical Emission Spectroscopy, rf-Phase-Resolved Optical Spectroscopy

The optical emission of low-pressure oxygen rf-plasmas in the spectral range from 300 to 900 nm shows two intense emission lines of atomic oxygen at 777 and 844 nm. Furthermore, weak emission of molecular and atomic oxygen ions is observed around 525 and 464 nm, respectively, as well as the emission of molecular

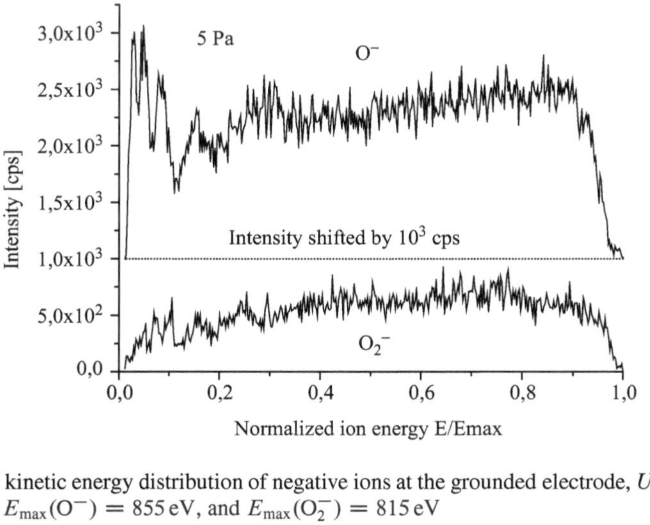

Fig. 11.9 Ion kinetic energy distribution of negative ions at the grounded electrode, $U_{SB} = 800$ V, $d_E = 25$ mm, $E_{max}(O^-) = 855$ eV, and $E_{max}(O_2^-) = 815$ eV

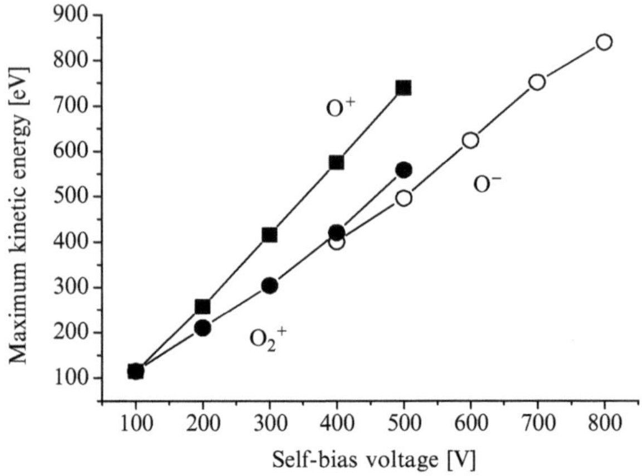

Fig. 11.10 Maximum energy of positive ions (powered electrode) and negative ions (grounded electrode) in low-pressure oxygen rf-plasma

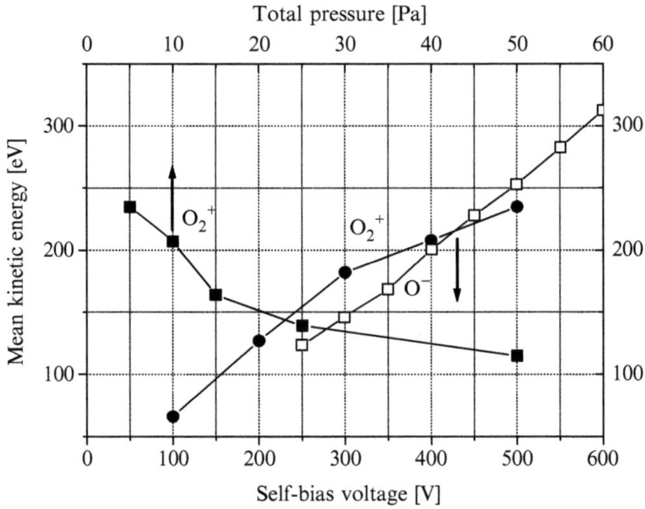

Fig. 11.11 Mean kinetic energy of positive and negative ions versus self-bias voltage at 5 Pa, and decreasing mean ion energy ($O_2{}^+$) with increasing total pressure ($U_{pp} = 1{,}000$ V)

oxygen from the atmospheric band at 760 nm. The rotationally resolved emission of the atmospheric band is often used to determine the rotational temperature of molecular oxygen by means of fitting to a Boltzmann distribution. The rotational temperature correlates often with the translation (gas) temperature. Besides this spectral range, in the VUV range the emission of atomic oxygen at 130 nm was found. Below 115 nm (window cutoff) now strong emission lines are expected from the oxygen rf-plasma.

In the following, the spatiotemporally resolved emission of atomic oxygen at 777 and 844 nm is in the focus of interest. The optical emission was imaged by use of a focusing lens and a 1 mm aperture diaphragm onto the entrance slit of the spectrometer which results in an axial resolution of 0.6 mm. The intensity was

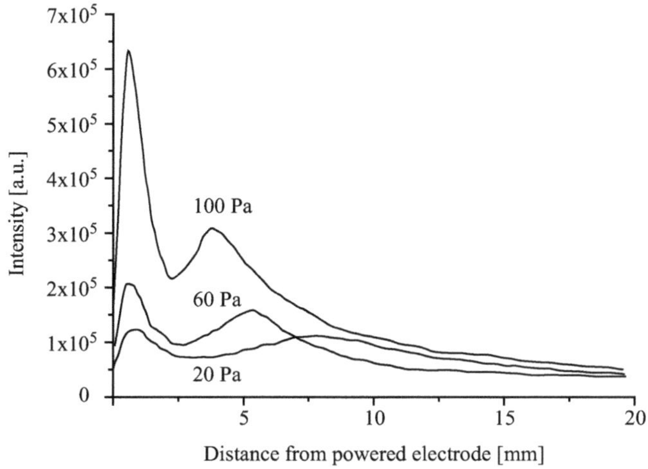

Fig. 11.12 Time averaged, axially resolved plasma-induced optical emission intensity of atomic oxygen at 844 nm

detected by means of an ICCD camera, mounted in the focal plane of a 0.5 m triple grating spectrometer. The detected emission intensity was averaged over the line of sight in radial direction.

The rf-cycle averaged axially resolved emission intensity of the asymmetric capacitively coupled rf-discharge provides two significant regions of atomic oxygen emission which are connected with the rf-plasma sheath in front of the powered electrode. The origin of this enhanced emission is on the one hand the electron heating due to the rf-sheath dynamics and electron impact excitation of oxygen with the intensity maximum at the position nearby the mean rf-sheath thickness, and on the other hand the emission due to excitation by heavy particle collision directly in front of the powered electrode surface (see Fig. 11.12) [14]. The latter one correlates with the energetic positive ion flux to the powered electrode. The apparent maximum at the axial position very close to the electrode surface is the result of a shadowing effect due to shearing the detection cone of the optical system with the discharge electrode. Indeed, the emission intensity should further increase toward the electrode surface. This could be verified by PIC–MCC simulations of the oxygen rf-discharge [22, 23].

Furthermore, the rf-cycle-resolved emission intensity was analyzed. In that case the spectrometer was replaced by an appropriate interference filter and the emission intensity was analyzed via a standard zoom objective by means of a fast ICCD camera with effective time resolution of 1.5 ns and repetition rate of 50 kHz. By triggering the digital delay/pulse generator with the rf-voltage signal from the powered electrode, the emission intensity was measured during the rf-cycle of about 74 ns by means of setting the gate of the ICCD camera to a fixed phase position. After intensity accumulation from few thousands rf-cycles the gate was shifted by 1 ns to the next rf-phase position and the procedure was done again for totally 74 time slots

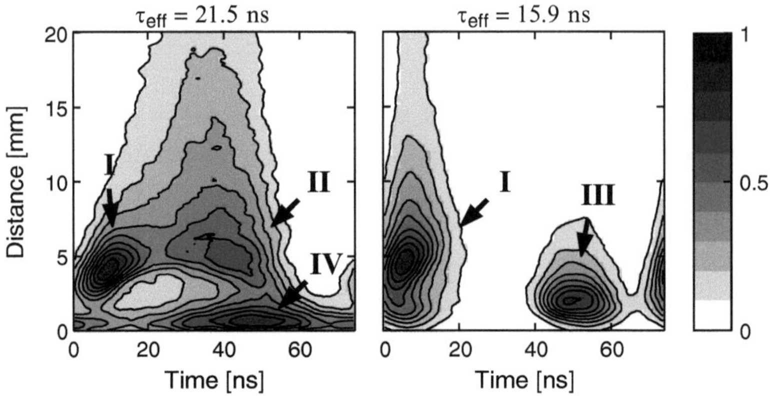

Fig. 11.13 Phase-resolved relative excitation rate of atomic oxygen due to electron heating during sheath expansion (I), secondary electrons (II), electron heating due to field reversal (III), and heavy species collisions (IV) for the transition at 844 nm (*left*, at $p = 80$ Pa, $P = 70$ W, $U_{SB} = -348$ V) and 777 nm (*right*, at $p = 100$ Pa, $P = 10$ W, $U_{SB} = -68$ V)

[14, 15]. From the measured time-resolved emission intensity, the relative excitation rate of atomic oxygen was determined by introducing the effective lifetime of the excited state which is determined by the natural lifetime (ca. 30 ns) and the quenching with the background gas. The axially and rf-cycle-resolved excitation rate of atomic oxygen is represented in Fig. 11.13 for two characteristic experimental conditions. The excitation pattern reveals four different regions of excitation:

(I) Electron impact excitation during the sheath expansion phase
(II) Electron impact excitation at full sheath expansion from secondary electrons including electrons from detached negative ions in the sheath region
(III) Electron impact excitation during the sheath collapse due to field reversal
(IV) Excitation by heavy particles in connection with the flux of the energetic positive ions to the powered electrode

Under all experimental conditions the pattern (I) is always observed due to the electron heating during the sheath expansion. The excitation pattern (II) by secondary electrons takes place at higher self-bias voltage and pressure, while the field reversal in oxygen rf-plasma is pronounced at low self-bias voltage (III). The emission pattern directly in front of the rf-electrode can only be explained by electronic excitation of atomic oxygen in heavy particle collisions. The dynamics of the excitation rate over the rf-cycle is characterized by harmonic behavior similar to the ion flux as shown by PIC–MCC simulations. This pattern is observed at high self-bias voltage with typical threshold at $U_{sb} = -100$ V. With raising total pressure the excitation rate increases significantly. The implementation of the atomic oxygen excitation by collisions of positive molecular oxygen ions with the neutral oxygen molecules of the background gas in PIC–MCC simulations and the appropriate fitting of the excitation cross section results in good agreement between experiment and simulation

[23, 24]. Nevertheless more detailed investigations have to be done to characterize the elementary processes leading to the oxygen excitation, for example, fast neutrals from charge exchange collisions or reflected fast neutrals coming from the powered electrode.

11.5.1.5 Atomic Oxygen Ground-State Density

The *two-photon absorption laser-induced fluorescence* (TALIF) spectroscopy technique was successfully applied to measure the density of the reactive atomic oxygen [17]. The oxygen transition $3p^3P_{2,1,0} \rightarrow 2p^3P_2$ was used to excite the oxygen atoms from the ground state due to the absorption of two photons at $\lambda = 225.6$ nm. The fluorescence light of the transition to the state $3s^3S_1^0$ at $\lambda = 844.6$ nm was detected. The calibration of the absolute atomic oxygen density in the ground state was performed by means of a TALIF reference measurement using pure xenon gas of known density. The two-photon resonance of the xenon atom $5p^{61}S_0 \rightarrow 5p^57p[3/2]_2$ at $\lambda = 225.4$ nm was taken into consideration which is close to the oxygen resonance. The xenon fluorescence line $7p \rightarrow 6s$ at $\lambda = 462$ nm was analyzed. The necessary laser radiation in the region of 225 nm was produced by means of frequency doubling of the dye laser radiation. The third harmonic of a Nd:YAG laser pumped the 1:1 dye mixture of coumarin 47 and coumarin 120 at $\lambda = 452$ nm. Passing through the BBO-I crystal, the desired UV radiation with pulse energy up to 350 μ J was generated and separated by the Pellin Broca prism. The fluorescence light perpendicular to the laser beam was measured by means of a photomultiplier after focusing and guiding through the interference filter (846 nm, 10 nm FWHM). The optical arrangement of the TALIF experiment was fixed and the complete parallel plate electrode system of the rf-discharge could be moved in axial and radial direction to achieve a space-resolved density distribution. In axial direction the minimum step of spatial resolution amounts to 0.5 mm. Figure 11.14 shows the characteristic axial distribution of the atomic oxygen density at different total pressure of the rf-oxygen plasma. An asymmetric distribution is observed with the maximum density nearby the plasma sheath edge in front of the powered electrode and a strongly decreasing density in direction to the powered electrode. The pressure dependence reflects the increasing sheath extension with decreasing pressure. The axial atomic oxygen distribution is mainly determined by the atomic oxygen production in connection with the rf-sheath dynamics, the diffusion and surface recombination. The total atomic oxygen density increases with raising total pressure. The absolute oxygen density determined in the discharge center is in the range of 10^{19}m^{-3} (see Fig. 11.15).

Next, we compare the normalized axial distributions of the time averaged intensity of excited oxygen atoms from optical emission spectroscopy with the ground-state atomic oxygen from TALIF experiment is shown in Fig. 11.16. The distribution of the plasma-induced atomic oxygen emission is mainly determined by the excitation processes in the rf-sheath, whereas the axial distribution of the ground-state oxygen density is significantly influenced by diffusion processes

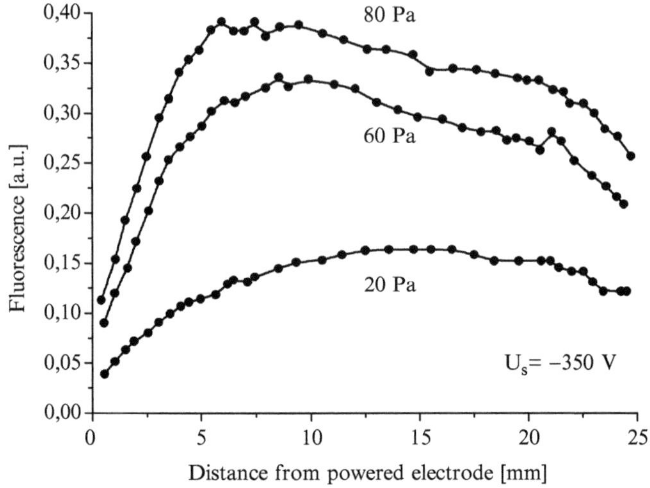

Fig. 11.14 Axial density distribution of ground-state atomic oxygen from two-photon absorption laser-induced fluorescence (TALIF) spectroscopy

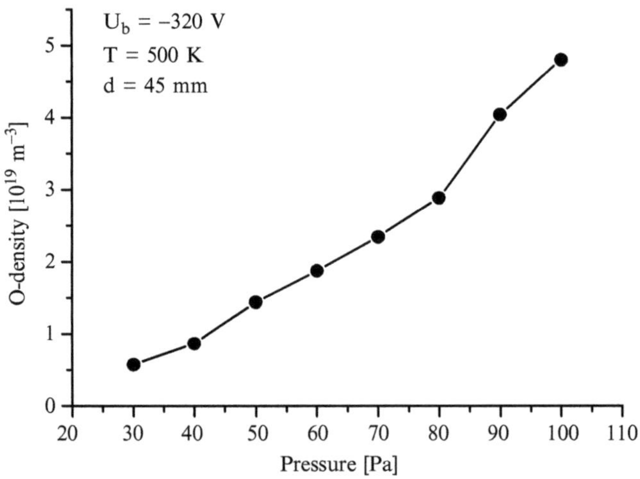

Fig. 11.15 Absolute density of ground-state atomic oxygen at the discharge center, calibrated by means of TALIF measurements in neutral xenon gas

and surface recombination. The dashed lines in Fig. 11.16 correspond to the time averaged rf-sheath thickness. Moreover, the strong increasing plasma-induced optical emission intensity in front of the powered electrode is not correlated with the ground-state density. In turn, it reveals the low influence of direct electronic excitation of atomic oxygen by electron impact.

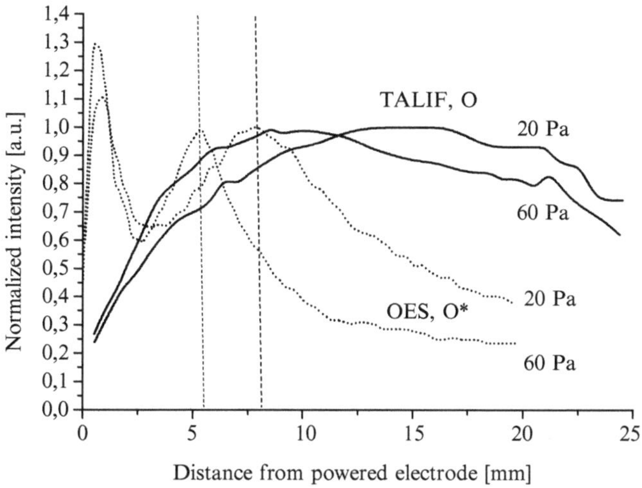

Fig. 11.16 Comparison between plasma-induced and laser-induced optical emission (TALIF) of atomic oxygen for the transition at 844 nm

11.5.2 Interaction of Oxygen Plasma with Polymers

The interaction of oxygen rf-plasma with polymer surfaces is exemplarily studied using the simplest synthetic hydrocarbon polymer, polyethylene (PE). In particular, the specific interaction mechanisms of the different plasma species and the physical and chemical modification of the polymer interface is of interest.

11.5.2.1 Fourier Transform Infrared Spectroscopy of Thin Polymer Films

The influence of the oxygen plasma on polyethylene was investigated by means of the special technique of the attenuated total reflection (ATR) [25, 26]. For the ATR measurements the internal reflection element (IRE) with 45° incident angle and 34 reflections was used. The IRE consists of chalcogenide glass. For the study of plasma chemical modification in a polymer interface, the polymer samples were prepared as thin films with thickness lower than 50 nm. These thin films are prepared on the IRE by means of a dip coating technique using a solution of additive-free low-density polyethylene (0.1–1%) in toluene at 70°C.

The IRE was integrated in the powered electrode inside the plasma apparatus and the thin PE film is faced to the rf-plasma. Therefore, the thin polymer film could be plasma treated from one side and probed by FTIR spectroscopy from the other side.

Figure 11.17 (top) shows the absorption spectrum of a thin PE film with characteristic absorption bands of the asymmetric and symmetric C–H stretching vibrations (2,950–2,700 cm^{-1}) as well as the C–H deformation vibration (1,460 cm^{-1}) of the $-(CH_2)_n-$ polymer groups. The observed absorption from 2,400 to 2,280 cm^{-1}

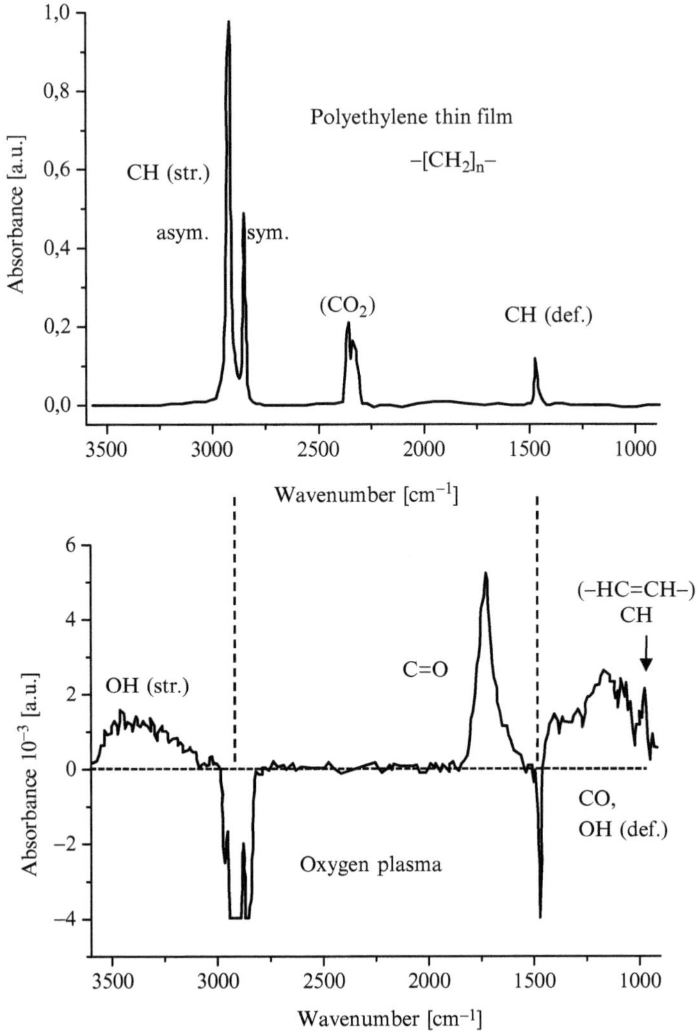

Fig. 11.17 Absorption spectrum of a thin polyethylene film (ca. 30 nm), taken by FTIR-(ATR) spectroscopy (*top*), and the ratio spectrum of PE thin film before/after 3 s oxygen plasma treatment (*bottom*). Etching/degradation of CH_2 polymer groups and formation of new functional groups (–OH, C=O, CO) as well as crosslinking (–CH=CH–)

has its origin from not compensated CO_2 outside the sample. The ratio spectrum in the Fig. 11.17 (bottom) generated from IR spectra before and after plasma treatment contains all the changes due to the plasma treatment. The strong polymer degradation/etching from the negative bands in the spectral range of the C–H vibrations is clearly seen. Additionally, different oxygen-containing structures were produced in the polymer interface. In the wave number range between 3,500 and 3,000 cm^{-1}

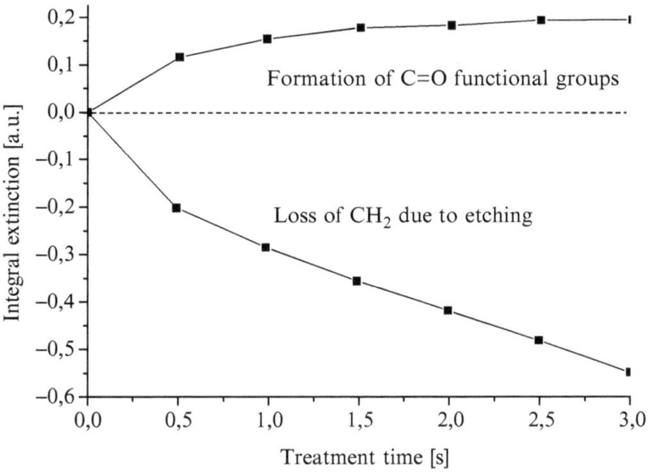

Fig. 11.18 Generation of functional groups (C=O) and etching of polyethylene on the powered electrode in oxygen rf-plasma. In steady state, a thin dynamic interface is formed with new chemical and physical properties, and the polymer material is etched continuously

O–H stretching vibrations of the hydroxyl groups appear broadened due to hydrogen bonds. The carbonyl group (C=O stretching) is observed at 1,800–1,690 cm^{-1}. The wide spectral range of the carbonyl group reveals a variety of C=O structures with slightly different characteristic frequencies. In the region from 1,400 to 1,000 cm^{-1} there are many overlapped bands associated with C–O stretching vibrations, O–H deformation in alcohols as well as C–O stretching vibrations in esters and ethers. The absorption around 960 cm^{-1} can be interpreted as C–H stretching vibration from a crosslinked (C=C) molecular structure.

The investigation of the temporal development of the observed molecular structures was performed by use of the integral extinction of the C–H and C=O stretching vibration in dependence on the total plasma treatment time (see Fig. 11.18). The polymer degradation/etching represented by the integral extinction of C–H vibrations shows after short treatment time a constant etching rate, while the carbonyl band (C=O) is formed and reaches a constant integral extinction. This behavior can be explained by the formation of a thin modified surface layer and polymer etching simultaneously. In steady state, a dynamic modified surface layer at nearly constant thickness is observed, while polymer material is permanently removed.

The penetration depth of energetic positive ions in polymer material was simulated by the TRIM code (transport of ions in matter) for the case of fast oxygen ions interacting with polyethylene. The depth distribution of energetic oxygen species depends on their kinetic energy received in the plasma sheath in front of the discharge electrode. The kinetic energy of oxygen ions in the order of few 100 eV relevant for the rf-plasma sheath provides a mean penetration depth of few nm only (see Fig. 11.19). Therefore, the modified surface layer should have a thickness of the same order of magnitude.

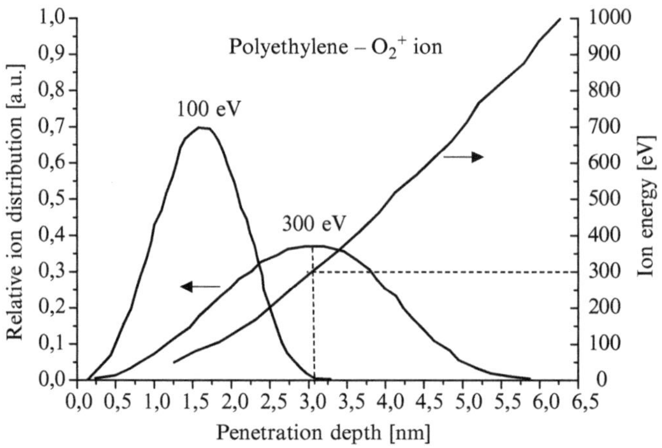

Fig. 11.19 TRIM simulation of the interaction of energetic oxygen ions with polyethylene

11.5.2.2 Spectroscopic Ellipsometry of Thin Plasma-Treated Polymer Films

The ellipsometry investigates the change of the polarization state of light reflected non-normally from a substrate [27]. At known substrate material, covered with thin transparent film, the polarization state contains information about the thickness of the film as well as the refractive index. In the last case multiple interferences have to be considered in the reflection coefficient which includes the optical path difference and the *Fresnel reflection coefficient* at the different interfaces. The shape of the measured polarization ellipse depends on the ratio of the complex reflection coefficients for parallel and perpendicular polarization which is related to the measured ellipsometric angles Ψ and Δ:

$$\rho = \frac{R_P}{R_S} = \tan\Psi \cdot \exp(i\Delta). \tag{11.26}$$

The ratio ρ is a transcendental function of the film thickness, the complex refractive index of the film, substrate and ambient, as well as of the angle of incidence and wavelength. Only in the case of substrate material it is possible to solve (11.26) analytically, while an iteration procedure is necessary if one-layer or multilayer models are used. The interpretation of the measured ellipsometric angles Ψ and Δ as optical relevant data (refractive index, thickness of thin film) strongly depends on the used optical model which has to fit the physical situation. Sufficient information about the material under investigation together with spectroscopic ellipsometry or measurements at different angles of incidence may contribute to obtain the physically relevant data.

The investigation of the plasma treatment of thin polymers was performed by the preparation of thin polyethylene films (thickness <30 nm) on a silicon wafer. The prepared samples were investigated by spectroscopic ellipsometry in the spectral

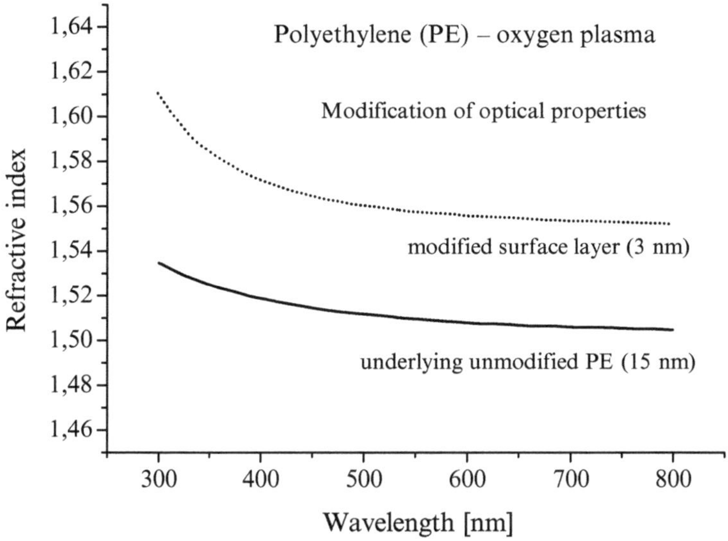

Fig. 11.20 Dispersion relation of thin polyethylene film. The ellipsometric measurements provide the modified thin surface layer due to the oxygen plasma treatment

range between 300 and 800 nm before and after the oxygen plasma treatment. The interpretation of the measured ellipsometric angles with the best fit analysis enforced the application of a two-layer model for the thin polymer film after the plasma treatment. For the dispersion behavior of the polymer film the *Cauchy dispersion relation* was taken into account. The polymer film has to be divided in a modified surface layer and a nearly unchanged underlying polyethylene film. The modified surface layer is characterized by a significantly higher refractive index (see Fig. 11.20). Furthermore, as expected from the FTIR analysis, the thickness of the modified surface layer is on the order of a few nm at comparable experimental conditions.

11.5.2.3 Mass Spectrometric Investigation of Reaction Products in Plasma/Gas Phase

The reaction products due to the interaction of oxygen rf-plasma with polyethylene were investigated by means of mass spectrometry analyzing the extracted neutrals and positive ions at the grounded discharge electrode. The polyethylene sample was placed either on the powered rf-electrode or at the grounded electrode. The mass spectra shown in Fig. 11.21 represent difference spectra. That means each spectrum is the difference of the mass spectra, with and without polyethylene on the considered discharge electrode. The oxidative decomposition of polyethylene on the powered and grounded discharge electrode is shown in both, the ion as well as neutral gas mass spectrum. The prominent peaks in the mass spectra have their

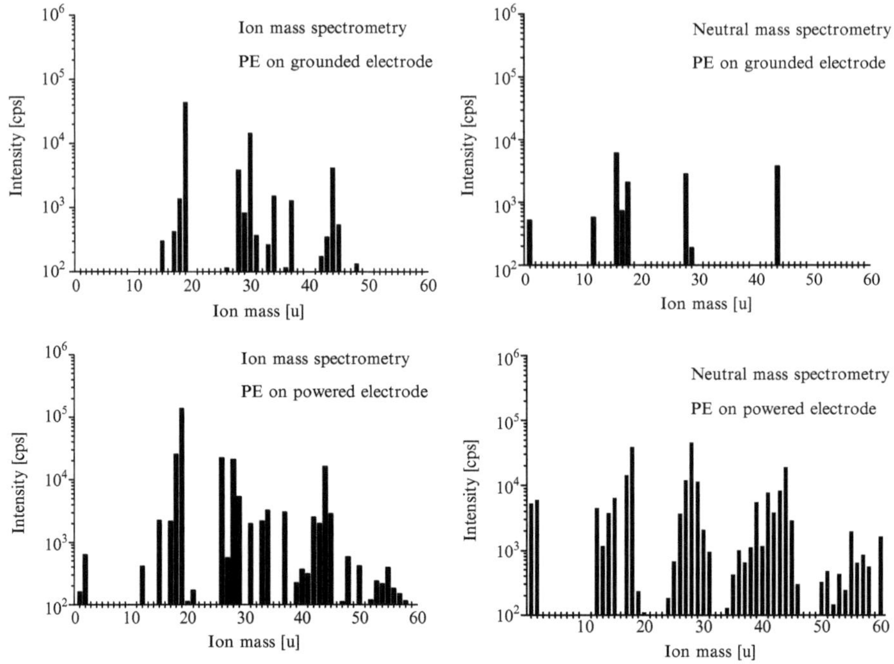

Fig. 11.21 Mass spectrometric analysis of reaction products at the grounded electrode due to polyethylene degradation in oxygen plasma ($p = 5\,\text{Pa}$, $U_{pp} = 600\,\text{V}$). The spectra are difference spectra (with/without polyethylene on the considered discharge electrode)

origin in the oxidative decomposition of the polyethylene, characterized by the ions H_2O^+(18u), H_3O^+ (19u), CO^+ (28u), and $CO_2{}^+$ (44u). Moreover, the ions H^+, $H_2{}^+$, and CHO^+ (29u) appear. In the case when the polymer sample is placed on the powered electrode the intensity of the extracted ions discussed above rises, due to the stronger ion bombardment. In the neutral mass spectrum with polyethylene on the powered electrode the typical groups of peaks are observed which have their origin in the fragmentation of hydrocarbons, but also hydrogen and carbon oxides.

References

1. S. Pfau, M. Tichy, in *Low Temperature Plasmas*, vol. 1, ed. by H. Hippler, K. Kersten, M. Schmidt, K.H. Schoenbach (Wiley-VCH, Weinheim, 2008)
2. U. Kogelschatz, J. Salge, in *Low Temperature Plasmas*, vol. 1, ed. by H. Hippler, K. Kersten, M. Schmidt, K.H. Schoenbach (Wiley-VCH, Weinheim, 2008)
3. K. Becker, K.H. Schoenbach, in *Low Temperature Plasmas*, vol. 1, ed. by H. Hippler, H. Kersten, M. Schmidt, K.H. Schoenbach (Wiley-VCH, Weinheim, 2008)
4. Y.P. Raizer, *Gas Discharge Physics* (Springer, Berlin, 1991)
5. V.M. Lelevkin, D.K. Otorbaev, D.C. Schram, *Physics of Non-Equilibrium Plasmas* (Elsevier, Amsterdam, 1992)

6. P.P.J.M. Schram, *Kinetic Theory of Gases and Plasmas* (Kluwer, Dordrecht, 1991)
7. L.S. Polak, Yu.A. Lebedev (eds.), *Plasma Chemistry* (Cambridge International Science Publishing, Cambridge, 1998)
8. D. Loffhagen, F. Sigeneger, R. Winkler, in *Low Temperature Plasmas*, vol. 1, ed. by H. Hippler, H. Kersten, M. Schmidt, K.H. Schoenbach (Wiley-VCH, Weinheim, 2008)
9. M.A. Lieberman, A.J. Lichtenberg, *Principles of Plasma Discharges and Material Processing* (Wiley, Hoboken, NJ, 2005)
10. J. Meichsner, M. Zeuner, B. Krames et al., Surf. Coat. Technol. **98**, 1565 (1998)
11. M. Krämer, B. Clarenbach, K. Kaiser, Plasma Sour. Sci. Technol. **15**, 332 (2006)
12. M. Zeuner, H. Neumann, J. Meichsner, J. Appl. Phys. **81**, 2985 (1997)
13. M. Zeuner, J. Meichsner, L.A. Rees, J. Appl. Phys. **79**, 9379 (1996)
14. K. Dittmann, D. Droszdov, B. Krames, J. Meichsner, J. Phys. D: Appl. Phys. **40**, 6593 (2007)
15. S. Nemschokmichal, K. Dittmann, J. Meichsner, IEEE Trans. Plasma Sci. **36**, 1360 (2008)
16. K. Niemi, V.S. van der Gathen, H. Döbele, J. Phys. D: Appl. Phys. **34**, 2330 (2001)
17. M. Geigl, S. Peters, O. Gabriel, B. Krames, J. Meichsner, Contrib. Plasma Phys. **45**, 369 (2005)
18. M. Nitschke, J. Meichsner, J. Appl. Polym. Sci. **65**, 381 (1997)
19. R. Rochotzki, M. Nitschke, M. Arzt, J. Meichsner, Phys. Stat. Sol. (a) **145**, 289 (1994)
20. C. Küllig, Diploma Thesis, University of Greifswald, Germany (2009)
21. C. Wild, P. Koidl, J. Appl. Phys. **69**, 2909 (1991)
22. F.X. Bronold, K. Matyash, D. Tskhakaya, R. Schneider, H. Fehske, J. Phys. D: Appl. Phys. **40**, 6583 (2007)
23. K. Matyash, R. Schneider, K. Dittmann, J. Meichsner, F.X. Bronold, D. Tskhakaya, J. Phys. D: Appl. Phys. **40**, 6601 (2007)
24. K. Dittmann, Ph.D. Thesis, University of Greifswald, Germany (2009)
25. N.J. Harrick, *Internal Reflection Spectroscopy* (Interscience, New York, 1975)
26. Y.J. Chabal, *Internal Reflection Spectroscopy: Theory and Application* (Dekker, New York, 1992)
27. R.M.A. Azzam, N.M. Bashara, *Ellipsometry and Polarized Light* (North-Holland, Amsterdam, 1977)

Chapter 12
Formation and Deposition of Nanosize Particles on Surfaces

Rainer Hippler, Satya R. Bhattacharyya, and Boris M. Smirnov

Abstract The processes in the cathode region of a magnetron discharge involving metal atoms and metal clusters are analyzed. The processes include formation of fast metal atoms as a result of cathode bombardment by accelerated buffer gas ions, thermalization of these metal atoms in collisions with buffer gas atoms, nucleation processes leading to the transformation of metal atoms into metal clusters, and attachment of metal atoms to the cathode and walls of the magnetron chamber as a result of their diffusion in a buffer gas. It is observed that nanosize clusters deposited on surfaces undergo solid–liquid phase transition at reduced melting temperatures and finally fragment into smaller clusters.

12.1 Introduction

The interest in so-called *dusty plasmas* containing small particles or clusters has grown enormously over the past decade [1, 2]. Much of this interest is due to applied research related to materials science and surface-processing technology and with respect to plasma diagnostics. There are numerous relations between dusty plasma physics and materials science. In dusty plasmas, the formation and growth of small particles (e.g., nanosize clusters) and the interaction of the plasma with the

---

R. Hippler (✉)
Institut für Physik, Ernst-Moritz-Arndt-Universität Greifswald,
Felix-Hausdorff-Str. 6, 17489 Greifswald, Germany
e-mail: hippler@physik.uni-greifswald.de

S.R. Bhattacharyya
Surface Physics Division, Saha Institute of Nuclear Physics, 1/AF Bidhan Nagar,
Kolkata 700 064, India
e-mail: satyar.bhattacharyya@saha.ac.in

B.M. Smirnov
Joint Institute for High Temperatures, Russian Academy of Sciences, Izhorskaya 13/19,
Moscow 125412, Russia
e-mail: bmsmirnov@gmail.com

M. Bonitz et al. (eds.), *Introduction to Complex Plasmas*, Springer Series on Atomic, Optical, and Plasma Physics 59, DOI 10.1007/978-3-642-10592-0_12,

particle's surface is the subject of investigations. One goal is the tailoring and usage of particles for specific purposes. Here one can think of deposition, etching, surface activation, modification, or separation of agglomerated grains in the plasma.

A cluster is a system of bound atoms or molecules and as such is an intermediate physical object between individual atoms or molecules and condensed bulk matter. Typical cluster sizes vary from a few atoms up to nanosize clusters (particles) with more than 10^6 atoms or molecules. Clusters due to their limited size have specific optical, magnetic, and electronic properties that are of interest from an applied point of view.

Cluster deposition onto a surface is of interest for contemporary nanotechnology [5, 6]. Solid metal clusters deposited on a target form an amorphous porous metal film that consists of individual grains. Such films are working as catalysts. Then a respondent molecule attaches to a grain thereby decreasing the activation energy of a chemical process and allowing for such chemical transformations which are impossible in a gaseous or liquid matter. Silver, for example, is an effective antibacterial remedy that kills microbes [7–9]. Recent studies show that the strongest anti-bacterial action is due to nonuniformities of the silver surface in the 1–10 nm size range [10] which is just the size of structural elements that result from deposition of solid clusters on a surface. This means that nanosize silver particles or clusters act as catalyst of biochemical reactions due to the nanometer-size nonuniformity of its surface. A surface with deposited metal clusters may be used in the same manner as a chemical catalyst; these deposition processes are thus of interest from a nanotechnological [11–13] and nanoelectronics point of view [14], and thus for applications in chemistry, biophysics, and medicine.

12.2 Magnetron Discharge

For the generation of nanosize clusters a magnetron discharge is a suitable device. The efficient cathode erosion in this discharge due to sputtering forms gaseous metal atoms, which are subsequently converted into clusters. The stability and uniformity of magnetron discharge allows to obtain high rates of atom formation in spite of small current densities.

The magnetron discharge was invented 70 years ago [17] and it received its contemporary form with two coaxial magnets (Fig. 12.1) in the 1960s [18]. Due to the cathode bombardment with fast ions and the subsequent cathode erosion by sputtering, a magnetron discharge is a source of metal atoms that determines its wide range of applications, for example, in the deposition of functional films on surface [4, 19]. Due to its unique properties the magnetron discharge may be also used as a source of metal clusters, and the first such scheme was proposed by Haberland et al. [20–24].

In essence, a magnetron discharge consists of a negatively biased cathode. Once the discharge is ignited, positive ions are accelerated toward the negatively biased cathode where they produce cathode secondary electrons and sputter cathode atoms producing a vapor of atoms from the cathode material. The secondary electrons are

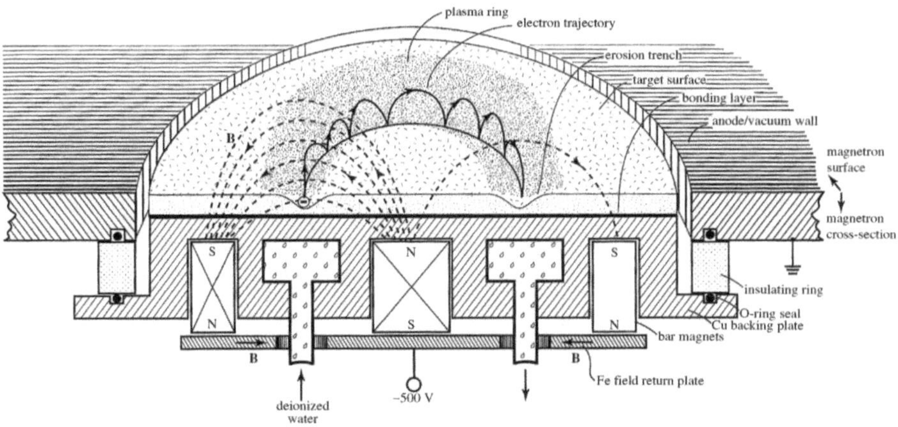

Fig. 12.1 Magnetron discharge [3]. Positive ions are accelerated toward the negatively biased cathode where they liberate secondary electrons and by sputtering of cathode material produce a vapor of cathode atoms. Secondary electrons are accelerated by the cathode potential and become trapped by the combined action of electric and magnetic field in the so-called *electron trap* region where they undergo a spiral motion losing most of their energy by ionizing collisions with buffer gas atoms (from [4])

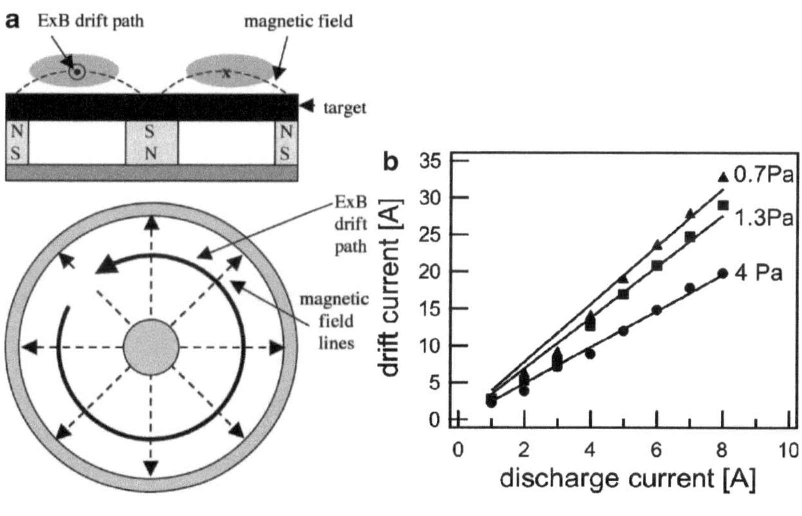

Fig. 12.2 (**a**) Top and side view of a planar magnetron, its electric $\boldsymbol{E}$ and magnetic $\boldsymbol{B}$ field lines (schematic), and the resulting $\boldsymbol{E} \times \boldsymbol{B}$ drift path [15]. (**b**) Measured drift current versus discharge current at different working pressures in an argon discharge [16] (from [4])

accelerated away from the cathode and become confined to the so-called *electron trap* region by the combined action of electric and magnetic fields (Figs. 12.1 and 12.2). The magnetic field is maintained by permanent magnets which are located underneath the cathode and whose magnetic fields penetrate into the space in front of the cathode.

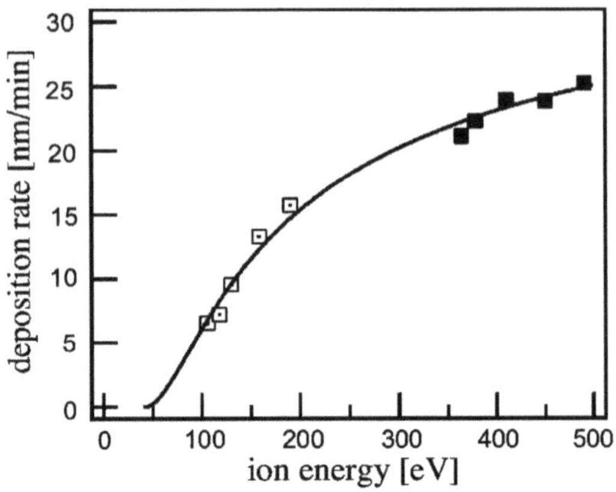

Fig. 12.3 ZnO deposition rate as as function of ion energy [25] (from [4])

Typical cathode voltages U are in the few hundred volts range so that positive ions impinging on the negatively biased cathode have gained a sufficient amount of kinetic energy $E \leq e_0 U$ to sputter atoms and ions from the cathode. The sputtering yield Y increases roughly linearly with cathode voltage, as does the deposition rate (see Fig. 12.3). Impinging ions also produce secondary electrons which are required to maintain the discharge; a typical secondary electron emission coefficient is $\gamma \approx 5\%$ [26]. Secondary electrons gain energy equal to the cathode potential (i.e., a few 100 eV) and become trapped in the region of perpendicular electric and magnetic fields. In order to maintain the discharge, each secondary electron thus has to produce as many as $1/\gamma \approx 20$ ions.

The magnetic field in a magnetron discharge is created by two coaxial magnets; it has a specific structure as is shown in Fig. 12.1. As a result, the total magnetic field has a maximum near the cathode and in this region is directed parallel to the cathode. Under the joint action of the magnetic field and the discharge electric field directed perpendicular to the cathode, electrons drift along a ring (Fig. 12.2). This gives a magnetic trap of a torus shape, and a quasineutral plasma of an enlarged density is formed in this region.

Because the plasma conductivity is proportional to the number density of electrons or ions of this quasineutral magnetron plasma, the conductivity and hence the current density in the electron trap region is rather large. Figure 12.2b shows the so-called *drift current* above the cathode in the tap region, exceeding the discharge current by factors of typically 3–7.

12.3 Nucleation Processes in a Magnetron Plasma

The central point of cluster generation in a magnetron plasma is the formation and growth of metal clusters in a rare buffer gas. The purpose of the buffer gas is twofold: (1) as working gas for maintaining the discharge and (2) to cool the atoms sputtered

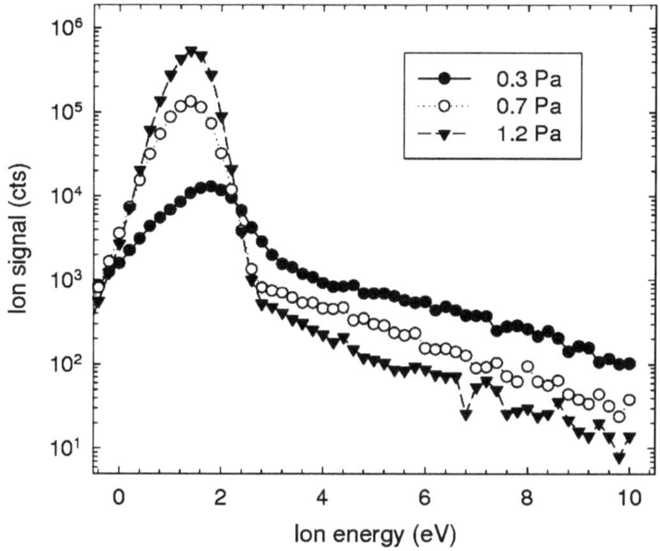

Fig. 12.4 Ion energy distribution of Ti^{+} ions versus pressure [27]

from the cathode. The energy distribution $Y(E)$ of sputtered atoms is approximately given by [26]

$$Y(E) \propto \frac{E}{(E+U)^3}, \tag{12.1}$$

where E is the kinetic energy and U the surface binding energy. Figure 12.4 shows the energy distribution of sputtered titanium ions measured 15 cm away from the cathode. The energy distribution is composed of a high-energy tail of Ti^{+} ions reflecting sputtered metal ions and thermalized Ti^{+} ions with low energies around 1 eV (this small energy reflects the plasma potential and thus corresponds to the energy gained by positive ions in the plasma sheath at walls where the energy measurement takes place). With increasing buffer gas pressure the density of cold (thermalized) ions increases while that of hot (nonthermalized) ions decreases [27]. Obviously, the sputtered atoms and ions lose kinetic energy through collisions with buffer gas atoms, and the higher the pressure the more efficient is the cooling process with buffer gas atoms.

In analyzing cluster growth we base ourselves on the following scheme [28–30]

$$2\mathrm{M} + \mathrm{A} \rightarrow \mathrm{M}_2 + \mathrm{A} \quad \text{and} \quad \mathrm{M}_n + \mathrm{M} \rightarrow \mathrm{M}_{n+1}. \tag{12.2}$$

Here M and A are metal atom and buffer gas atom, respectively. The first three-body process leads to formation of a diatomic molecule M_2, with a buffer gas atom taking away the energy excess. Next, starting from a diatomic molecule, we assume the energy excess resulting from formation of a bound state to be taken in collisions with buffer gas molecules. This means the following character of the second stage of of the process (12.2)

$$\mathrm{M}_n + \mathrm{M} \to \mathrm{M}^*_{n+1} \quad \text{and} \quad \mathrm{M}^*_{n+1} + \mathrm{A} \to M_{n+1}, \tag{12.3}$$

where M^*_{n+1} is an excited cluster. Next, we use the liquid drop model for the cluster [31, 32], so that a cluster is similar to a spherical drop with an atom density equal to that a macroscopic system of these atoms. We also use the hard sphere model [31, 32] for cluster collisions with atoms. This means that the cluster radius is large compared to the interaction radius of atomic forces during the collision of atoms with cluster surfaces. For simplicity, we further assume that each contact of a metal atom with the cluster surface leads to atom attachment. It is valid if the binding energy per metal atom in a cluster is large compared to the thermal energy of an incident metal atom. Since we are guided by typical temperatures ~1,000 K, and the binding energy per atom for metal clusters is of the order of several eV, this assumption is reasonable.

We also neglect evaporation processes. Note that heat balance of the growing cluster includes its heating as a result of atom attachment and release of the binding energy, and its cooling in collisions with buffer gas atoms. Since the rate of the first process is proportional to the number density of free metal atoms, and the rate of the second process is proportional to the number density of buffer gas atoms, the cluster temperature is determined by the concentration of metal atoms in a buffer gas, and at small concentrations cluster heating is weak and thus negligible. The rate of evaporation is expressed through the rate of atom attachment on the basis of the principle of detailed balance [31, 34–36], and under typical conditions the rates of atom attachment to the cluster surface and cluster evaporation are equal at the buffer gas temperatures of several thousand Kelvin.

Within the framework of the hard sphere model, the rate constant of the second process (12.2) for a large cluster consisting of n atoms is [28, 31]

$$k_n = k_0 n^{2/3} \quad \text{with} \quad k_0 = \pi r_\mathrm{W}^2 \sqrt{\frac{T}{2\pi m}}, \tag{12.4}$$

where r_W is the Wigner–Seitz radius for a given metal, T is the gas temperature expressed in energy units, and m is the mass of a metal atom. Under typical conditions of cluster growth we have a large value of the parameter [31, 37]

$$G = \frac{k_o}{K N_\mathrm{a}}, \tag{12.5}$$

where K is the rate constant of the three body (nucleation) process in reaction (12.2), N_a is the number density of buffer gas atoms. Because the rate constant of the three body process is between 10^{-33} and 10^{-32} cm^6/s for typical metals and buffer gases [19, 28, 31, 38], the parameter G is large for $N_\mathrm{a} \ll 10^{21}$ cm^{-3}.

A large value of the parameter G leads to a specific character of cluster growth. In fact, the first process of (12.2) proceeds slowly, while the subsequent cluster growth is fast. As a result, at any time the main contribution to the size distribution function of clusters gives large n. This justifies the liquid drop model under consideration. In addition, the condition $G \gg 1$ provides a large size of clusters in the end of the cluster growth process, that is, $n \sim G^{3/4}$ [28, 31, 32]. In particular, in the case of a magnetron plasma, where $N_a \approx 10^{15}$ cm^{-3} we have in the end $n \approx 10^4$. In addition, an average time of transformation of an atomic metal vapor located in a buffer gas into clusters is [37]

$$\tau_{\mathrm{nuc}} = \frac{3.2G^{1/4}}{N_{\mathrm{m}} k_o}, \tag{12.6}$$

where N_{m} is the number density of metal atoms. One can use the above formulas under conditions of experiment for a titanium vapor in argon as a buffer gas at pressure of 1 mbar [33], where $k_0 = 5.8 \times 10^{-11}$ cm^3/s at room temperature [31], $K = 2.5 \times 10^{-33}$ cm^6/s, $N_a = 4 \times 10^{15}$ cm^{-3} giving $G = 6 \times 10^6$. According to (12.6) under the experimental conditions ($N_{\mathrm{m}} \approx 8 \times 10^{11}$ cm^{-3}) [33], we have for the nucleation time $\tau_{\mathrm{nuc}} \approx 3$ s.

The above analysis is related to a low buffer gas temperature, where nucleation of thermal metal atoms is the stronger the higher their concentration. Hence, one can expect that increasing the discharge current will lead to an increase of the cluster formation efficiency and of the cluster flux. But this takes place as long as the temperature of the buffer gas is sufficiently low and the supersaturation of the metal vapor is large. At higher temperatures clusters evaporate and the efficiency of cluster generation decreases. Therefore one can find an optimal discharge current that provides the maximum efficiency of cluster formation for a given construction of magnetron discharge.

12.4 Nanosize Cluster Deposition

Nanosize silver clusters have been produced inside a gas aggregation nanocluster source followed by deposition on a Si(100) substrate with a native oxide layer. Figure 12.5 shows schematically the experimental setup which is equipped with a NC 200 nanocluster source (Oxford Applied Research) inside a vacuum chamber [33, 39]. Target atoms are sputtered by direct current (dc) magnetron sputtering. Clusters form by the attachment of free atoms and through coagulation processes inside the liquid nitrogen (LN_2) cooled aggregation tube of variable length. By increasing the length (thereby the volume) of the aggregation tube the residence time of the growing clusters increases, which helps the clusters to grow to bigger sizes. The clusters flow together with the buffer gas (Ar) through a variable orifice with a typical diameter of 0.3 cm. An optional mass filter can be used to select a particular cluster size/mass. Finally these clusters are deposited on a Si(100) substrate kept at a distance of about 53 cm from the exit of the magnetron chamber.

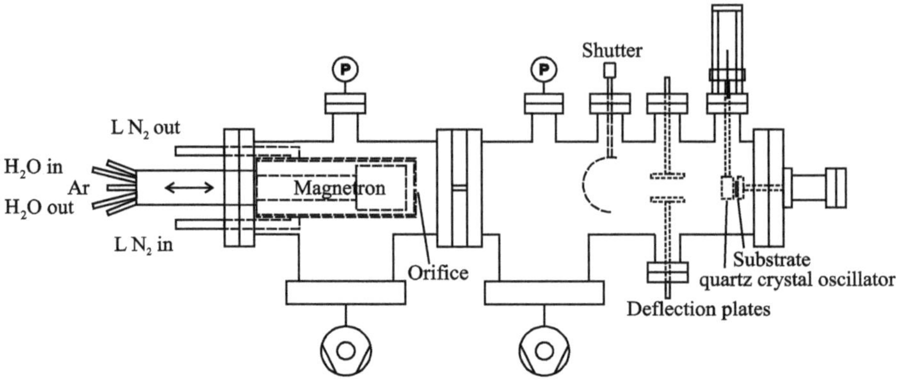

Fig. 12.5 Experimental setup for nanosize cluster production [33]

Fig. 12.6 AFM picture of deposited silver clusters showing aggregation of clusters due to coagulation of neighboring clusters (deposition time 15 s) [39]

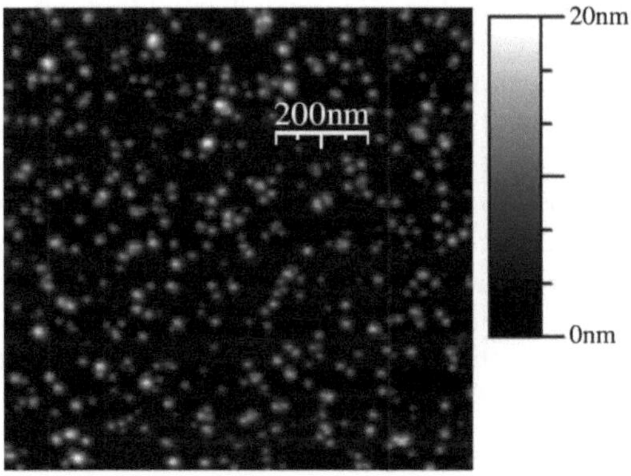

Atomic force microscopy (AFM) and grazing incidence diffractometry (GIXD) techniques are used to analyze the deposited cluster films. AFM is done in tapping mode, with high resolution tips. The sewing tip approach is used in order to avoid tip damage (Fig. 12.6).

An AFM picture of deposited silver clusters and the extracted mean heights is shown in Fig. 12.6. A closer inspection of the AFM picture of Ag clusters deposited for 15 s shows both isolated and aggregated clusters on Si(100) wafer. Statistical analysis of Fig. 12.6 reveals that about 40% of the clusters are completely isolated, most clusters are very close to each other forming two-dimensional aggregates. Very few aggregates ($< 3\%$) show growth in the third dimension (cluster on top of each other). Formation of larger nanoparticles thus occurs as a result of coagulation of two or more neighboring clusters because of surface diffusion as well as the possibility of clusters landing on top of each other far before the first layer is completed. For longer deposition times more and more clusters form large aggregates (see Sect. 12.6).

12.5 Melting Temperature and Lattice Parameters of Ag Clusters

Grazing incidence diffractometry (GIXD) in off-specular reflectivity geometry is performed using a $\theta - \theta$ diffractometer [40]. An integrated high-temperature chamber allows for in situ measurements at temperatures up to 1,000°C.

The X-ray pattern of deposited Ag-clusters shows characteristic peaks of face-centered-cubic (fcc) silver lattice. Ag(111) reflection is the most intensive peak in the X-ray pattern of randomly oriented poly-crystalline Ag. To collaborate the theory of melting and plastic deformation due to cluster-surface collisions we carried out annealing experiments using in situ high-temperature GIXD. Figure 12.7 shows diffractometry measurements of the Ag(111) reflection at different temperatures [39]. While heating up the samples a peak shift caused by thermal expansion is observed. Above a certain temperature no crystalline silver is found anymore, indicating cluster melting. Further, no traces of Ag are found after annealing and with the sample cooled to room temperature (RT) which indicates that evaporation took place. The melting of the samples occurs in the temperature range between 853 and 883 K depending on the cluster size (Fig. 12.8), while the melting point of bulk silver is 1,233 K. The cluster sizes are obtained from GIXD measurements. The very small melting point of samples is because of the small grain size.

Theoretical considerations about the size effect on the melting point suggest a depression of the melting point for the measured clusters sizes. The decrease of the melting point ΔT for small crystallites of radius r can be calculated from the Gibbs-Thomson equation which follows from thermodynamic principles [39]

$$\Delta T = \frac{4\sigma M T_{\text{bulk}}}{\Delta H_{\text{m}} r \rho} \qquad (12.7)$$

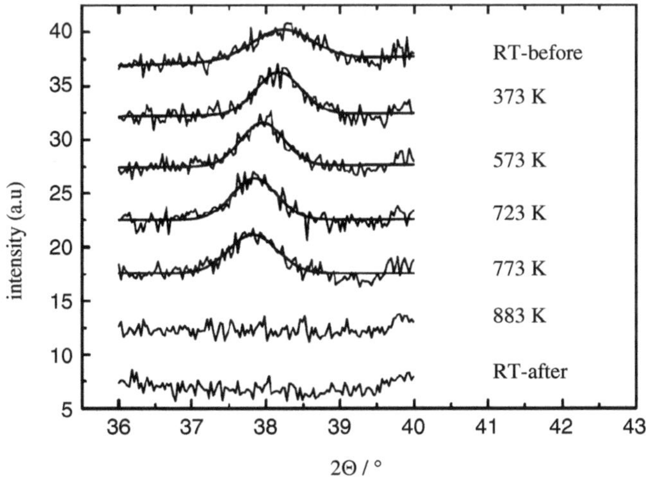

Fig. 12.7 Examples of in situ GIXD measurements during annealing of a sample with 15 nm cluster size [39]

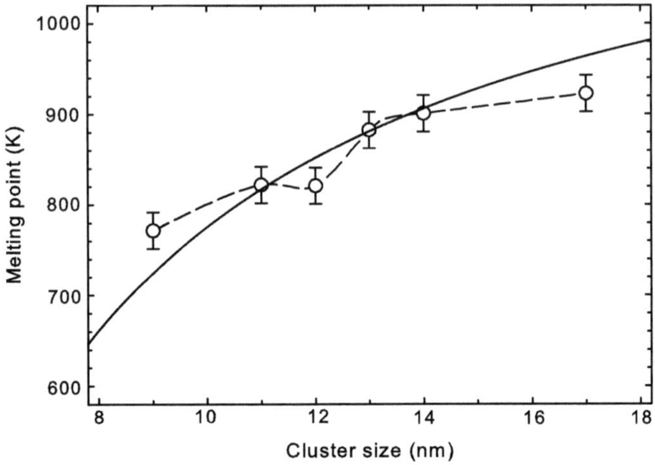

Fig. 12.8 Melting temperature versus cluster size [39]

with $\sigma = 1.02$ J/m^2 the surface energy, $M = 107.9$ g/mol the molar mass, $T_{\text{bulk}} =$ 1,233 K the melting point of the bulk material, $\Delta H_{\text{m}} = 11.3$ kJ/mol the melting enthalpy and $\rho = 10.5$ g/cm^3 the density of the material. Our experimental results confirm the theoretical considerations. The good agreement between theoretically calculated and the experimentally observed melting points also demonstrates the capability of GIXD technique as a powerful tool for cluster size measurements.

The width of the diffraction peak (Fig. 12.7) is related to the cluster or domain size of the sample. Qualitatively, a larger peak width corresponds to a smaller domain size. For a quantitative analysis, the peak profile is described by a Stokes Fourier series. The Fourier coefficients $F(L)$ of the physical line profile contain information about particle size. In the traditional Warren-Averbach plot, the linear part of the graph $\ln F(L)/L$ versus L then allows for the domain size determination (see [40]). Figure 12.9 shows results of diffractometry measurements for the Ag(111) reflection at different cluster sizes. The measured lattice constants change with Ag nanocluster sizes: it first shows an increase from 9 nm to 12 nm size followed by a decrease and already reaching at 17 nm the lattice constant of bulk silver ($a_0 = 0.4086$ nm). For all investigated clusters strains due to residual tensile stresses were observed. It is worth noticing that strain is the result of lattice disorders and structural disorder.

12.6 Rapid-Thermal Annealing (RTA) of Deposited Cluster Films

Mass-selected silver cluster ions were deposited on silicon wafers with a native oxide (SiO_2) layer. The deposited films were examined by a scanning electron microscope (SEM) with energy dispersive X-ray analysis (EDX). The SEM was

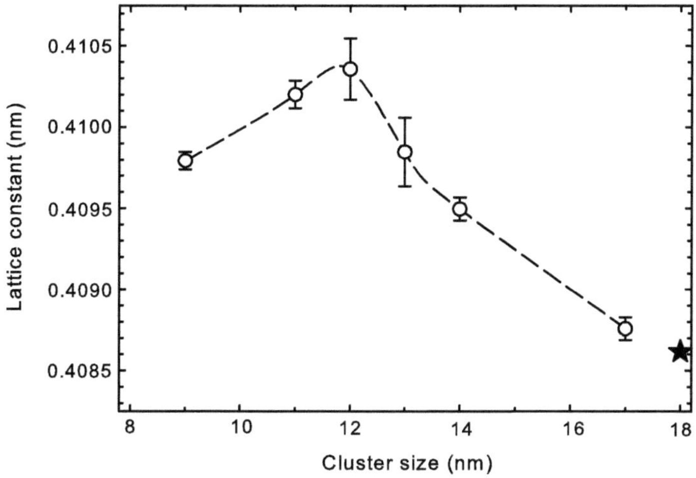

Fig. 12.9 Lattice constant versus cluster size (see text) [39]. *Star*: lattice constant of bulk silver

operated at an electron acceleration voltage of 10 kV. In order to get information about the height of the clusters, the films were also examined by AFM in tapping mode at a resonance frequency of about 428 kHz under ambient conditions. One of the deposited films was further treated by rapid-thermal annealing (RTA) at temperatures of 473, 673, 873, and 1,073 K for 3 min each in a nitrogen environment. The RTA chamber was first evacuated by a turbo molecular pump to a pressure of about 2×10^{-6} mbar and then pure (99.99%) nitrogen gas was introduced into the chamber. The annealed samples were examined by SEM/EDX [41,42].

Figure 12.10a shows the morphology of mass-selected clusters deposited on the Si substrates. The film was deposited for 6 min on the Si substrates without removing the native oxide chemically. The morphology shows irregular islands or aggregates with varied size distributions. The average size, that is, mean diameters of the aggregates, is found to be approximately 35 nm. Obviously the size of supported aggregates is much bigger than the size of incident mass-selected nanoclusters (5.6 nm). This means that nanoclusters once incident on a surface merge to form bigger agglomerates as compared to individual nanoclusters. In the following, hence, the term nanocluster will refer to an individual cluster and their assembly will be termed as aggregates or islands.

Figures 12.10b–e show the SEM images of the mass-selected Ag-nanocluster film (deposition time 6 min) after it underwent sequential rapid-thermal annealing at temperatures of 473, 673, 873, and 1,073 K for 3 min in a N_2 environment; the EDX spectrum of the sample annealed at 1,073 K is shown in Fig. 12.10f. At the annealing temperature of 473 K, the surface morphology does not change appreciably as compared to the as-deposited film; only the number density is reduced and some aggregates are joined to form bigger shapes. At 673 K, a closer look reveals that the features have a tendency to become spherical and the sizes of the features is also reduced. During annealing at 673 K, the silver nanoclusters presumably reach

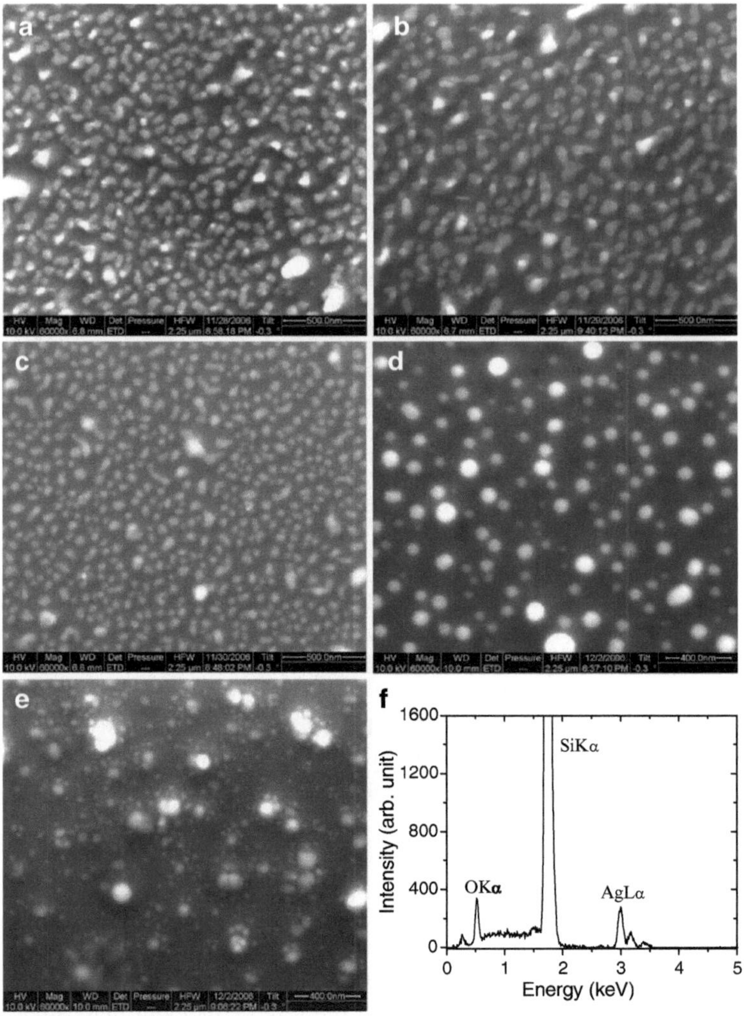

Fig. 12.10 Morphology of mass-selected silver nanosize clusters of 5.6 nm sizes of (**a**) as-deposited and of the same sample after rapid-thermal annealing at (**b**) 473 K, (**c**) 673 K, (**d**) 873 K, and (**e**) 1,073 K in a N_2 environment for 3 min in each case. SEM operating voltage 10 kV, magnification 60,000. (**f**) EDX spectrum of the film annealed at 1,073 K, showing lines from oxygen, silicon, and silver [41,42]

a molten liquid state. To minimize the free surface energy such a liquid phase must take a spherical shape which is particularly evident from Fig. 12.10d. We know that the melting point of bulk silver is 1,233 K. The present results again show that Ag in nanocluster phase has a much lower melting temperature compared to its bulk phase. Annealing at 873 K leads to spherical clusters of even bigger sizes than at 673 K (see Fig. 12.10c). Finally, at the annealing temperature of 1,073 K, the cluster

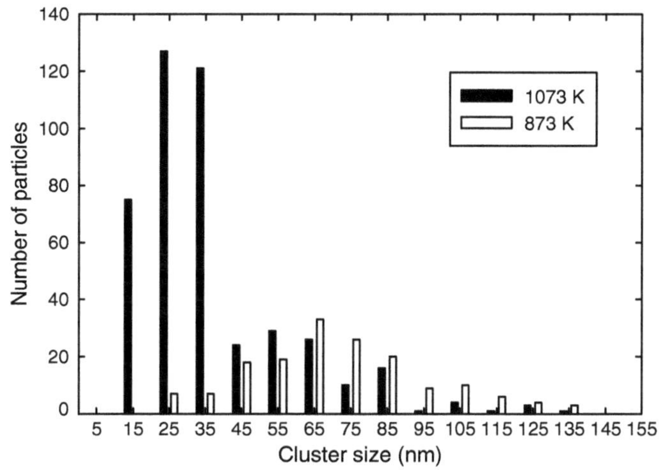

Fig. 12.11 Histogram of mass-selected silver nanosize clusters of 5.6 nm size after rapid-thermal annealing (RTA) to 873 and 1,073 K in a N_2 environment for 3 min in each case

sizes are again reduced. This time the reduction in the cluster sizes and the increase in particle numbers are due to fragmentation of a large cluster into a few smaller clusters (Fig. 12.11).

In the process of nanocluster deposition on a surface, three physical phenomena arise [43], namely, deposition of nanoclusters, diffusion of nanoclusters on the surface and interaction of the deposited nanoclusters. Initially when nanoclusters are incident on a substrate, surface coverage is low, and gradually with time, the nanoclusters diffuse on the surface to form islands or aggregates. It is likely that Ag clusters form a continuous film with the distribution of nanocluster or aggregate sizes after possible restructuring due to diffusion and interaction of the nanoclusters.

In our case of low-energy impact on a surface the deposited nanoclusters do not embed into the substrate, and after deposition they can still move over the substrate surface. Therefore, at small nanocluster beam intensity, the character of nanocluster attachment to a surface is similar to the formation of fractal structures when solid particles attach to a surface and can move over it. This mechanism of joining of solid particles on a surface, in accordance with the DLA model (diffusion limited aggregation) [44–46] or the DDA model (deposition diffusion aggregation) [47], accounts for diffusive motion of particles on the surface. According to the DLA model, a cluster is assembled by adding individual particles to it. A definite number of particles are initially introduced into the formation region and the particles coalesce when they collide with each other. This will initially result in the appearance of a larger number of smaller nanoclusters, and subsequent collisions will lead to aggregation and formation of large agglomerates. In fact, there is always a certain probability of joining two adjacent particles, resulting in a decrease in the diffusion coefficient of a grown aggregate on the surface. As far as the DDA model [11] is concerned, a highly porous thick film is formed after diffusion and coalescence of granular nanostructures that move on a substrate in a random manner. In considering the real character of particle joining, we note that the reconstruction of a joined nanocluster proceeds slowly after their contact and this reconstruction is concerned with a region near the particle surface. In contrast to liquid clusters which merge initially in a single drop and finally form a compact film on the surface, in the case of solid particles a porous film is formed and deposited solid clusters conserve much of their individuality. In the present case the film has a typical pore size of the order of the cluster size, quite in contrast to the fractal structure of a film at low intensities of the nanocluster beam, where the typical pore size exceeds the nanocluster size [41, 42].

12.7 Evaporation of Clusters

Heating of a deposited film leads to its partial evaporation, and the analysis of this process allows to determine the parameters for cluster evaporation at high temperature. In the analysis, we assume the surface film to consist of individual clusters, that is, that the surface of cluster contacts to be small compared to the total area of the cluster surface. In addition, the total area of contacts between clusters and the surface is also relatively small. We can then consider the cluster behavior at high temperature in the framework of the liquid drop model [31], and the binding energy of atoms in clusters may be determined from experimental results.

Indeed, in the framework of the liquid drop model for the cluster, taking a liquid cluster to consist of $n \gg 1$ atoms (and, hence, its surface energy is small), we define the total binding energy E_b in terms of both the volume and the surface cluster energy as [31]

$$E_b = \epsilon_0 n - A n^{2/3}. \tag{12.8}$$

In the case of silver, the specific binding energy of bulk silver is $\epsilon_0 = 2.87$ eV, and the specific surface energy is $A = 2$ eV [31]. The rate constant of atom attachment to a cluster is given by (12.4), and with a Wigner–Seitz radius for liquid silver of $r_W = 0.166$ nm and for $T = 600$ K we get $k_0 = 7.4 \times 10^{12}$ cm^3/s. The rate of cluster evaporation $\nu_{ev}(T)$ is

$$\nu_{ev}(T) = k_0 n^{2/3} N_{sat}(T) \exp(\Delta\epsilon / T n^{1/3}) \tag{12.9}$$

where $N_{sat}(T) \propto \exp(-\epsilon_0/T)$ is the number density of atoms for saturated vapor, and where $\Delta\epsilon = 2A/3$ accounts for a decrease in the atom binding energy due to surface energy.

We apply the above formulas to experimental conditions of evaporation of silver and silver clusters. In particular, at temperatures $T = 600$, 700, and 800 K, the saturated number density of atoms N_{sat} is, respectively, equal to 15 cm^{-3},

1.3×10^4 cm^{-3}, and 1.2×10^7 cm^{-3}, respectively. From this, we have the balance equation for a decrease in the cluster size due to cluster evaporation

$$\frac{dn}{dt} = -\nu_{ev}(T). \tag{12.10}$$

Its solution gives the total time τ_0 of cluster evaporation

$$\tau_0 = \frac{9Tn^{2/3}}{2Ak_0 N_{sat}(T)} \exp\left(-\frac{2A}{3Tn^{1/3}}\right). \tag{12.11}$$

In particular, for the cluster radius $r = 1$ nm, at temperatures of $T = 600$, 700, and 800 K, we have evaporation times of $\tau_0 = 2 \times 10^5$ h, 1,000 h, and 0.8 h. A twofold increase in the cluster radius leads to an increase in τ_0 by an order of magnitude. We also evaluate the cluster evaporation time at $T = 873$ K and 1,073 K for the cluster radius $R = 25$ nm according to the data in Fig. 12.10 ($n = 3 \times 10^6$). In this case, the exponential in (12.11) is unity, $N_{sat}(873\text{K}) = 6 \times 10^7$ cm^{-3}, $N_{sat}(1{,}073K) = 7 \times 10^{10}$ cm^{-3}, and $k_0 = 1 \times 10^{11}$ cm^3/s. Then (12.11) gives the total evaporation time to be about 2 h at $T = 1{,}073$ K and three orders of magnitude higher at $T = 873$ K. This means that under the experimental conditions, several percent of silver should evaporate at $T = 1{,}073$ K, and the evaporation process is not significant at $T = 873\ K$.

12.8 Conclusions

In conclusion, magnetron plasmas are efficient tools for production of nanosize particles (clusters) and deposition of nanosize particle layers on surfaces. Deposited nanosize clusters have many interesting features which make them of interest for scientific and technological purposes.

Acknowledgments This work was supported by the Deutsche Forschungsgemeinschaft DFG through SFB/TRR 24 and by RFBR, Russia (09-02-00108a). We like to thank Prof. C.A. Helm, Dr. H. Wulff, Prof. T.K. Chini, Dr. D. Datta, O. Ivanova, P.V. Khastanov, Dr. A. Majumdar, Dr. M. Quaas, Dr. I. Shyjumon, and Dr. V. Stranak.

References

1. A. Melzer, J. Goree, in *Low Temperature Plasmas*, vol. 1, ed. by R. Hippler, H. Kersten, M. Schmidt, K.H. Schoenbach (Wiley-VCH, Berlin, 2008), p. 129
2. R. Hippler, H. Kersten, in *Low Temperature Plasmas*, vol. 2, ed. by R. Hippler, H. Kersten, M. Schmidt, K.H. Schoenbach (Wiley-VCH, Berlin, 2008), p. 787
3. D.L. Smith, *Thin Film Deposition: Principles and Practice* (McGraw-Hill, New York, 1995)

4. K. Ellmer, in *Low Temperature Plasmas*, vol. 2, ed. by R. Hippler, H. Kersten, M. Schmidt, K.H. Schoenbach (Wiley-VCH, Berlin, 2008), p. 675
5. H. Gleiter, Nanostruct. Mater. **1**, 1 (1992)
6. H. Gleiter, Nanostruct. Mater. **6**, 3 (1995)
7. S.Y. Liau, D.C. Read, W.J. Pugh, J.R. Furr, A.D. Russell, Lett. Appl. Microbiol. **25**, 279 (1997)
8. A. Gupta, S. Silver. Nat. Biotechnol. **16**, 888 (1998)
9. K. Nomiya, A. Yoshizawa, K. Tsukagoshi, N.C. Kasuga, S. Hirakawa, J. Watanabe, J. Inorg. Biochem. **98**, 46 (2004)
10. J.R. Morones, J.L. Elechiguerra, A. Camacho, K. Holt, J.B. Kouri, J.P. Ramirez, M.J. Yacaman, Nanotechnology **16**, 2346 (2005)
11. A. Perez, P. Melinon, V. Dupuis, P. Jensen, B. Prevel, J. Tuaillon, L. Bardotti, C. Martet, M. Treilleux, M. Broyer, M. Pellarin, J.L. Vaille, B. Palpant, J. Lerme, J. Phys. D **30**, 709 (1997)
12. C. Binns, Surf. Sci. Rep. **44**, 1 (2001)
13. K. Shintani, Y. Taniguchi, S. Kameoka, J. Appl. Phys. **95**, 8207 (2004)
14. T.H. Lee, C.R. Hladik, R.M. Dickson, Appl. Phys. Lett. **84**, 118 (2004)
15. S.M. Rossnagel, IBM J. Res. Dev. **43**, 163 (1999)
16. S.M. Rossnagel, H.R. Kaufman, J. Vac. Sci. Technol. A **5**, 88 (1987)
17. F.M. Penning, Physica **3**, 873 (1936)
18. E.K. Kay, J. Appl. Phys. **34**, 760 (1963)
19. P.V. Kashtanov, B.M. Smirnov, R. Hippler, Phys. Usp. **50**, 455 (2007)
20. H. Haberland, B. von Issendorff, J. Yufeng, T. Kolar, Phys. Rev. Lett. **69**, 3212 (1992)
21. H. Haberland et al., J. Vac. Sci. Technol. A **10**, 3266 (1992)
22. H. Haberland et al., Mater. Sci. Eng. B **19**, 31 (1993)
23. H. Haberland, B. von Issendorff, J. Yufeng, T. Kolar, G. Thanner, Z. Phys. D **26**, 8 (1993)
24. H. Haberland, M. Mall, M. Moseler, Y. Qiang, T. Reiners, Y. Thurner, J. Vac. Sci. Technol. A **12**, 2925 (1994)
25. G. Pribil, Z. Hubicka, R.J. Soukup, N.J. Janno, J. Vac. Sci. Technol. A **19**, 1571 (2001)
26. R. Hippler, in *Low Temperature Plasmas*, vol. 1, ed. by R. Hippler, H. Kersten, M. Schmidt, K.H. Schoenbach (Wiley-VCH, Berlin, 2008), p. 71
27. R. Hippler, S. Wrehde, V. Stranak, O. Zhigalov, H. Steffen, M. Tichy, M. Quaas, H.Wulff, Contrib. Plasma Phys. **45**, 1 (2005)
28. B.M. Smirnov, Phys. Usp. **46**, 589 (1994)
29. B.M. Smirnov, A.Yu. Strizhev. Phys. Scripta **49**, 615 (1994)
30. B.M. Smirnov, I. Shyjumon, R. Hippler, Phys. Rev. E **77**, 066402 (2007)
31. B.M. Smirnov, *Clusters and Small Particles in Gases and Plasmas* (Springer, New York, 1999)
32. B.M. Smirnov, *Principles of Statistical Physics* (Wiley-VCH, Berlin, 2006)
33. I. Shyjumon, M. Gopinadhan, C.A. Helm, B.M. Smirnov, R. Hippler, Thin Solid Films **500**, 41 (2006)
34. B.M. Smirnov, Plasma Chem. Plasma Process. **13**, 673 (1993)
35. B.M. Smirnov, Phys. Usp. **36**, 933 (1993)
36. B.M. Smirnov, R.S. Berry, *Phase Transitions of Simple Systems* (Springer, Berlin, 2007)
37. B.M. Smirnov, I. Shyjumon, R. Hippler Phys. Scripta **73**, 288 (2006)
38. V.N. Kondrat'ev, *Rate Constants of Gas-Phase Reactions* (Nauka, Moscow, 1971) (in Russian)
39. I. Shyjumon, M. Gopinadhan, O. Ivanova, M. Quaas, H. Wulff, C.A. Helm, R. Hippler, Eur. Phys. J. D **37**, 409 (2006)
40. H. Wulff, H. Steffen, in: *Low Temperature Plasmas*, vol. 1, ed. by R. Hippler, H. Kersten, M. Schmidt, K.H. Schoenbach (Wiley-VCH, Berlin, 2008), p. 329
41. S.R. Bhattacharyya, D. Datta, I. Shyjumon, B.M. Smirnov, T.K. Chini, D. Ghose, R. Hippler, J. Phys. D: Appl. Phys. **42**, 035306 (2009)
42. S.R. Bhattacharyya, T.K. Chini, D. Datta, R. Hippler, I. Shyjumon, B.M. Smirnov, JETP **107**, 1009 (2008)
43. P. Melinon et al., Int. J. Mod. Phys. B **9**, 339 (1995)
44. T.A. Witten, I.M. Sander, Phys. Rev. Lett. **47**, 1400 (1981)
45. P. Meakin, Phys. Rev. A **27**, 604 (1983)
46. M. Sahimi, M. McKarnin, I. Nordahl, M. Tirrell, Phys. Rev. A **32**, 590 (1985)
47. P. Jensen, A.L. Barabási, H. Larralde, S. Havlin, H.E. Stanley, Phys. Rev. B **50**, 15316 (1994)

Chapter 13
Kinetic and Diagnostic Studies of Molecular Plasmas Using Laser Absorption Techniques

Jürgen Röpcke, Richard Engeln, Daan Schram, Antoine Rousseau, and Paul B. Davies

Abstract Within the last decade, mid-infrared absorption spectroscopy between 3 and 20 μm – known as infrared laser absorption spectroscopy (IRLAS) and based on tunable semiconductor lasers, namely lead salt diode lasers, often called tunable diode lasers (TDLs), and quantum cascade lasers (QCLs) – has progressed considerably as a powerful diagnostic technique for in situ studies of the fundamental physics and chemistry of molecular plasmas. The increasing interest in processing plasmas containing hydrocarbons, fluorocarbons, and organosilicon compounds has led to further applications of IRLAS because most of these compounds and their decomposition products are infrared active. IRLAS provides a means of determining the absolute concentrations of the ground states of stable and transient molecular species, which is of particular importance for the investigation of reaction kinetics. Information about gas temperature and population densities can also be derived from IRLAS measurements. A variety of free radicals and molecular ions have been detected, especially using TDLs. Since plasmas with molecular feed gases are used in many applications such as thin film deposition, semiconductor processing, surface activation and cleaning, and materials and waste treatment, this has stimulated the adaptation of infrared spectroscopic techniques to industrial requirements. The recent development of QCLs offers an attractive new option for the monitoring and control of industrial plasma processes as well as for highly time-resolved studies on the kinetics of plasma processes.

J. Röpcke (✉)
INP-Greifswald, Felix-Hausdorff-Str. 2, 17489 Greifswald, Germany
e-mail: roepcke@inp-greifswald.de

R. Engeln and D. Schram
Eindhoven University of Technology, P.O. Box 513, 5600 MB, Eindhoven, The Netherlands
e-mail: r.a.h.engeln@tue.nl; d.c.schram@tue.nl

A. Rousseau
LPP, Ecole Polytechnique, CNRS, 91128 Palaiseau Cedex, France
e-mail: Antoine.Rousseau@lpp.polytechnique.fr

P.B. Davies
University of Cambridge, Cambridge CB2 1EW, UK
e-mail: pbd2@cam.ac.uk

M. Bonitz et al. (eds.), *Introduction to Complex Plasmas*, Springer Series on Atomic, Optical, and Plasma Physics 59, DOI 10.1007/978-3-642-10592-0_13,

The aim of the present contribution is threefold (1) to review recent achievements in our understanding of molecular phenomena in plasmas including interactions with solid surfaces, (2) to report on selected studies of the spectroscopic properties and kinetic behavior of radicals, and (3) to review new applications of QCLs and to describe the current status of advanced instrumentation for QCLAS in the mid-infrared.

13.1 Introduction

Low-pressure, nonequilibrium molecular plasmas are of increasing interest not only in fundamental research but also in plasma and vapor processing technology. Molecular plasmas and vapors are used in a variety of applications such as thin film deposition, semiconductor processing, surface activation and cleaning, and in materials and waste treatment. The investigation of plasma physics and chemistry in situ requires detailed knowledge of plasma parameters, which can be obtained by appropriate diagnostic techniques. The need for a better scientific understanding of plasma physics and chemistry has stimulated the improvement of established diagnostic techniques and the introduction of new ones.

The methods of absorption spectroscopy are of great importance in plasma diagnostics because they provide a means of determining the population densities of species in both ground and excited states. The spectral line positions provide species identification while line profiles are often connected with gas temperature and relative intensities provide information about population densities and thus give information on the excitation dynamics. An important advantage of absorption spectroscopy (AS) over optical emission spectroscopy (OES) methods is that only relative intensities need to be measured to determine absolute concentrations, avoiding the problems of complete instrument calibration inherent in the OES methods. Absorption spectroscopy has been applied across the spectrum from the vacuum ultraviolet (VUV) to the far infrared (FIR). A wide variety of light sources, dispersive elements, detectors, and data acquisition methods can be used. The classic absorption spectroscopy experiment uses a continuous light source in combination with a narrow bandwidth frequency filter, in particular a spectrograph, and a detector suitable for the spectral range of interest. Besides continuous emission lamps (e.g., the Xe lamp for the VIS and NIR, and the D_2 lamp for the UV), tunable narrowband light sources (e.g., tunable dye lasers, diode lasers) can be used as external light sources as well.

Two principle cases need to be distinguished (I) the measurement of the absorption of light emitted by the plasma itself (*self-absorption* or *reabsorption*) and (II) absorption of an external light source. In the case of self-absorption measurements, the light from the plasma is measured with and without a retroreflector or compared with the direct light emitted from an identical plasma. This ensures identical line profiles for the emitted and the absorbed light. This method is particularly important for checking the optical thickness of the plasma. The probing light intensity of an external light source in case (II) has to be low to avoid saturation effects of the exited states of the species under investigation.

The change in light intensity $\mathrm{d}I_\nu$ when a beam passes through a transparent homogeneous plasma zone of thickness $\mathrm{d}l$ is given by the net balance of the intensity due to absorption and emission within the layer $\mathrm{d}l$:

$$\mathrm{d}I_\nu = (\varepsilon(\nu) - \kappa(\nu)I_\nu)\mathrm{d}l, \tag{13.1}$$

where $\varepsilon(\nu)$ is the emission coefficient and $\kappa(\nu)$ is the absorption coefficient per unit length. The absorption coefficient $\kappa(\nu)$, describing the light absorption in an infinitesimally thin layer of a plasma, is given by

$$\kappa(\nu) = \sum_i N_i \sigma_i(\nu), \tag{13.2}$$

where the summation is over all absorbing species and states. N_i are the species population densities and $\sigma_i(\nu)$ are their absorption cross sections at frequency ν. The line profile P_ν is normalized by

$$\int_{\text{line}} P_\nu \mathrm{d}\nu = 1. \tag{13.3}$$

Assuming the upper level k is not populated, then the absolute value of $\kappa(\nu)$ results:

$$\int_{\text{line}} \kappa(\nu)\mathrm{d}\nu = \frac{h\nu}{c_0} N_i B_{ik}. \tag{13.4}$$

The line absorption coefficient is then given by

$$\kappa(\nu) = \frac{h\nu}{c_0} N_i(\alpha, \beta, \gamma, \ldots) B_{ik} P_\nu, \tag{13.5}$$

where $N_i(\alpha, \beta, \gamma, \ldots)$ is the population density of the ith level, which depends on different plasma parameters $(\alpha, \beta, \gamma, \ldots)$, and B_{ik} is the Einstein coefficient for the transition between the levels i and k.

In the case where (I) the external light source has much higher intensity than that of the plasma itself and (II) the absorption is continuous over the whole distance l ($\kappa(\nu)$ = constant), the absorption of radiation can be described by the Beer–Lambert law:

$$I_\nu(l) = I_\nu(0)\exp(-\kappa(\nu)l), \tag{13.6}$$

where $I_\nu(0)$ and $I_\nu(l)$ are the fluxes of the radiation entering and leaving the plasma, l is the length of the absorbing (homogeneous) plasma zone, and $\kappa(\nu)$ is the absorption coefficient. Figure 13.1 illustrates this situation [1]. It should be noted that the definition of the effective absorption length of the plasma zone and the assumption of a constant absorption coefficient over the complete distance l are two nontrivial questions in plasma spectroscopy. Inhomogeneities, anisotropies, and temperature gradients caused by local changes in the plasma can lead to serious errors.

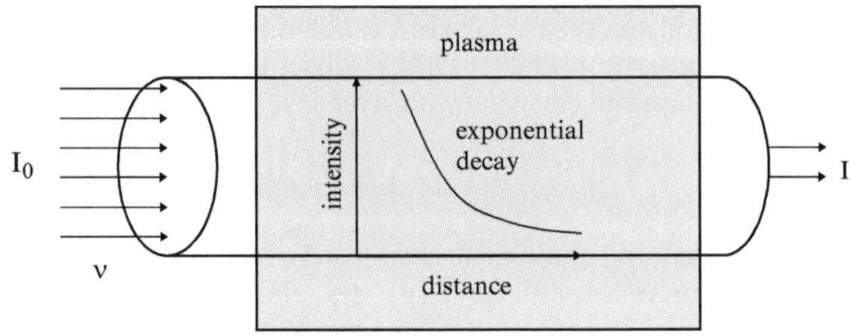

Fig. 13.1 Absorption of external radiation in a plasma according to the Beer–Lambert law [1]

With the development of tunable, narrowband light sources such as tunable dye lasers and infrared diode lasers, these have been substituted for continuous light sources in absorption spectroscopy experiments. These narrowband laser sources have the advantage of high spectral intensity, narrow bandwidth, and continuous tunability over the absorption profile.

The increasing interest in processing plasmas containing hydrocarbons, fluorocarbons, or organosilicon compounds has led to further applications of infrared absorption spectroscopy techniques because most of these compounds and their decomposition products are infrared active. Fourier transform infrared (FTIR) spectroscopy has been used for in situ studies of molecular plasmas for a number of years, but it is generally insufficiently sensitive for detecting free radicals or ions in processing plasmas. Tunable diode laser absorption spectroscopy (TDLAS) is increasingly being used in the spectral region between 3 and 20 μm for measuring the concentrations of free radicals, transient molecules, and stable products in their electronic ground states. TDLAS can also be used to measure neutral gas temperatures [2] and to investigate dissociation processes of molecular low-temperature plasmas [3–6]. The main applications of TDLAS until now have been for investigating molecules and radicals in fluorocarbon-etching plasmas [2, 5, 7], in plasmas containing hydrocarbons [6, 8–16] and nitrogen, hydrogen, and oxygen [17–20]. A wide variety of low molecular weight free radicals and molecular ions have been detected by TDLAS in purely spectroscopic studies, for example, Si_2^- [21] and SiH_3^+ [22] in silane plasmas. Most of these spectroscopic results have yet to be applied in plasma diagnostic studies.

Molecular plasmas are increasingly being used not only for basic research but also, due to their favorable properties, for materials processing technology. These fields of application have stimulated the development of infrared spectroscopic techniques for industrial requirements. To exploit the capabilities of infrared TDLAS for effective and reliable on-line plasma diagnostics and process control in research and industry, compact and transportable tunable infrared multicomponent acquisition (IRMA, TOBI) systems have been developed [17, 18]. These systems are mainly focused on (1) high-speed detection of stable and transient molecular species in plasmas under nonstationary excitation conditions and (2) sensitive (sub-ppb) trace gas detection with the aid of multipass absorption cells.

The main disadvantage of TDLAS systems, based on lead salt diode lasers, is the necessary cryogenic cooling of the lasers (and also of the detectors), because they operate at temperatures below 100 K. Systems based upon lead salt diode lasers are typically large in size and require closed cycle refrigerators and/or cryogens like liquid nitrogen. The recent development and commercial availability of quantum cascade lasers (QCLs) offers an attractive new option for infrared absorption spectroscopy.

The present chapter is intended to give an overview of recent achievements which have led to an improved understanding of phenomena in nonequilibrium molecular plasmas based on the application of infrared laser absorption spectroscopy (IRLAS) techniques. The chapter is divided into three main sections. In Sect. 13.2, special attention is devoted to recent achievements in our understanding of molecular phenomena in plasmas including interactions with solid surfaces using TDLs. Section 13.3 describes selected studies of the spectroscopic properties and kinetic behavior of radicals. Section 13.4 concerns an overview over selected new applications of QCLs in the mid-infrared including the current status of advanced spectroscopic instrumentation.

13.2 Plasma Chemistry and Reaction Kinetics

13.2.1 Studies of $Ar/H_2/N_2/O_2$ Microwave Plasmas

Low-temperature plasmas, in particular microwave and radio-frequency (RF) plasmas, have high potential for applications in plasma technology. In molecular low-temperature plasmas, the species and surface conversion is frequently governed by high degrees of dissociation of the precursor molecules and by the high amounts of chemically active transient and stable molecules present [23–25]. To gain further insight into plasma chemistry and kinetics, the study of (mainly) electron-induced plasma reactions is an important objective. These processes lead to entire series of different secondary chemical reactions involving all the source gas molecules. In all cases, the monitoring of transient or stable plasma reaction products, in particular the measurement of their ground-state concentrations, is the key to improved understanding of fundamental phenomena in molecular nonequilibrium plasmas which can in turn be applied to many other aspects of plasma processing. Radicals containing carbon and oxygen are of special interest for fundamental studies and for applications in plasma technology.

In recent years, several types of microwave discharge with different precursor gases have been at the center of interest. The most recent applications of TDLAS for plasma diagnostic purposes include studies in which many different species have been monitored under identical plasma conditions [24, 26]. This experimental data have frequently been used to model plasma chemical phenomena.

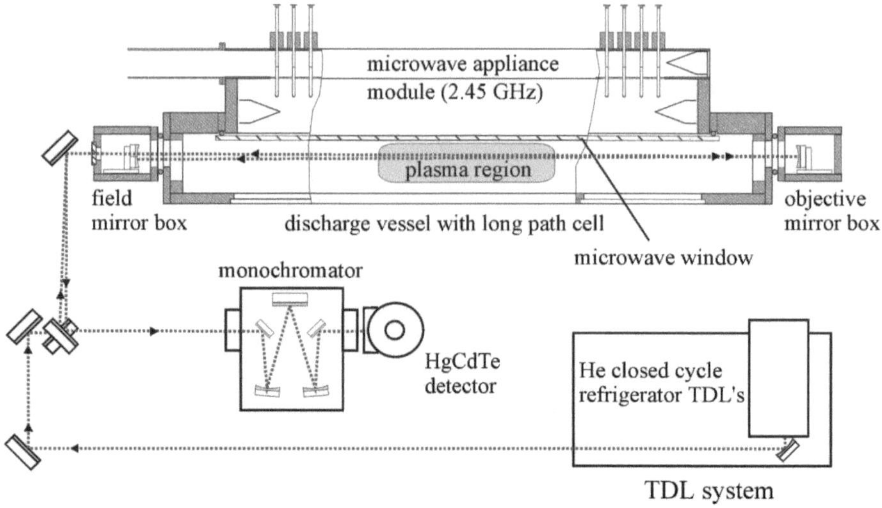

Fig. 13.2 Experimental arrangement of a planar microwave plasma reactor (*side view*) used in plasma diagnostics with a White-type multiple pass optical arrangement, and tunable diode laser (TDL) infrared source. The path of the diode laser beam is indicated by *dotted lines* [24]

Although hydrogen- and hydrocarbon-containing plasmas with admixtures of oxygen and nitrogen have been extensively studied [1, 15, 24], there is still a lack of experimental data concerning the absolute densities of radicals in these discharges. The hydroxyl radical is known to be one of the main oxidizing radicals. So far, only a few studies have been reported on absolute OH concentrations in plasmas. Mostly, they have been monitored and measured in the UV spectral region [27–29]. In the present study, quantitative measurements of OH in plasmas by means of TDLAS at 530 cm^{-1} are reported. The measurements were performed in a planar microwave reactor in $Ar/H_2/N_2/O_2$ discharges. The reactor was equipped with multipass cell optics (White cell) to increase the absorption length (Fig. 13.2). A mirror spacing of 1.5 m and 40 passes gave an effective absorption length of 60 m. The pressure was kept constant at 1.5 mbar during the experiments.

With a typical input power of 1.5 kW and 420 sccm Ar, the following gas mixture was used: (0–20) sccm H_2, 10 sccm N_2, (0–20) sccm O_2. For an Ar/H_2 plasma H_2O and – depending on the oxygen flow – NH_3 or NO turned out to be the most abundant species (Fig. 13.3). The mole fraction of water was typically in the order of 1%. Further details on the experimental setup, data acquisition, and data processing can be found elsewhere [6, 31–33].

It can be seen in Fig. 13.3 that the ammonia concentration is quite sensitive to even small amounts of oxygen in the plasma, and as its concentration decreases the NO and OH concentrations increase. This behavior can be qualitatively understood in the light of three reactions (13.7)–(13.9) which are most likely to occur [34]. NH_3 is effectively destroyed due to the impact of O and OH radicals (13.7) and (13.8). When varying the hydrogen flow in the system (Fig. 13.4) the drop in $[NH_3]$ is not

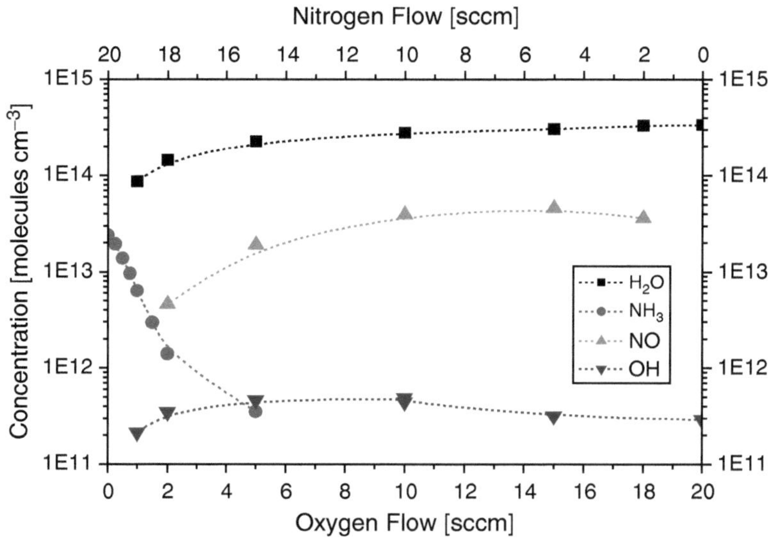

Fig. 13.3 Concentrations of the most abundant species and of OH in an $Ar/H_2/N_2/O_2$ microwave plasma (420 sccm Ar, 10 sccm H_2, (0–20) sccm N_2+O_2; 1.5 mbar). [NH_3] was below the detection threshold for oxygen flows higher than 6 sccm, $p = 1.5$ mbar, $P = 1.5$ kW [30]. It can be seen that the ammonia concentration is quite sensitive to even small amounts of oxygen in the plasma, and as its concentration decreases the NO and OH concentrations increase

so pronounced for a higher hydrogen content as for a lower one. There is still also the possibility of forming NH_3 from NH_2, which is more probable for higher H_2 flows.

$$NH_3 + O \rightarrow NH_2 + OH \tag{13.7}$$

$$NH_3 + OH \rightarrow NH_2 + H_2O \tag{13.8}$$

$$N + OH \rightarrow NO + H \tag{13.9}$$

The longer NH_3 is present in the discharge the longer OH can build up. Finally, a slight maximum of the OH concentration was found for H_2/O_2 ratios of 1:1 (Fig. 13.4). For higher oxygen flows NH_3 can no longer act as a source for OH, especially at hydrogen flows lower than 20 sccm. So the decrease of [OH] should be mainly attributed to the loss channel (3) which may also explain the ambiguous behavior of NH_3 and NO when increasing the oxygen content in the system [30].

Recently based on new experiments, it has been shown that not only gas phase reactions but also surface reactions play an important role for the formation of NO. Plasma-produced radicals may adsorb on the walls and can pick up species from the gas phase or react with each other heterogeneously. When changing the N_2/O_2 ratio the number of N and O radicals adsorbed on the surface is altered. Increasing the oxygen amount acts therefore in two directions: it prevents NH_3 formation on

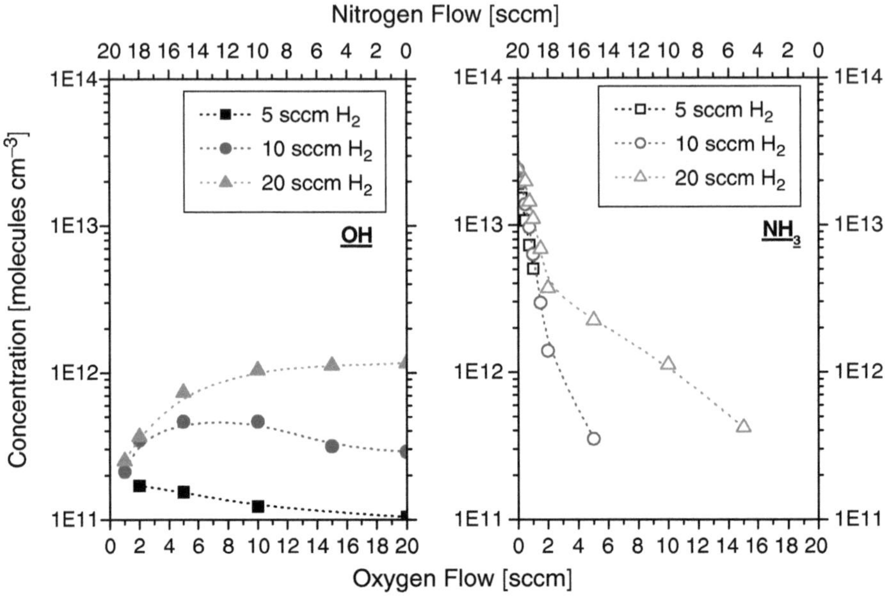

Fig. 13.4 Concentrations of OH (*left panel*) and NH_3 (*right panel*) for $Ar/H_2/N_2/O_2$ microwave plasmas (420 sccm Ar, (0–20) sccm H_2, (0–20) sccm $N_2 + O_2$; 1.5 mbar) plasmas [30]. The decrease of [OH] is mainly attributed to the loss channel (3). The concentration of NH_3 is reduced via channels (1, 2)

the surface and destroys the NH_3 present in the gas phase. On the other hand, NO can be formed more efficiently and to a substantial extent via surface reactions (see Sect. 13.2.2) [35, 36].

13.2.2 On the Importance of Surface Association to the Formation of Molecules in a Recombining N_2/O_2 Plasma

The interaction of reactive species, like atomic or molecular radicals, with a surface is a very general phenomenon. In plasma processing it is this interaction that leads to the modification of surfaces, that is, deposition or etching [37–40]. But also in the plasma-assisted conversion of gases the surface plays an important role [41–43]. In interstellar space, it has been recognized that ice or dust particles act as third body in association processes in which not only hydrogen molecules are formed, but also larger molecules [44]. How to control the very fast erosion of the carbon tiles due to the plasma–surface interaction in the divertor region of Tokamaks is one of the most important research questions still to be answered in thermonuclear fusion research [45]. Here, we unravel the contribution of surface related processes to the formation of new molecules in a low-pressure, recombining plasma. We report on the study of plasmas created from mixtures of N_2 and O_2 admixed to an argon plasma expanding from a thermal plasma source (see also [46,47]). In this plasma, only a small number of different types of new molecules can be created, and the gas phase chemistry is well known. This enables us to describe the surface processes with a relatively small set of surface reactions. The reported values of the wall loss probabilities of O and N atoms tend to vary strongly with (1) type of radical, (2) surface material, (3) wall temperature, (4) plasma composition, and (5) radical fluxes toward the surfaces. This implies that the formation of new molecules from these plasma-produced radicals also varies strongly with these effects. The surface material determines the typical bond strengths of the radicals to the surface. The contribution of plasma–surface interactions of a remote, recombining, low-pressure plasma ($p = 20$ Pa), produced by a cascaded arc, was studied (Figs. 13.5 and 13.6) in [35]. The electron temperature is low ($T_e = 0.3$ eV) which implies that no direct electron impact excitation, dissociation, or ionization occurs. The molecules that are admixed to the argon expansion are efficiently dissociated by means of charge exchange and subsequent dissociative recombination [48,49].

Association of radicals into new molecules presumably occurs mainly at the walls of the reactor. Mixtures of N_2 and O_2 are admixed to the argon expansion, and the resulting steady-state mole fractions of the stable molecules in the system (Ar, N_2, O_2, NO, N_2O, and NO_2) are measured by means of a combination of mass spectrometry and tunable diode laser absorption spectroscopy. A simulation of our plasma system has been developed in CHEMKIN, a commercially available chemical kinetics simulation program. With this type of simulation, it is possible to concentrate on the kinetics in the plasma and or on the kinetics of plasma–surface

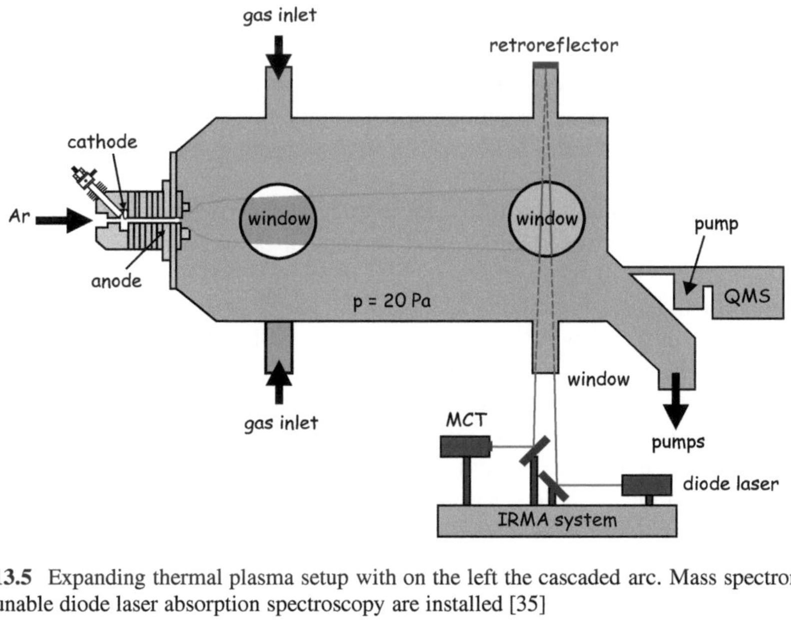

Fig. 13.5 Expanding thermal plasma setup with on the left the cascaded arc. Mass spectrometry and tunable diode laser absorption spectroscopy are installed [35]

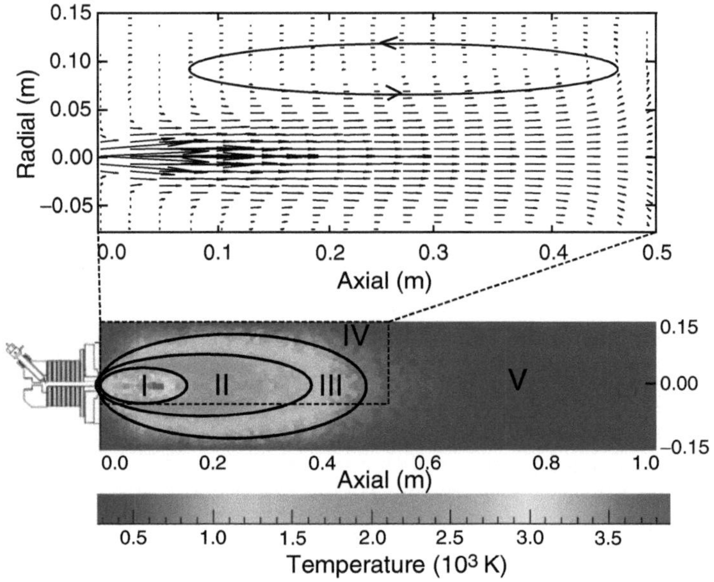

Fig. 13.6 The calculated temperature distribution in the reactor and the division of the reactor into five different regions: (I) the supersonic expansion; (II) the subsonic expansion; (III) the hot part of the background; (IV) the cold background, including the walls of the reactor; and (V) the background downstream in the reactor, where the measurements are performed. The calculated velocity profile of the first half of the reactor is also sketched. Note that the IR measurements are performed at an axial position of 0.8 m [35]

interactions. Furthermore, the calculations can be performed very fast. Although the CHEMKIN simulation consists of zero-dimensional components, flow effects in the system are estimated and implemented. By comparing the results of these simulations with the measurements, the contributions of surface related processes are unraveled. Furthermore, indications on the sticking coefficients and activation energies are obtained.

Although the simulation is based on a small set of parameters to describe the surface reactions, it gives a fairly good agreement between measured and calculated mole fractions of the stable molecules in the plasma (Fig. 13.7). The kinetics can be summarized in four steps (1) charge exchange and subsequent dissociative recombination initiated by Ar^+ ions; (2) dissociation of feedstock gases and surface-produced species with N and O atoms; (3) sticking of N, O, and NO to the walls of the reactor; and (4) formation and desorption of species at the walls of the reactor. It can be deduced from the simulation that the surface related processes contribute greatly to the total kinetics in the plasma. Without surface interactions taken into account, the calculated mole fractions of NO_2 and N_2O are three to four orders of magnitude lower than the measured ones. When the surface interactions are taken into account, the contribution to the total kinetics can be determined. The rate of surface production is up to a factor of 2 bigger than the primary rate of production

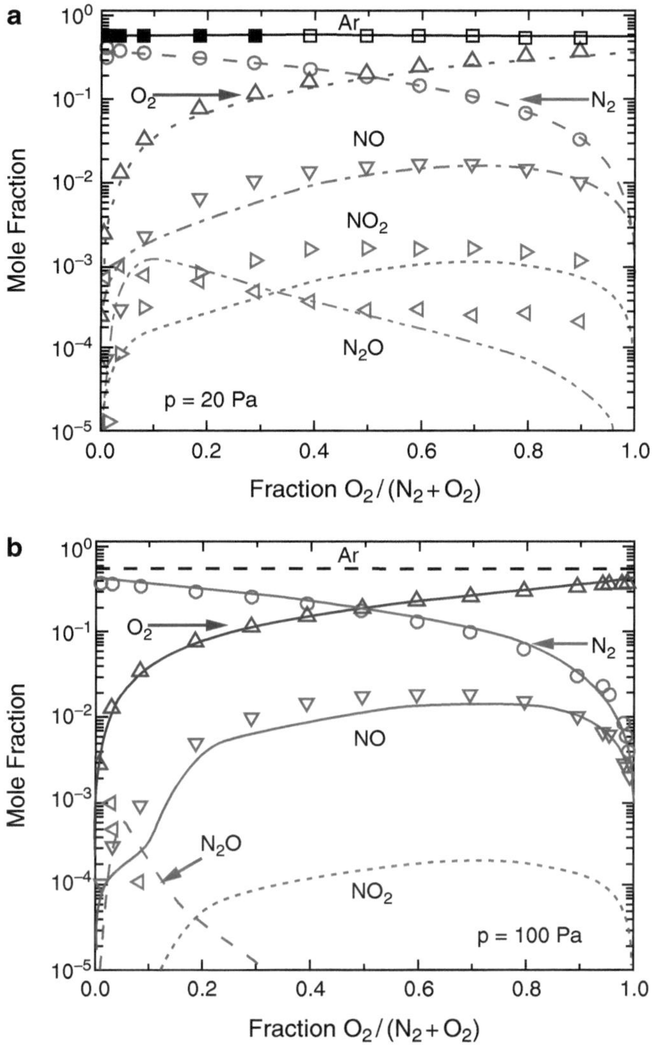

Fig. 13.7 *The solid lines* (Ar, N_2, O_2, and NO), the *dashed lines* (N_2O), and the *dotted lines* (NO_2) represent the results of the simulation in which surface interactions are taken into account. The *symbols* represent the measured mole fractions of the different types of molecules. A flow of 3,000 sccm Ar and 1,800 sccm of mixtures of N_2 and O_2 is injected at two different pressures: (**a**) $p = 20$ Pa and (**b**) $p = 100$ Pa [35]

of radicals by charge exchange with Ar^+ ions, followed by dissociative recombination. The rate of surface production of NO_2 is significantly higher than of N_2O, even though the observed abundances of these types of molecules is in the same order of magnitude (Fig. 13.7). This illustrates that follow-up reactions between NO_2 and gas phase radicals are very important in the production of N_2O. It was therefore expected and confirmed by measurements and simulations that the same experiment performed at a lower pressure reflects more clearly the ratio of surface-produced flows, particularly of NO_2 and N_2O. Finally, the best agreement between calculations and measurements was found for low activation energies and desorption energies of the surface processes and low calculated surface coverage. These results indicate that the surface processes take place on a very mobile surface layer [35]. These results reflects in general the value of radical studies to deduct how important surface reactions are, then the radical densities are smaller as they are used at the surface to produce new molecules.

13.3 Kinetic Studies and Molecular Spectroscopy of Radicals

13.3.1 Line Strengths and Transition Dipole Moment of CH_3

13.3.1.1 The ν_2 Fundamental Band

This section describes a new measurement of μ_2 for the ν_2 fundamental band of the methyl radical in order to resolve the differences between earlier experimentally measured values and between experiment and theory. The method used for determining the absolute methyl radical concentrations was the same as that used by Yamada and Hirota [8]. However, integrated intensities and many more methyl radical lines were used. Furthermore, the kinetic conditions were more precisely specified and the temperature determined more exactly. The resulting value of μ_2 is now in much better agreement with theory.

The methyl radical has no electric dipole allowed rotational transitions because of its D_{3h} symmetry and so IR spectroscopy is one of the few suitable methods for its detection. The determination of methyl radical concentrations in terrestrial and astronomical sources using IR spectroscopy relies on the availability of accurate line strengths and transition dipole moments. The ν_2 band of CH_3 is the strongest of its IR active fundamentals and particularly useful for quantitative measurements. The need for a more accurate and precise value of μ_2 has been highlighted by the measurements of CH_3 in the atmospheres of Saturn [50], Neptune [51], and in the interstellar medium [52].

The experimental setup of the planar microwave plasma reactor with the optical arrangements used for the methyl transition dipole moment study is comparable to that shown in Fig. 13.2. Details of the diode laser spectrometer, IRMA, and discharge absorption cell have been reported elsewhere [40,53]. The methyl radical was produced in mixtures of tertiary butyl peroxide ($[(CH_3)_3CO]_2$) and argon at total pressure of 1 mbar. Two kinds of experiments were performed (a) time-dependent measurement of the decay of the absorption coefficient when the discharge was turned off, to obtain absolute methyl concentrations and (b) measurements of the absorption coefficients of different rovibronic lines. In total, ten lines were studied

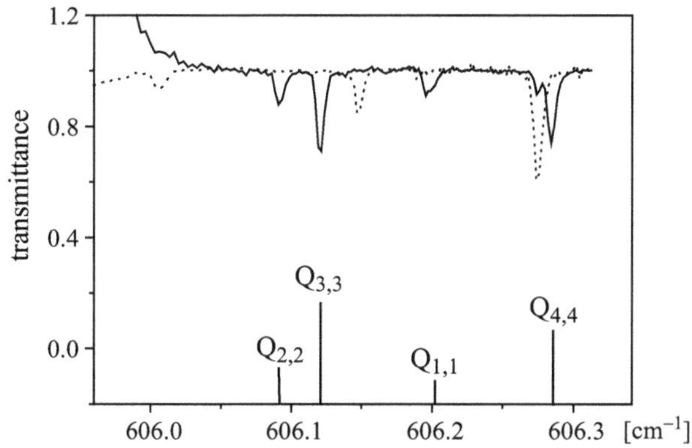

Fig. 13.8 Survey spectrum showing several ($J = K$) Q-branch lines of the ν_2 fundamental of the CH_3 free radical around the band origin. The spectrum represented by the *dashed line* is a calibration spectrum from N_2O and CO_2 [54]

in the fundamental band, seven in the first hot band, and one from the second hot band. A survey spectrum of the Q-branch region of the ν_2 fundamental band is shown in Fig. 13.8.

To derive accurate line strengths and the transition dipole moment, it is necessary to obtain the absolute concentration of the methyl radical and its temperature in the discharge. The decay method was the experimental approach for methyl radical concentration measurements. The plasma was switched on and off for periods of 10 s and the decay of the methyl radical signal measured during the off period with ms time resolution [55, 56]. The absolute concentration was obtained from the decay of the integrated absorption coefficient and the recombination rate constant. It is well known from numerous kinetic studies that the main loss channel under the conditions used here is self-recombination via a three body reaction. Hence by measuring the integrated absorption coefficient as a function of time and knowing the value of the recombination rate constant, the absolute concentration of the methyl radical can be obtained.

The rate constant k_1 for the self-recombination reaction of methyl radicals has been extensively investigated in experimental and theoretical work [53, 57–63]. The selected value for k_1 was based on the compilation of Baulch et al. and was appropriate for the specific temperature and argon concentration [54, 64]. The translational, rotational, and vibrational temperatures of the methyl radical were measured. A near similarity of T_{trans} and T_{rot} was observed. Based on experimental results, the vibrational temperature was found to be in equilibrium with the translational and rotational temperature within experimental uncertainties, that is, $T_{vib} = 600$ K. For details, see [54, 65].

Figure 13.9 shows as an example of the temperature dependence of the line strengths of several transitions from the lower energy levels. The gas temperature was found to be about 800 K near the deposition substrate, which was essentially the same as the temperature measured by a thermocouple probe. The line strengths

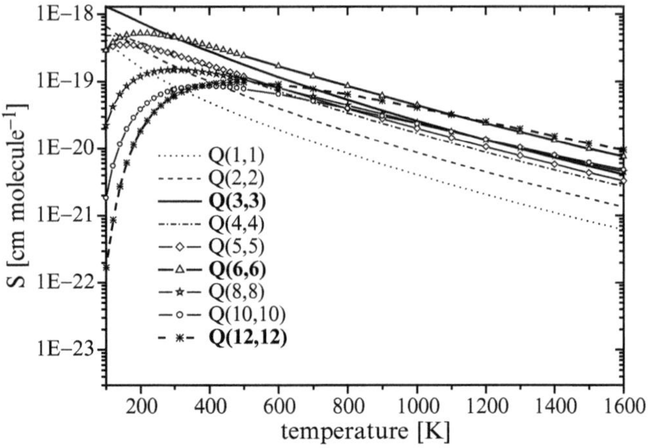

Fig. 13.9 Line strengths, S, for different Q-branch transitions of CH_3 as a function of temperature. Values calculated from the reference temperature values at 296 K [54]

of the nine Q-branch lines in the ν_2 fundamental band of the methyl radical in its ground electronic state were used to derive a more accurate value of the transition dipole moment of this band: $\mu_2 = 0.215(25)$ Debye (D). Improved accuracy over earlier measurements of μ was obtained by integrating over the complete line profile instead of measuring the peak absorption and assuming a Doppler line width to deduce the concentration; and the derivation of more accurate temperatures by examining a large number of lines. In addition a more precise value for the rate constant for methyl radical recombination than available earlier was employed. The new value of μ_2 is in very good agreement with high-quality ab initio calculations. Figure 13.10 shows the chronological summary of calculated and measured values of μ_2 for the transition dipole moment of the ν_2 fundamental band [54].

13.3.1.2 The ν_2 First Hot Band

The infrared laser absorption spectra of the methyl radical recorded in electric discharges consist not only of fundamental band lines but also of weaker hot band lines, suggesting the possibility of determining the transition dipole moments of the hot bands for comparison with theory. This section describes the determination of the transition dipole moment of the first hot band of the ν_2 band using the same experimental method as used for the fundamental band in Sect. 13.3.1.1.

A survey spectrum of some of the Q-branch lines of the first hot band is shown in Fig. 13.11. The required methyl recombination rate coefficient in the presence of argon and at the experimental temperatures is available from the literature. The measured line strengths for the ν_2 first hot band Q-branch transitions are given in [66]. The translational, rotational, and vibrational temperatures are required to

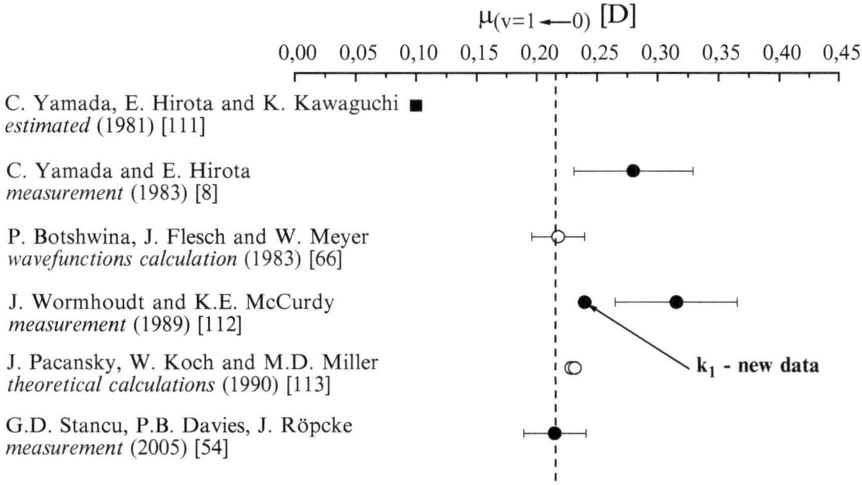

Fig. 13.10 Chronological summary of calculated and measured values of μ_2 for the transition dipole moment of the ν_2 fundamental band of CH_3 [54]. The new value of μ_2 is in very good agreement with high-quality ab initio calculations

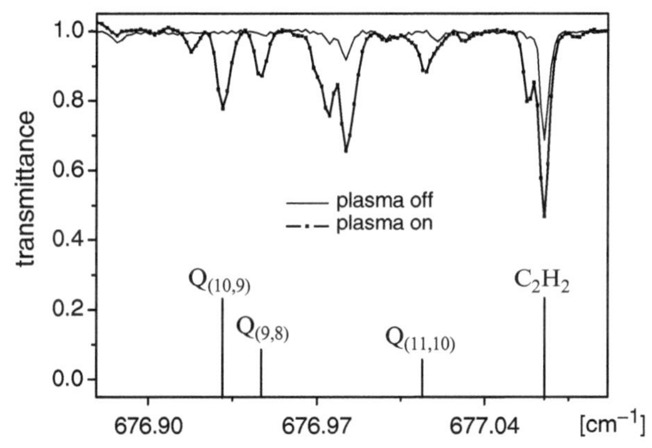

Fig. 13.11 Survey spectrum showing Q-branch lines from the first hot band. Several lines of stable species are observed both inside and outside the plasma [66]

calculate the corresponding partition functions. Although the partition functions were deduced in the earlier study [54] on the fundamental band, they were also evaluated here from the hot band transitions and using a more extended analysis.

The translational temperature was obtained by fitting the line profiles to a Gaussian function since the line width is determined by Doppler broadening at the pressures used here. The net broadening due to the intrinsic Doppler broadening of the line corresponded to $T = 630\,\mathrm{K}$. The rotational temperature was obtained using $Q(4,2)$, $Q(9,8)$, $Q(10,9)$, $Q(11,10)$, and $Q(12,11)$. Experimentally, the vibrational temperature is obtained from measured integrated absorption coefficients

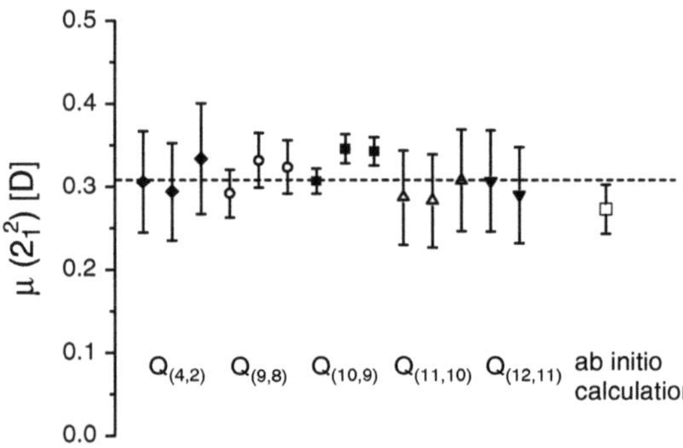

Fig. 13.12 The transition dipole moment of the ν_2 first hot band, measured values compared with the ab initio calculation. Multiple determinations based on five measured transitions are shown [66]. This value is in satisfactory agreement with the value of 0.27(3) D from a high-precision ab initio calculation

of resolved vibration rotation transitions in the fundamental and the hot bands. The transition dipole moment of the first hot band was determined to be 0.31(6) D (Fig. 13.12). This value is in satisfactory agreement with the value of 0.27(3) D from a high-precision ab initio calculation using the self-consistent electron pair (SCEP) method reported by Botschwina, Flesch, and Meyer [66, 67].

13.3.2 Molecular Spectroscopy of the CN Radical

The CN radical is of fundamental importance in laboratory spectroscopy and in astrophysics. Electronic emission spectra arising from the red ($A^2\Pi - X^2\Sigma^+$) and violet ($B^2\Sigma^+ - X^2\Sigma^+$) band systems excited in flames and discharges have been studied in the laboratory over decades while CN spectra have been detected in the atmospheres of stars and in the interstellar medium. Most recently, the electronic band systems have been very extensively measured and analyzed in emission [68, 69] using high-resolution Fourier transform spectroscopy.

Rotationally resolved spectra of the fundamental band of the CN free radical in four isotopic forms have been measured using TDLAS [70]. The source of the radical was a microwave discharge in a mixture of isotopically selected methane and nitrogen diluted with argon. The lines were measured to an accuracy of 5×10^{-4} cm^{-1} and fitted to the formula for the vibration rotation spectrum of a diatomic molecule, including quartic distortion constants. The band origins of each of the isotopomers from the five parameter fits were found to be $^{12}C^{14}N$: 2,042.42115(38) cm^{-1}, $^{13}C^{14}N$: 2,000.08479(23) cm^{-1}, $^{12}C^{15}N$: 2,011.25594(25) cm^{-1}, and $^{13}C^{15}N$: 1,968.22093(33) cm^{-1} with one standard deviation from the fit given in parenthesis. Some of the lines showed a resolved splitting due to the spin rotation interaction. This was averaged for fitting purposes. The average equilibrium internuclear distance derived from the $\nu = 0$ and $\nu = 1$ rotational constants of the four isotopomers is 1.171800(6) Å which is in good agreement with the value determined from microwave spectroscopy. Figure 13.13 shows a stick diagram of all the lines measured in the four isotopic forms and their intensities calculated for a rotational temperature of 950 K [70].

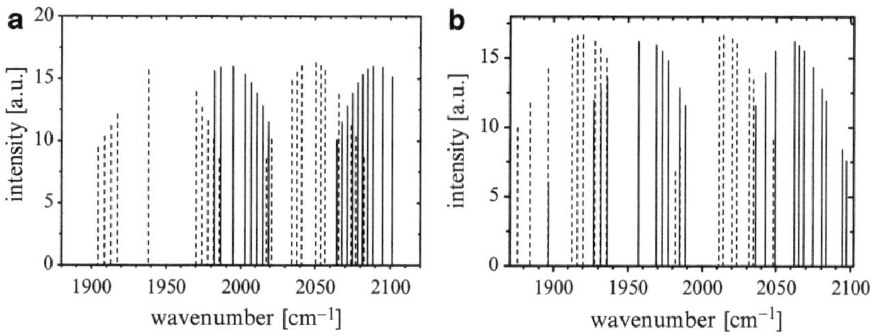

Fig. 13.13 Detected absorption lines from the fundamental band of CN and their intensities at 950 K: (**a**) $^{12}C^{14}N$ and $^{13}C^{14}N$ (*dotted lines*) and (**b**) $^{12}C^{15}N$ and $^{13}C^{15}N$ (*dotted lines*) [70]

13.4 Quantum Cascade Laser Absorption Spectroscopy for Plasma Diagnostics and Control

13.4.1 General Considerations

The recent development of pulsed and continuous wave (cw) QCLs and their commercial availability offer promising new possibilities for infrared absorption spectroscopy [71–73]. QCLs are unipolar semiconductor injection lasers based on intersubband transitions in a multiple quantum-well heterostructure. QCLs are designed using band-structure engineering and grown by molecular beam epitaxy. They are able to emit mid-infrared radiation at near room temperature operation. Compared to lead salt laser systems, QCL systems are very compact mid-infrared sources characterized by narrow line width combining single-frequency operation and considerably higher powers, that is, tens of mW.

The output power is sufficient to combine them with thermoelectrically cooled infrared detectors, which permits a decrease of the apparatus size and provides an opportunity to design compact cryogen-free mid-infrared spectrometer systems. The positive features of quantum cascade laser absorption spectroscopy (QCLAS) opens up new fields of application in research and industry. Recently, a compact quantum cascade laser measurement and control system (Q-MACS) has been developed (Figs. 13.14 and 13.15) for time-resolved plasma diagnostics, process control, and trace gas monitoring which can be used as platforms for various applications of QCLAS [1, 75].

Nowadays, QCLAS has been used to detect atmospheric trace constituents or trace gases in exhaled breath. Furthermore, it has already been successfully applied to the study of plasma processes, for example, of microwave and RF discharges [76, 77]. Of special interest for all hydrocarbon-based processes is the capacity to detect transient molecules like the CH_3 radical, as the supposed key growth species, by means of QCLAS [74].

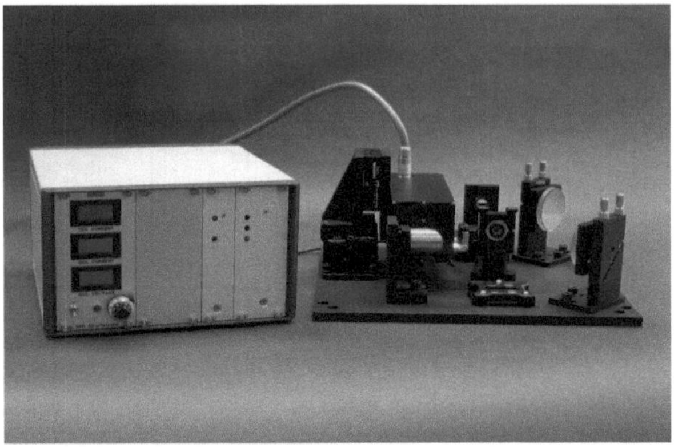

Fig. 13.14 Q-MACS Basic with laser head, supply unit, connection cable, and optical collimation module [74]

Fig. 13.15 New generation of compact QCLAS equipment with laser head in TO-8 configuration (*left-hand side*), reference gas cell (*middle*), and detector (*right-hand side*). The laser beam is guided by two Out of Axis Paraboloids [74]

Using pulsed QCL the scan through an infrared spectrum is commonly achieved by two different methods. In the *interpulse* mode, a bias DC ramp is applied to a series of short laser pulses of a few 10 ns [56, 78]. Another option is the *intrapulse* mode, that is, the scanning in single, longer pulses acquiring an entire spectrum [79]. Since this scan is performed in tens up to a few hundred nanoseconds a time resolution below 100 ns has become possible for quantitative in situ measurements of molecular concentrations in plasmas for the first time. Therefore it fits very well to measurements of rapidly changing chemical processes.

13.4.2 Time-Resolved Study of a Pulsed DC Discharge: NO and Gas Temperature Kinetics

One of the key issues when studying plasma processing for gas treatment is to make sure that no undesirable byproduct results from the process. Among them, NO_x components are readily produced in air plasmas. Over the past few years mid-infrared absorption techniques are increasingly being used for quantifying studies of phenomena related to NO_x production and removal in plasmas.

A high time resolution, up to the nanosecond scale, could open up a new approach for studying kinetic phenomena in molecular plasmas in real time and in situ. However, the extension of real time measurements on a submicrosecond scale to the mid-infrared has been hampered due to the lack of suitable light sources. QCLs offer an attractive new option for time-resolved infrared absorption spectroscopy [71–73, 75].

Based on this new approach for fast in situ plasma diagnostics, an experiment to study the time decay of NO in single discharge pulses of 1 ms duration has been designed. At the center of interest was the kinetics of the destruction of NO in a pulsed DC discharge formed in an Ar–N_2 mixture containing 0.91% of NO. The NO time dependence in N_2–O_2 pulsed discharges has already been measured or calculated using LIF [80, 81], mass spectrometry [82], or simplified models [82, 83]. However, the aims of these previous studies were different from the present case, namely identifying the NO(A) excited state formation [80–83] and describing a modulated glow air discharge by a simplified kinetic model [82]. In addition to the NO measurements in the present study, the temporal behavior of the electric field, the applied voltage, and the injected current during the pulse were also monitored. The comparison of the time evolution of the NO concentration under static and flowing conditions combined with simplified model calculations enabled the analysis of the dynamics of the plasma heating to be achieved.

The kinetics of the destruction of NO has been studied under static and flowing conditions, that is, in a closed and open discharge tube ($p = 266$ Pa) in a pulsed DC discharge of an Ar–N_2 mixture containing 0.91% of NO. Figure 13.16 shows the experimental setup. The time decay of the NO concentration was measured in single discharge pulses of 1 ms duration. The comparison of the time evolution of the NO concentration under static and flowing conditions and simplified model calculations enabled an analysis of the dynamics of the plasma heating to be made. The temperature increase during the pulse is below 40 K, but it has a strong influence on the line strength of the NO absorption line. The apparent decrease in the NO concentration in a single pulse, of about 20%, is due to the heating of the gas which in turn makes the line strength vary while the concentration remains almost constant for several successive pulses. Therefore, the QCLAS measurements, combined with model calculations, are a powerful noninvasive temperature probe with a remarkably high time resolution approaching the submicrosecond time scale (Fig. 13.17) [84].

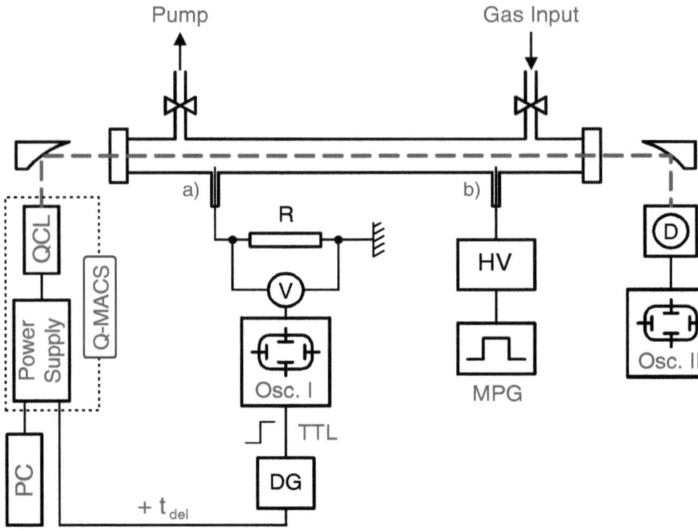

Fig. 13.16 Experimental setup incorporating the QCL system (*left*), the optical configuration and the pyrex DC discharge tube used for single pass absorption (*dashed line*), and the thermoelectrically cooled detector (*MPG* master pulse generator, DG delay generator, Osc. oscilloscope I and II, D detector) [84]

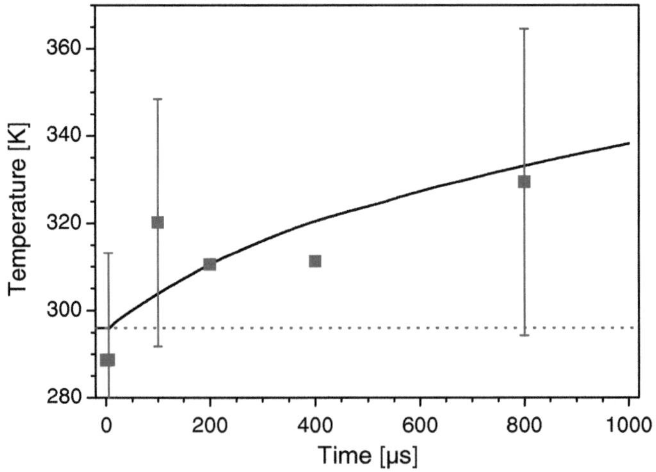

Fig. 13.17 Comparison between modeled gas heating dynamics and the experimentally derived temperature data. The *dashed line* represents the room temperature. *Solid line* is the modeled temperature [84]. The QCLAS measurements, combined with model calculations, are a powerful noninvasive temperature probe with a remarkably high time resolution approaching the submicrosecond time scale

13.4.3 Trace Gas Measurements Using Optically Resonant Cavities

The development of compact and robust optical sensors for molecular detection is of interest for an increasing number of applications, such as environmental monitoring and atmospheric chemistry [85–87], plasma diagnostics and industrial process control [1, 77], combustion studies, explosive detection, and medical diagnostics [88–92]. Absorption spectroscopy in the mid-infrared spectral region using lasers as radiation sources is an effective method for monitoring molecular species. In principle, path lengths up to several kilometers can be achieved by using optical resonators for cavity ring-down spectroscopy (CRDS) [93], cavity-enhanced absorption spectroscopy (CEAS) [94], or integrated cavity output spectroscopy (ICOS) [95,96]. The majority of cavity-based methods have used sources of radiation in the ultraviolet and visible regions. For many years, the infrared spectral range could not be employed either for CRDS or for the CEAS or ICOS techniques, because of the lack of suitable radiation sources with the required power and tunability, but this situation has now changed. Near-infrared (NIR) applications have profited from developments in telecommunications where cheap and compact light sources became available [97, 98] in the 1990s whereas similar lasers were not available in the mid-infrared range (MIR, 3–20 μm). Recent advances in semiconductor laser technology, in particular the advent of intersubband quantum cascade lasers (QCLs) and interband cascade lasers (ICLs) [71, 73, 99, 100], provides new possibilities for highly sensitive and selective trace gas detection using MIR absorption spectroscopy in combination with multiple pass cells [85, 101–107] and enabled sensitivities of $5 \times 10^{-10}\,\mathrm{cm}^{-1}\,\mathrm{Hz}^{-1/2}$ to be accomplished but at the expense of large sample volumes [56]. Distributed feedback (DFB)-QCLs combine single-frequency operation with tunability over several wave numbers, and average powers over a mW; hence, they are superior to lead salt lasers. The relatively high output power of the QCL permits the use of optical cavities with high finesse. With the help of such cavities, the effective path length of the laser beam in the absorbing medium can essentially be increased to more than the 200 m limit usually available from conventional optical multipass cells [101, 108] while keeping the sample and pumped volume small. While pulsed QCLs working at room temperature have been commercially available for several years, room temperature cw QCLs have only recently been introduced. Thermoelectrically cooled QCLs have been combined with high finesse optical resonators (Fig. 13.18) in order to profit from their enlarged path lengths at reasonably small sample volumes in combination with the high absorption cross section in the infrared molecular fingerprint region. Two different approaches have been investigated. Firstly, pulsed QCLs at 7.42 and 8.35 μm were used to perform CRDS and ICOS experiments. The spectra measured at pressures between 100 and 300 mbar were normally characterized by a broadening of the absorption lines and reduced absorption in comparison with theoretical expectations, which makes an absolute calibration necessary. The resulting decrease in sensitivity, that is, to $\sim 5 \times 10^{-7}\,\mathrm{cm}^{-1}\,\mathrm{Hz}^{-1/2}$, means that long path cell configurations of

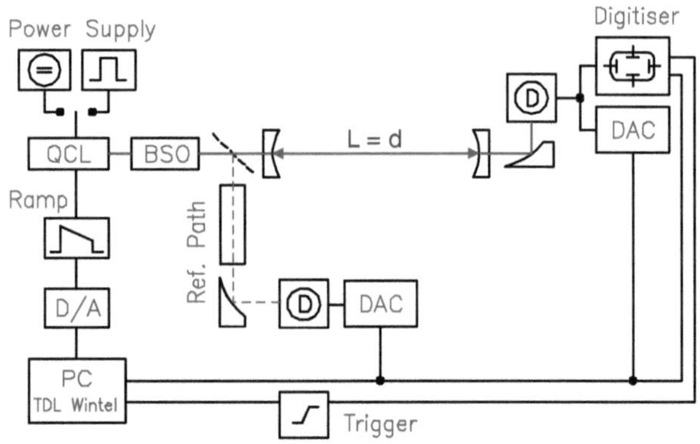

Fig. 13.18 Apparatus used for both CRDS and CEAS experiments. Beam-shaping optics (BSO) provided efficient light transmission to the resonator. The detector (D) signals were recorded with data acquisition cards (DACs). The quantum cascade lasers (QCLs) were driven by either a pulsed or a cw source, and the tuning ramp for the laser was generated via the digital-to-analog (D/A) converter [109]

similar sample volumes have superior sensitivity to either CRDS or ICOS. It transpires that the frequency-down chirp inherent to pulsed QCLs sets the fundamental limit.

Due to the chirp the effective line width of the QCL is much broader than the narrow molecular absorption features and the rather fast chirp rate does not allow for an efficient build-up of the laser field in the cavity. CRDS using pulsed QCLs therefore has only a limited number of useful applications, for example, for the determination of the reflectivity of the cavity mirrors in preliminary experiments or for the detection of complex and broad molecular absorptions at higher pressures. Secondly, a cw QCL at 7.66 μm has been combined with an unstabilized and unlocked cavity. With this straightforward arrangement, comprising only a TE-cooled detector, an effective path length of 1,080 m has been achieved. The main limit to sensitivity, of 2×10^{-7} cm^{-1} Hz$^{-1/2}$, was the cavity mode noise which may be reduced by off-axis alignment. Such mode noise is absent in conventional long path cell spectrometers. With a 20 s measurement interval detection limits for N_2O and CH_4 of 6×10^8 and 2×10^9 molecules cm^{-3}, respectively, could be achieved at 2.2 mbar indicating that sub-ppb levels could easily be measured at higher pressures (see Fig. 13.19 and Table 13.1). Furthermore, the small sample volume of 0.3 l used here, and consequently the reduced pumping requirement, is an advantage over long path cells with much larger volumes than 0.5 l. The detection limits achieved here are relevant both for the detection of radicals in plasma chemistry and for trace gas measurements with field-deployable systems. Additionally, trace gas measurements may also be carried out without pre-concentration procedures. Radicals with small abundances in the gas phase might now be detectable via QCL-based spectroscopy. The choice of an appropriate method depends on whether the important criterion is ultimate sensitivity or a more compact system. In the former case, a multipass

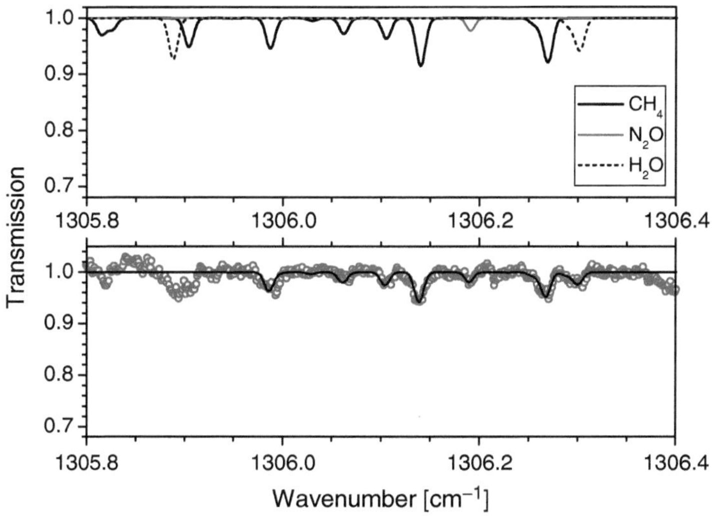

Fig. 13.19 Calculated spectrum (*upper*) for air containing CH_4, N_2O, and H_2O. Experimental conditions: $p = 2.2$ mbar, $L_{\mathrm{eff}} = 1,080$ m, 0.005 cm^{-1} instrumental broadening. The corresponding CEAS spectrum (*lower*) was observed with a TE-cooled detector (*open circles*) and fitted (*solid line*) to determine the actual concentrations [109]. The detection limits achieved here are relevant both for the detection of radicals in plasma chemistry and for trace gas measurements with field-deployable systems

Table 13.1 Species detected in ambient laboratory air at 2.2 mbar [109]

Molecule	ν(cm^{-1})	Line strength (cm per molecule)	Mix. ratio	Number density (molecules cm^{-3})	MDND (molecules cm^{-3})
CH_4	1,305.987	4.78×10^{-20}	1.72 ppm	9.2×10^{10}	2.5×10^{9}
CH_4	1,306.062	2.41×10^{-20}	1.82 ppm	9.8×10^{10}	4.9×10^{9}
CH_4	1,306.105	3.19×10^{-20}	1.68 ppm	9.0×10^{10}	3.7×10^{9}
CH_4	1,306.140	7.60×10^{-20}	1.66 ppm	8.9×10^{10}	1.6×10^{9}
N_2O	1,306.191	1.12×10^{-19}	346 ppb	1.9×10^{10}	6.4×10^{8}
H_2O	1,306.29	6.08×10^{-24} [a]	1.1%	5.8×10^{14}	1.8×10^{13}

MDND minimum detectable number density
[a] Effective line strength from two unresolved lines

cell spectrometer would be preferable because of its better signal-to-noise characteristics but this configuration would exclude the detection of processes on short time scales or certain types of in situ measurements. Sophisticated locked cavity schemes might overcome such restrictions at the expense of complex spectrometer geometries. To achieve a compact system a small volume cavity-based spectrometer employing cw QCLs would be more appropriate. This configuration would also be of special interest for applications where the pressure cannot arbitrarily be chosen in order to adapt the absorption line width to the laser line width, for example, in low-pressure plasmas. Moreover, for these applications in situ measurements are essential because multipass cell sampling ex situ is not an option.

13.4.4 In Situ Monitoring of Plasma Etch Processes with a QCL Arrangement in Semiconductor Industrial Environment

During the last 40 years, plasma etching has become a fundamental feature for processing integrated circuits. The optimization of the plasma chemistry in etch processes includes the identification of the mechanisms responsible for plasma-induced surface reactions combined with the achievement of uniformity in the distribution of molecules and radicals for homogeneous wafer treatment.

In etch plasmas used for semiconductor processing concentrations of the precursor gas NF_3 and of the etch product SiF_4 were measured on-line and in situ using a new experimental arrangement, designated the Q-MACS Etch system, which is based on QCLAS. In addition, the etch rates of SiO_2 layers and of the silicon wafer were monitored including plasma-etching endpoint detection. For this purpose the Q-MACS Etch system worked in an interferometric mode. The experiments were performed in an industrial dual frequency capacitively coupled magnetically enhanced reactive ion etcher (MERIE), which is a plasma reactor developed for dynamic random access memory (DRAM) technologies (Fig. 13.20). The quantum cascade laser system Q-MACS Etch consists of a pulsed infrared QCL source with the laser wavelength tunable in the range $1,027$–$1,032\,cm^{-1}$, optical components, detectors, and data acquisition cards controlled by a PC (Fig. 13.21). On board the IR beam is split into two channels using two IR transparent ZnSe beam splitters. The main part of the beam is coupled into an IR fiber using an off-axis parabolic mirror, and then collimated into the plasma reactor. In the second channel a reference spectrum of C_2H_4 is measured for line locking. Industrial requirements, such as (1) no open optical path and (2) the availability of just one optical access port makes coupling the infrared beam into the reactor a challenging task. The solution realized in the Q-MACS Etch is based on the use of (a) mid-infrared fibers and (b) internal reflections in the reactor chamber. The capability of Q-MACS Etch for SiF_4 concentration monitoring via reactor side access and for monitoring the etch rates via reactor top access is shown in Fig. 13.20.

This first application of a QCL arrangement for monitoring of industrial etch processes has opened up a challenging new option for control of demanding semiconductor production applications. Focused on sensitive and fast concentration measurements of key molecular components, while ensuring compactness, robustness, and long-term stability, this new class of process control equipment has the potential to become implemented into other fields of technology [77, 110].

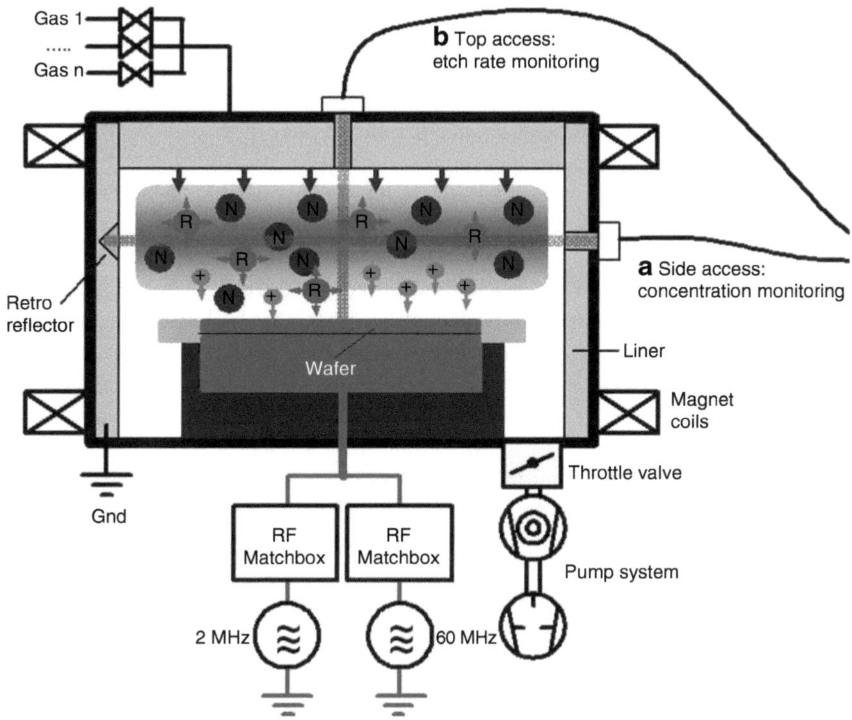

Fig. 13.20 Experimental arrangement of the MERIE plasma-etching reactor: (**a**) side access and (**b**) top access monitoring of the Q-MACS Etch system [77]

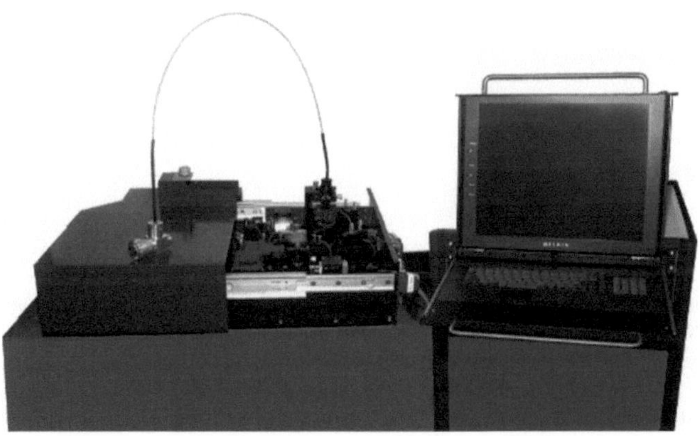

Fig. 13.21 Q-MACS Etch with the optical module and the separate control unit in a rack [110]

13.5 Summary and Conclusions

During the past few years, a variety of phenomena in molecular plasmas in which many short-lived and stable species are produced have been successfully studied based on diode laser absorption techniques in the mid-infrared spectral range, with which the present chapter is concerned. It has been possible to determine absolute concentrations of ground states using spectroscopy thereby providing a link with chemical modeling of the plasma, the ultimate objective being better to understand the chemical and reaction kinetic processes occurring in the plasma. The other essential component needed to reach this objective is to determine physical parameters of the plasma, as for example, temperatures, degrees of dissociation, and dynamics of reaction kinetic processes, and the present chapter discusses methods for achieving this. The need for a better scientific understanding of plasma physics and chemistry has stimulated the application of IRLAS, which has been proven to be one of the most versatile techniques for studying molecular plasmas. Based on the recent development of QCLs, the further spread of this method of high-resolution mid-infrared spectroscopy to industrial applications has become a reality.

Acknowledgments This work was partly supported by (a) the Deutsche Forschungsgemeinschaft, Sonderforschungsbereich Transregio 24, and Transferbereich 36; (b) the Bundesministerium für Bildung und Forschung, FKZ 13N7451/8; and (c) the Deutscher Akademischer Austauschdienst and EGIDE as part of the French–German PROCOPE Collaboration Program (Project 04607QB).

Furthermore, it was supported within the scope of a technology development by the EFRE fund of the European Community and by funding of the State Saxony of the Federal Republic of Germany.

The authors give sincere thanks to all present and former members of the laboratories involved in Greifswald, Paris, Eindhoven, and Cambridge for permanent support and a stimulating scientific climate. In particular, the authors are indebted to all coauthors of former papers whose contributions made the present chapter possible. K.-D. Weltmann is gratefully acknowledged for his continuous encouragement and general support.

References

1. J. Röpcke, G. Lombardi, A. Rousseau, P.B. Davies, Plasma Sour. Sci. Technol. **15**, S148 (2006)
2. M. Haverlag, E. Stoffels, W.W. Stoffels, G.M.W. Kroesen, F.J. de Hoog, J. Vac. Sci. Technol. A **14**, 380 (1996)
3. P.B. Davies, P.M. Martineau, Adv. Mater. **4**, 729 (1992)
4. S. Naito, N. Ito, T. Hattori, T. Goto, Jpn. J. Appl. Phys. **34**, 302 (1995)
5. M. Haverlag, E. Stoffels, W.W. Stoffels, G.M.W. Kroesen, F.J. de Hoog, J. Vac. Sci. Technol. A **12**, 3102 (1994)
6. J. Röpcke, L. Mechold, M. Käning, W.Y. Fan, P.B. Davies, Plasma Chem. Plasma Process. **19**, 395 (1999)
7. M. Haverlag, E. Stoffels, W.W. Stoffels, G.M.W. Kroesen, F.J. de Hoog, J. Vac. Sci. Technol. A **14**, 384 (1996)
8. C. Yamada, E. Hirota, J. Chem. Phys. **78**, 669 (1983)
9. S. Naito, M. Ikeda, N. Ito, T. Hattori, T. Goto, Jpn. J. Appl. Phys. **32**, 5721 (1993)

10. S. Naito, N. Ito, T. Hattori, T. Goto, Jpn. J. Appl. Phys. **33**, 5967 (1994)
11. M. Ikeda, N. Ito, M. Hiramatsu, M. Hori, T. Goto, J. Appl. Phys. **82**, 4055 (1997)
12. G.M.W. Kroesen, J.H.W.G. den Boer, L. Boufendi, F. Vivet, K. Khouli, A. Bouchoule, F.V. de Hoog, J. Vac. Sci. Technol. A **14**, 546 (1996)
13. C. Busch, I. Möller, H. Soltwisch, Plasma Sour. Sci. Technol. **10**, 250 (2001)
14. A. Serdioutchenko, I. Möller, H. Soltwisch, Spectrochim. Acta A **60**, 3311 (2004)
15. R.A.B. Zijlmans, O. Gabriel, S. Welzel, F. Hempel, J. Röpcke, R. Engeln, D.C. Schram, Plasma Sour. Sci. Technol. **15**, 564 (2006)
16. A. Rousseau, O. Guaitella, L. Gatilova, M. Hannemann, J. Röpcke, J. Phys. D **40**, 2018 (2007)
17. A. Rousseau, A. Dantier, L.V. Gatilova, Y. Ionikh, J. Röpcke, Y.A. Tolmachev, Plasma Sour. Sci. Technol. **14**, 70 (2005)
18. Y. Ionikh, A.V. Meshchanov, J. Röpcke, A. Rousseau, Chem. Phys. **322**, 411 (2006)
19. L.V. Gatilova, K. Allegraud, J. Guillon, Y.Z. Ionikh, G. Cartry, J. Röpcke, A. Rousseau, Plasma Sour. Sci. Technol. **16**, S107 (2007)
20. J.H. Van Helden, W. Wagemans, G. Yagci, R.A. Zijlmans, D.C. Schram, R. Engeln, G. Lombardi, G.D. Stancu, J. Röpcke, J. Appl. Phys. **101**, 043305 (2007)
21. Z. Liu, P.B. Davies, J. Chem. Phys. **105**, 3443 (1996)
22. P.B. Davies, D.M. Smith, J. Chem. Phys. **100**, 6166 (1994)
23. J. Röpcke, P.B. Davies, F. Hempel, B.P. Lavrov, in *Low Temperature Plasmas – Fundamentals, Technologies and Techniques*, vol. **1**, ed. by R. Hippler, H. Kersten, M. Schmidt, K.H. Schönbach (Wiley-VCH, Berlin, 2008), p. 215
24. F. Hempel, P.B. Davies, D. Loffhagen, L. Mechold, J. Röpcke, Plasma Sour. Sci. Technol. **12**, S98 (2003)
25. A. Cheesman et al., J. Phys. Chem. A **110**, 2821 (2006)
26. G. Lombardi, G.D. Stancu, F. Hempel, A. Gicquel, J. Röpcke, Plasma Sour. Sci. Technol. **13**, 27 (2004)
27. A. Ershov, J. Borysow, J. Phys. D: Appl. Phys. **28**, 68 (1995)
28. R. Sankaranarayanan, B. Pashaie, S.K. Dhali, Appl. Phys. Lett. **77**, 2970 (2000)
29. Z. Falkenstein, J. Appl. Phys. **81**, 7158 (1997)
30. S. Welzel, A. Rousseau, P.B. Davies, J. Röpcke, J. Phys.: Conf. Ser. **86**, 012012 (2007)
31. J. Röpcke, L. Mechold, M. Käning, J. Anders, F.G. Wienhold, D. Nelson, M. Zahniser, Rev. Sci. Instrum. **71**, 3706 (2000)
32. J.B. McManus, D. Nelson, M. Zahniser, L. Mechold, M. Osiac, J. Röpcke, A. Rousseau, Rev. Sci. Instrum. **74**, 2709 (2003)
33. A. Ohl, J. Phys. IV **8**, 83 (1998)
34. NIST Kinetics Database (http://kinetics.nist.gov/index.php)
35. R.A.B. Zijlmans, Ph.D. Thesis, TU Eindhoven, 2008
36. J. Röpcke, S. Welzel, N. Lang, F. Hempel, L. Gatilova, O. Guaitella, A. Rousseau, P.B. Davies, Appl. Phys. B **92**, 335 (2008)
37. M.C.M. van de Sanden, R.J. Severens, W.M.M. Kessels, R.F.G. Meulenbroeks, D.C. Schram, J. Appl. Phys. **84**, 2426 (1998)
38. L. Martinu, D. Poitras, J. Vac. Sci. Technol. A **18**, 2619 (2000)
39. J. Benedikt, R.V. Woen, S.L.M. van Mensfoort, V. Perina, M.C.M. van de Sanden, Diamond Relat. Mater. **12**, 90 (2003)
40. R. Förch, Z. Zhan, K. Wolfgang, Plasma Process. Polym. **2**, 351 (2005)
41. S. Tanaka, H. Uyama, O. Matsumoto, Plasma Chem. Plasma Process. **14**, 491 (1994)
42. H. Kim, Plasma Process. Polym. **1**, 91 (2004)
43. A. Rousseau, A.V. Meshchanov, J. Röpcke, Appl. Phys. Lett. **88**, 021503 (2006)
44. G. Vidali, J.E. Roser, L. Ling, E. Congiu, G. Manico, V. Pirronello, Faraday Discuss. **133**, 125 (2006)
45. Y. Shimomura, J. Nucl. Mater. **363–365**, 467 (2007)
46. G.J.H. Brussaard, K.G.Y. Letourneur, M. Schaepkens, M.C.M. van de Sanden, D.C. Schram, J. Vac. Sci. Technol. B **21**, 61 (2003)
47. M.F.A.M. van Hest, J.R. Haartsen, M.H.M. van Weert, D.C. Schram, M.C.M. van de Sanden, Plasma Sour. Sci. Technol. **12**, 539 (2003)

48. R.F.G. Meulenbroeks, A.J. van Beek, A.J.G. van Helvoort, M.C.M. van de Sanden, D.C. Schram, Phys. Rev. E **49**, 4397 (1994)
49. R.F.G. Meulenbroeks, R.A.H. Engeln, M.N.A. Beurskens, R.M.J. Paffen, M.C.M. van de Sanden, J.A.M. van der Mullen, D.C. Schram, Plasma Sour. Sci. Technol. **4**, 74 (1995)
50. B. Bézard, H. Feuchtgruber, J.I. Moses, T. Encrenaz, Astron. Astrophys. **334**, L41 (1998)
51. B. Bezard, P.N. Romani, H. Feuchtgruber, T. Encrenaz, Astrophys. J. **515**, 868 (1999)
52. H. Feuchtgruber, F.P. Helmich, E.F. van Dishoeck, C.M. Wright, Astrophys. J. **535**, L111 (2000)
53. M.T. Macpherson, M.J. Pilling, M.J.C. Smith, Chem. Phys. Lett. **94**, 430 (1983)
54. G.D. Stancu, J. Röpcke, P.B. Davies, J. Chem. Phys. **122**, 014306 (2005)
55. M.S. Zahniser, D.D. Nelson, C.E. Kolb, in *Applied Combustion Diagnostics*, ed. by K. Kohse-Hoinghaus, J. Jeffries (Taylor & Francis, New York, 2002), p. 648
56. D.D. Nelson, J.H. Shorter, J.B. McManus, M.S. Zahniser, Appl. Phys. B **75**, 343 (2002)
57. G.A. Laguna, S.L. Baughcum, Chem. Phys. Lett. **88**, 568 (1982)
58. R.J. Cody, W.A. Payne, R.P. Thorn, F.L. Nesbitt, M.A. Iannone, D.C. Tardy, L. Stief, J. Phys. Chem. A **106**, 6060 (2002)
59. D. Walter, H.H. Grotheer, J.W. Davies, M.J. Pilling, A.F. Wagner, in *Proceedings of the 23rd Symposium on Combustion*, The Combustion Institute, vol. 107, 1990
60. M.T. Macpherson, M.J. Pilling, M.J.C. Smith, J. Phys. Chem. **89**, 2268 (1985)
61. I.R. Slagle, D. Gutman, J.W. Davies, M.J. Pilling, J. Phys. Chem. **92**, 2455 (1988)
62. A.F. Wagner, D.M. Wardlaw, J. Phys. Chem. **92**, 2462 (1988)
63. A.B. Callear, M.P. Metcalfe, Chem. Phys. **14**, 275 (1976)
64. D.L. Baulch, C.J. Cobos, R.A. Cox, P. Frank, G. Hayman, T.h. Just, J.A. Kerr, T. Murrells, M.J. Pilling, J. Troe, R.W. Walker, J. Warnatz, J. Phys. Chem. Ref. Data **23**, 980 (1994)
65. G.D. Stancu, Ph.D. Thesis, University of Greifswald, Germany, 2004
66. G.D. Stancu, J. Röpcke, P.B.J. Davies, Phys. Chem. A **112**(28), 6285 (2008)
67. P. Botschwina, J. Flesch, W. Meyer, Chem. Phys. **74**, 321 (1983)
68. B.D. Rehfuss, M.H. Suh, T.A. Miller, V.E. Bondybey, J. Mol. Spectrosc. **151**, 437 (1992)
69. C.V.V. Prasad, P.F. Bernath, C. Frum, R. Engleman, J. Mol. Spectrosc. **151**, 459 (1992)
70. M. Hübner, M. Castillo, P.B. Davies, J. Röpcke, Spectrochim. Acta A Mol. Biomol. Spectrosc. **61**, 57 (2005)
71. J. Faist, F. Capasso, D.L. Sivco, C. Sirtori, A.L. Hutchinson, A. Cho, Science **264**, 553 (1994)
72. C. Gmachl, D.L. Sivco, R. Colombelli, F. Capasso, A.Y. Cho, Nature **415**, 883 (2002)
73. M. Beck, D. Hofstetter, T. Aellen, J. Faist, U. Oesterle, M. Ilegems, E. Gini, H. Melchior, Science **295**, 301 (2002)
74. S. Welzel, A. Rousseau, P.B. Davies, J. Röpcke, J. Phys.: Conf. Ser. **86**, 012012 (2007)
75. F. Hempel, S. Glitsch, J. Röpcke, S. Saß, H. Zimmermann, in *Plasma Polymers and Related Materials*, ed. by M. Mutlu (Hacettepe University Press, Turkey, 2005), p. 142
76. A. Cheesman et al., J. Phys. Chem. A **110**, 2821 (2006)
77. G.D. Stancu, N. Lang, J. Röpcke, M. Reinicke, A. Steinbach, S. Wege, Chem. Vap. Deposition **13**, 351 (2007)
78. K. Namjou et al., Opt. Lett. **23**, 219 (1998)
79. T. Beyer et al., J. Appl. Phys. **93**, 3158 (2003)
80. S. De Benedictis, G. Dilecce, M. Simek, J. Phys. D: Appl. Phys. **30**, 2887 (1997)
81. S. De Benedictis, G. Dilecce, J. Phys. III **6**, 1189 (1996)
82. M. Castillo, V.J. Herrero, I. Mendez, I. Tanarro, Plasma Sour. Sci. Technol. **13**, 343 (2004)
83. G. Cartry, L. Magne, G. Cernogora, J. Phys. D: Appl. Phys. **32**, 1894 (1999)
84. S. Welzel, L. Gatilova, J. Röpcke, A. Rousseau, Plasma Sour. Sci. Technol. **16**, 822 (2007)
85. D.D. Nelson, B. McManus, S. Urbanski, S. Herndon, M.S. Zahniser, Spectrochim. Acta A **60**, 3325 (2004)
86. J.H. Steinfeld, S.N. Pandis, *Atmospheric Chemistry and Physics: From Air Pollution to Climate Change* (Wiley, New York, 1998)
87. S.S. Brown, Chem. Rev. **103**, 5219 (2003)
88. H. Gupta, L.S. Fan, Ind. Eng. Chem. Res. **42**, 2536 (2003)
89. C. Bauer, P. Geiser, J. Burgmeier, G. Holl, W. Schade, Appl. Phys. B **85**, 251 (2006)

90. M.L. Silva, D.M. Sonnenfroh, D.I. Rosen, M.G. Allen, A. O'Keefe, Appl. Phys. B **81**, 705 (2005)
91. M.W. Todd, R.A. Provencal, T.G. Owano, B.A. Paldus, A. Kachanov, K.L. Vodopyanov, M. Hunter, S.L. Coy, J.I. Steinfeld, J.T. Arnold, Appl. Phys. B **75**, 367 (2002)
92. M.R. McCurdy, Y. Bakhirkin, G. Wysocki, R. Lewicki, F.K. Tittel, J. Breath Res. **1**, 014001 (2007)
93. A. O'Keefe, D.A.G. Deacon, Rev. Sci. Instrum. **59**, 2544 (1988)
94. R. Engeln, G. Berden, R. Peeters, G. Meijer, Rev. Sci. Instrum. **69**, 3763 (1998)
95. A. O'Keefe, J.J. Scherer, J.B. Paul, Chem. Phys. Lett. **307**, 343 (1999)
96. A. O'Keefe, Chem. Phys. Lett. **293**, 331 (1998)
97. D. Romanini, A.A. Kachanov, F. Stoeckel, Chem. Phys. Lett. **270**, 538 (1997)
98. G. Berden, R. Peeters, G. Meijer, Chem. Phys. Lett. **307**, 131 (1999)
99. R.Q. Yang, C.J. Hill, B.H. Yang, C.M. Wong, R.E. Muller, P.M. Echternach, Appl. Phys. Lett. **84**, 3699 (2004)
100. R.Q. Yang, C.J. Hill, B.H. Yang, Appl. Phys. Lett. **87**, 151109 (2005)
101. C. Roller, A.A. Kosterev, F.K. Tittel, K. Uehara, C. Gmachl, D.L. Sivco, Opt. Lett. **28**, 2052 (2003)
102. D.D. Nelson, J.B. McManus, S.C. Herndon, J.H. Shorter, M.S. Zahniser, S. Blaser, L. Hvozdara, A. Muller, M. Giovannini, J. Faist, Opt. Lett. **31**, 2012 (2006)
103. R. Lewicki, G. Wysocki, A.A. Kosterev, F.K. Tittel, Opt. Express **15**, 7357 (2007)
104. A. Grossel, V. Zeninari, B. Parvitte, L. Joly, D. Courtois, Appl. Phys. B **88**, 483 (2007)
105. A.A. Kosterev, Y.A. Bakhirkin, F.K. Tittel, Appl. Phys. B **80**, 133 (2005)
106. A.A. Kosterev, R.F. Curl, F.K. Tittel, C. Gmachl, F. Capasso, D.L. Sivco, J.N. Baillargeon, A.L. Hutchinson, A.Y. Cho, Appl. Opt. **39**, 4425 (2000)
107. B. Tuzson, M.J. Zeeman, M.S. Zahniser, L. Emmenegger, Infrared Phys. Technol. **51**, 198 (2008)
108. J.B. McManus, P.L. Kebabian, M.S. Zahniser, Appl. Opt. **34**, 3336 (1995)
109. S. Welzel, G. Lombardi, P.B. Davies, R. Engeln, D.C. Schram, J. Röpcke, J. Appl. Phys. **104**, 093115 (2008)
110. N. Lang, J. Röpcke, A. Steinbach, S. Wege, in *OPTAM*, Düsseldorf 2008, vol. 2047, VDI-Berichte, 2008, p. 11
111. C. Yamada, E. Hirota, K. Kawaguchi, J. Chem. Phys. **75**, 5256 (1981)
112. J. Wormhoudt, K.E. McCurdy, Chem. Phys. Lett. **156**, 47 (1989)
113. J. Pacansky, W. Koch, M.D. Miller, J. Am. Chem. Soc. **113**, 317 (1991)

Chapter 14
X-Ray Diagnostics of Plasma-Deposited Thin Layers

Harm Wulff

Abstract Grazing incidence X-ray diffractometry (GIXD) and X-ray reflectometry (XR) have been introduced as well-suited tools for investigations of plasma deposited thin films as well as plasma exposed surface layers. These X-ray methods have a number of advantages compared to other commonly used techniques. A combination of GIXD and XR can give a range of interesting information about chemical, physical and crystallographic properties of thin films. Conclusions can be drawn how plasma deposition techniques and plasma parameters influence film formation and film growth. In this chapter the basic principles of GIXD and XR were introduced and the efficiency of these X-ray methods will be discussed. By means of examples the phase analysis of thin films will be presented, the influence of plasma parameters on film properties will be discussed as well as the defect structure analysis of plasma-deposited films and plasma chemical reactions in surface layers.

14.1 Introduction

The phenomena associated with plasma surface interactions involve an interesting mix of plasma physics, ion–solid collision physics, surface chemistry, and materials science.

The deposition of films by plasma techniques as well the plasma treatment of solid walls are well known and widely used methods. Nevertheless, the fundamental mechanisms of plasma–wall interaction are not yet understood in detail.

To understand the complexities involved in plasma–wall interaction it is necessary to take a close look at the substrate area. The substrate area is subjected to plasma radiation and especially to fluxes of energetic and reactive charged and neutral particles in various excited states. Chemical reactions, diffusion processes, particle deposition, defect formation, and sputter processes can take place and show a very complex behavior during plasma surface interactions.

H. Wulff (✉)
Institut fr Biochemie, Ernst-Moritz-Arndt-Universität Greifswald, Felix-Hausdorff-Str. 4, 17487 Greifswald, Germany
e-mail: wulff@chemie.uni-greifswald.de

M. Bonitz et al. (eds.), *Introduction to Complex Plasmas*, Springer Series on Atomic, Optical, and Plasma Physics 59, DOI 10.1007/978-3-642-10592-0_14,

Hence, a specific challenge in plasma science is the characterization of thin surface layers as a response of external plasma parameters like plasma power, gas pressure, gas composition and substrate bias. Otherwise, characterization of thin films or thin surface layers is very important to improve device quality to the level acceptable to the design engineers. The properties of plasma-deposited films and plasma treated surfaces decide finally whether and for which purposes films or surfaces can be used.

Of particular interest is the analysis of film properties in the nanometer range. Grazing incidence diffractometry (GIXD) and X-ray reflectometry (XR) have been established as well-suited tools for investigation of chemical, physical and crystallographic properties of thin films and surface layers. Besides the X-ray methods also other analytical techniques can be used to give information on surfaces and deposited films.

Table 14.1 gives a survey about X-ray techniques and other analytical methods to analyze thin films after plasma exposure. In this chapter the basic principles of the

Table 14.1 Survey of X-ray and alternative methods for thin film characterization

Film property	X-ray method	Alternatives
Phase composition	GIXD: Bragg angle, intensity	Transmission electron microscopy (TEM)
Chemical composition (concentration depth profile)	GIXD: Bragg angle	Energy dispersive X-ray spectroscopy (EDX), X-ray photoelectron spectroscopy (XPS), Rutherford back scattering (RBS), elastic recoil detection analysis (ERDA)
Macrostress	GIXD: Bragg angle	Substrate curvature, laser optics
Grain size	GIXD: line profile, line width	TEM, scanning electron microscopy (SEM)
Microstrain	GIXD: line profile	
Preferred orientation	GIXD: intensity, polfigure	
Crystal structure	GIXD: Rietveld analysis, structure refinement	
Thickness	GIXD: intensity XR: Kiessig fringes	Interferometry, ellipsometry, TEM
Density	XR: critical angle for total reflection	Ellipsometry
Surface roughness interface roughness	XR: amplitude of Kiessig fringes	SEM, ellipsometry, atomic force microscopy (AFM)
Diffusion behavior	GIXD, thermal and time resolved: intensity	Secondary ion mass spectroscopy (SIMS), Auger electron spectroscopy (AES), combined with sputtering
Crystallization rate melting point	GIXD, thermal and time resolved: intensity	

X-ray techniques will be presented. By means of examples the efficiency of the X-ray methods will be demonstrated. These involve the characterization of ITO films, deposited on Si(100), especially the influence of oxygen gas flow and substrate voltages and studies of the alumina (Al_2O_3) formation during microwave plasma treatment of aluminum films.

14.2 X-Ray Analytical Methods

14.2.1 Grazing Incidence X-Ray Diffractometry, Asymmetric Bragg Case

X-ray diffraction is a powerful tool for material characterization. Various measurement techniques have been developed for analyzing different material properties. Thin polycrystalline films can be studied with advantage in a highly asymmetric Bragg case to obtain information from the near-surface region of a sample.

In this technique the diffraction volume can be increased by decreasing the angle of incidence. In the schematic diagram of grazing incidence X-ray diffractometry, the optical path in GIXD can be seen (Fig. 14.1). Due to small angle of incident X-ray beam its penetration depth normal to the film surface is reduced and interfering substrate peaks are eliminated. The diffraction vector, perpendicular to the reflecting planes and parallel to the bisectrix of incident and diffracted beam defines the direction along which structural information, such as lattice parameters or crystallite size, is obtained. In the asymmetric Bragg case it changes its direction with Bragg angle Θ, whereas on the normal Bragg-Brentano (BB) geometry it is fixed and perpendicular to the sample surface. Only the crystallites of the thin film that are oriented with a specific set of lattice planes perpendicular to the diffraction vector contribute to the intensity diffracted from these planes. Therefore, only polycrystalline films with a random or nearly random distribution of grain orientations may diffract X-rays from all major lattice planes.

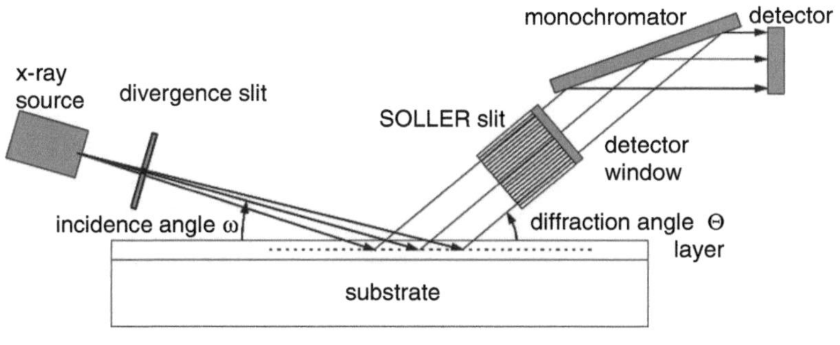

Fig. 14.1 Grazing incidence X-ray diffractometry (GIXD), asymmetric Bragg case, Bragg angle Θ, and angle of incidence ω

However, many thin films have strong preferred orientation of crystallites. In many cases a low-indexed set of lattice planes is parallel to the surface, e.g., in epitaxial films or in films with fiber texture. To get sufficient diffracted intensity from these planes we must choose an angle of incidence $\omega \approx \Theta$.

The measured X-ray intensity I originates from the volume element YA, with the information depth Y according to (14.1) and (14.2) and A the irradiated sample surface.

In the analysis of layers, the information depth Y of X-rays is an important factor, in particular, if gradients of structure parameters occur in the films. The film information depth Y strongly depends on the thickness x_0, the mean absorption coefficient μ and of course on the incidence angle ω.

$$Y = \frac{1}{\mu z}\left(1 - \exp\left(-\mu x_0 z\right)\right), \tag{14.1}$$

with

$$z = \frac{1}{\sin \omega} + \frac{1}{\sin\left(2\Theta - \omega\right)}. \tag{14.2}$$

A further advantage of GIXD compared to the normally used BB geometry is that the information depth is independent of the Bragg angle 2Θ.

14.2.2 GIXD, Bragg Case, Specular Reflected

X-rays can also be specularly reflected from surfaces at angles close to the critical angle for total reflection. Marra et al. [1] have demonstrated that strong diffracted X-ray beams can also be observed under these conditions.

Figure 14.2 shows the geometry of the in plane grazing incidence diffraction method. The primary X-ray beam, monochromated and focused at the sample irradiates the sample under an angle close to the critical angle for total reflection (<0.5°), thus enabling intense specular reflection. Executing $\Theta - 2\Theta$ scan strong diffracted beams can be observed exiting from the surface under very shallow angles with diffraction vectors parallel to the surface.

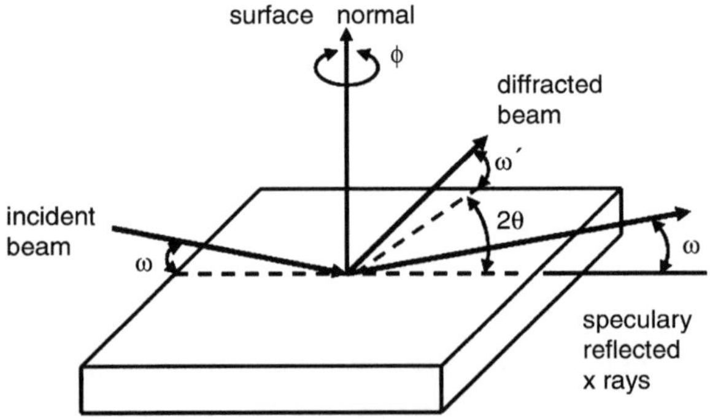

Fig. 14.2 Geometry of the grazing incidence diffraction method, in plane geometry

The penetration depth can be changed by varying the angle of incidence. For angles of incidence below the critical angle ω_c the penetration depth perpendicular to the surface is in the order of a few nanometers, determined by the evanescence of the electrical field. This GIXD is an in-plane scattering geometry combining the Bragg condition with the conditions of X-ray total external reflection from crystal surfaces.

Unfortunately both, the very asymmetric Bragg case reflection with a small incidence angle ω as well as the technique with specularly reflected X-rays are labeled "grazing incidence X-ray diffractometry." There is no clear conceptual separation between these two techniques in the literature.

A disadvantage of X-ray diffraction methods is the fact that only crystalline materials can be investigated. X-ray amorphous or amorphous materials do not give information on chemical or crystallographic properties.

14.2.3 X-Ray Reflectometry

X-ray reflectometry around the critical angle for total reflection allows the determination of film thickness, mass density, as well as surface and interface roughness irrespective of the crystalline structure. X-ray reflectometry is equally well applicable to crystalline, polycrystalline and amorphous materials. It only requires a sufficient flat sample. The basic principle is shown in Fig. 14.3.

In case of thin films on a substrate constructive interference occurs between the beam reflected at the surface and the beams reflected at the interfaces. Constructive interference results in intensity maxima called *Kiessig fringes*, whose angular spacing is characteristic for the thickness of a layer. The value for the critical angle of the total reflection ω_c can be used to determine the mass density ρ of a deposited film [2]:

$$\omega_c = \sqrt{2\delta}, \quad \delta = \sqrt{\frac{N_A r_0 \rho \sum_i (Z_i + f_i)}{2\pi \sum_i A_i}}, \tag{14.3}$$

where N_A is the Avogadro constant, r_0 the classical electron radius, $(Z_i + f_i)$ the scattering factor of an atom i, and A_i the atomic weight.

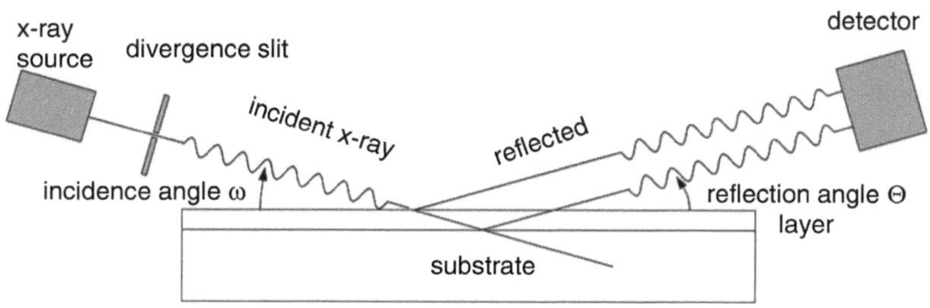

Fig. 14.3 X-ray reflectometry

X-ray reflectometry measurements can be performed with commercial X-ray diffractometers using the normal Bragg-Brentano geometry.

14.3 Examples

14.3.1 Characterization of ITO Films

Over the past years, there has been a growing interest in both fundamental research and device applications based on transparent conductive oxides (TCOs). This unique class of fascinating materials [3] is gaining a variety of applications in the fields of solar cells, gas sensors, organic light-emitting diodes, liquid crystal, and high definition displays, as well as for electro chromic and smart windows. Classical TCO layer systems are based on semiconducting oxides like indium oxide (In_2O_3). Indium tin oxide (ITO) is the major material for TCO at present. The doping material should cause only small lattice distortion and the deposition procedure should form TCO films with only small amounts of imperfections such as interstices, dislocations, grain boundaries [4, 5].

Plasma methods are characterized by a set of free parameters to optimize the deposition process. Plasma power, working pressure, oxygen partial pressure, and substrate bias voltage can influence deposition rate, film thickness [6], film stoichiometry, defect concentration and therefore the optoelectronic properties. An attractive technique for the deposition of TCO material is the so called reactive sputtering, where a metal or a metallic alloy target is sputtered in an atmosphere of argon admixed with the reactive gas oxygen. An advantage of plasma-enhanced deposition techniques is the low thermal load for the substrates, so TCO films can be deposited on organic foils, too.

In this example we present a study on influence of oxygen gas flow and low energy ion bombardment to DC magnetron sputtered ITO film growth processes and film properties like composition and structure degradation. Figure 14.4 shows the device used to deposit metallic In/Sn as well as ITO films on silicon (100) wafers by DC planar magnetron sputtering. A metallic In/Sn (90/10) target was used. Power, pressure and Ar-flow were kept constant. The standard deposition conditions were DC power: 30 W; inert gas flow: 15 sccm Ar; working pressure: 5.6×10^{-3} mbar. The oxygen flow was varied between 0 and 2 sccm. Substrate bias voltages between 0 and -100 V were used. The film thickness was approximately 50 nm for all samples. The films were deposited without substrate heating.

Film Properties Determined by GIXD

During deposition, the oxygen flow influences chemical and phase composition. Figure 14.5 shows the X-ray patterns of as-deposited layers. Samples deposited at

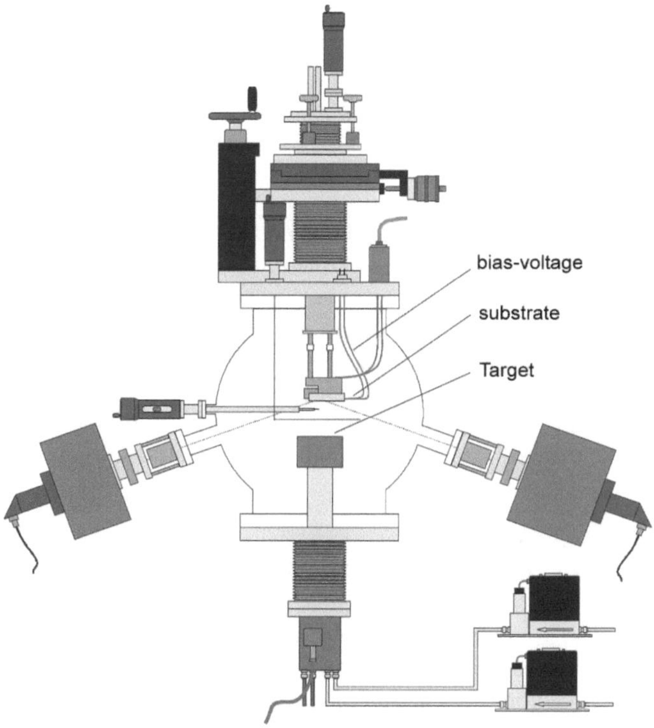

Fig. 14.4 DC magnetron device

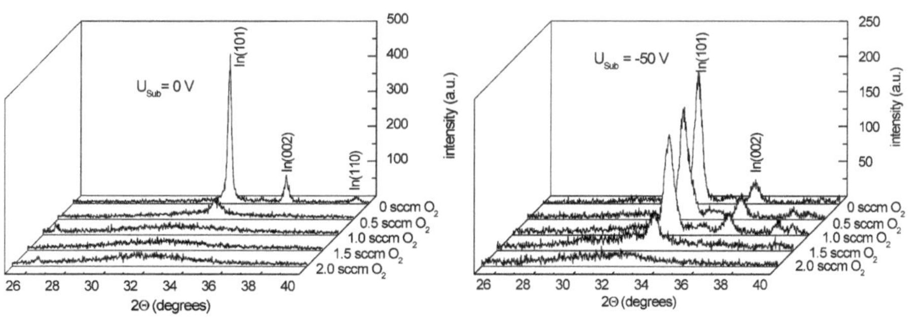

Fig. 14.5 Diffraction patterns of ITO films deposited at different oxygen flows, $U_{\text{bias}} = 0\,\text{V}$ (*left*), $U_{\text{bias}} = -50\,\text{V}$ (*right*)

low oxygen flow show reflections attributable to polycrystalline metallic indium [7]. No preferred orientation was observed. The intensity ratios are similar to those of the X-ray pattern of polycrystalline bulk In. No crystalline indium oxide phases were detected.

The amount of crystalline metallic In/Sn decreases with increasing oxygen flow. In case of $U_{\text{bias}} = 0\,\text{V}$ the films apparently become X-ray amorphous for oxygen flows more than 1 sccm. Appling a negative bias voltage $U_{bias} = -50\,V$ results in partly crystalline metallic films at higher oxygen flows. It appears that a negative bias voltage has the same effect as a diminished oxygen flow to the substrate.

Postannealing of Deposited Indium Tin Oxide, GIXD In Situ Phase Analysis

To extract further information the as-deposited films have been postannealed for several hours at 200 °C. During heat treatment the phase transformation of metallic In/Sn to crystalline indium tin oxide can be observed (Fig. 14.6). The appearance of a crystalline indium oxide phase is due to a slow diffusion of oxygen into the film followed by a fast crystallization process [8].

The new peaks can be identified as reflections of In_2O_3 in the bixbiyte (Mn_2O_3) structure type. Postdeposition treatment is often used to improve the crystallinity and therefore the desired properties as high conductivity and high transparency in TCO films.

Deposition Rate and Film Density Determined by XR

Film thickness and roughness in dependence of the oxygen flow in the deposition chamber can be evaluated from X-ray reflectometry measurements (Fig. 14.7). For these X-ray reflectometry measurements the deposition time was kept constant (30 s). Without oxygen the film thickness is 15.1 nm, with 1 sccm 11.5 nm and with 2 sccm oxygen the thickness is 7.1 nm.

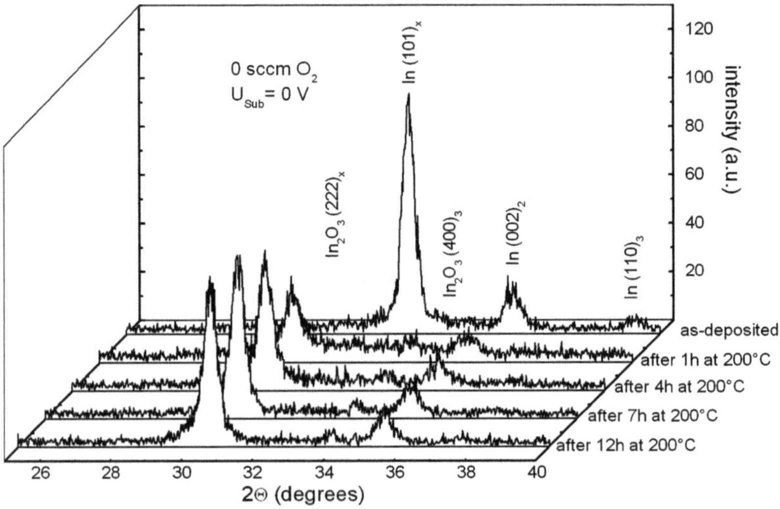

Fig. 14.6 Annealing process – in situ high-temperature diffractometry. Phase transformation of metallic In/Sn into crystalline ITO at 200 °C

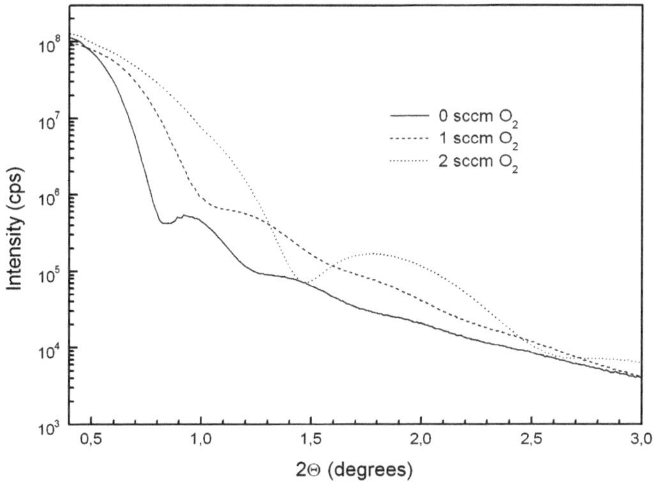

Fig. 14.7 X-ray reflectivity measurements of films deposited at different oxygen flows, $U_{\mathrm{bias}} = 0\,\mathrm{V}$: 0 sccm O_2, thickness 15.1 nm, roughness 1.52 nm; 1 sccm O_2, thickness 11.5 nm, roughness 0.98 nm; 2 sccm O_2, thickness 7.1 nm, roughness 0.75 nm

Fig. 14.8 Density (*triangle*) and deposition rate (*circle*) of samples deposited at 0 V substrate voltage versus oxygen flow

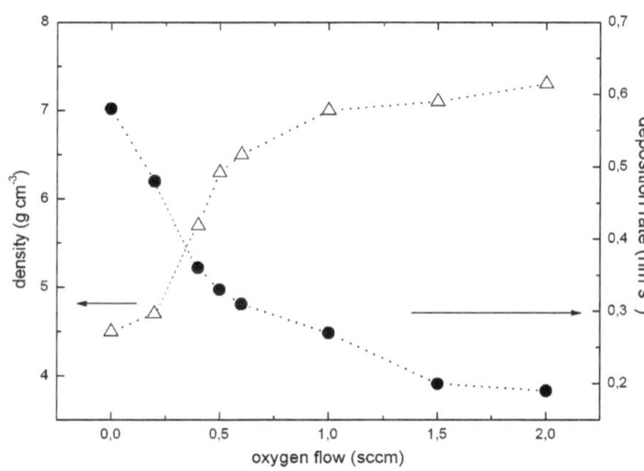

The thickness decreases with increasing oxygen flow and this also holds for the surface roughness of the films. The calculated deposition rate and the film density derived from the XR investigation are displayed in Fig. 14.8. The deposition rate decreases with increasing oxygen partial pressure. The film densities increase with raising oxygen flow. Oxygen flow more than 1 sccm leads to film densities similar to the X-ray density of indium oxide ($7.12\,\mathrm{g\,cm^{-3}}$). The low film density at low oxygen flows in comparison to indium bulk value suggests a high amount of voids in the more metallic films. Increasing oxygen flows effect more compact layers. From XR measurements (see Fig. 14.7) and AFM images (Fig. 14.9) additionally a drop

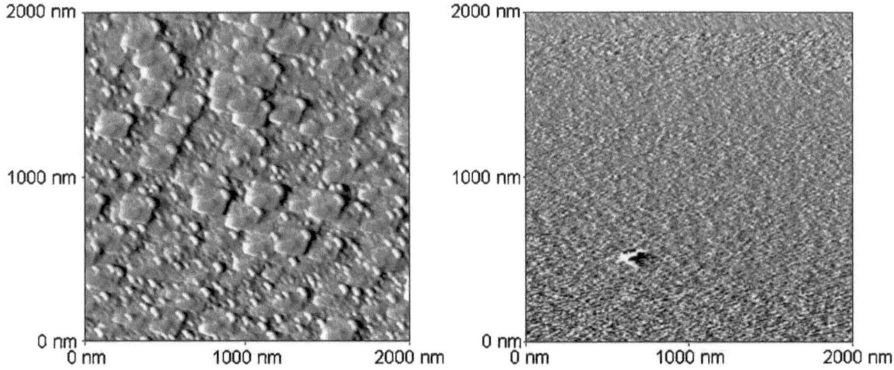

Fig. 14.9 AFM micrographs, samples deposited (*left*) without O_2 and (*right*) at 1.5 sccm O_2; $U_{bias} = 0$ V, deposition time 30 s

of the substrate roughness follows. The metallic films show large particles forming a rough film surface. Increasing oxygen flow causes a smooth surface without clearly observable particles.

The layer growth is determined by the sputter yield on the target and by the reaction of indium/tin and oxygen on the substrate, too. As the oxygen flow is increased the cathode is partially covered by an oxide layer. This results in a drop of the indium/tin sputter yield. Another reason for the observed drop of the deposition rate is the crystal growth. Indium grows in crystallites which include a lot of voids particularly in the initial stage. In_2O_3 grows presumably amorphous (X-ray amorphous), i.e., in very small particles which are closely packed [8, 9]. AFM micrographs of these films (Fig. 14.9) clearly demonstrate that grains become smaller with increasing oxygen flow. The metallic film shows large grain sizes forming a rough film surface. Deposition with higher oxygen partial pressure causes a smooth surface, grains are not clearly observable.

Substrate Bias Influences Film Properties and Film Growth

To obtain further information on the effect of energy flux due to ion energy the substrate voltage U_{bias} was varied from 0 to -100 V at the same oxygen flows. The additional energy flux to the growing films due to increased ion energy can be responsible for changes observed in diffraction patterns (Fig. 14.5) and also in film properties. An increased amount of metallic In/Sn in the films was detected in films deposited at -50 V substrate voltage although the supply of oxygen in the discharges remains the same as in 0 V experiments. Film density and growth rate show a different behavior compared to films deposited at 0 V. In principle the slopes of density and deposition rate curves (Fig. 14.8) are similar. However, the densities are smaller and the growth rates are higher for -50 and -100 V substrate

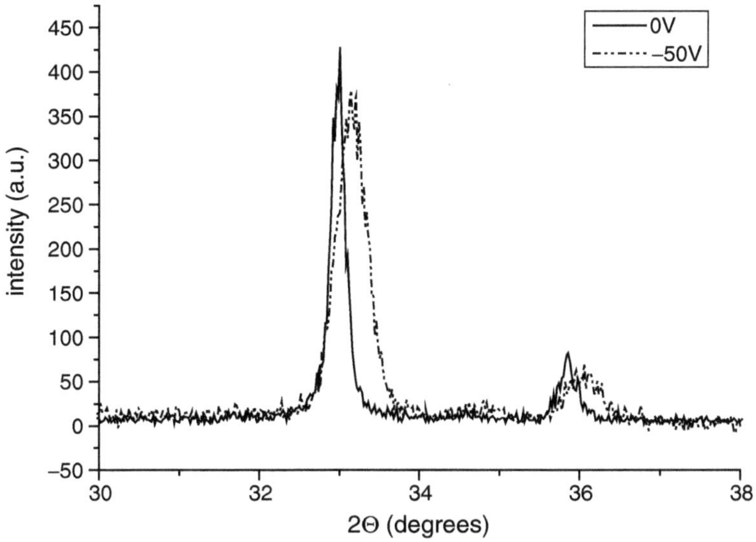

Fig. 14.10 X-ray pattern of In/Sn. line shape and line position are influenced by negative substrate voltage

voltages. There are noticeable differences in the profiles and peak positions and therefore in the microstructure between the 0 V samples and films deposited at −50 V (Fig. 14.10).

Lattice Defects in Thin Surface Layers Influence X-Ray Data

If we compare the indium X-ray profiles at 0 V substrate voltage and the X-ray profiles of a film deposited at −50 V substrate voltages we can observe that the peak profiles are broadened and the peak position is shifted to higher angles (Fig. 14.10). Apparently, the increasing ion energy affects the X-ray profiles.

Lattice defects in thin films can influence X-ray data. Microstructural data are partially or wholly manifested by diffraction intensity, line shape, or the change in line shape with respect to the diffraction angle 2Θ. Imperfections of the first type, such as point defects, displacement disorders or substitution disorders, shift the line position; imperfections of the second type such as domain sizes or dislocations act on the diffraction line shape. The shape analysis of diffraction peaks essentially comprises three problems:

- Extraction of the pure physical line profile from the experimental profile
- Unraveling of the contributions of various types of lattice imperfections to the physical line profile
- Quantitative estimation of microstructural parameters

X-Ray Profile Analysis

The experimentally observable diffraction line profile of an X-ray reflection $h(x)$ is the convolution of a physical line profile $f(y)$ caused by lattice disorder of the second type and an instrumental line profile $g(x)$:

$$h(x) = \int \mathrm{d}y f(y) g(x - y). \tag{14.4}$$

The profile $g(x)$ was determined with a standard material that does not contain defects, strain or particle size broadening. In the described experiments LaB_6, a SRM (standard reference material) from NIST was used. The apparatus function $g(x)$ is separated from the experimental measured profile $h(x)$ applying the Stokes Fourier series [10]. Warren and Aberbach [11] introduced a procedure for separating the contributions from crystallite size and microstrain using the method of Stokes and Wilson [10]. They assumed that the cosine coefficients A_n of the line profile caused by imperfections are simply the product of A_S and A_P, the "distortion" (strain) coefficients and the particle size coefficients. A detailed description of the Warren–Averbach method is given in [12]. On the base of the Warren–Averbach theory [13] Klimanek developed a single profile method [14, 15]. The Fourier coefficients $F(L)$ of the resulting physical line profile $f(y)$ contain information on particle size effect P, mean strains effect S, which are constant within a crystallite or a subgrain, and a dislocation effect D:

$$F(L) = A_P(L)\, A_S(L)\, A_D(L), \tag{14.5}$$

where L is the domain of definition of the experimental line profile

$$L = \frac{\lambda}{2(\sin\Theta_2 - \sin\Theta_1)}, \tag{14.6}$$

with λ the used wavelength and Θ_1 and Θ_2 the lower and upper limit of the observed reflection (hkl).

From

$$-\frac{\ln F(L)}{nL} = \frac{1}{T} + K\left\langle e^2(L)\right\rangle L, \tag{14.7}$$

the linear part of a graph $\ln F(L)/nL$ versus nL allows the determination of the mean particle size T (intercept) and the microstrain $\left\langle e^2(L)\right\rangle^{1/2}$ (slope). The dislocation density can be estimated from the linear branch of a Krivoglaz-Wilkens plot [16, 17]:

$$-\frac{\ln F(L)}{L^2} = \frac{1}{T}L + \left(K\left\langle e^2(L)\right\rangle + B\ln L_0\right) - B\ln L, \tag{14.8}$$

with T the effective particle size, B a factor proportional to the mean total dislocation density, L_0 a length proportional to the core radius r_0 of the strain field of a dislocation and $\langle e^2 (L) \rangle^{1/2}$ the mean square microstrain due to internal stress.

In/Sn Films: Results of the Line Profile Analysis

Figure 14.11 shows typical Warren–Averbach plots $\ln F (L) / nL$ versus nL of three In/Sn films, deposited at 0, −50, and −100 V. There are appreciable differences in the microstructure between the 0 V sample and the films deposited at negative U_{bias}. From 0 to −100 V domain sizes and microstrains increase considerably, attaining a maximum at −50 V and diminish again at −100 V.

Table 14.2 summarizes the results of the line profile analysis. As expected from the X-ray reflection profiles, there is a strong influence of negative substrate voltage on film microstructure. It is generally accepted that domain sizes and interplanar spacings are affected by changes in particle and energy flux to the substrate [18].

In the −50 V samples the crystallite growth is strongly disturbed, domain sizes T are small and microstrain $\langle e^2 (L) \rangle^{1/2}$ and dislocation density ρ_V are high. The shift of the interplanar spacings d to smaller values can be attributed to contraction of the unit cell due to an increasing concentration of vacancies in these films. The defect concentration of the films deposited without negative substrate voltage and at $U_{\text{bias}} = -100$ V are similar, but significantly smaller than that of the −50 V samples.

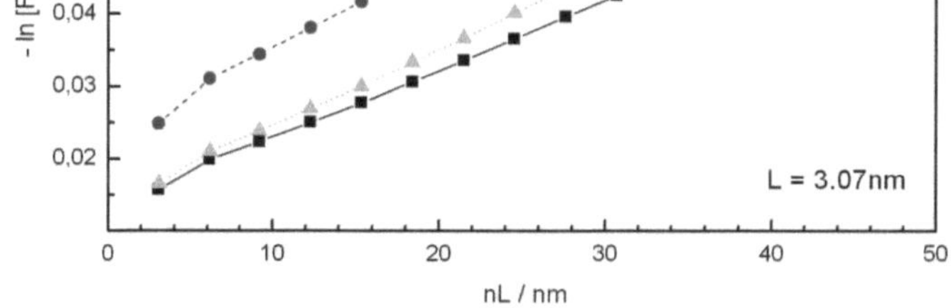

Fig. 14.11 Warren–Averbach plot of In (101)-reflection

Table 14.2 Physical parameters of In/Sn films deposited at various substrate voltages

U_{bias} (V)	T (nm)	$\langle e^2 \rangle^{1/2}$	ρ_V (cm^{-2})	d_{101} (Å)	$\Delta V/V$
0	74	1.87×10^{-3}	0.56×10^{11}	2.718	Reference
−50	43	2.14×10^{-3}	1.10×10^{11}	2.714	−0.00455
−100	71	1.98×10^{-3}	0.64×10^{11}	2.717	−0.00149

The development of the defects in the −100 V samples is not quite clear. We assume that the increasing energy flux causes a higher mobility of the atoms. Thus more often regular lattice positions can be occupied during film growth process.

14.3.2 *Study of Al_2O_3 Formation During Microwave Plasma Treatment of Al Films in Ar–O_2 Gas Mixtures*

The desire to control the physical and chemical properties of very thin metal oxide films by tailoring oxide film growth, morphology, and chemical composition has increased interest in the oxidation of metal surfaces in a variety of application areas such as microelectronics, catalysis, and surface coatings. While thermal oxidation and several chemical vapor deposition (CVD) techniques have been used for many years in aluminum oxide coating production, alternative methods were applied recently to reduce the temperature during the oxidation process. Microwave-induced plasma oxidation is one possibility to form homogeneous alumina layers [19].

The next example concerns the kinetics of aluminum oxide formation during a microwave plasma treatment of Al-films in different Ar/O_2 gas mixtures and different plasma powers. In the center of interest are two questions:

1. How do plasma activated gases affect the structure and the composition of the coatings?
2. How do plasma activated species influence the kinetics of the formation or modification of the layers?

The Al-coatings used in these experiments were prepared by thermal evaporation of Al on Si(100) wafers. The typical thickness of the Al-layers was determined by X-ray reflectometry to about 50 nm. The plasma treatment experiments were carried out in microwave plasma chamber. The Al-samples were laid onto a substrate holder, which is centered in the quartz tube of the plasma reactor. Plasma power, gas composition, and exposure time were varied.

The process gases were two different Ar/O_2 gas mixtures and pure oxygen. By means of a potentiometric oxygen sensor we could determine the partial pressure of oxygen for Ar/O_2 (1) $p(O_2) = 1.5 \times 10^{-6}$ bar, Ar/O_2 (2) $p(O_2) = 2.6 \times 10^{-6}$ bar and for pure O_2 $p(O_2) = 6.1 \times 10^{-6}$ bar at total gas flow of 20 sccm.

The X-ray patterns of a 50 nm thick untreated Al-layer at different incident angles confirm the existence of polycrystalline Al films. These patterns are typical for polycrystalline fcc Al. There is no preferred orientation. Figure 14.12 shows the decrease of the integral intensity of the Al(111)-reflection after each 1h plasma oxidation in pure oxygen (700 W). Also in Ar/O_2-gas with only small amounts of oxygen the intensity of the Al reflection decreases, and the whole film thickness increases to more than 50 nm. Therefore we drew the conclusion that also in Ar/O_2-gas mixtures with small oxygen concentrations chemical reactions take place. The formed Al_2O_3 is X-ray amorphous. XPS and FTIR measurements confirm the formation of Al_2O_3.

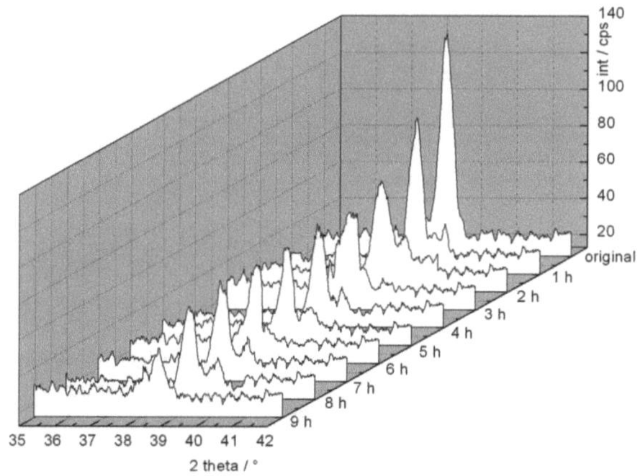

Fig. 14.12 Decrease of the integral intensity of the Al(111)-reflection after each 1 h plasma oxidation (O_2, 700 W)

X-Ray Reflectometry Measurements, Reflectivity Curves

Four reflectivity curves are shown in Fig. 14.13. The values in the legend demonstrate the film parameters obtained from the simulation with the simulation program REFSIM [20]. The best fit is obtained with the assumption of four layers. The two top layers describe the chemical reaction of Al (aluminum) to Al_2O_3 (alumina). The whole thickness in the as-deposited film is 51 nm, the whole thickness after 1h plasma oxidation is 57 nm. The increasing thickness is due to the differences in molecular volume between metallic aluminum and aluminum oxide.

Using these values the maximum amount of stoichiometric Al_2O_3 which could have formed is easy to calculate. We only need the densities of Al and Al_2O_3 from the X-ray reflectivity simulation:

$$\frac{\rho_{\mathrm{Al}}}{M_{\mathrm{Al}}} x_{\mathrm{Al}} = \frac{\rho_{\mathrm{Al_2O_3}}}{M_{\mathrm{Al_2O_3}}} x_{\mathrm{Al_2O_3}}, \tag{14.9}$$

with x the film thickness, ρ the densities, and M the molecular weight. Taking into account (14.9) the oxidation of 1 nm metallic aluminum produces 1.75 nm Al_2O_3.

Formation of Alumina

From the measured thickness of the untreated Al-film (from X-ray reflectometry) and from the integral intensity of the Al(111) peak (GIXD) the thickness of residual Al was calculated after each oxidation step. Together with total thickness of the Al/Al_2O_3 coating, this enabled us to estimate the produced amount of aluminum oxide (Fig. 14.14, dashed line).

From the balance of the number of moles, with consideration of measured oxide densities (from X-ray reflectometry) it is possible to calculate the amount of oxide which is theoretically expected (Fig. 14.14, full line).

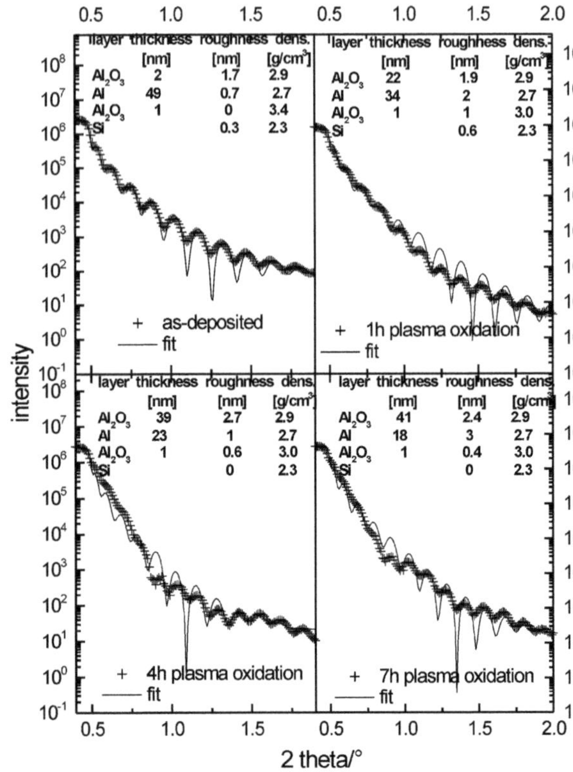

Fig. 14.13 Overview of the Al samples, oxidized in plasma for 1, 4 and 7 h in comparison to the as-deposited layer. The values in the legend present the film parameters

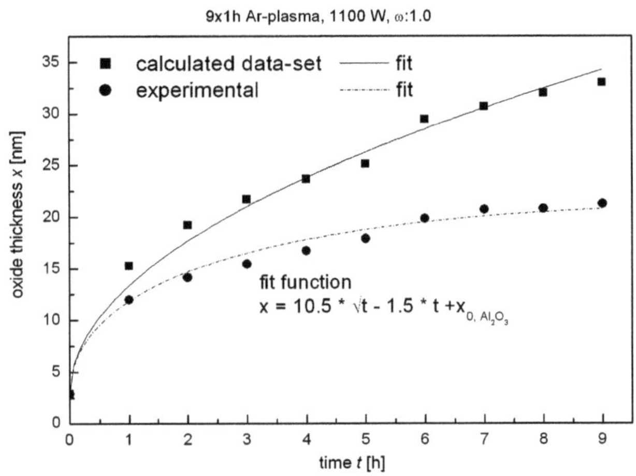

Fig. 14.14 Calculated and experimentally determined thickness of the Al_2O_3 formation. The difference between the curves is due to the sputtering process

The fit-function for theoretical values of oxide thickness x is

$$x = b\sqrt{t}, \tag{14.10}$$

referring to a diffusion limited process [21]. The deviation of the experimental results from the theoretical data-set is attributed to a sputtering process. The experimental data follow the function

$$x = b\sqrt{t} - at, \tag{14.11}$$

with b the diffusion rate constant and a the sputter rate. From the experimental results we draw the conclusion that the formation of aluminum oxide is founded on three processes:

- Formation of plasma activated oxygen
- Diffusion of activated oxygen into the Al-layer to form Al_2O_3
- Simultaneous sputtering of the aluminum oxide which reduces the thickness of the oxide layer

Mathematical Model for the Oxidation Process

On the base of Fick's second law and a linear sputter process we have developed a mathematical model for calculation of the diffusion coefficient D and the sputter coefficient S. The basic idea is that the integral intensity of an X-ray reflection depends on the concentration of the respective crystalline phase within a volume element which is a function of time during the oxide formation. The mean absorption coefficient changes with plasma operation time due to chemical reaction. Because the formed aluminum oxide is X-ray amorphous, its quantification can only be determined indirectly by the decrease of the integral intensities in the X-ray patterns of the Al films (see Fig. 14.12). The time depending of the decrease of the X-ray integral intensity of a particular reflection after each oxidation step was used for the determination of the diffusion coefficient and the sputter rate. We take into account an inhomogeneous layer with the thickness d which consists of a material n (polycrystalline Al) and m (formed Al_2O_3) (Fig. 14.15). The phase transformation to phase m raises the total layer thickness. The thickness of phase m increases, the thickness of phase n decreases. Parts of the formed phase m will be sputtered during plasma modification process, reducing the thickness of phase m and thus the total thickness of the layer.

For the calculation of the concentration profile, Fick's law for one-dimensional diffusion has to be solved:

$$\frac{\partial C(x,t)}{\partial t} = D\,\frac{\partial^2 C(x,t)}{\partial x^2}. \tag{14.12}$$

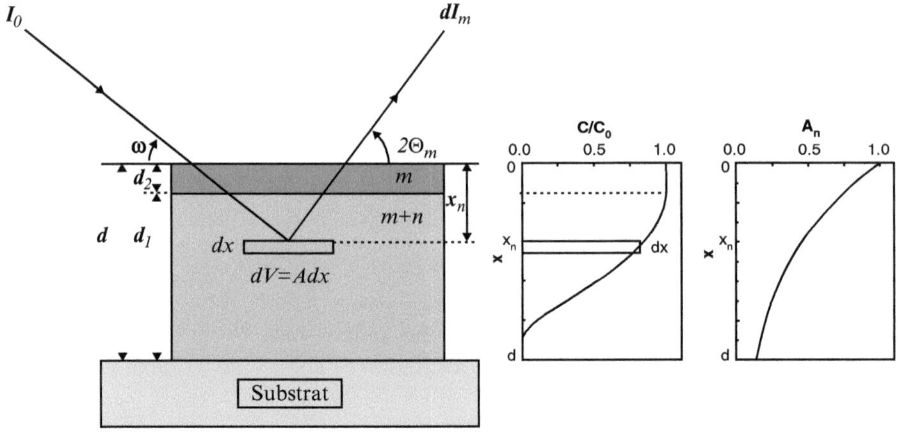

Fig. 14.15 Schematic model for the kinetic of the plasma-enhanced oxidation process

Applying the following conditions:

$$C\,(x,0) = 0, \qquad \text{for } t = 0, \qquad \text{and } 0 \le x < \infty, \tag{14.13a}$$

$$\frac{\partial C\,(x,t)}{\partial x} = 0, \qquad \text{for } x = d_1, \qquad \text{and } 0 \le t < \infty, \tag{14.13b}$$

$$C\,(0,t) = C_0, \qquad \text{for } x = 0, \qquad \text{and } 0 \le t < \infty, \tag{14.13c}$$

the solution is:

$$\frac{C'\,(x,t)}{C'_0} = 1 - \frac{4}{\pi}\sum_{n=1}^{\infty}\frac{(-1)^{n-1}}{2n-1}\cos\left(\frac{2n-1}{2}\pi\frac{d_1-x}{d_1}\right)\exp\left\{-\left(\frac{2n-1}{2}\pi\right)^2\frac{Dt}{d_1^2}\right\}, \tag{14.14}$$

with D the diffusion coefficient .

The molecular volume of Al_2O_3 and Al is different. Then the total thickness of the layer can be written as a function of time:

$$d_1\,(t) = d_{\mathrm{Al}}\,(t) + d_{\mathrm{Al_2O_3}}\,(t) = d_{\mathrm{Al}}\,(t) + K\left(d^0_{\mathrm{Al}} - d_{\mathrm{Al}}\,(t)\right), \tag{14.15}$$

K considers the molvolumina ratio Al_2O_3/Al of the layers. The sputtering process is introduced by

$$d = d_1 + d_2 - S\,t, \tag{14.16}$$

with S, the sputter rate and d_2 the oxide layer thickness. The contribution of a volume element dV in the depth x_n to the measured X-ray integral intensity of a reflection of phase n is given by

$$I_n\,(t) = B_n K_0 I_0\,A\left(1 - \frac{C\,(x,t)}{C_0}\right)A_n\,\mathrm{d}x, \tag{14.17}$$

with B_n the theoretical intensity factor, K_0 the apparatus constant and I_0 the intensity of the incident X-ray beam. A is the irradiated area of the film, C is the concentration of phase m (X-ray amorphous Al_2O_3) in the fraction of phase n in dV. The absorption correction factor A_n is given by

$$A_n = \mathrm{e}^{-\bar{\mu}(x,t)\, z_n x_n}, \tag{14.18}$$

where z corresponds to (14.2). $\bar{\mu}$ is the mean linear absorption coefficient, averaged over all materials in the irradiated film:

$$\bar{\mu} = \mu_n + (\mu_m - \mu_n)\,\frac{1}{x_n}\int_0^{x_n} \mathrm{d}x\,\frac{C\,(x,t)}{C_0}. \tag{14.19}$$

From the known intensity at the beginning $I_n^0(t = 0)$ and the thickness of the oxide at time t the constants in (14.17) can be determined:

$$B_n K_0 I_0 A = \frac{I_n^0 \mu_n z_n\, \mathrm{e}^{\mu_m z_n d_2^0}}{1 - \mathrm{e}^{-\mu_n z_n d_1^0}}. \tag{14.20}$$

By employing (14.20) into (14.17) follows:

$$I_n\,(t) = \frac{I_n^0 \mu_n z_n\, \mathrm{e}^{\mu_m z_n d_2^0}}{1 - \mathrm{e}^{-\mu_n z_n d_1^0}}\,\mathrm{e}^{-\mu_m z_n d_2}\int_0^{d_1} \mathrm{d}x\,\left(1 - \frac{C\,(x,t)}{C_0}\right)\mathrm{e}^{-\bar{\mu}(x,t) z_n x}. \tag{14.21}$$

By comparing the measured and the calculated peak intensities as well as the total thicknesses, the D and S values were adjusted to get the best fit between the curves.

Diffusion Coefficients and Sputter Rates

Figure 14.16 shows the diffusions coefficients and the sputter rate S in dependence on the microwave plasma and the gas mixtures. At constant power D increases as the amount of oxygen in the plasma is increased and D also increases with increasing

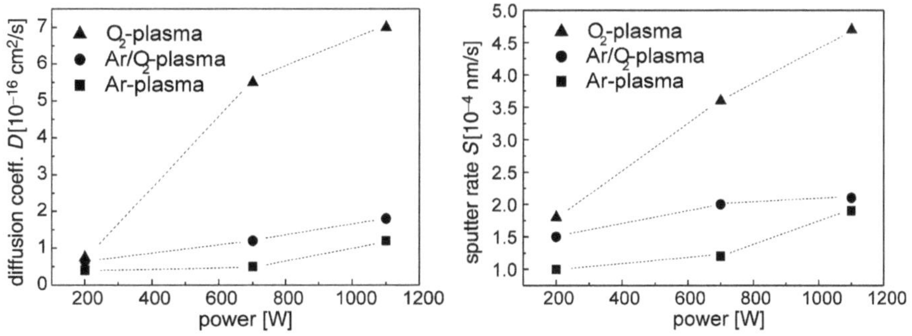

Fig. 14.16 Diffusion coefficient D and sputter rate S in dependence on the incoupled microwave power. D and S increase with incoupled microwave power as well as with increasing oxygen in the plasma

microwave power. Because of the effect of microwave power on the kinetic energy of the plasma particles an increasing number of defects is incorporated in the aluminum films and the oxide layers. This causes a subsequent penetration of activated oxygen trough the oxide film to the Al-film and oxide can be produced more quickly. The sputtering process limits the oxide formation [19]. The sputter rate S depends on plasma power and plasma gas mixture. Both, the kinetic energy of plasma particles and the amount of activated oxygen, obviously influence the sputter process.

Temperature Dependence of D and S

The activation energy E_A for the diffusion process in microwave plasma was calculated in usual way from the Arrhenius plot [21]. Figure 14.17 shows the Arrhenius plot for different gas mixtures and plasma powers in a temperature range from 150 to 300 °C. The values of activation energies $E_A = -23.4\,\mathrm{kJ\,mol^{-1}}$ (O_2 plasma, 700 W), $E_A = -23.2\,\mathrm{kJ\,mol^{-1}}$ (O_2 plasma, 200 W) and $E_A = -19.0\,\mathrm{kJ\,mol^{-1}}$ (Ar/O_2 (1) plasma, 700 W) point to a similar diffusion mechanism irrespective of the microwave power and the used plasma gas mixtures. The values are clearly smaller than the activation energy for thermal-induced oxidation of aluminum. For thermal-induced diffusion in fcc aluminum the activation energies are $E_A = -131\,\mathrm{kJ\,mol^{-1}}$ for lattice diffusion and $E_A = -69.1\,\mathrm{kJ\,mol^{-1}}$ for grain boundary diffusion [22]. The different reaction path compared to a thermal stimulated process can be assigned to atomic or ionic species even at comparatively low temperatures. As expected, the sputter rates are independent of temperature in the temperature range under investigation [23].

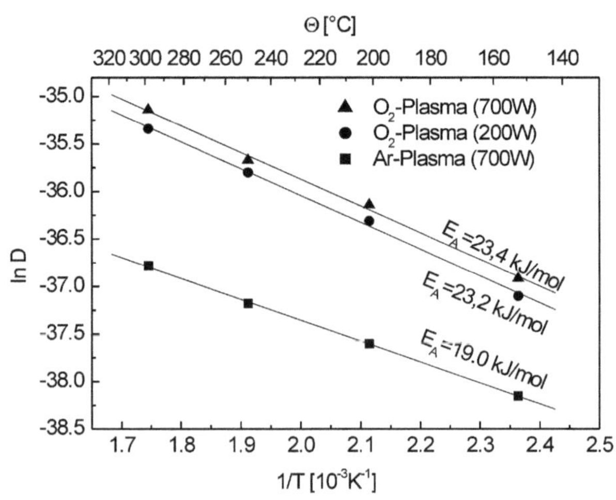

Fig. 14.17 Arrhenius-plot, determination of activation energy E_A in dependence on gas mixture and power

14.4 Summary

Grazing incidence X-ray diffractometry and X-ray reflectometry were used to study *non destructive* the microstructure of thin films as well as the influence of plasma parameters on microstructure and film formation processes.

The examples were aimed to demonstrate the effectiveness of X-ray methods to investigate plasma treated surfaces.

Four types of diagnostic methods have been discussed, which are of interest to investigations of plasma treated surfaces, the phase analysis, the defect structure analysis, film formation processes under plasma exposure and the study of growth kinetic in thin films.

Acknowledgments This work was supported by the Deutsche Forschungsgemeinschaft DFG through SFB198 and SFB TR24. I would like to acknowledge the contributions of many colleagues to the work described in this chapter, in particular the contributions of Dr. M. Quaas, Dr. A. Quade, and Dr. H. Steffen.

References

1. W.C. Marra, P. Eisenberger, A.Y. Cho, J. Appl. Phys. **50**, 6927 (1979)
2. B. Lengeler, Microchim. Acta **1**, 455 (1987)
3. B. Szyska, Vak. Forsch. Praxis **1**, 38 (2001)
4. M. Quaas, C. Eggs, H. Wulff, Thin Solid Films **332**, 277 (1998)
5. M. Quaas, H. Wulff, H. Steffen, R. Hippler, Surf. Sci. **540**, 337 (2003)
6. W. Beyer, J. Huepkes, H. Siebig, Thin Solid Films **516**, 147 (2007)
7. International Centre for Diffraction Data (ICDD), PDF-2 (1999)
8. H. Wulff, M. Quaas, H. Steffen, Thin Solid Films **355–356**, 395 (1999)
9. M. Quaas, H. Steffen, R. Hippler, H. Wulff, Surf. Sci. **454–456**, 790 (2000)
10. R. Stokes, Proc. Phys. Soc. **61**, 382 (1948)
11. B.L. Averbach, B.E. Warren, J. Appl. Phys. **20**, 885 (1949)
12. B.E. Warren, *X-Ray Diffraction* (Dover, New York, 1991), p. 251
13. B.E. Warren, Prog. Metal Phys. **9**, 147 (1969)
14. P. Klimanek, Freib. Forsch. B **265**, 76 (1988)
15. P. Klimanek, Mater. Sci. Forum **79–82**, 73 (1991)
16. M.A. Krivoglaz, *X-Ray and Neutron Diffraction in Nonideal Crystals* (Springer, Berlin, 1996)
17. M. Wilkens, Phys. Stat. Sol. (a) **2**, 359 (1970)
18. E.J. Mittemeijer, P. Scardi (eds.), *Diffraction Analysis of the Microstructure of Materials* (Springer, Berlin, 2004)
19. A. Quade, H. Wulff, Thin Solid Films **377–378**, 626 (2000)
20. S. Grassel, D. Fuchs, Program Package Win-Refsim, Siemens AG (1992)
21. K. Meyer, *Physikalisch-chemische Kristallographie* (Verlag fuer Grundstoffindustrie, Leipzig, 1968)
22. M. Ohring, *The Material Science of Thin Films* (Academic, San Diego, 1992), p. 358
23. P. Sigmund, Phys. Rev. **184**, 383 (1969)

Chapter 15
The Use of Nonthermal Plasmas in Environmental Applications

Kurt H. Becker

Abstract Nonthermal plasmas are widely used environmental applications covering a broad and diverse range of topics, which have been studied extensively in laboratory, bench-scale investigations, particularly during the past two decades. However, only a few research breakthroughs were successfully translated into economically viable technologies for large-scale industrial applications. Those include ozonizers using dielectric barrier discharges and electrostatic precipitators using corona discharges. In this chapter, we describe the current status of ozonizers and electrostatic precipitators. We also try to identify some of the bottlenecks that prevent other nonthermal plasma-based environmental technologies from becoming widely applicable. As an example, we look at the use of nonthermal plasmas for the treatment of gaseous and (to a lesser extent) liquid waste streams, with an emphasis on waste streams containing volatile organic compounds (VOCs). Bottlenecks preventing or slowing down the rapid and efficient translation of successful bench-scale studies to widely used industrial technologies include the identification, quantification, and control of byproducts produced in the plasma chemical reactions and establishing carbon closure (i.e., the accounting of the ultimate fate of each initial carbon atom that is exposed to the plasma).

15.1 Introduction

The use of nonthermal ("cold") plasma for environmental applications has been studied extensively for a long time in the context of water purification and disinfection using ozone (O_3) generated in dielectric barrier discharges (DBDs) as well as in the context of electrostatic precipitators (ESPs) using corona discharges. Plasma-assisted, plasma-initiated, and plasma-enabled remediation of gaseous (and

K.H. Becker (✉)
Department of Physics, Polytechnic Institute of New York University, Six MetroTech Center, Brooklyn, NY 11201, USA
e-mail: kbecker@poly.edu

M. Bonitz et al. (eds.), *Introduction to Complex Plasmas*, Springer Series on Atomic, Optical, and Plasma Physics 59, DOI 10.1007/978-3-642-10592-0_15,

to a lesser extent liquid) waste streams has been the subject of intense laboratory studies for a long time. A summary of the work performed in this field before 1993 can be found in the book by Penetrante and Schultheiss [1]. Areas of application for nonthermal plasmas in environmental remediation include primarily NO_x and SO_x decomposition and the destruction of volatile organic compounds (VOCs). High-energy electron beams, corona discharges, dielectric barrier discharges, and various surface-type discharges, sometimes in conjunction with a packed bed of ferroelectric pellets, have been the most widely utilized discharge configurations for the generation of nonthermal plasmas for environmental applications [1]. From a practical point of view, plasmas that can be generated and maintained at atmospheric pressure are more desirable than low-pressure plasma, since they do not require costly vacuum chambers, pumps, and other vacuum equipment.

Self-sustaining, diffuse discharge plasmas at high pressures, where single collision conditions no longer prevail and collisional and radiative processes beyond binary collisions involving ground-state species become important, are difficult to maintain because of their inherent instability and susceptibility to filamentation and transition to an arc [2]. One promising route to the generation and maintenance of stable, high-pressure plasmas uses spatial confinement of the plasma to a spatial region, where at least one characteristic dimension is smaller 1 mm, so-called *microplasmas* [3, 4]. The energy and current densities that can be realized in microplasmas are much higher than those in large-volume plasmas at atmospheric pressure.

While much research has been carried out using laboratory-scale plasma-based remediation systems, the transition to industrial prototypes and commercial pilot units has been slow outside of ozonizers and electrostatic precipitators. This is due, on one hand, to engineering challenges such as choice of materials for long-term operation, energy, and cost efficiency of plasma-based methods relative to competing technologies and to problems associated with the scale-up of laboratory reactors with gas flow rates of liters per minute or less to industrial size units for flow rates of many cubic meters per minute. On the other hand, there are also remaining science issues that have not been adequately addressed such as the complete characterization and control of byproduct formation, dependence of destruction efficiencies on contaminant concentration, reactor volume and geometry, and residence time. Many of these issues are related to the fact that the plasma chemistry involved in the destruction processes is not well understood and that we lack a detailed understanding of how the discharge physics is coupled to the plasma chemistry. Ideally, the destruction of VOCs would lead to CO_2 and H_2O as the only ultimate reaction products, that is, full mineralization of the VOC. However, that goal is rarely, if ever, achieved in practice.

The remaining part of this chapter is organized as follows. Ozonizers and electrostatic precipitators are discussed in Sect. 2. Section 3 discusses various aspects of plasma-assisted degradation of VOCs in low-flow environments including byproduct formation and carbon closure. Section 4 is devoted to a brief summary of oxidant formation and organics destruction in water using pulsed electrical discharges.

15.2 Commercially Viable, Large-Scale Plasma-Based Environmental Applications

To date, there are only two commercially viable, that is, large-scale industrial plasma-based environmental applications: electrostatic precipitators and ozonizers. An excellent summary of the state of the art of both applications can be found in a recently published book [5]. Here, we will only give a brief overview of both applications.

15.2.1 Ozonizers

15.2.1.1 Historical Background

Industrial ozone (O_3) generation is the classical application of nonequilibrium air plasmas at atmospheric pressure. Low temperature is mandatory because ozone molecules decay fast at elevated temperature. At the same time, a relatively high pressure is required because ozone formation is a three-body reaction involving an oxygen atom, an O_2 molecule, and a third collision partner, O_2 or N_2. The DBD (or silent discharge) originally proposed by Siemens in 1857 [6] for "ozonizing air" is ideally suited for this purpose. Ozone is an extremely effective oxidant. Siemens, in collaboration with Ohlmüller, tested the effect of ozone exposure on cholera, typhus, and coliform bacteria. The result was complete sterility after ozone treatment [7]. Subsequently, industrial ozone production started for applications in small water treatment plants followed by installation in major drinking water plants all over the world in the early twentieth century. Since then, ozone generating technology has been closely linked to the development of water purification processes. In many countries, ozonation in water treatment was later replaced by more cost-effective processes using chlorine or chlorine compounds, which are not only cheaper, but also more soluble in water than ozone. Recent concerns about potentially harmful disinfection byproducts have reversed this trend back toward the use of ozone. To date, more than 250 operating water purification plants in the US use ozone.

15.2.1.2 Ozone Properties and Ozone Applications

O_3 is a colorless gas with a characteristic odor. It condenses at $-112°C$ to an indigo blue liquid, which is highly explosive. Below $-193°C$, ozone forms a deep blue–violet solid. Because of the explosion hazard, O_3 is used only in diluted form in gas or water streams. Due to its oxidizing power, it can be used as a potent germicide and viricide as well as a bleaching agent. Ozone is increasingly used to replace other oxidants such as chlorine that present more environmental problems and safety hazards. Because strong oxidants are chemically active species, their storage, handling, and transportation involve substantial hazards. Ozone, in particular, is inherently unstable, so that only controlled in situ, on-demand generation at a rate controlled by its consumption in the process is practical in most applications.

The most important application of ozone is still the treatment of water, because O_3 is capable of oxidizing many organic and inorganic compounds in water. Ozone chemistry in water involves the generation of hydroxyl free radicals, which react rapidly with chemicals such as aromatic hydrocarbons, unsaturated compounds, aliphatic alcohols, and formic acid.

15.2.1.3 Ozone Formation in Electrical Discharges

Ozone can be generated in different types of gas discharges in which the electron energy is high enough to dissociate O_2 and in which the gas temperature can be kept low enough for the O_3 molecules to survive without undergoing thermal decomposition. Nonequilibrium discharges meet these requirements, above all corona discharges, and dielectric barrier discharges.

Ozone formation in both positive and negative corona discharges has been extensively investigated and is reasonably well understood, although the process is rarely used on an industrial scale. The preferred discharge type for industrial ozone generators has always been the DBD (or silent discharge). In recent years, industrial ozone generation benefited very much from a better understanding of the discharge properties and of the ozone formation process [8–10]. Operating in air or O_2 at pressures between 1 and 3 bar, at frequencies between 0.5 and 5 kHz, and using gap spacings in the mm range, the discharge is always a filamentary DBD. Major improvements were obtained by tailoring microdischarge properties in air or in oxygen in such a way that recombination of oxygen atoms is minimized and ozone formation is optimized. Cylindrical as well as planar electrode configurations have been used. The majority of commercial ozone generators use cylindrical electrodes forming narrow annular discharge spaces of 0.5–1 mm radial width. The outer electrode is normally a stainless steel tube, which is at ground potential and which is water-cooled. These tubes have a length of 1–4 m. The coaxial inner electrode is a glass or ceramic tube, closed at one side, and having an inner metal coating as a high-voltage electrode (Fig. 15.1), or a closed steel cylinder covered by a dielectric layer (ceramic and enamel). The feed gas is streaming in the axial direction through the annular discharge region between the inner and outer tube. Each volume element of the flowing gas is subjected to the action of many microdischarges and the gas leaves the ozonizer enriched with ozone.

Any electric discharge in air or oxygen causes chemical changes induced by reactions of electrons or ions with N_2, O_2 and trace elements like H_2O and CO_2 and subsequent free radical reactions. Extensive lists of possible reactions have been collected and reliable sets of rate coefficients have been established [11]. It is noteworthy to state that ion reactions play only a minor role and that the main reactions are triggered by the electron impact dissociation of O_2 and N_2 and of a few excited molecular states. In pure oxygen, which is used in most commercial ozonizers, O_3

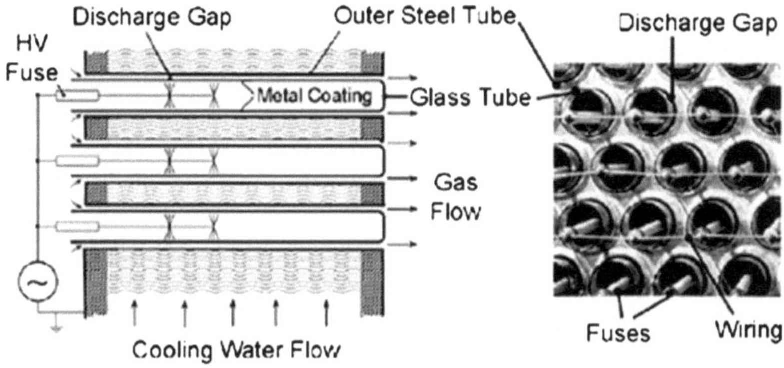

Fig. 15.1 Water-cooled discharge tubes in an O_3 generator

formation is a straightforward process. O_3 always originates from a three-body reaction of an oxygen atom reacting with two O_2 molecules:

$$O + O_2 + O_2 \rightarrow O_3^* + O_2 \rightarrow O_3 + O_2 \tag{15.1}$$

where O_3^* stands for a transient excited state in which the O_3 molecule is initially formed. The time scale for O_3 formation in atmospheric-pressure oxygen is a few microseconds. Fast side reactions (15.2a)–(15.2d) converting O atoms or O_3 into O_2 compete with the ozone formation and pose an upper limit on the O atom concentration or the degree of O_2 dissociation tolerable in the DBD:

$$O + O + O_2 \rightarrow 2O_2 \tag{15.2a}$$

$$O + O_3 + O_2 \rightarrow 3O_2 \tag{15.2b}$$

$$O(^1D) + O_3 \rightarrow 2O_2 \tag{15.2c}$$

$$O + O_3^* + O_2 \rightarrow 3O_2 \tag{15.2d}$$

Computations with large reaction schemes show that complete conversion of O to O_3 can only be expected, if the concentration ratio $[O]/[O_2]$ stays below 10^{-4}. If, on the other hand, the energy density in a microdischarge and consequently also the degree of dissociation is too low, a considerable fraction of the deposited energy is dissipated by ions (up to 50%). Since ions do not appreciably contribute to the O_3 formation, this situation has to be avoided. A reasonable compromise between excessive energy losses due to ions and best use of O atoms for O_3 formation is found when the relative oxygen atom concentration in a microdischarge reaches about 2×10^{-3} in the microdischarge channel. This concentration can be obtained at an energy density of about $20\,\mathrm{mJ/cm^3}$ [12]. At zero O_3 background concentration, this leads to a maximum energy efficiency of ozone formation corresponding to roughly 25%.

In a conventional ozonizer, the O_3 concentration in the gas stream passing through the generator increases along the path of the gas flow due to the accumulated action of a large number of microdischarges. With increasing O_3 concentration, back reactions gain importance, which finally leads to a situation where each additional microdischarge destroys as much O_3 as it generates. The attainable saturation concentration defined by this equilibrium depends strongly on the pressure and on gas temperature, but is typically around 20%.

The O_3 formation in air is more complicated and will not be discussed here. We refer the reader to [5] for details of the ozone formation processes in the presence of N_2. We note that the maximum attainable O_3 concentration in dry air is lower than in pure O_2. The situation is further complicated, if water vapor is present in the feed gas (humid air) [5].

15.2.1.4 Technical Aspects of Large Ozone Generators

Large ozone generators use several hundred discharge tubes and can produce up to 100 kg ozone per hour. Typically, several O_3 generators are installed. Figure 15.2 shows a photograph of the entrance section of a large ozonizer. Depending on the feed gas, ozone concentration up to 5 wt% (in air) and up to 18 wt% (in pure O_2) can be obtained. Advanced water treatment processes utilize O_3 at concentrations up to 12 wt%. Depending on the desired O_3 concentration, the energy required to produce 1 kg of O_3 ranges from 7.5 to 10 kWh in O_2 and from 15 to 20 kWh in air. Such a level of performance of the ozonizer requires that stringent design criteria and operating conditions are met (alignment of the tubes relative to one another,

Fig. 15.2 Large O_3 generator at the Los Angeles Aqueduct Filtration Plant

constant gap width, pressure, temperature of the coolant, purity and preparation of the feed gas, thermal and electrical properties of the dielectric, type and stability of the power source, etc.).

15.2.1.5 Future Prospects of Industrial Ozone Generation

A better understanding of microdischarge properties in nonequilibrium DBDs and advances in power semiconductors resulted in improved performance and reliability of industrial ozone generation in recent years. Higher O_3 generation efficiencies and drastically reduced sizes of ozonizers helped lower the manufacturing and operating costs. Today, O_3 can be produced at a total cost of about 2 US$/kg. Further progress can be expected through engineering efforts aimed at developing dielectrics with superior properties, better flow control, and improved thermal management. Rapid advances in power semiconductor design resulting in improved gate turn-off thyristors and insulated gate bipolar transistors, which are now capable of switching 1,000 A at 5 kV will have a major impact. Soon bulky step-up transformers will no longer be required and almost arbitrary excitation wave forms can be generated with high efficiency. Investigations into homogeneous self-sustained volume discharges may lead to more favorable plasma condition for O_3 formation [5, 13].

A novel approach to operate a DBD for ozone generation, the so-called *intelligent gap design*, was recently proposed by Vezzu [14, 15]. Here, the discharge tube is segmented into three or four parts and different types of microdischarges are generated in the different segments using geometric considerations (e.g., a conical gap) and changing the dielectric properties of each segment. As a consequence, strong microdischarges are maintained in the first segment (where the background O_3 concentration is low) generating 50–70% of the final O_3 concentration. In the subsequent segments (where there is an appreciable O_3 background concentration), the microdischarges are progressively weaker in an effort to minimize the back reactions that favor some of the O_3 destruction reactions. Ozone generation concentrations exceeding 20 wt% at the ozonizer exit (in pure oxygen) have been routinely obtained and maintained over long periods of operation [14, 15].

15.2.2 Electrostatic Precipitation

15.2.2.1 Historical Background

The earliest account of the influence of electric discharges on smoke, fumes, and suspended particles dates back to 1600 [16] when Gilbert published De Magnete (on The Magnet), a comprehensive review of what was then known about electrical and magnetic phenomena. In 1824, Hohlfeld [17] reported an experiment of clearing smoke in a jar by applying a high voltage to a corona wire electrode. Lodge [18] was the first to systematically investigate this effect and to put it to test on a large scale in lead smelters to suppress the white lead fume escaping from the chimney. A number of important industrial applications followed and resulted in an improved precipitator, an electrical device, which could collect dusts and fumes as well as acid mists and fogs. Cottrell [19] was the first to realize that for precipitation the negative corona discharge was superior to the positive corona. Within a few years, commercial applications evolved for collecting sulfuric acid mists, for zinc and lead fumes, for cement kiln dust, for gold and silver recovery from electrolytic copper slimes, and for alkali salt recovery from waste liquors in paper–pulp plants [19]. Soon, the first use of ESPs collecting fly ash from a pulverized coal-fired power plant was reported and, very quickly, this became by far the largest single application of ESPs. Investigations by Deutsch [20, 21] and Seeliger [22] brought new insight in the physical processes involved in electrostatic precipitation and a first quantitative formulation of precipitator performance.

The main advantages of ESPs are that various types of dust, mist, droplets, etc., can be collected under both dry and wet conditions, and also that submicron size particles can be collected with high efficiency. ESPs can handle very large air or flue gas streams, typically at atmospheric pressure, with low power consumption and low-pressure drop. These properties led to a number of large-scale commercial applications in steel mills, nonferrous metal processing, cement kilns, pulp/paper plants, power plants and waste incinerators, sulfuric acid plants, and in petroleum refineries for powder catalyst recovery. Much smaller ESPs of different designs are used for indoor air cleaning in homes and offices.

15.2.2.2 Main Physical Processes Involved in Electrostatic Precipitation

Electrostatic precipitation is a physical process in which particles suspended in a gas flow are charged electrically by ions produced in a corona discharge, separated from the gas stream under the influence of an electric field, and transported to collecting plates, from which they can be removed periodically mechanically (dry ESP) or continuously by washing (wet ESP). Typical configurations are corona wires centered in cylinders or wires mounted at the center plane between parallel plates forming ducts (Fig. 15.3). The electrodes can be simple weighted wires, barbed wires, helical wires, or rods, serrated strips and many other kinds. They all have in common that they have parts with a small radius of curvature or sharp edges to facilitate corona formation. The particle laden gas flow is channeled to pass through many cylinders or ducts either in the vertical (cylinders) or in the horizontal (ducts) direction. In larger precipitators, negative coronas are used almost exclusively because they have a larger stability range and can be operated at higher voltages. Electrode plate distances of 0.2–0.4 m and voltages in the range 50–110 kV are common. Small ESPs for indoor air cleaning normally use positive coronas, because they produce less ozone, a matter of great concern for indoor applications.

The active corona region where electrons and positive and negative ions are generated is restricted to a very thin layer around the corona electrodes. Typically, ionization occurs only in a layer extending a fraction of 1 mm into the gas volume.

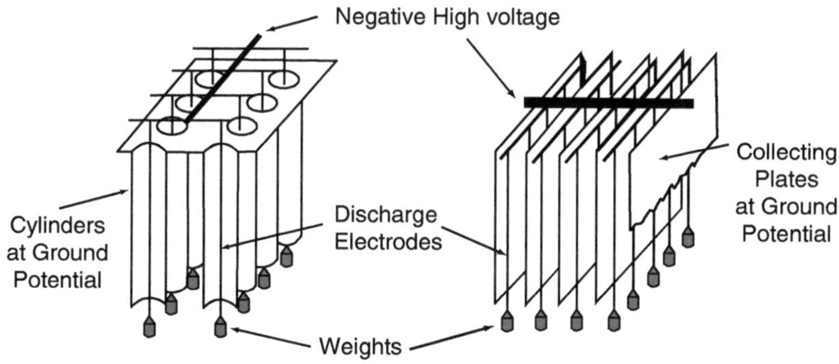

Fig. 15.3 Precipitator configurations with weighted wire corona electrodes

Positive ions travel only a short distance to the negative electrode, while electrons and negative ions start moving toward the collecting surface at ground potential. In air or flue gas mixtures at atmospheric pressure, dissociative electron attachment to O_2, CO_2, or H_2O form negative ions, which fill most of the space in the duct. They are utilized to charge dust particles, so that these can be subjected to electrical forces in order to separate the dust from the gas stream.

The physical processes involved in corona charging of powders and droplets have been studied in great detail. Apart from precipitators, these phenomena are utilized in electrophotography [23], copying machines, printers, liquid spray guns, and in powder coating [24]. Solid particles or droplets entering a precipitator pass many corona zones, undergo collisions with ions resulting in charge accumulating, and are subjected to Coulomb forces in the electric field and to drag forces in the viscous flow. The charging process of solid particles or droplets has two main contributions, the relative importance of which depends on particle size. Field charging is the dominating process for particles of diameter of about 2 mm or more. Charging stops when a saturation charge is reached and additional approaching ions will be deflected in the electric field of the previously accumulated charges on the particle. Under typical precipitator conditions, a 5 mm particle may accumulate several thousand elementary charges. For very small particles, field charging becomes a slow process and another charging process, diffusion charging, which depends on the Brownian motion of the ions, takes over [25]. Diffusion charging does not depend on the electric field and does not reach a saturation charge.

15.2.2.3 Large Industrial Electrostatic Precipitators

Industrial precipitators can be very large installations. As an example, the precipitator at the exit of a pulverized coal-fired utility boiler of a 500 MW power plant consumes about 200 tons of coal per hour resulting in 20–80 tons of fly ash per hour, depending on the origin and quality of coal. At the exit of the boiler, the fly ash particles, which range from 0.1 to 10 mm in size, are dispersed in a flue gas stream of about 2,500,000 m^3/h with a mass concentration of about 20 g/m^3. To meet acceptable output concentrations of 20 mg/m^3, the precipitator has to reach a weight collection efficiency of 99.9%, which can be achieved routinely (and even a 99.99% collection efficiency is not out of the question). One of the major achievements of modern precipitator technology is the fact that these performance goals can be reached with a very low power consumption of 0.1% of the generated power and a pressure drop of only 1 mbar. To handle such a large gas flow, the flue gas is slowed down to about 1 m/s, channeled into many parallel ducts of 15 m height, up to 15 m length, and 0.3–0.4 m width. Such large ESPs are subdivided into segments of about 5 m length. About 110–150 such ducts add up to a total width of 45 m and a total collection area of 60,000 m^2 is realized.

Up until recently, ESPs were designed on the basis of the quantitative investigations of Deutsch in the early part of the twentieth century [20, 21, 26], which, for the first time, established a quantitative relation between the collection efficiency of a precipitator and some operational and geometry parameters. Today, a better understanding of the pertinent physical processes and the availability of fast computers and advanced computational tools allows sophisticated modeling and simulation to determine the paths of individual particles through a large precipitator. This approach requires a sufficiently accurate description of the field distribution and ion production, of the charging process, of the flow field, and of the particle motion. Because of the strong interaction between the different processes, the differential equations describing the various processes have to be solved simultaneously with appropriate boundary conditions. All computations show that there is a particle size range between 0.1 and 1 mm diameter that is difficult to collect. Larger particles are more efficiently collected because they accumulate sufficient charge in the corona zones and are subjected to strong electric forces. Very small particles are also easily collected because they experience less flow resistance when particle diameters approach the mean free path of the gas molecules.

Other effects may limit the performance of ESPs, for example, large amounts of fine dust entering the precipitator, which causes the corona current in the entrance sections to drop to a small fraction of its value in the absence of the dust (corona quenching). The collected material on the collecting plates can also cause performance limitations. If particles have a very low electrical resistivity, for example, metal particles, they do not stick to the collecting plates and cannot be collected efficiently. On the other hand, if the dust resistivity is very high, the deposited dust layer will eventually limit the current flow and stop the corona. Normally, a different phenomenon, called back corona, occurs instead. When the deposited dust forms a porous layer of growing thickness, gas breakdown in interstices and on particle surfaces can occur, which causes the corona current to increase. Now positive ions, generated by the back corona inside the dust, travel toward the center electrodes and counteract the charging process, which results in a "bipolar" corona. Obviously, only unipolar ion flow guarantees optimum charging conditions. Back coronas are observed in precipitators serving boilers using low sulfur coal and also in powder coating, where high resistivity polymer particles and pigments are deposited.

A significant improvement in the ESP performance can be obtained by replacing direct-current (dc) high-voltage excitation by intermittent alternate-current (ac) excitation, in which a certain number of half-cycles or even several cycles of the exciting voltage are suppressed. Thus, peak voltages higher than those achievable with dc excitation and lower average voltages and average currents are obtained. This results in energy savings as well as in an improved performance of the precipitator, if back coronas are a problem. Even better results can be obtained with pulsed excitation. In a pulsed corona, the duration of the applied voltage pulse is shorter than the ion transit time from the discharge electrode to the collecting plate (typically of the order 1 ms in a large precipitator). In practice, periodic short high-voltage pulses (pulse widths from 1 to 300 ms and pulse repetition rates of 30–300 s^{-1}) are superimposed on a dc high voltage. This increases the uniformity of the corona along the discharge electrodes and on the collecting plates and helps suppress the occurrence of back coronas in the collection of high resistivity dust.

15.2.2.4 Summary

Electrostatic precipitation is the leading and most versatile method for high-efficiency collection of solid particles, fumes, and mists escaping from industrial processes. It presents by far the most important application of industrial air pollution control. About 100 years of practical experience with various kinds of dust, a growing understanding of the physical processes involved, and more recently, the use of advanced computational tools simulating the particle charging, particle motion, and particle collection process have led to its present supremacy.

15.3 Decomposition of Volatile Organic Compounds in Microplasmas

As an example for the plasma-assisted remediation of gaseous waste streams, data are presented on the use of atmospheric-pressure microplasmas in the decomposition of selected prototypical VOCs in low gas flow environments such as advanced life support (ALS) systems. Issues associated with byproduct characterization, quantification, and control – which represent obstacles in the rapid technological commercialization of these plasma processes – will be discussed briefly.

15.3.1 Experimental Details

All results described in this chapter were obtained with plasma reactors based on the capillary plasma electrode (CPE) discharge. The novel aspect of the CPE discharge is a novel electrode design [3, 27], which uses dielectric capillaries that cover one or both electrodes of a discharge device. In many other aspects, the CPE discharge looks similar to a conventional DBD. However, the CPE discharge exhibits a mode of operation that is not observed in a DBD, the so-called *capillary jet mode*. Here, the capillaries serve as plasma sources and produce jets of high-intensity plasma at high pressure, which emerge from the end of the capillary and form a "plasma electrode" for the main plasma. The CPE discharge displays two distinct modes of operation when excited by pulsed dc or ac. When the frequency of the applied voltage pulse is increased above a few KHz, one observes first a diffuse mode. When the frequency reaches a critical value (which depends strongly on the length-to-diameter, L/D, ratio and the feed gas), the capillaries "turn on" and a bright, intense plasma jet emerges from the capillaries. When many capillaries are placed in close proximity to each other, the emerging plasma jets overlap and the discharge appears uniform. This "capillary" mode has been characterized in a rudimentary way for several laboratory-scale research discharge devices in terms of its characteristic electric and other properties [28–32]. Using a Monte Carlo modeling code [33], the existence of the threshold frequency, which depends critically on the L/D ratio of the capillaries and the dielectric material, has been verified [28, 29]. The model also predicts relatively high average electron ,energies of 5–6 eV in the "capillary" mode, which have been verified experimentally [33].

Figure 15.4 shows schematically three plasma reactors that were used in the VOC destruction experiments carried out in our group. The top diagram in Fig. 15.4 shows a rectangular reactor with solid-pin electrodes, which is operated in the cross-flow regime (CF-CPE reactor). This design maximizes contaminant exposure to the plasma and thus improves the energy efficiency. The reactor consists of two parallel dielectric plates, one of which is perforated with 100 capillaries of 0.4 mm diameter each. The plates are separated by a 3.2 or 1.6 mm gap through which the contaminated gas stream passes (cross-flow) resulting in a plasma volume of about 50 or 25 cm^3. The cathode consists of metallic pins, which are partially inserted into the capillaries. The use of metallic pins minimizes energy losses in the system and improves plasma generation and stability.

The center diagram in Fig. 15.4 shows a rectangular plasma reactor with hollow-pin electrodes, which is operated in the flow-through regime (FT-CPE reactor). The gas stream is introduced through the hollow-pin electrodes and the capillaries. This maximizes the exposure of the contaminants to the region of highest plasma density. The design employed 37 pins of 0.8 mm diameter in an effort to minimize the development of a pressure drop through the system when the gas flows through the hollow pins. Depending on the gap size between the two parallel plates, the effective reactor volume was 3.5 and 6.8 cm^3, respectively.

The bottom diagram in Fig. 15.4 shows an annular (A-CPE) reactor, which was designed to simulate plug-flow conditions. The reactor body is made of Pyrex glass. One of the electrodes consists of an Al screen that jackets the Pyrex glass cylinder, whereas the other electrode is made of an aluminum bar placed concentrically inside the glass cylinder surrounded by a perforated alumina silicate dielectric material. The contaminant streams are fed axially into the reactor. This mode of operation allows the contaminated gas stream to flow through the entire annulus

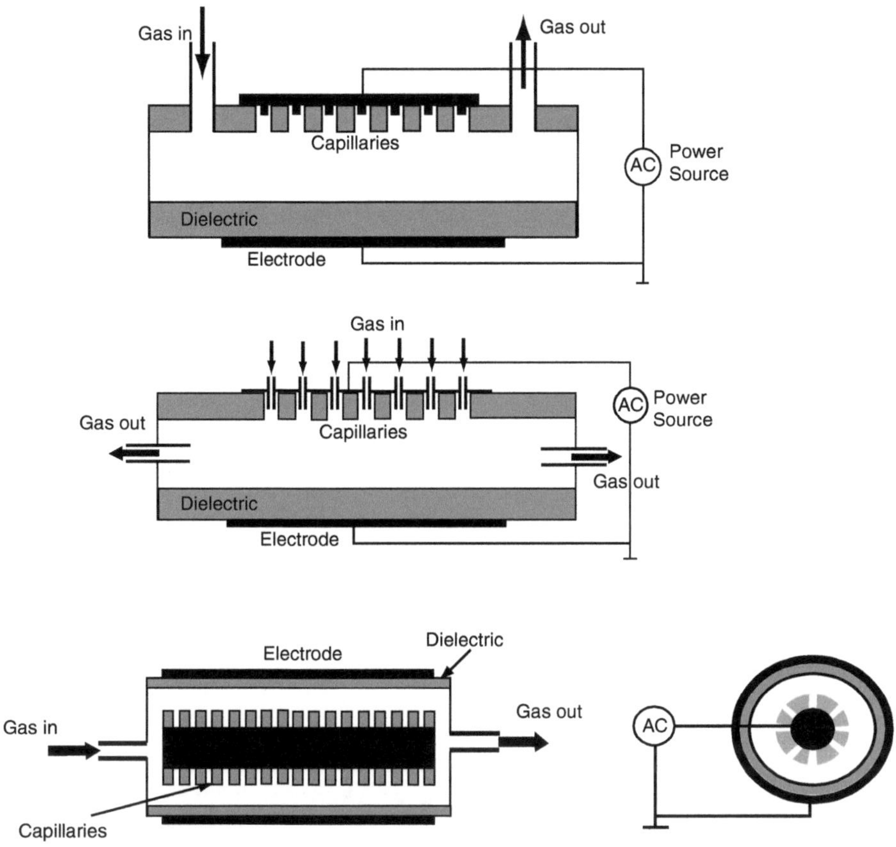

Fig. 15.4 Three capillary plasma electrode (CPE) discharge configurations. *Top*: cross-sectional view of a cross-flow CPE (CF-CPE) reactor with solid-pin electrodes. *Center*: cross-sectional view of a flow-through CPE (FT-CPE) reactor with hollow-pin electrodes. *Bottom*: two cross-sectional views of an annular CPE (A-CPE) reactor with a solid center electrode

of the reactor where exposure to the plasma occurs. The overall dimensions of the reactor were 254 mm in length with a 38 mm outer diameter and a plasma volume of about 20 cm^3.

15.3.2 VOC Destruction Efficiency

The results of comprehensive studies of VOC destruction using the three CPE plasma reactors schematically shown in Fig. 15.4 have been reported recently [34–36] in the context of exploring the use of nonthermal plasmas generated in these reactors for use in ALS system designs for controlled cabin environments (submarine, aircraft, spacecraft, and space station). VOCs studied included prototypical aliphatic (ethylene, heptane, and octane) and aromatic (benzene, toluene, ethylbenzene, and xylene) compounds as well as ammonia. Initial contaminant concentrations varied between 75 and 1,500 ppm(v) (parts per million in volume) in ambient air with gas flow rates between 1 and 10 l/min, which are typical for ALS systems. Destruction efficiencies were determined as a function of plasma energy density, initial contaminant concentration, residence time in the plasma volume, reactor volume, and single contaminant versus contaminant mixture and a kinetic model was developed to determine the relevant rate constants for contaminant destruction. Input powers ranging from 10 to 100 W were used to generate plasma volumes between 3 and 20 cm^3. This resulted in energy densities (J/cm^3), defined as electrical input power (W) divided by the gas flow rate entering the plasma reactor (cm^3/s), between 0.5 and 10 J/cm^3 and residence times in the range of 0.2–2 s. For all compounds, operating conditions could be identified that guaranteed destruction efficiencies of 95% and higher; in many cases, complete destruction (within the detection limits) could be obtained. Figure 15.5, as an example, shows the destruction efficiency for benzene in two CPE reactors, an annular reactor and a cross-flow, solid-pin electrode reactor. As can be seen, essentially complete destruction is achieved in the annular reactor for specific energies of 3 J/cm^3 and above. On the other hand, specific energies approaching values of up 10 J/cm^3 are required for comparable destruction efficiency in the cross-flow reactor. This indicates that optimization of the reactor geometry is a critical component in achieving maximum destruction efficiency.

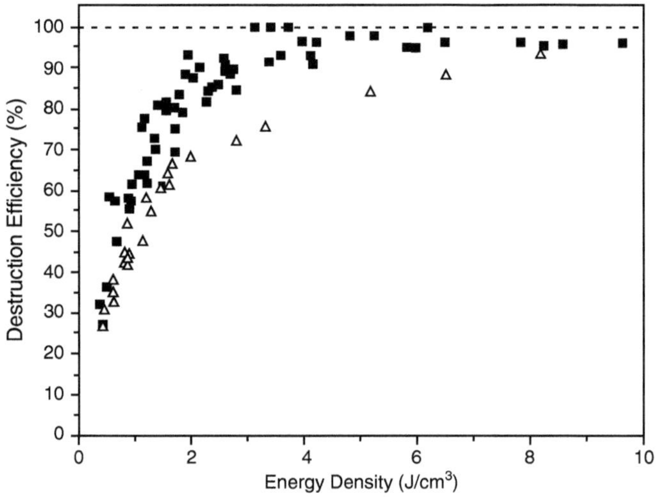

Fig. 15.5 Benzene destruction efficiency in an annular CPE plasma reactor (*solid squares*) and in a cross-flow CPE plasma reactor (*open triangles*). For both reactors, date are shown for initial contaminant concentrations varying between 200 and 1,200 ppm(v) and flow rates varying between 2 and 8 l/m; in all cases, the residence was chosen to be sufficiently long to ensure maximum destruction efficiency

15.3.3 Byproduct Formation

The maximum destruction efficiency of a given compound in a particular reactor, however, is only one aspect of the VOC destruction process. Byproduct formation is another important consideration, particularly in the context of the practical implementation of plasma-based processes. Ultimately, one needs to measure the concentrations of all identified byproducts and the gaseous carbon oxides, CO and CO_2. In air plasmas, the formation of byproducts such as the nitrogen oxides (N_2O, NO, NO_2, and N_2O_5, which in the presence of moisture will be converted rapidly to nitric acid, HNO_3) and ozone are of great concern. In our studies [35, 36], we analyzed the plasma-treated gas stream leaving the reactor using a variety of standard techniques and instruments including UV/visible emission spectroscopy, UV/visible spectrophotometry, FTIR spectroscopy, gas chromatography with mass spectrometry (GC–MS), gas chromatography with a thermal conductivity detector (GC–TCD), a Foxboro Miran 980 nondispersed infrared (ND-IR) spectrophotometer, and an ENERAC 2000E Electrochemical Combustion Gas Analyzer for NO_x, CO, and CO_2. In addition to the byproducts mentioned earlier, we also found that the plasma-initiated degradation of essentially all aliphatic and aromatic compounds studied here led to the formation of formaldehyde as well as several alkanes, alkenes, alcohols, and aldehydes, which were impossible to identify uniquely and to quantify.

"Carbon closure" can be determined by evaluating the carbon mass balances for the reactants and all identified products. If contaminants and products are measured in ppm(v), the contribution of each component to the carbon concentration (measured in $\mu g/cm^3$) can be expressed as

$$C_{C(i)} = [(pM_i/RT)]C_i r_C, \tag{15.3}$$

where $C_{C(i)}$ denotes the carbon concentration contributed by component i, p refers to the pressure, M_i is the molar mass of component i, R is the universal gas constant, T is the absolute temperature, C_i refers to the concentration of compound i, and $r_C = M_C/M$ denotes the carbon mass content per mole of compound i. The resulting product carbon is obtained by adding the carbon concentrations of the unreacted hydrocarbons, CO, and CO_2 using the above equation. Carbon closure is then computed by taking the ratio of the product carbon to the influent carbon. Carbon closure for benzene and ethylene in our studies was essentially 100%, which indicates that all influent carbon is accounted for in the form of identified final products, CO, and CO_2. Carbon closure for toluene, heptane, octane, and ethylbenzene ranged from 70 to 90%. Although no attempt was made [8–10] to identify and quantify all byproducts, the gap between reported carbon closure values and 100% is most likely due to the formation of low-mass chain hydrocarbons, alcohols, and aldehydes [37–39].

15.3.4 Kinetic Studies

A limited number of kinetic studies on the degradation of organic and inorganic contaminants have been performed. Penetrante [38] used an empirical relationship to successfully fit degradation data for carbon tetrachloride. According to their model, the final contaminant concentration is directly proportional to the initial concentration and the exponential of the negative input energy density. The model uses two empirical fitting parameters that can be determined experimentally. A double-exponential kinetic model has been proposed by Yamamoto [40]. However, the complexity of this model and the number of parameters that must be determined experimentally limit severely its use as a tool for reactor design and operation. Mok [41] used an assumed (not experimentally or computationally verified) plug-flow concept and a second-order expression with respect to contaminant concentration and power density to arrive at a pseudo-first-order model that is fairly easy to use for reactor design and performance prediction.

Here, we summarize (1) the development of a kinetic model to describe the degradation of selected contaminants in one of our plasma reactors [36] and (2) the results of a series of kinetic studies aimed at determining the relevant apparent plasma chemical degradation rate constants and comparing them to the model predictions. Three model compounds were selected for this purpose, toluene, ethylbenzene, and *m*-xylene, primarily because of their chemical similarity.

We first summarize the kinetic model that was developed for the rectangular plasma reactor operated in the flow-through regime (Fig. 15.4, center diagram) [36]. Flow-through plasma reactors operate in a way that exposes the gas stream to comparatively short plasma residence times of typically only a fraction of a second, which makes the experimental determination of the flow regime in the reactor difficult, if not impossible [35, 42]. Alternatively, computational methods can be employed for this purpose. The Reynolds (N_{Re}) and dispersion (N_{d}) numbers can provide information on the flow conditions in the reactor. The calculated Reynolds numbers for our reactor range from about 15 to more than 120 for flow rates between 1 and 10 l/min corresponding to residence times from as long as 0.5 s to as short as 0.05 s, so that laminar flow conditions prevail in the reactor over a wide range of operating conditions. The relative effect of dispersion in the reactor can be evaluated from calculated dispersion numbers obtained from the dispersion coefficients, the flow-through velocity, the reactor length, and the contaminant molecular diffusivity of compound A in air (B) [43]. For the three prototypical VOCs studied here, molecular diffusivities of $8.0 \times 10^{-6}\,\mathrm{m^2/s}$ (toluene) and $7.5 \times 10^{-6}\,\mathrm{m^2/s}$ (ethylbenzene and *m*-xylene) were calculated. Using these values and flow rates in the range of 1–10 l/min, the dispersion numbers, N_{d}, for all three target compounds were found to be in the range of 0.14–0.93. When these values are entered in dispersion nomographs available in the literature [43] along with an assumed final to initial concentration ratio $C/C_0 = 0.1$ (corresponding to 90% contaminant removal), we find that plug-flow conditions prevail in the reactor. Thus, the reactor used in tour studies can be modeled as a plug-flow reactor subject to the following assumptions:

(a) Plug-flow conditions prevail throughout the reactor.
(b) The temperature remains constant; thus, the density of the gaseous influent and effluent streams remains constant.

As discussed in detail in [36], a second-order rate expression appears to be the natural way to describe the plasma chemical degradation of gaseous contaminants (assuming that the plasma chemical degradation of environmental contaminants depends on both the concentration of the reactive plasma species such as radicals, charged, and neutral species and the concentration of the contaminant)

$$r = -k_2 C_{\mathrm{PRS}} C, \tag{15.4}$$

where C_{PRS} and C are the concentration of the plasma reactive species and the contaminant, respectively, and k_2 is the second-order reaction rate constant. The concentration of the generated plasma reactive species can be considered directly proportional to the power input per unit reaction volume:

$$C_{\mathrm{PRS}} = k_{\mathrm{p}} P_{\mathrm{W}} / V, \tag{15.5}$$

where k_{p} is a proportionality factor, P_{W} denotes the power input, and V refers to the reactor volume. This yields

$$r = -k_{\mathrm{d}}(P_{\mathrm{W}}/V)C, \tag{15.6}$$

where k_{d} is an "apparent" plasma chemical degradation rate constant, defined as the product of k_{p} and k_2. Inserting (15.6) into the definition of the contaminant residence time, τ

$$\tau = \int_{C_0}^{C} \mathrm{d}C/r \tag{15.7}$$

and integrating over the range of influent/effluent concentrations yields

$$\ln(C/C_0) = -k_{\mathrm{d}}(P_{\mathrm{W}}/V)\tau = -k_{\mathrm{d}} E_{\mathrm{s}}, \tag{15.8}$$

where the energy density E_{s} in the reactor is defined as the ratio of the power input P_{W} and flow rate Q, $E_{\mathrm{s}} = P_{\mathrm{W}}/Q$. This yields

$$C = C_0 \exp(-k_{\mathrm{d}} E_{\mathrm{s}}). \tag{15.9}$$

The above equation models the plasma chemical degradation of organic contaminants as a pseudo-first-order process with respect to the energy density and we finally arrive at an expression for the treatment efficiency E of the form

$$E = \frac{C_0 - C}{C_0} = \frac{1 - C}{C_0} = 1 - \exp(-k_{\mathrm{d}} E_{\mathrm{s}}). \tag{15.10}$$

Equations (15.9) and (15.10) allow the determination of the apparent plasma chemical destruction rate constant from plots of the concentration or degradation efficiency versus the energy density.

Kinetic studies were carried out to verify the above model and to determine the relevant apparent plasma chemical degradation rate constants for the three model compounds mentioned earlier (toluene, ethylbenzene, and *m*-xylene). Tight control was exercised in these studies on keeping the influent flow rate and contaminant concentration constant. For each contaminant, five sets of experiments were carried out corresponding to plasma powers of 20, 30, 40, 50, and 75 W. At each power, a number of runs were carried out by varying the influent stream flow rate in the range from 2.0 to 8.0 l/s. For all experimental runs, the influent target compound concentration were kept constant (to better than 3%) at 265, 270, and 155 ppm(v) for, respectively, toluene, ethylbenzene, and *m*-xylene. Effluent concentration data for the three target contaminants are plotted versus energy density and the exponential decay was fitted to the data by using a nonlinear regression routine (Fig. 15.6). The values of the correlation coefficient (R^2) indicate excellent quality of the fit for ethylbenzene and toluene (not shown here) with $R^2 > 0.95$ and a very good fit for

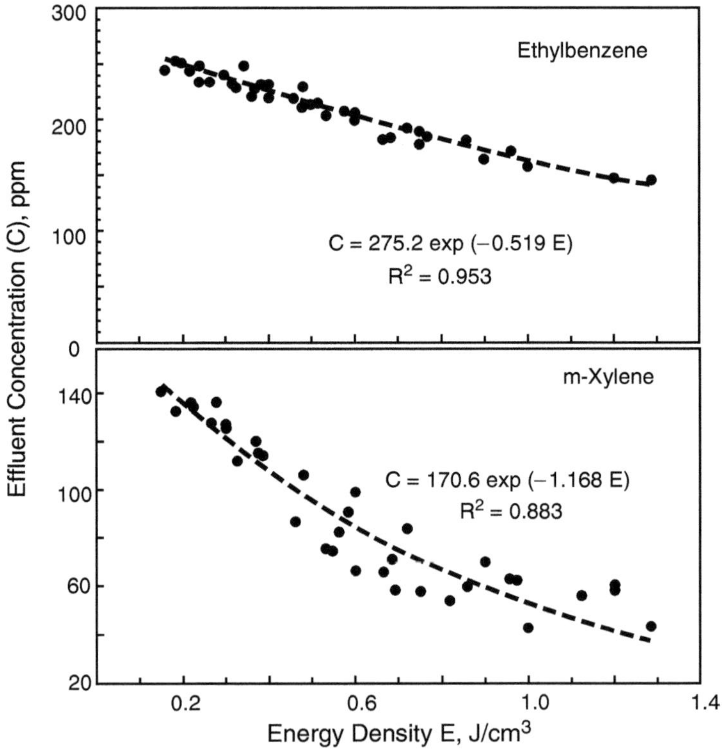

Fig. 15.6 Evaluation of kinetic data for the plasma chemical degradation of ethylbenzene and *m*-xylene

m-xylene ($R^2 > 0.88$). The error in the determination of the apparent degradation rate was also small (3–7%). The experimental values for the initial concentration and those derived from the model are in very good agreement. A comparison of the apparent degradation rate constants for the three model compounds shows that the rate constants are of the same order of magnitude. The apparent degradation rates decrease from m-xylene to toluene to ethylbenzene.

15.3.5 Summary

The results of the extensive parametric and kinetic studies reported in [34–36] can be summarized as follows:

1. Maximum VOC destruction efficiencies between 95 and 100% were obtained for all compounds studied.
2. The VOC destruction efficiency increased with the specific energy initially, but leveled off at values of the specific energy that are compound dependent.
3. The destruction efficiency increased with increasing initial contaminant concentration, that is, dilute waste streams are more difficult to treat.
4. The VOC destruction efficiency increased with increasing residence time initially, but leveled off for higher values of the residence time; the minimum residence time to achieve maximum destruction efficiency was compound dependent.
5. The VOC destruction efficiency increased with reactor volume as long as the plasma density did not fall below a certain threshold; when the plasma density was below that level, an increase in reactor volume did not result in a higher destruction efficiency.
6. For chemically similar compounds, for example, aromatic hydrocarbons, the maximum destruction efficiency was found to be inversely related to the ionization energy of the compound and directly related to the degree of chemical substitution; this suggests that chemical substitution sites are the places of highest plasma-induced chemical activity.
7. Byproduct formation was found to be an important issue that slows down the rapid translation of plasma-based VOC degradation processes that are successful in the research laboratory to their implementation on a large scale in industry. The identification, quantification, and control of various undesirable byproducts remain a major challenge and bottleneck along with the full accounting of the fate of each C atom (carbon closure).
8. For one of the plasma reactors used in this study, a model was developed that treats the reactor as a plug-flow reactor following pseudo-first-order reaction kinetics with respect to the specific energy; model predictions for plasma chemical degradation rates are in good agreement with experimentally determined rates and were found to decline from m-xylene to toluene to ethylbenzene.

15.4 Pulsed Electrical Discharges in Water

The topic of (pulsed) electrical discharges generated in water, aqueous solutions, and liquids for disinfection purposes is very broad and it is impossible to discuss this subject in depth within the constraints of this chapter. Therefore, no attempt is made in this section to provide comprehensive coverage of all aspects relating to the physics and chemistry of pulsed electrical discharges in "liquids." Instead, a brief overview of the basic mechanisms involved in the generation of pulsed electrical discharges in a liquid is presented. Subsequently, the results of some specific experiments aimed at elucidating the mechanisms that lead to the generation of reactive species (OH, O_3, O, and H_2O_2) by a pulsed discharge in water are discussed (for further details, we refer the reader to [44–51]). Lastly, we will briefly introduce some results for the disinfection and decontamination of water using pulsed electrical discharges, a process that has the advantage that it does not require the transfer or disposal of chemicals, which is a necessity in conventional disinfection processes using O_3 or Cl_2.

15.4.1 Background

The generation of a discharge in water requires electric field strengths of about 2 MV/cm (depending on the conductivity of the water) and proceeds through a reduction in the local density by field or heat related mechanisms (i.e., by forming gaseous bubbles in the liquid phase) thus creating conditions favorable for the development of electron and ionization avalanches [44, 52]. This process consumes a significant amount of power and converts a large fraction of the input energy into heat. Producing electrical discharges in water in the presence of externally introduced gas bubbles improves the energy balance, with more energy used to produce chemically active species [53–55]. The discharge is initiated in the gas bubble, so that no energy has to be consumed in order to form the bubbles in the liquid. Operating at electric field strengths of about 30 kV/cm limits the discharges to the gaseous bubbles. It is also possible to create a discharge above the water surface by placing one electrode outside the water [50, 51], in which case the discharge occurs mostly in the gas phase above the water.

Different mechanisms are possible for creating an electrical discharge in the gas bubbles. Investigations using mesh and wire electrodes report discharges that initiate at points of contact between the wire, the solid insulation, air, and possibly water and proceed along the surface of the bubble [54]. Some experiments have used bubbles passing through the water relatively far from the electrodes. In other arrangements, bubbles spend long periods of time attached to the electrodes and the bubble may be in contact with one or both electrodes. In cases where the gas bubble is in direct contact with both electrodes, the discharge forms via an electron avalanche and a transition to an arc must be prevented by the outside electrical circuit or by using short voltage pulses. If water covers one or both electrodes, the discharge current is limited, particularly for short voltage pulses. As a consequence, dielectric barrier discharges or corona discharges may be expected [45,50–56]. The study of discharges in bubbled water using wire-to-plane, wire-to-wire, or mesh electrode configurations is complicated by the flow of bubbles [57, 58]. The discharge can occur in various bubbles at the same time and hence the field in the bubbles is difficult to estimate and the properties of the discharge are difficult to study. Single-bubble systems allow detailed studies of the discharge process in the bubble, which is stationary and can be subjected to repeated voltage pulses.

15.4.2 Experimental Systems

Two experimental systems that have been used in our laboratory will be discussed here briefly. The first system allows a discharge to be created in a single stationary bubble in water [55]. This system consists of a water chamber constructed from PVC to reduce the impedance, a Blumlein pulse forming network, a gas delivery system, and various diagnostic tools. The chamber has three ports (for the fiber optic cable, the needle electrode with gas inlet, and the movable disk electrode) and two windows at 90° to each other (see Fig. 15.7a). A gas flow controller delivers gas to a microsyringe, which is used to form a gas bubble at the tip of a hypodermic needle inserted into the water chamber. The amount of gas injected can be controlled up to a microliter and a pressure release valve ensures that the overall pressure remains at about one atmosphere. The syringe is used to control the size of the bubble. A digital camera is used to record the size of the bubble and a UV transparent optical fiber is used to collect optical emissions from the bubble. Argon and oxygen bubbles have been used in this study. This experimental arrangement allows the study of a single stationary bubble with just the tip of the needle extending into the bubble (see Fig. 15.7b). The hypodermic needle, which is mostly covered by the insulating material of the chamber with only a 0.5 mm tip protruding into the bubble,

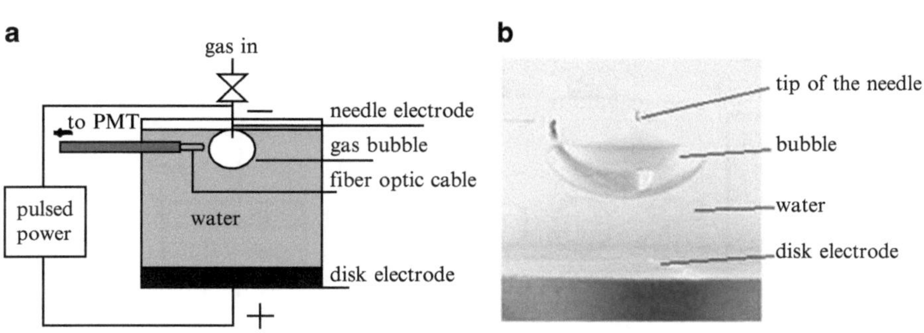

Fig. 15.7 (**a**) Experimental arrangement showing the bubble around the needle electrode with the disk electrode immersed in water. Gas is fed through the needle and a stationary bubble forms around the tip of the needle. (**b**) Bubble and the electrodes

serves as the negative electrode. The second electrode is a stainless steel disk. The distance between the electrodes can be varied between about 10 and 25 mm. The electrical power delivery system consists of a dc high-voltage power supply and a pulse shaping network. A 60 kV, 10 mA power supply charges the pulse generator. A Thyratron switch delivers the voltage pulse. The Thyratron is triggered manually delivering single pulses at intervals from 30 s to several minutes. The Blumlein pulse shaping network delivers a 1 μs long rectangular pulse in the 3–30 kV range. Optical emissions of the OH band (309 nm) and the Ar lines (765 nm) from inside the bubble are sent to a cooled photomultiplier tube (PMT) with a temporal resolution of 2 ns. All current, voltage, and PMT data are recorded by a digital oscilloscope. Deionized water was used with a conductivity of 1.2–1.5 μScm, a pH of 5.0, and a temperature of 21°C. Experimental results reported here include measurements of optical emissions and the discharge current for various sizes of Ar and O_2 bubbles.

The second experimental setup, shown in Fig. 15.8, is a reactor that was developed for water disinfection applications. This system includes a gas flow controller and water pump. The power supply employs two rotating spark-gap switches to generate high-voltage pulses and a 40 kV dc power supply. This reactor employs a set of mesh electrodes in plane-to-plane configuration connected sequentially either to high voltage or to ground. A forced oxygen bubble flow passes vertically through the electrode system along with the flow of contaminated water from a raw water tank. The treated water is collected in a reservoir. In addition to monitoring the

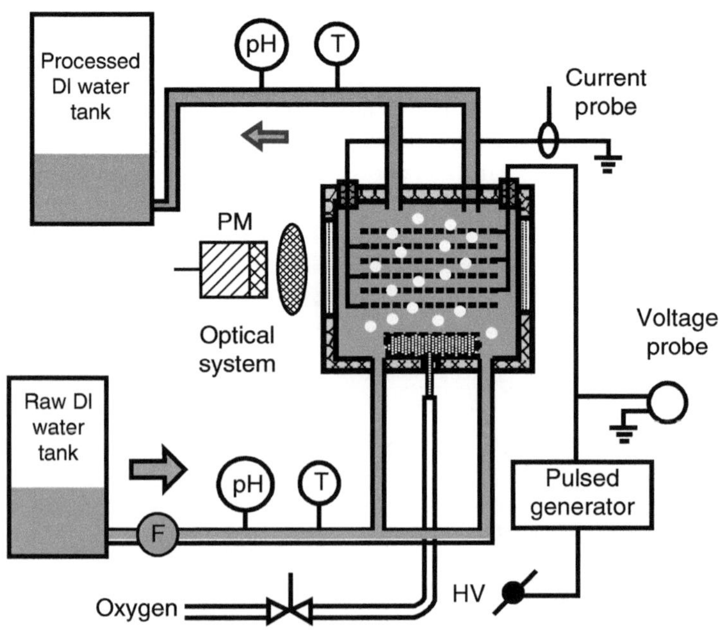

Fig. 15.8 Water treatment system

current, voltage, and optical data, the system is also equipped with diagnostics that allow the measurement of radical concentrations in the water and the concentration of selected contaminants in the water prior to and after plasma treatment.

15.4.3 Selected Experimental Results

Experiments using the single-bubble system (Fig. 15.7) were conducted with Ar and O_2 bubbles of various sizes and we used several distances between electrodes. Saturated water vapor is an important component in all bubbles, since the gas is injected into the water and the bubble resides in the water during the entire experiment. Sharp current and light emission peaks are observed in all experiments except for those situations when a gas bubble covers the entire gap between the electrodes. In this case, the discharge occurs essentially in an atmospheric-pressure gas between two metal electrodes and has the electrical signature of a spark (increasing current and decreasing voltage). In all situations where the bubble does not cover the entire space between the two electrodes, the discharge resembles a dielectric barrier discharge with the needle electrode surrounded by a mixture of gases at atmospheric pressure and the disk electrode covered by a layer of water. Typical current, voltage, and optical emission pulses observed for voltages above a certain threshold, which depends on several factors such as the electrode separation, the gas in the bubble (Ar or O_2), and the conductivity and pH of the water, are shown in Fig. 15.9 for an Ar bubble.

Individual or series of concurrent current and light emission pulses are seen with no observable distortion of the voltage during these events. The rise and decay times of these pulses are less than 10 ns and 50–80 ns, respectively. The optical pulses are slightly broader than the corresponding current pulses. Two sets of peaks are generally observed: a set of peaks at arbitrary times during the application of the voltage pulse and another peak in both current and optical signal during the shut-off phase of the pulse. The amplitude, the number of peaks, and the time delay for the onset of the discharge depend on the applied voltage and on the size of the bubble. The shapes of the current and the PMT traces do not change significantly from pulse to pulse for a fixed voltage. A subsequent increase in the applied voltage leads to a decrease in the time delay to the onset of the discharge. In addition, the number of peaks observed during the time the voltage is applied increases with increasing voltage. The intensity of the peak that occurs during the shut-off part of the applied pulse also increases for higher applied voltages. The origin of the (reverse) discharge peak that occurs when the applied voltage is turned off is a reverse electric field caused by charges that accumulate on the inside surface of the bubble during the regular discharge pulses and trigger a discharge in the opposite direction when the external voltage is switched off. Experiments conducted with Ar and O_2 bubbles do not show any significant differences in the optical emission signal using the 309 nm OH filter, indicating that similar concentrations of OH radicals are produced in both Ar and O_2 bubbles.

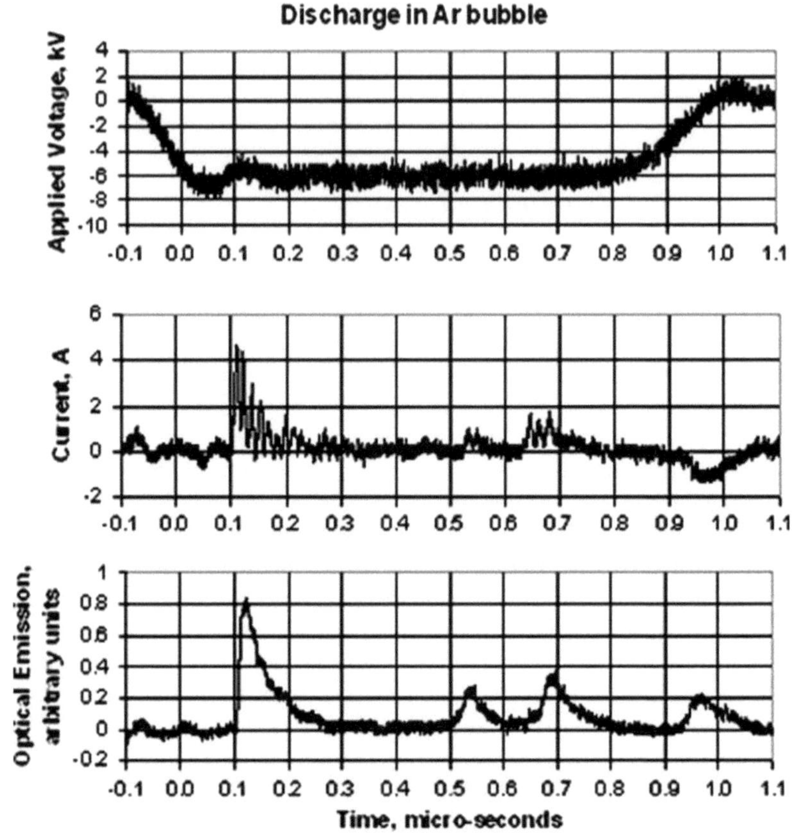

Fig. 15.9 Typical traces for the applied voltage, the current, and the optical emission (OH band emission) for an 8 mm Ar bubble

Experiments using the water treatment reactor (Fig. 15.8) were carried out with the objective to (1) ascertain the potential of pulsed electric discharges in water to generate reactive oxidants such as OH, O_3, O, and H_2O_2 and (2) quantify the potential to decompose organic contaminants in the water such as phenol and rhodamine as well as residues of energetic compounds (explosives) such as TNT, HMX, and CL20 [59]. Figure 15.10 shows the production of hydrogen peroxide (H_2O_2) as a function of the size of the mesh electrodes and the O_2 gas flow rate under otherwise constant operating conditions. Since the discharge produces a significant amount of hydrogen peroxide, the Fenton reaction, that is, the reaction of H_2O_2 with Fe^{2+} to form Fe^{3+} along with OH^- and OH can be used to increase the decomposition rate of organic materials. Experiments were performed with the objective to assess the potential of pulsed electrical discharges in bubbled water to decompose the energetic compound HMX. This was done at a reduced pH 3, which is known to increase the reaction rate [60]. The reaction was initiated by adding $FeSO_4$ to the water to yield a [Fe^{2+}] concentration of 0.3 mmol/l. We found that roughly 50% of the initial

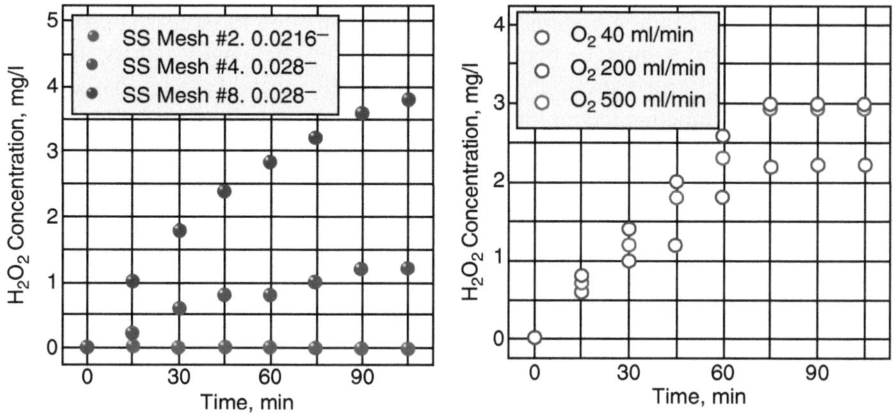

Fig. 15.10 H_2O_2 concentration for various mesh electrode sizes at a fixed O_2 flow rate and fixed operating conditions (*left*) and various O_2 gas flow rates at a fixed mesh size and fixed operating conditions (*right*)

concentration of the energetic compound HMX is decomposed in about 1 h when O_2 bubbles were used. The decomposition efficiency was significantly lower when Ar bubbles were used.

15.4.4 Summary

This brief section touched on some aspects of the physical and chemical processes that are initiated by pulsed electrical discharges in bubbled water. As an example of how water decontamination applications can benefit from a technology based on the production of reactive species by pulsed electrical discharges in water, results for the decomposition of the energetic compound HMX in an aqueous solution by pulsed discharges in water were presented.

15.5 Conclusion

This chapter reviews the status of the use of nonthermal ("cold") plasma for environmental applications. Water purification and disinfection using ozone (O_3) generated in DBDs and ESPs using corona discharges are the only two applications that have achieved full-scale industrial implementation. The plasma-assisted, plasma-initiated, and plasma-enabled remediation of gaseous (and to a lesser extent liquid) waste streams has been the subject of intense laboratory studies for more than 20 years with a focus on the decomposition of NO_x and SO_x and the destruction of VOCs. Much research has been carried out in this area using laboratory-scale plasma-based remediation systems; however, the transition to industrial prototypes and commercial pilot units has been slow. Engineering challenges such as choice of materials for long-term operation, energy, and cost efficiency of plasma-based methods relative to competing technologies and to problems associated with the scale-up of laboratory reactors as well as unresolved science issues such as the complete characterization, quantification, and control of byproduct formation, dependence of destruction efficiencies on contaminant concentration, reactor volume and geometry, and residence time are the main reason for this slow transition. Many of these issues are related to the fact that the plasma chemistry involved in the destruction processes is not well understood and that we lack a detailed understanding of how the discharge physics couples to the plasma chemistry. Lastly, laboratory studies of pulsed electrical discharges induced in externally introduced gas bubbles in water and aqueous solutions indicate that these discharges may represent a convenient way for the in situ, on-demand generation of oxidants in these liquids for disinfection and decontamination purposes.

Acknowledgments I would like to acknowledge the contributions of many colleagues to the work described in this chapter, in particular the contributions of Profs. Christos Christodoulatos, Agamemnon Koutsospyros, Erich Kunhardt, and Abraham Belkind as well as Drs. Shu-Min Yin, Sophia Gershman, and Oksana Mozgina. I also gratefully acknowledge many insightful discussions with Dr. U. Kogelschatz, Dr. Guido Vezzu, Prof. Jose Lopez, and Prof. Weidong Zhu. Partial financial support for some of the work discussed here was provided by NASA, NSF, and the US Army (Picatinny Arsenal).

References

1. B.M. Penetrante, S.E. Schultheiss (eds.), *Non-Thermal Plasma Technologies for Pollution Control*, NATO-ASI Series, vol. 34A/B (Plenum, New York, 1993)
2. E.E. Kunhardt, IEEE Trans. Plasma Sci. **28**, 189 (2000)
3. K.H. Schoenbach, K. Becker, E. Kunhardt, in *Non-Equilibrium Air Plasmas at Atmospheric Pressure*, ed. by K.H. Becker, U. Kogeschatz, K.H. Schoenbach, R.J. Barker (IOP, Bristol, 2004). Chapter 6 "DC and Low-Frequency Air Plasma Sources" and references therein
4. K.H. Becker, K.H. Schoenbach, J.G. Eden, J. Phys. D **39**, R55 (2006)
5. U. Kogelschtz, in *Non-Equilibrium Air Plasmas at Atmospheric Pressure*, ed. by K.H. Becker, U. Kogeschatz, K.H. Schoenbach, R.J. Barker (IOP, Bristol, 2004). Chapter 9 "Applications of Atmospheric-Pressure Air Plasmas" and references therein
6. W. Siemens, Poggendorff's Ann. Phys. Chem. **102**, 66 (1857)
7. W. Ohlmüller, *Über die Einwirkung des Ozons auf Bakterien* (Springer, Berlin, 1891)
8. U. Kogelschatz, in *Process Technologies for Water Treatment*, ed. by S. Stucki (Plenum, New York, 1988), p. 87
9. U. Kogelschatz, B. Eliasson, in *Handbook of Electrostatic Processes*, ed. by J.S. Chang, A.J. Kelly, J.M. Crowley (Dekker, New York, 1995), p. 581
10. G. Pietsch, V.I. Gibalov, Pure Appl. Chem. **70**, 1169 (1998)
11. J.T. Herron, D.S. Green, Plasma Chem. Plasma Process. **21**, 459 (2001) and references therein to earlier publications
12. B. Eliasson, U. Kogelschatz, in *Proceedings of the 8th International Symposium on Plasma Chemistry (ISPC-8)*, Tokyo, Japan, vol. 2 (1987), p. 736
13. M. Kogoma, S. Okazaki, J. Phys. D: Appl. Phys. **27**, 1985 (1994)

14. G. Vezzu, Private Communication (2006)
15. G. Vezzù, J.L. Lopez, A. Freilich, K.H. Becker, *Optimization of Large-Scale Ozone Generators*, IEEE Trans. Plasma Sci. **37**, 890 (2009)
16. W. Gilbert, *Tractatus, sive Physiologia de Magnete, Magnetisque corporibus mango Magnete tellure, sex libris comprehensus* (Excudebat Petrus Short, London, 1600)
17. M. Hohlfeld, Arch. f. d. ges. Naturl. **2**, 205 (1824)
18. O.J. Lodge, J. Soc. Chem. Ind. **5**, 572 (1886)
19. F.C. Cottrell, J. Ind. Eng. Chem. **3**, 542 (1911)
20. W. Deutsch, Z. Tech. Phys. **6**, 423 (1925)
21. W. Deutsch, Ann. Phys. **76**, 729 (1925)
22. R. Seeliger, Z. Tech. Phys. **7**, 49 (1926)
23. J.M. Crowley, in *Wiley Encyclopedia of Electrical and Electronic Engineering*, ed. by J.G. Webster (Wiley-Interscience, New York, 1998), p. 719
24. M.K. Mazumder, in *Wiley Encyclopedia of Electrical and Electronic Engineering*, ed. by J.G. Webster (Wiley-Interscience, New York, 1999), p. 15
25. N.A. Fuchs, *The Mechanics of Aerosols* (Pergamon, Oxford, 1964)
26. W. Deutsch, Ann. Phys. **68**, 335 (1922)
27. E.E. Kunhardt, K. Becker, U.S. Patents 5,872,426 (1999), 6,005,349 (2000), 6,147,452 (2004), and 6,879,103 (2005)
28. E.E. Kunhardt, K. Becker, L.E. Amorer, in *Proceedings of the 12th International Conference on Gas Discharges and Their Applications*, Greifswald, Germany, 1997, p. I-374
29. E.E. Kunhardt, K. Becker, L.E. Amorer, L. Palatini, Bull. Am. Phys. Soc. 42, 1716 (1997)
30. E.E. Kunhardt, G.P. Korfiatis, K. Becker, C. Christodoulatos, in *Proceedings of the 4th International Conference on Protection and Restoration of the Environment*, Halkidiki, Greece, 1998, ed. by G.P. Korfiatis
31. L. Moskwinski, P.J. Ricatto, N. Abramzon, K. Becker, G.P. Korfiatis, C. Christodoulatos, in *Proceedings of the 14th Symposium on Applications of Plasma Processes (SAPP)*, Jasna, Slovakia, 2003, p. 17
32. N.S. Panikov, A. Paduraru, R. Crowe, P.J. Ricatto, C. Christodoulatos, K. Becker, IEEE Trans. Plasma Sci. **30**, 1424 (2002)
33. L.E. Amorer, Ph.D. Thesis, Stevens Institute of Technology (1999) (unpublished data)
34. S.-M. Yin, C. Christodoulatos, K. Becker, A. Koutsospyros, in *Proceedings of the International Conference on Environmental Systems (ICES)* (SAE International, Colorado Springs, 2003). Paper No. 2003-01-2501, on CD ROM
35. A. Koutsospyros, S.-M. Yin, C. Christodoulatos, K. Becker, Int. J. Mass Spectrom. **233**, 305 (2004)
36. A. Koutsospyros, S.-M. Yin, C. Christodoulatos, K. Becker, IEEE Trans. Plasma Sci. **33**, 42 (2005)
37. C.M. Nunez, G.H. Ramsey, W.H. Ponder, J.H. Abbott, L.E. Hamel, P.H. Kariher, Air Waste **43**, 242 (1993)
38. B.M. Penetrante, M.C. Hsiao, J.N. Bardsley, B.T. Merrit, G.E. Voglin, A. Kuthiz, C.P. Burkhart, J.R. Bayless, Plasma Sour. Sci. Technol. **6**, 251 (1997)
39. S. Futurama, A. Zhang, G. Prieto, T. Yamamoto, IEEE Trans. Ind. Appl. **34**, 967 (1998)
40. T. Yamamoto, J. Electrostat. **42**, 227 (1997)
41. Y. S. Mok, C. M. Nam, M. H. Cho, I.-S. Nam, IEEE Trans. Plasma Sci. 30, 408 (2002)
42. H. Qiu, K. Martus, W. Y. Lee, K. Becker, Int. J. Mass Spectrom. **233**, 19 (2004)
43. O. Levenspiel, *The Chemical Reactor Omnibook* (Oregon State University Press, Corvallis, 1993)
44. M. Sato, T. Ohgiyama, S. Clements, IEEE Trans. Ind. Appl. **32**, 106 (1996)
45. B. Sun, M. Sato, J.S. Clements, J. Electrostat. **39**, 189 (1997)
46. S. Ihara, M. Miichi, S. Satoh, C. Yamabe, IEEE Int. Pulsed Power Conf. 5451, 1291 (1999)
47. H. Akiyama, IEEE Trans. Diel. Electr. Insul. **7**, 646 (2000)
48. A.M. Anpilov, E.M. Barkhudarov, Yu.B. Bark, J. Phys. D **34**, 993 (2001)
49. A. Abou-Ghazala, S. Katsuki, K. H. Schoenbach, F.C. Dobbs, K.R. Moreira, IEEE Trans. Plasma Sci. **30**, 1449 (2002)

50. P. Lukes, A.T. Appleton, B. Locke, IEEE Trans. Ind. Appl. **40**, 60 (2004)
51. P. Lukes, B. Locke, Ind. Eng. Chem. Res. **44**, 2921 (2005)
52. P. Sunka, V. Babicky, M. Clupek, P. Lukes, M. Simek, J. Schmidt, M. Cernak, Plasma Sour. Sci. Technol. **8**, 258 (1999)
53. T. Miichi, S. Ihara, S. Satoh, C. Yamabe, Vacuum **59**, 236 (2000)
54. C. Yamabe, F. Takeshita, T. Miichi, N. Hayashi, S. Ihara, Plasma Process. Polym. **2**, 246 (2005)
55. S. Gershman, O. Mozgina, A. Belkind, K. Becker, E. Kunhardt, Contrib. Plasma Phys. **46**, 19 (2007)
56. C.J. Hochanadel, J. Phys. Chem. **6**, 587 (1952)
57. B. Sun, M. Sato, J. Clemens, J. Phys. D **32**, 1908 (1999)
58. Y. Wen, H. Liu, W. Liu, X. Jiang, Plasma Chem. Process. **5**, 137 (2005)
59. O Mozgina, S. Gershman, K. Becker, C. Christodolatos, A. Belkind, in *Atomic and Surface Physics and Related Phenomena*, ed. by V. Grill, T.D. Märk (University of Innsbruck Press, Innsbruck, Austria, 2006), p. 136
60. K.-D. Zoh, M.K. Stenstrom, Water Res. **36**, 1331 (2002)

Chapter 16
Complex (Dusty) Plasmas: Application in Material Processing and Tools for Plasma Diagnostics

Holger Kersten and Matthias Wolter

Abstract Formation, trapping, and modification of powder particles in laboratory received growing interest during the last decade in research and technology with novel and unique properties. Applications of complex (dusty) plasmas are numerous, most of them emerging in modern material science. For the optimizations of these processes, a detailed knowledge of plasma–particle interaction is needed. On the other hand, due to the interaction between small (test) particles and the surrounding plasma information on the electric field in front of surfaces and the energy fluxes in the plasma can be obtained. Dust particles can be used as a diagnostic tool, for example, by observing position and motion of the particles in dependence on the discharge parameters. In a certain sense, one can state that "complex (dusty) plasmas" are a rapidly expanding field of research at the border between plasma physics, material processing, and diagnostics, for example, for synthesis and modification of powder particles.

16.1 Introduction

Powder formation, modification, and trapping in laboratory discharges have received growing interest during the last decade. The unique possibility of dust particle confinement and control in the gas phase makes plasmas excellent media for particle handling and treatment. The applications of dust particles in plasma are numerous, most of them emerging in modern material science. However, to optimize the processes, a detailed knowledge of plasma–particle interaction is needed. On the other hand, the interaction between small (test) particles and the surrounding plasma can be used as a diagnostic tool, for example, by observing position and motion of the particles in dependence on the discharge parameters. For example, information on the electric field in front of surfaces and on the energy fluxes in the plasma can be obtained.

H. Kersten (✉) and M. Wolter
Institut für Experimentelle und Angewandte Physik, Christian-Albrechts-Universität Kiel, Leibnizstr. 19, 24098 Kiel, Germany
e-mail: kersten@physik.uni-kiel.de; wolter@physik.uni-kiel.de

M. Bonitz et al. (eds.), *Introduction to Complex Plasmas*, Springer Series on Atomic, Optical, and Plasma Physics 59, DOI 10.1007/978-3-642-10592-0_16,

In a certain sense, one can state that "complex (dusty) plasmas" are a rapidly expanding field of research on the border between plasma physics, material processing, and diagnostics. Some examples for this statement will be provided in this contribution:

- *Synthesis and modification of powder particles in plasmas*
  [–] Application and perspectives of nano- and microparticles
  [–] Particle growth in reactive plasmas
  [–] Modification of particles by process plasma
- *Dust particles as electrostatic microprobes*
  [–] In front of floating walls and in front of (biased) electrodes
  [–] In plasma process sources
- *Dust grains as thermal probes*
  [–] Energy balance of particles/substrates in plasma processing
  [–] Special (phosphor) particles for temperature measurement

16.2 Disturbing Side Effects of Dust Particles in Plasma Processing

The fact that plasma processes heavily implied in semiconductor device processing, where dust formation yields to serious problems, has been one of the major inputs for the development of many basic and technological studies devoted to complex (dusty) plasmas [1, 2]. Especially in plasma-enhanced dry etching of semiconductors, the industry has always to fight against problems of surface contamination by dust particles. It is obvious that small particles, which fall onto the wafer surface during plasma treatment and stick in the wafer trenches, can cause defects and, hence, make the integrated circuit chips useless. The fact that the size of plasma-etched structures has decreased down to the submicrometer range and is reduced to below 100 nm at present implies the importance for avoiding "killer particles" down to the nanometer region. The problems become even more important by using modern processes as inductively coupled plasma (ICP) where the electric fields in front of the substrates are rather weak and the particles can easily reach the surfaces. It is not surprising that the main goal of early dusty plasma investigations was to obtain a good control of contamination in plasma-processing reactors, either by eliminating dust particles from the gas phase or by preventing them from getting into contact with the surface. As a result of numerous elaborated studies [3–6] dust contamination by relatively large (>100 nm) particles is at present under control.

The sources of particle contamination during the plasma–surface processes are:

- Formation of large molecules, mesoscopic clusters, and particles in the plasma by chemically reactive gases

- Formation of macroscopic particles at surfaces by means of plasma–wall interaction

In the first case, the plasma process itself promotes the particle formation by excitation, dissociation, and reaction of the involved species in the gas phase. Typical examples are plasma polymerization [7] and thin film deposition in silane-containing plasma-enhanced chemical vapor deposition (PECVD) processes [7, 8]. The different stages of the particle growth in the gas phase can be observed by mass spectrometry [9], laser-induced evaporation [10], photodetachment [8, 11], IR absorption [3], microwave cavity measurements [8], Mie scattering [12], and self-excited electron resonance spectroscopy (SEERS) [13, 14]. The generated particles are often analyzed by transmission electron microscopy (TEM).

Examples for the generation of particulates from the surrounding surfaces are reactive-ion etching (RIE) [15, 16], surface sputtering of targets [17, 18], vacuum arc deposition [19–21], and hollow cathode processes [22].

Formation of dust particles will be illustrated shortly for RIE processes with halogen containing species: An originally small roughness which might exist at the substrate surface can result in local deposition of polymer films. Those parts will be etched slower than the clean surface. In the anisotropic RIE etch process caused by directed ion flux, small columnary etch residues are formed (Fig. 16.1). As a result of slight underetching, after a certain time the columns become thinner at their base and, thus, unstable. Since the structures are negatively charged, Coulomb repulsion from the surface causes them to break off and to be ejected into the discharge. The splitted and charged etch residues are finally trapped in the glow by the force balance acting onto the particles [23].

Very similar mechanisms are also responsible for particle formation and contamination in fusion devices [25–27]. Dust particles are likely to be formed during the

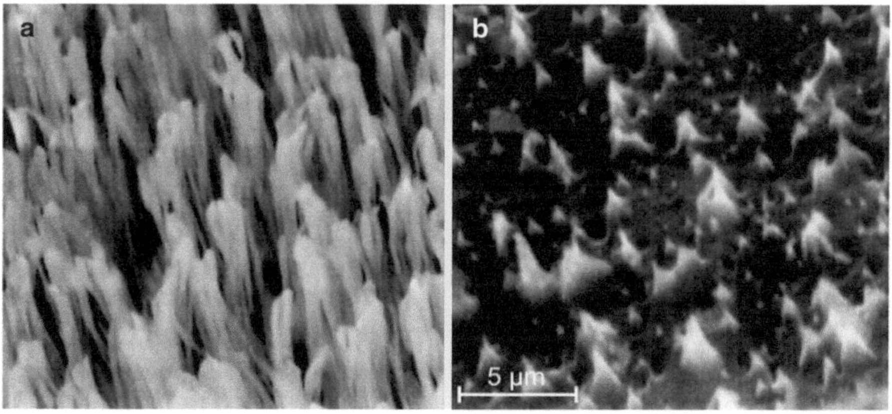

Fig. 16.1 (**a**) A scanning electron microscope (SEM) photograph of a silicon wafer treated in a pure CCl_2F_2 plasma. (**b**) The wafer was later exposed to a pure argon plasma. The surface structure of etching residues (so-called *silicon grass*) developed in the CCl_2F_2 discharge is sputtered away in argon [24]

glow discharge cleaning and during thin film deposition for wall conditioning. The particles will fall to the bottom at the end of a fusion plasma discharge and, in particular, the lighter ones can be reinjected into the plasma either by magnetic effects or by electrostatic charging when they are in contact with the edge plasma. Due to thermophoretic forces and due to repetitive evaporation and condensation the particles may accumulated at cold areas of the device. They may block spacings and fill gaps which were introduced for engineering reasons. An additional danger is that dust particles may retain a large fraction of hydrogen which will lead to considerable tritium inventories.

Dust appearance seems to be a natural limitation in plasma–surface treatment. To minimize the influence of dust particles in thin film deposition and etching, it is important to develop either new processes avoiding dust generation or to develop process cycles in complex (dusty) plasmas without contamination at the relevant substrate regions, which are sensitive against dust fallout. This means an active influence on the collection of particles as well as their trapping behavior and their movement, respectively.

There exist several chances of such active influences, for instance:

- Intelligent arrangement of electrodes and substrates
- Construction of special electrode shapes (grooved electrodes) [4,5,28],
- Square wave plasma modulation [29,30]
- Fast transport regimes of the reactive species
- Additional electrostatic forces by an external potential supply [31,32]
- Additional other forces based on neutral drag (gas flow) or thermophoresis (temperature gradient) by external heating [33, 34], or photophoresis (laser irradiation) [35,36]

For example, the basic idea for the introduction of square wave plasma modulation with "on–off" cycles is that the small and negatively charged dust precursors clusters are not allowed to grow in size and concentration during the "on" sequence and leave the plasma volume during the "off" sequence.

The introduction of special electrode shapes and additional forces will result in changes of the equilibrium planes where the dust particles are trapped. By means of these peculiarities the particles can be actively pushed toward regions in the reactor where their presence is not dangerous [31].

16.3 Formation and Modification of Powder Particles in Plasmas for Various Industrial Applications

In contrast to the disadvantages of dust particles in plasmas, particles which are produced and/or modified in plasma can also have valuable properties for specific applications. In particular, the increased knowledge and ability to control particles in a plasma environment has led to new lines of technological research, namely the tailoring of particles with desired surface properties.

Present and potential applications of such plasma-treated particles are numerous:

- Treatment of soot and aerosols for environmental protection [37]
- Powder particle synthesis in high-pressure [38] and low-pressure plasmas
- Illumination technology, cluster lamp [39]
- Enhancement of adhesive, mechanical, and protective properties of powder particles for sintering processes in metallurgy [40]
- Fragmentation of powder mixtures in order to sort them [31]
- Improvement of thin film properties by incorporation of nanocrystallites for amorphous solar cells [41] and hard coatings [42, 43]
- Coating of lubricant particles [44]
- Application of tailored powder particles for chemical catalysis [45]
- Functionalization of microparticles for pharmaceutic and medical application
- Production of color pigments for paints
- Improvement of surface protection against corrosion of fluorescent particles [46]
- Tailoring of optical surface properties of toner particles [47], etc.

The different methods for particle production, modification, and application as well as their various possibilities of employment are schematically summarized in Fig. 16.2.

From the point of view for environmental protection, gas cleaning is required in flue gas processing and diesel engine technology. Electrostatic and plasma-based cleaning tools are often applied. For example, the charging behavior of small soot particles in a plasma can be used for their removal. So far, corona discharges are already used as electrostatic filters. These discharges are favored for particle charging because they operate under atmospheric pressure [37].

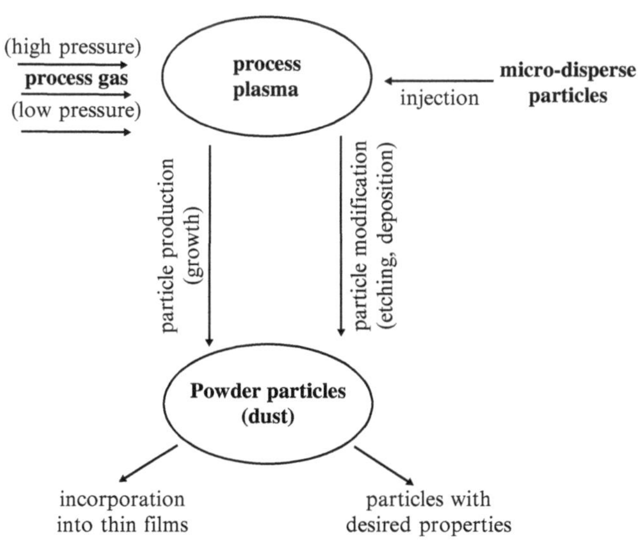

Fig. 16.2 Different ways of particle treatment in process plasmas [24]

Production of large amounts of powders in high-pressure thermal plasma jets is also state of the art in metal and ceramic industry. However, the size distribution and uniformity of such produced particles is very poor. A solution is the use of low-pressure plasmas for powder generation and modification. But, unfortunately, particle formation in low-pressure discharges is not efficient due to the low density of active species as a result of the quite limited trapping capacity. In addition, external handling, like injection of particles into the vacuum system and collection of processed particles, gives always a lot of problems. Thus, as far as mass production is concerned, low-pressure technology cannot compete with high-pressure thermal methods, for example, thermal plasmas or ovens.

The major advantage of low-pressure plasmas is their nonequilibrium chemistry, which creates the unique possibility for fine surface treatment, without damage and thermal overload. Hence, using low-pressure technologies, one can fabricate small amounts of particles with a high added value. Synthesis of particles with very sharp size distributions was demonstrated by Tachibana et al. [48], who injected small carbon soot precursors into a methane plasma in order to obtain perfectly spherical objects. Vivet et al. [49] investigated an integral process of particle growth and coating in order to obtain grains with catalytic properties. In their experiments, small silicon carbide particles were produced in a SiH_4/CH_4 plasma. In the next step, the grains were in situ coated using evaporated palladium wires. Such plasma-produced particles have been shown to be excellent carriers for catalytic coatings with large and active surfaces [49]. Another interesting field of application for low-pressure plasma-produced particles is their incorporation into thin films. The enhancement of the layer properties of such composite materials is expanding rapidly. For example, photoluminescent properties of films seeded with nanometer size silicon particles are essential in optoelectronic applications. In solar cell technology, crystalline silicon particles included in an amorphous matrix act as precursors in the recrystallization of the film. Microcrystalline silicium (MC-Si) films have been produced by P. Roca i Cabarrocas et al. [41] which have a high degree of microstructure. They could show that by running the plasma in a regime close to the formation of powder, it is possible to produce a new type of silicon thin films (polymorphous silicon) which result from the simultaneous deposition of radicals and nanoparticles. Despite being heterogeneous, polymorphous silicon films exhibit improved transport properties and stability with respect to a-Si:H. Moreover, their implantation in single junction p–i–n solar cells has resulted in stable efficiencies close to 10%. Finally, nanostructured films may have excellent mechanical properties, like high plastic hardness and high elastic modulus [42].

Furthermore, it is possible to produce composite coatings, where the properties of various materials are combined. An example is the deposition of a wear resistant self-lubricating coating. In this process, which is schematically shown in Fig. 16.3, small lubricating MoS_2 particles are included in a hard titanium nitride matrix [44]. The resulting layer will have the hardness and chemical stability of a TiN-layer. However, as the layer wears off during its lifetime, the embedded MoS_2 particles will emerge at the surface and form a lubricating layer. This is the principle of a self-lubricating hard coating.

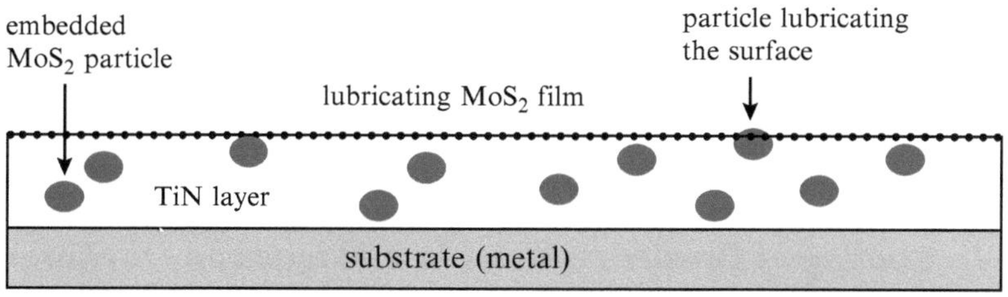

Fig. 16.3 The idea of a self-lubricating coatings. Small grains of lubricant are included in the hard matrix. When the surface is exposed to friction and wear, small amounts of lubricant are released to form a thin protective film over the surface. Such hybrid coatings are both effectively lubricated and environmentally clean [24, 44]

Nanostructured materials exhibit properties that are not observed for macroscopic feature sizes. For example, Nanocomposites containing metallic nanoparticles in a dielectric matrix are of particular interest because of their functional properties. This include electronic conductivity ranging from single electron hopping and tunneling to percolation, surface plasmons giving rise to characteristic optical absorption, ferromagnetic single domain behavior and superparamagnetism, granular giant magnetoresistance, and enhancement of catalytic activity [50]. As a consequence nanomaterials have not only been of strongly increasing fundamental interest during the last decade but also of great technology importance.

Since a lot of these materials are prepared under low-pressure PECVD conditions, the optimization of applications for nanocomposite materials need the knowledge gained in the studies of dusty plasmas. Nanometer size particles are extremely difficult to handle outside the processing reactor. Therefore, integrated process of dust particle fabrication in the plasma and codeposition in the layer is the most promising technology for the production of composite material [50].

Low-pressure discharges can also be used solely as trapping media to process externally injected particles. The plasma itself can contain chemically active species, which will cause surface modification of powders. For example, in new ceramic materials fabricated from sintered powders, pretreatment of the particle surface is desired. For such application, Kitamura et al. [51] described coating of zinc oxide particles with SiO_x, by injecting ZnO grains into TEOS/Ar/O_2 plasma. Tetraethoxysilane (TEOS) is a common silicon-containing organic precursor in reactive plasma deposition. Advantages of particle coatings are obvious in sintering of those particles which led to better quality material in terms of endurance and shrinkage properties or for the improvement of surface stability against corrosion. The latter issue is also important for the protection of fluorescent particles [46, 52].

Moreover, processing of externally injected particles in the plasma can yield unique objects, like coated or modified grains with desired surface structure, color, and/or fluorescent properties. For instance, such particles are useful as toners in copying machines [47] or in optical devices. Often, the conditioning of stable emulsions and pastes out of polymer powders is an important prestage process step in polymer particle technology [53]. However, modification of the grains can be a complex, multistep process, involving various chemical reactions and the support of external devices. In the following, some examples for coating of powder particles in complex plasma environment will be presented.

16.3.1 Coating of Powder Particles in a Magnetron Discharge

Thin metallic films (Al, Ti, and Cu) were deposited by means of dc-magnetron sputtering on silicon oxide particles (18 μm in diameter) suspended in an argon rf-discharge (5 Pa and 5 W) [54]. The coated particles were examined by scanning electron microscopy (SEM). Photographs of the particles are shown in Fig. 16.4.

The structure of the metallic films was different depending on the target material. While the Al- and Cu-films were deposited smoothly, the Ti-films showed a distinct island formation. This can be explained by the lower sputter yield of the titanium target and also by the different surface adhesion between the particles and the deposited material [55]. The surface energy of titanium (1.39 $\mathrm{J\,m^{-2}}$) is higher than that of aluminum (0.91 $\mathrm{J\,m^{-2}}$) and copper and, therefore, the wetting of the particle surface is worse for titanium.

The deposition rates have been determined as a function of the magnetron power. To obtain the film thickness the metals were sputtered pre-experimentally onto glass substrates. During the deposition process, the transmission of the sample was determined and, by taking into account the optical relationships, the thickness, and, thus, the growth rate could be estimated. In conclusion, from the parameter study of the deposition rate at relevant experimental conditions (magnetron power between 30 and 100 W), one can expect growth rates of 0.5–2.5 $\mathrm{nm\,s^{-1}}$ for aluminum and titanium layers and growth rates of 2.5–15 $\mathrm{nm\,s^{-1}}$ for copper on the powder grains. This expectation is also supported by observation of the surrounding films on the powder particles (see, e.g., Fig. 16.4d). For an optically thick aluminum coating of about 60 nm, the particles have to be confined for 1 min, which is not a problem because the particles can be trapped for a rather long time.

The difference between the original and powder-coated particles is obvious. The particles are completely covered with close and quite thick metal layers. The coated particles show a rough and cauliflower-like shape which makes them attractive, especially for catalytic applications. The modified particles also differ from the originals in their contrast. If the deposited films were thinner, the layer surrounding the particles could also be smoother. In such cases, the surface structure of the coated particles looked golf ball-like which might be interesting for optical applications.

In the experimental setup used it was only possible to trap about 10^5 particles per cubic centimeter, corresponding to a mass of powder particles of about 0.5 mg. The particles were treated in the magnetron discharge for about 5 min. The metallic layers were deposited evenly on all the particles. This would give a yield of coated particles of about 50 $\mathrm{mg\,h^{-1}}$ if it were possible to extract the particles directly from the reactor by a load-lock system. This throughput is not very much but can probably be enhanced using a scaled setup.

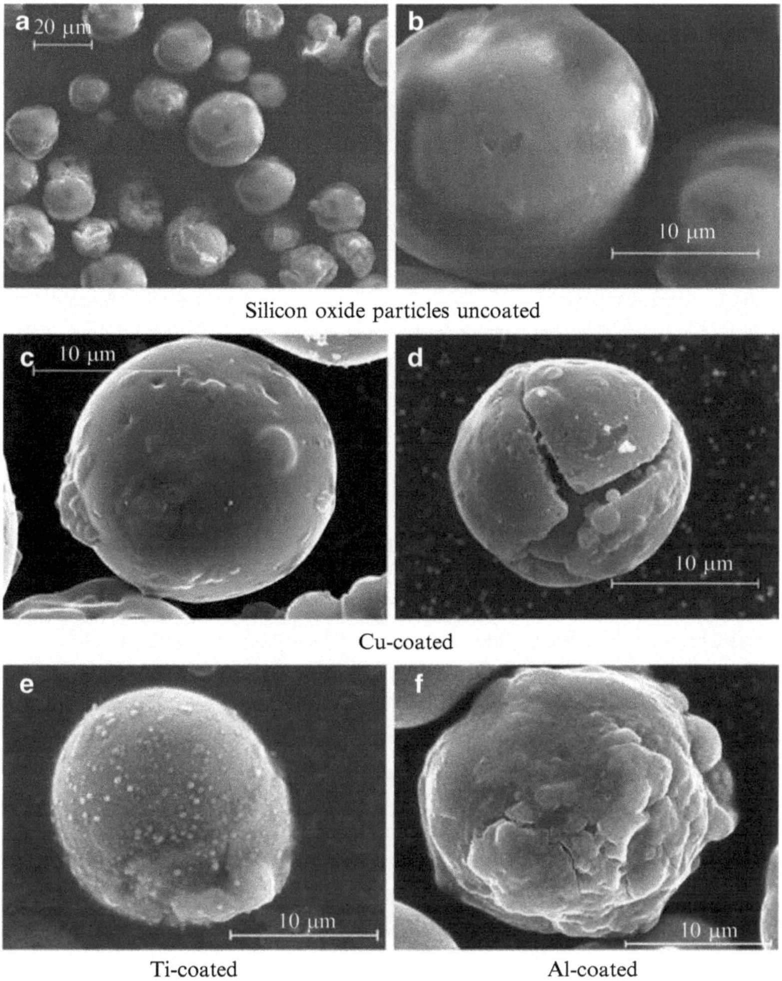

Fig. 16.4 SEM photograph of silicon oxide particles (**a, b**) which are confined in a rf-plasma and coated by magnetron sputtering [54]. The particles have been coated by copper (**c, d**) as well as by titanium (**e**) and aluminum (**f**)

During the magnetron experiments, an interesting observation was made: When the magnetron was turned on, the entire particle cloud began to rotate (see Fig. 16.5).

After the magnetron was turned off, the rotation ceased due to the neutral drag. Similar observations of particle motion under the influence of a permanent magnetic field were made by Konopka et al. [56]. They explain the rotation with the azimuthal component of the ion drag force as being due to $\boldsymbol{E} \times \boldsymbol{B}$-drift of the ions in the perpendicular radial electric and vertical magnetic fields.

However, we observed particle cloud rotation only under the influence of magnetron discharge. This means that for this effect, the interaction of the rf-plasma and the magnetron discharge is of significance.

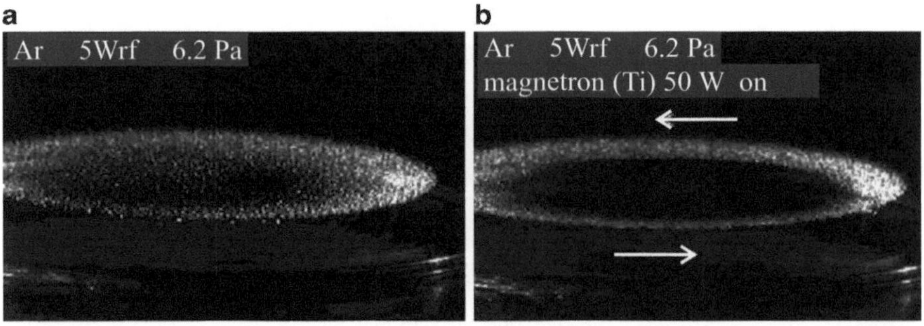

Fig. 16.5 Particle cloud in front of the rf-electrode (**a**), the discharge power was 5 W_{rf}. After switching on the dc-magnetron (50 W), which is outside the photograph, the particle cloud starts to rotated (**b**)

The magnetic field of the permanent magnets in the magnetron is already very weak for the particle distance ($B_{z,\mathrm{Mag}} \sim 10^{-4}$ T). It alone has no impact on the particles. Thus, we believe that the electrons (which exist at high density due to the magnetron effect in front of the target cathode) must be responsible for the ion drift. With these circulating electrons an additional magnetic field is formed and an additional electron current in the direction of the opposite rf-electrode is supplied. This current was observed in earlier experiments.

We ascribe the rotation of the particle cloud to an $\boldsymbol{E} \times \boldsymbol{B}$-drift of the argon ions which interact by momentum transfer with the particles. The perpendicular fields responsible for the ion drift are the vertical component of the permanent magnetic field, $B_{z,\mathrm{Mag}}$, of the magnetron and the radial components of the electric field above the rf-electrode which is changed significantly by the influence of the magnetron plasma particularly by the flow of electrons from the magnetron plasma (Fig. 16.6). Finally, the momentum transfer of the ions (ion drag) due to the tangential Lorentz force causes the grains to rotate. Even if the ion drag is only just sufficient in a small circular region, the other dust particles are coupled to each other by Coulomb interaction [35, 58]. This also explains why the entire particle cloud rotates like a rigid solid with the same frequency.

The motion of the particle cloud was studied using different parameters. Figure 16.7 shows the rotation frequency as a function of the magnetron power. It increases linearly with increasing magnetron power accounting for the increase in the charge carrier density in the magnetron discharge. This observation supports our supposition on the influence by interaction with the electron ring current, since the density of the charge carriers increases with magnetron discharge power.

The dependence on the gas pressure was also studied (Fig. 16.8). The rotation frequency f_{rot} decreases as the pressure p increases which can be explained by the increasing neutral friction force F_{N} (neutral drag). For the rotation frequency, the force balance between the ion drag F_{ion} and the neutral drag F_{N} is important [59]

$$F_{\mathrm{N}} = \xi p v_{\mathrm{s}} = F_{\mathrm{ion}}, \tag{16.1}$$

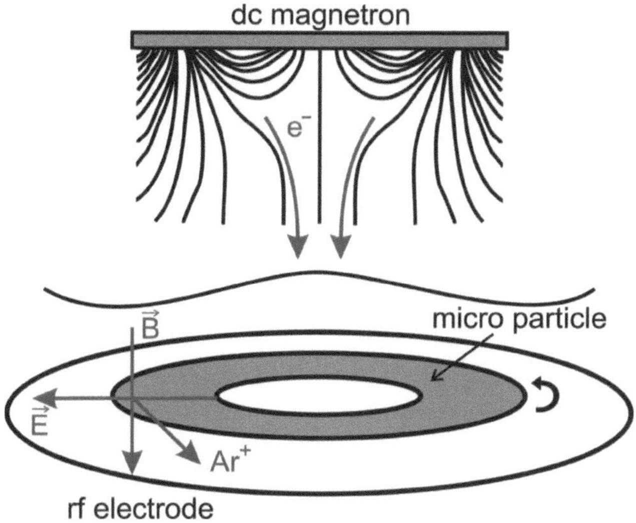

Fig. 16.6 Principle sketch of the fields which cause (by the tangential ion drag) the rotation of the grains with magnetron operation [57]

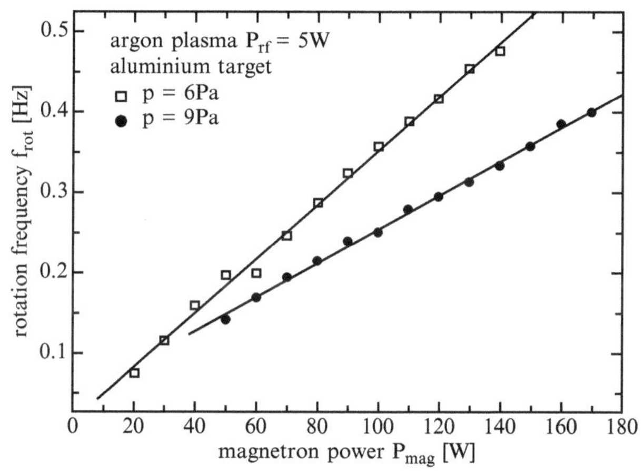

Fig. 16.7 Rotation frequency of the particle cloud as a function of the magnetron power at different gas pressures

where p is the gas pressure, $v_s = 2\pi r_s f_{rot}$ is the rotation velocity of the particle cloud, r_s is the mean radius of the rotating particle ring, and ξ is a constant.

With this, we can derive the following expression for the rotation frequency [60, 61]:

$$f_{rot} = \frac{(\sigma_{coll} + \sigma_{coul})\, n_i\, (m_{ar} v_{di}) \sqrt{v_{di}^2 + v_{th,i}^2}}{2\pi \xi r_s} \frac{1}{p}, \tag{16.2}$$

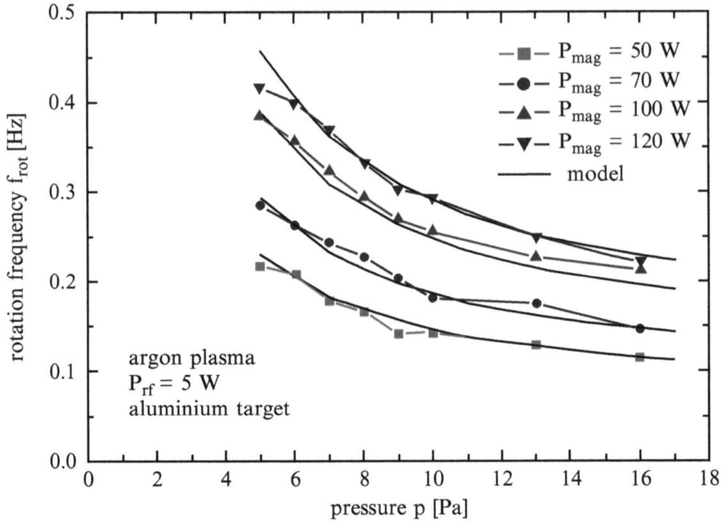

Fig. 16.8 Pressure dependence on the particle cloud rotation frequency at different powers

where σ_{coll} is the collision cross section for ion momentum transfer, σ_{coul} is the cross section for Coulomb interaction, n_i is the ion density, m_{ar} is the ion mass, v_{di} is the drift velocity of the ions, and $v_{th,i}$ is the thermal velocity of the ions.

Figure 16.8 shows a comparison of the experimental data points and the model. The curve yields a friction force of about 5×10^{-13} N which is of the same order as the confining forces.

An additional approach for coating of injected particles has been demonstrated in [47, 62], where again an argon radio-frequency plasma was employed to charge and confine particles, while the coating has been performed by means of a separate magnetron sputter source. The aim was to coat ferromagnetic iron particles with an optically thick white layer (aluminum), to make them suitable as toners for color copying machines. Particles of about 2-μm diameter are introduced through a sieve. The aluminum vapor was supplied by the magnetron sputter source, located above the particle cloud. In Fig. 16.9, a SEM photograph of processed particles is shown. The process time was about 300 s which resulted in Al-layer thickness of 150–200 nm. One of the major drawbacks of the magnetron source is its influence on the operation of the rf-plasma and the particle confinement. Moreover, high-energy aluminum atoms are introduced during the sputter process. This leads to heating of all surfaces, including grain particle surface [62].

At present, coating is still a batch process. It should be mentioned again that by means of the procedures as described earlier the yield of processed powder particles is rather low. A much higher yield may be obtained by continues processing in fluidized bed reactors at higher pressures as suggested by several authors [63, 64]. But this technology is only suitable for particle mixtures which trickle very well. For agglomerations of strongly adhesive particles, as it is often the case, there is almost

Fig. 16.9 SEM photograph of Fe particles coated with Al-layer (after [47]). The mean diameter of the iron particles is about 2 μm; however, smaller particles can also be observed due to size distribution

no other chance than the use of low-pressure plasmas. In those cases, separation is readily achieved in the plasma, because negative charging of a conglomerate leads to its "explosion." The separated particles are confined above the powered electrode (PE). Since electrostatic trapping force is exceeded by the gravitation force when the particle size is larger than about 10 μm, the electrostatic confinement is possible only for single, separated particles. If conglomerates are too heavy, they fall onto the electrode surface.

16.3.2 Deposition of Protective Coatings on Individual Phosphor Particles

Another example of technological powder treatment in process plasma is the deposition of alumina coatings on individual phosphor particles by a PECVD process. The deposited layers protect the particles against degradation and ageing during plasma and UV irradiation in fluorescent lamps. Thin films of alumina (Al_2O_3) are chemically stable and allow a substantially full light transmission at the excitation wavelength of mercury (254 nm) and in the visible range.

As original material an aluminum-organic precursor (ATI) was used. The ATI molecules are dissociated in the rf-plasma volume and the radicals contribute to the formation of Al-oxide on the small phosphor particles substrates. Fragmentation and deposition essentially depend on the plasma parameters and the process cycle.

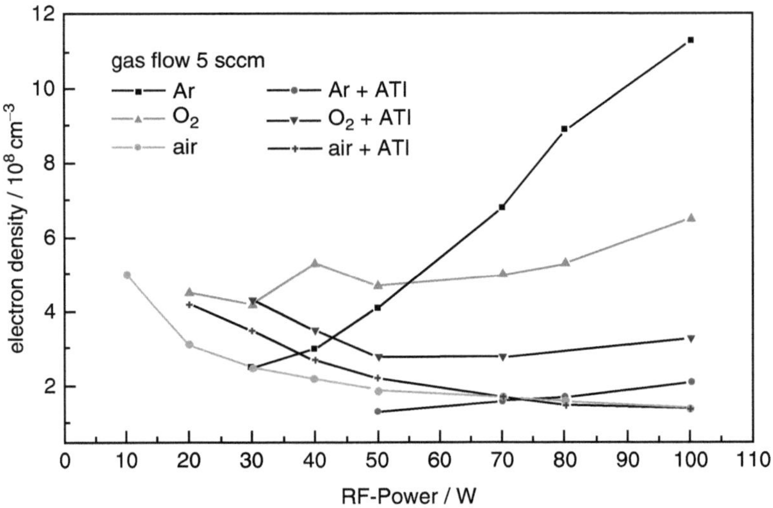

Fig. 16.10 Electron density as measured by SEERS with respect to dependence on discharge power [54]

Owing to the fact that the ATI fragments tend to form negative ions, the electron density shows only a weak variation with the power, whereas in a pure argon plasma a strong increase could be observed (Fig. 16.10).

This observation is due to the electron attachment by the radicals which contribute to the film growth at the luminophore particles. In a pure argon plasma the electron density increases with increasing power (30–100 W) from 2×10^8 to $12 \times 10^8\,\mathrm{cm}^{-3}$, while the density in an Ar/ATI plasma is about $10^8\,\mathrm{cm}^{-3}$. In the argon plasma, the degree of ionization increases with the discharge power and, hence, the density of the Ar^+ ions and electrons also increases. In ATI or oxygen plasma, respectively, the ionization also increases as a function of the discharge power. However, to a certain extent, the electrons are attached at the radicals to form negative ions. As a result, the density of the free electrons almost does not change (see Fig. 16.10).

The different order of magnitude and the tendency in the electron densities for different process conditions allow us to make conclusions about dissociation and ionization mechanisms and reveals information about the thin film deposition on the fluorescent grains. The original phosphor particles, which are not coated by a protective alumina layer, show a remarkable decrease in their light intensity after argon plasma treatment, which simulates the process conditions in a fluorescent lamp (Fig. 16.11). Irradiating the plasma for 1 h at 50 W causes an ageing effect and a decrease in the light intensity of the luminophores by a factor of 10. Therefore, the particles have been coated by the protective alumina layer. The success of particle deposition was proved by X-ray photoelectron spectroscopy (XPS). It could be demonstrated that the luminophore grains show CH-groups from the glue before deposition, whereas almost no CH-groups could be detected by XPS on the particles

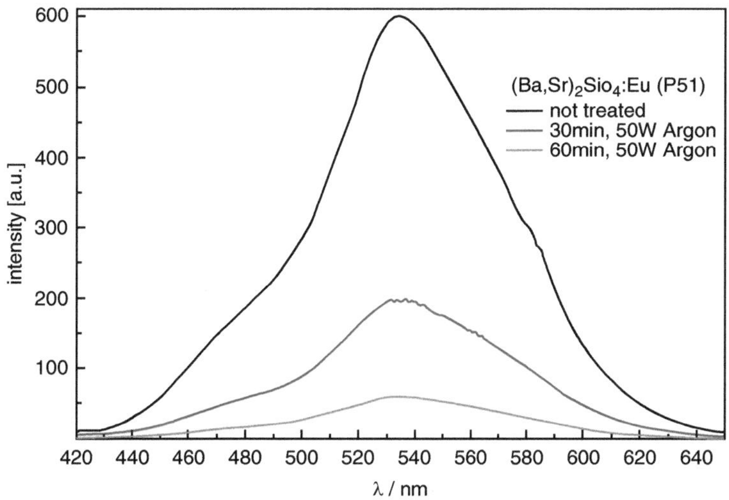

Fig. 16.11 Spectral light intensity of nontreated luminophores after plasma irradiation

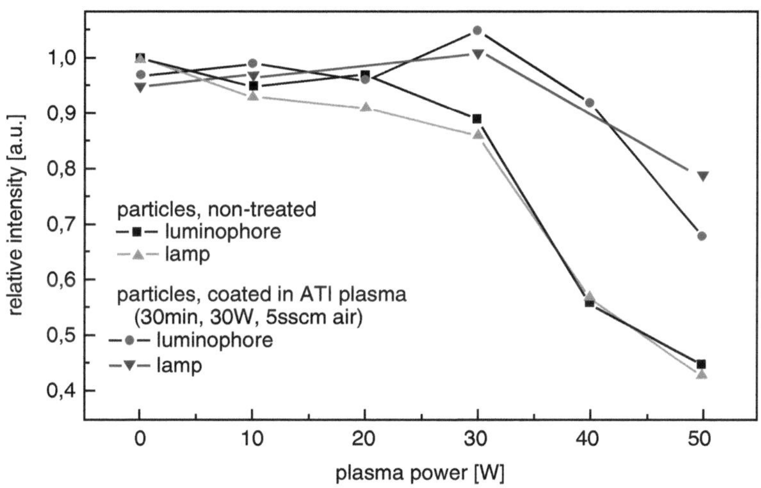

Fig. 16.12 Comparison of light intensities of noncoated and by PECVD-coated luminophore particles after plasma irradiation

after deposition of the protective layer. In comparison to the nontreated particles, PECVD decomposition of ATI in the rf-plasma giving transparent Al_2O_3-films onto phosphor particles results in a much higher stability against plasma irradiation, for example, against UV radiation and ion bombardment at low energies. Whereas the light intensity of nontreated luminophores decreases with high-power plasma irradiation, the light intensity of coated luminophores remains stable even for high power (see Fig. 16.12).

As the fluorescent properties of the grains should be preserved, it is important that there is no change in the emission spectra of the particles with respect to the Al_2O_3 protective layers. It could be shown that the emission spectra of alumina-coated $BaMg_2Al_{16}O_{27}$:Eu^{2+} do not differ from that of the uncoated materials. An additional advantage of plasma treatment under optimized conditions (e.g., 30 W, 5 sccm air/ATI, 30 min treatment time) is the decomposition of the glue material, which makes any additional annealing process unnecessary.

16.3.3 Particles as Microsubstrates

On the surface of microparticles in a process plasma, not only charging processes are of importance but also chemical processes like molecule association and radical reactions. With suitable diagnostics like scattering spectroscopy and Raman spectroscopy, one can obtain information about the mechanisms at the surfaces of microscopic solids, which are of great interest, for example, for catalyst research and particle coating.

Cavity-enhanced spectroscopy is a powerful diagnostic technique for the characterization of micron-sized spheres. It has been used very successfully for aerosol droplets [65]. The feasibility for applying this technique to solid micron-sized particles levitated in an rf-plasma and has been shown in [66].

A pulsed laser is used to excite cavity resonances (whispering gallery modes) in individual microparticles, leading to enhanced Raman scattering at characteristic wavelengths. This noninvasive method gives direct access to the size and also the chemical composition of the microspheres, and is potentially a very interesting tool for the characterization of growing layers deposited on microparticles in molecular plasmas.

Cavity-Enhanced Spectroscopy

If a droplet of a liquid or a microparticle is struck by a laser beam, under suitable incident conditions and at particular resonant wavelengths, the light can undergo total internal reflection and become trapped inside the droplet for long periods. The term "long" means the order of nanoseconds, during this time the light travels a few meters inside the particle. The trapped laser light leads to stimulated Raman scattering at particular resonant wavelengths within the Raman spectrum and an enhanced Raman scattering signal can be detected. The resonant behavior can be accurately described by Mie scattering theory [67]. Thus, microparticles can act as optical cavities, greatly enhancing light whose wavelength is coincident with resonant modes of the cavity [68].

Figure 16.13 shows a spectrum of a water droplet illuminated by a laser at a wavelength of 590 nm. Sharp, equally spaced peaks appear at a Raman shift of approximately 3,400 cm^{-1}, corresponding to the OH stretching band of water. Those

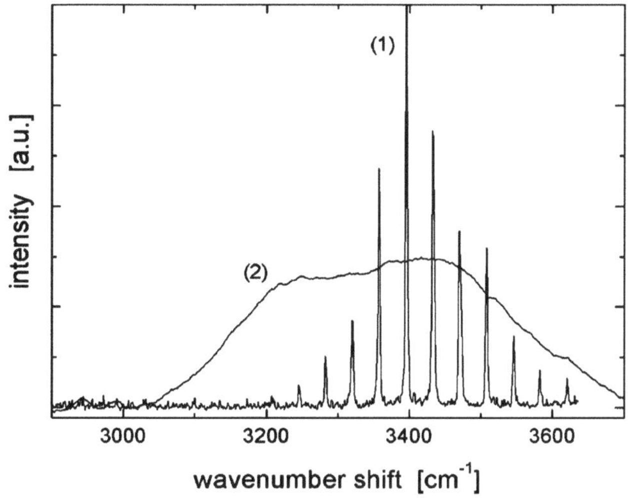

Fig. 16.13 Cavity-enhanced Raman scattering of a water droplet (1) in comparison to the spontaneous Raman scattering of bulk water (2)

resonant modes can be thought of as the light forming standing waves within the particle. They can be assigned a mode number, which corresponds to the number of wavelengths in the standing wave, and a mode order, which corresponds to the number of radial intensity maxima. Resonances of the same mode order and polarization are approximately equally spaced. From the line spacing, we readily arrive at the particle size. The following formula gives an approximation for the particle radius r [65]:

$$r \sim \frac{\tan^{-1}\left(\sqrt{m^2-1}\right)}{\sqrt{m^2-1}} \frac{1}{2\pi \Delta k}. \tag{16.3}$$

Here, m is the refractive index of the material and k the line spacing between resonances of consecutive mode numbers. Particle sizes can be calculated from (16.3) with an accuracy of a few hundred nanometers.

Optical Measurements on Microparticles

The diagnostic method which was successfully used for water droplets has been adapted to polymethylmethacrylate (PMMA) particles confined in a rf-plasma [66]. To excite cavity-enhanced scattering, a pulsed tuneable dye laser centered at 590 nm with pulse energies of about 1.5 mJ per pulse, pumped by a Nd:YAG-laser (532 nm), was employed. The spectra were resolved and captured by a spectrograph and an ICCD camera. The signal was sent to the entry optics of the spectrograph via an optical fiber that could be positioned at a variable angle with respect to the laser beam.

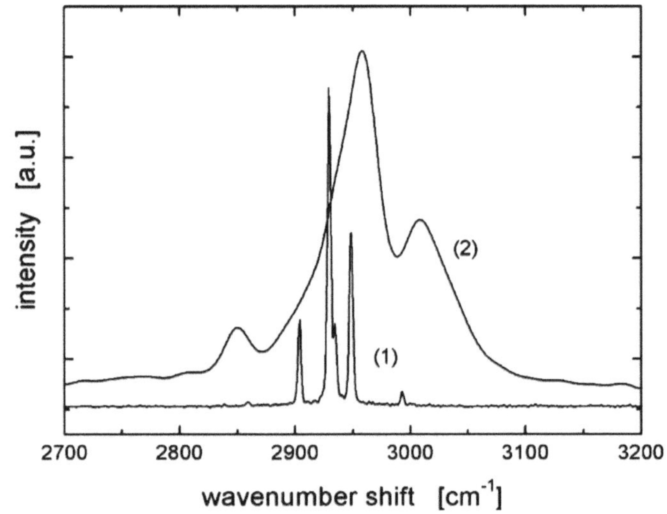

Fig. 16.14 Cavity-enhanced Raman scattering of PMMA particles (1) and their corresponding spontaneous Raman scattering (2)

Before investigating solid microparticles in the plasma, experiments were performed on particles in air, where they could be manipulated more easily. Spherical particles of different sizes and chemical composition were investigated. Cavity-enhanced Raman scattering was successfully detected from polymer spheres. Figure 16.14 shows a spectrum measured from a PMMA particle with a diameter of 50 μm. Clear peaks appear at a Raman shift of 2,900 cm^{-1}, which is in the C–H stretching region. Calculating the particle size from the spectrum gives a value of 51 μm, which is in good agreement with the value specified by the sphere manufacturer (Microparticles GmbH).

To test whether it is possible to detect a cavity-enhanced signal only from the particle surface, particles were investigated which are coated with a fluorescent dye (Rhodamine B). In an analogous way to cavity-enhanced Raman scattering, the fluorescence signal is enhanced at wavelengths corresponding to cavity resonances. The resulting spectra show peaks superimposed on the broad fluorescent band, where the signal is enhanced through coupling into cavity resonances.

Figure 16.15 shows a spectrum from a Rhodamine B-coated PMMA particle where both the cavity-enhanced Raman signal from the particle material and the fluorescence signal originating from the surface can be seen. These results demonstrate that cavity-enhanced spectroscopy is a surface-sensitive technique, making it suitable for the characterization of growing layers in molecular plasmas.

The particles are trapped at the position in the plasma sheath where all acting forces balance. They appear to the eye to be very stationary, but if viewed with a camera with high spatial and temporal resolution it becomes clear that they move around their equilibrium position by distances at least of the order of their diameter, which complicates the alignment of the laser beam to the particles.

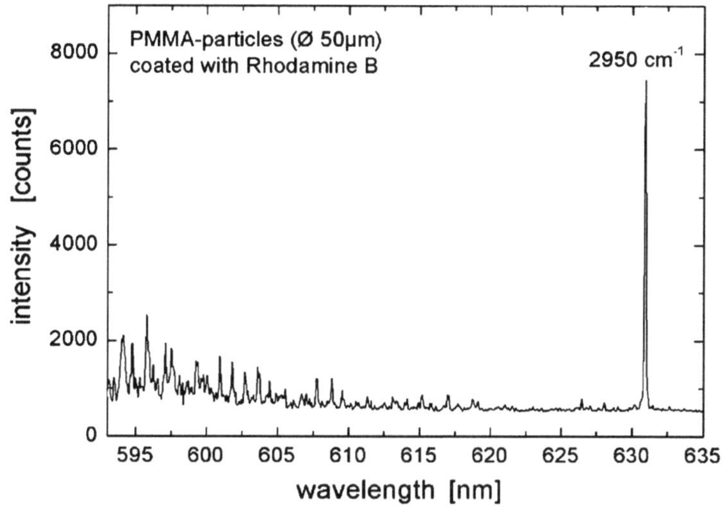

Fig. 16.15 Spectrum of a 50 μm-PMMA sphere coated with Rhodamine B. Both cavity-enhanced Raman scattering and cavity-enhanced fluorescence can be observed

The first cavity-enhanced spectra were obtained from melamine formaldehyde (MF) particles that were doped with Rhodamine B. The particles had a diameter of 9.4 μm and a very smooth surface.

Figure 16.16 shows the fluorescence spectrum obtained from such a particle levitated in an argon rf-plasma. The signal level is not much above the noise level, but it shows a clear periodicity of approximately 10 nm. This becomes much more obvious if we look at the autocorrelation function of the signal (Fig. 16.16b). The periodicity that can be determined experimentally (9.3 nm) is in satisfactory agreement with the theoretically expected separation between cavity resonances of consecutive mode numbers for microparticles of this size. One can conclude that cavity-enhanced spectroscopy of particles confined in a plasma yields valuable information on their size, chemical composition, and surface [66].

An extensive discussion of spectroscopic methods for the diagnostics of molecular plasmas using laser absorption techniques can be found in Chap. 13.

16.4 Particles as Electrostatic Probes

An interesting aspect in the research of complex (dusty) plasmas is the experimental study of the interaction of microparticles with the surrounding plasma for diagnostic purposes [23, 69]. The issue of structure and dynamics of dust clouds in plasmas as well as strongly coupled dusty plasmas has been extensively described in Chaps. 6–8 and 10.

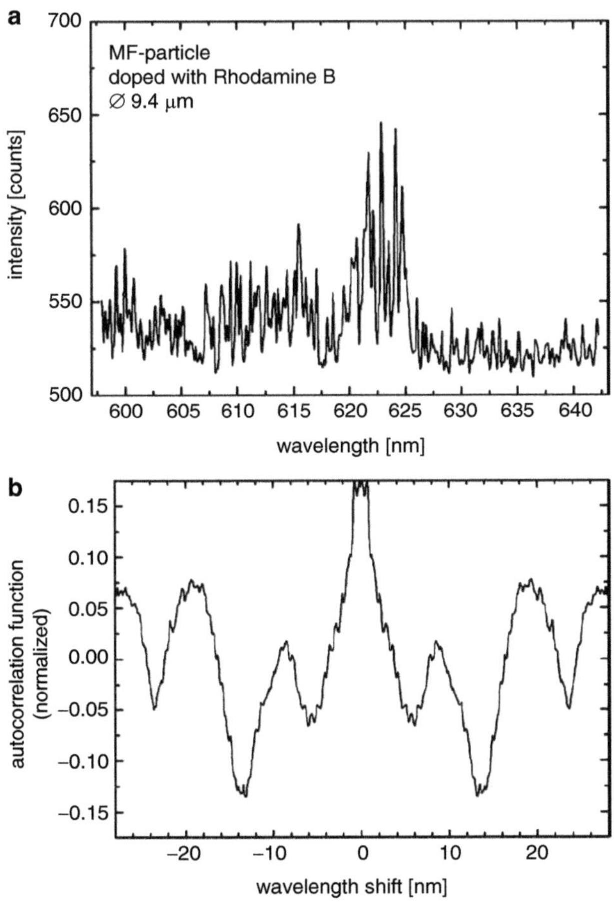

Fig. 16.16 (**a**) Cavity-enhanced fluorescence spectrum of a Rhodamine B-doped MF particle levitated in an argon plasma. (**b**) Autocorrelation function of the signal

Since in many cases of applications in plasma technology it is of great interest to describe the electric field conditions in front of floating or biased surfaces, the confinement and behavior of test particles are studied in front of floating walls inserted into a plasma as well as in front of additionally biased surfaces. For the latter case, the behavior of particles in front of an adaptive electrode (AE), which allows for an efficient confinement and manipulation of the grains, has been experimentally studied in dependence on the discharge parameters and on different bias conditions of the electrode [70].

The effect of the partially biased surface (dc, rf) on the charged microparticles has been investigated by particle-drop experiments as well as by particle oscillation experiments [66, 70]. Tracking the position and movement of the particles in dependence on the discharge parameters, information on the electric field has been obtained. These experiments have been performed in front of the AE which allows for an efficient confinement and manipulation of the grains [70, 71]. The electric field strength at particle trapping position could be determined to be in the order of several 10^3 V m^{-1}. The experimental results for the obtained electric field in the sheath basically agree with results from independent PIC simulations. If an additional negative bias potential is suddenly applied to the electrode, the behavior of a levitated dust particle depends on the magnitude of the bias voltage. At lower voltages, the dust particle oscillates around a new equilibrium position. For sufficiently large voltages, the particle is accelerated toward the pixel surface by gravity, ion drag, and electrostatic forces due to its positive charge, which the particle acquires in absence of electrons [70].

Since the microparticles can be observed in the plasma sheath easily, they can serve, in particular, as electrostatic probes for the characterization of the potential surfaces and electric fields in this region [23, 72, 73]. Usually, the plasma sheath – which is an important zone of energy consumption and, hence, often the essential part of a discharge for applications – is difficult to monitor by common plasma diagnostics such as Langmuir probes or optical spectroscopy. By monitoring the dependence of the position and movement of the particles on the discharge parameters, information can be obtained on the electric field in front of electrodes and substrate surfaces where other plasma diagnostic methods fail. If dust microscopic particles are injected into a plasma, they become negatively charged up to the floating potential V_{fl} by electron and ion currents (j_{e} and j_{i}) toward the particles, and can be confined in the discharge. The spatial distribution and movement of the dust particles in a low-temperature plasma is a consequence of several forces acting on the particles [1, 23, 74]. The charged particles interact with the electric field in front of the electrode or wall, respectively, where the electrostatic force has to be balanced by various other forces in order to confine the dust grains. These forces, which have been discussed extensively by several authors [74, 75], are gravitation, neutral and ion drag, thermophoresis, and photophoresis. In a variety of process plasmas the electrostatic force, which is proportional to the electric field strength in the sheath, is the dominant force in comparison with the others. Hence, the use of charged microparticles to obtain additional information on the sheath structure has been successfully demonstrated in front of the PE of capacitively coupled rf-discharges [23, 73, 76].

In dusty plasma experiments, fine particles usually levitate in the horizontal plane above a metal electrode and show a spatial distribution, which depends on the electric field structure above the electrode. Under some conditions, vortices appear and the microparticles move in the plasma [36, 57, 58, 77]. These motions are often generated by surfaces of different potential. Commonly, microparticles are negatively charged in plasmas. Then the surface of the negatively biased electrode pushes away the particles. When another surface in the plasma region is biased less negatively or even positively with respect to the floating potential, the grains move toward these surfaces.

According to the balance of gravitational force ($\boldsymbol{F}_{\mathrm{g}}$), electrostatic force ($\boldsymbol{F}_{\mathrm{el}}$), ion drag ($\boldsymbol{F}_{\mathrm{i}}$), neutral drag ($\boldsymbol{F}_{\mathrm{n}}$), thermophoresis, and Coulomb interaction, microparticles disperse only in a relatively small region of the plasma sheath depending on their size and charge. But only some of these forces will play a role in laboratory complex plasmas under certain conditions. Commonly, the electrostatic and gravitational forces are important. Superposition of the two forces results in a harmonic potential trap around an equilibrium position [73, 74]. The knowledge and the manipulation of the spatial particle distribution is of great interest, for example, for sorting of particles and surface modification of powders [13, 78, 79]. However, the interaction of the plasma with injected dust particles is of interest not only regarding their spatial distribution. Vice versa, from the particle behavior, conclusions about the surrounding plasma and sheath properties can be obtained, for example, field strength and structure.

Since dust grains are small isolated substrates in a plasma environment, they always attain the floating potential. As electrons are much more mobile than ions, the grain surface collects a negative charge, repelling electrons and attracting positive ions until a stationary state is reached. As a result, the net charge $Q = -Ze_0$ of a micron-sized particle can be on the order of a few thousands elementary charges e_0. In principle, the charge Q on a microparticle can be obtained by equating the fluxes of electrons and positive ions toward the particle surface and their recombination [23, 80, 81]. Several authors have studied both theoretical and experimental aspects of charging the dust grains in capacitively coupled rf-discharges, (see, e.g., [1, 23, 82, 83]).

The charged particles interact with the electric field in front of the electrodes or other surfaces and are often observed as levitated dust clouds forming rings or domes in the boundary regions of the plasma. For several applications in plasma technology such as etching or deposition of thin films, modification of powder or composite materials the microparticles are not confined in front of powered electrodes but in front of surfaces which are additionally biased. Therefore, in the following sections, the particle behavior and its use for sheath diagnostics in front of floating surfaces, which are inserted into the plasma, of dc-biased surfaces and of rf-biased surfaces will be shown.

The superposition of different forces, caused by the self-biased electrode and the floating glass wall, influences the shape of the particle cloud. A glass wall, which extends into the plasma and which is at floating potential, causes an attractive force on the particles toward its surface. As a result, the surrounding sheath and the trapped particles show a kind of wetting behavior. This comparison with the capillarity of fluids comes into mind when looking at Fig. 16.17 from this point of view [84]. In photograph 1 in Fig. 16.17, the common situation for low discharge power (3 W) can be seen. The rf-electrode is at the bottom and a lot of particles lie on it. At the left part of the image, the glass box stays on the electrode. Above the electrode a nearly 2D particle cloud is levitated by the balance between gravitation and electrostatic force due to the electric field induced by the dc self-bias of the rf-electrode. If the discharge power and, hence, the potential and field strengths in front of the electrode change with respect to the potential and field at the glass wall (photographs 2–6 in Fig. 16.17), the balance between gravitation and the superposed electrostatic field forces shifts, too. Since the particles are only confined in regions where the force balance is fulfilled, the changes in the shape of the particle cloud and its wetting behavior can be explained qualitatively.

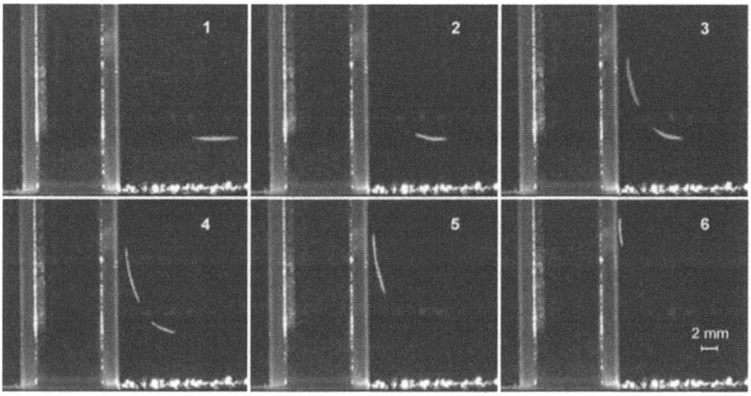

Fig. 16.17 Wetting of a glass wall by particles due to superposition of electrostatic fields (argon plasma with $p = 6$ Pa and a rf-power $P = 3, 6, 8, 10, 15$, and 44 W from picture 1 to 6)

16.4.1 Dust Particles in Front of an Adaptive Electrode

To influence the confined particles and to simulate different electrostatic surface conditions, a segmented AE [71, 85] has been used as an essential part of the experimental setup (see Fig. 16.18).

This device is well suited for such investigations because the sheath structure can be manipulated locally. Furthermore, the influence of additional plasma sources (e.g., external ion beam source or sputter magnetron) on the behavior of microparticles in the rf-plasma can be investigated [57, 86].

A typical asymmetric, capacitively coupled rf-plasma in argon ($p = 0.1$–100 Pa) is employed to charge the particles which are spherical MF particles of 0.5, 1, 5, and 10 μm in diameter. The reactor possesses an upper PE and the lower AE with 101 square pixels, which can be biased independently by dc-voltages.

The rf-power (5–100 W) is supplied by the upper electrode at a frequency of 13.56 MHz and amplitude up to 250 V. Depending on the discharge conditions, electron densities of 10^9–10^{11} cm^{-3}, electron temperatures of 0.8–2.8 eV, and plasma potentials with respect to the ground of 20–30 V for the pristine plasma [87] are measured.

The plasma is bounded to the surrounding surfaces by the self-organizing structure of the sheath, which has a characteristic potential slope and charge carrier profile. This region in front of the AE has finally been probed by the microparticles.

The particles are illuminated by a laser fan ($\lambda_{laser} = 532$ nm); their positions and movements have been observed by a fast CCD camera and video recording, which we employed to investigate the distribution of the powder particles in the plasma sheath (see Fig. 16.18).

At low pressure ($p < 1$ Pa) the sheath is collision less, whereas at higher pressure ($p > 10$ Pa) it is dominated by collisions. Most experiments have been carried out just in the transition regime, which is of interest for plasma processing. The

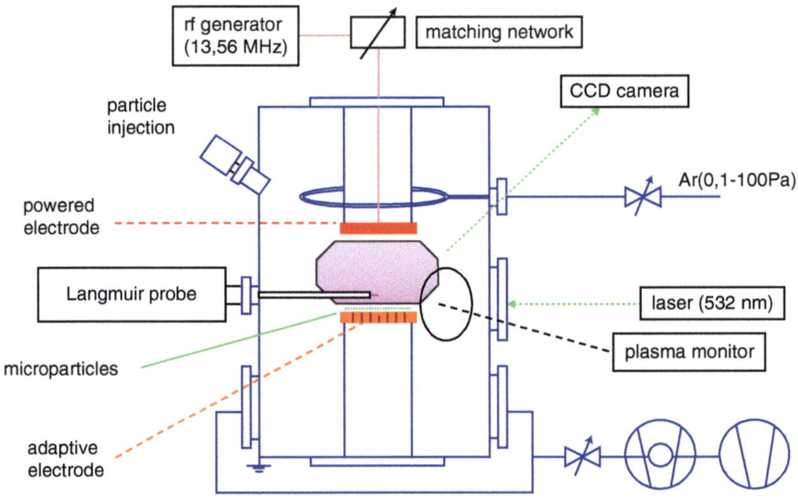

Fig. 16.18 Experimental setup. The central part is the adaptive electrode

transition behavior can be clearly observed in the ion energy distribution function at the grounded surface [66, 87]. The maximum ion energy varies between nearly thermal energy due to thermalization and charge exchange by collisions at high pressure and nearly plasma potential ($\approx$25 eV) due to collision less transfer from the plasma bulk to the surface. If the boundary wall (electrode and substrate) is without any external potential (e.g., floating), the electron and ion fluxes toward the surface are equal and the floating potential V_{fl} reflects the internal plasma properties. An external biasing of selected surface elements (pixels) results in a local change of the sheath and the plasma. This idea is just the basis for the concept of the AE [71, 86] by which a spatial and temporal manipulation of the plasma sheath is possible.

The 101 identical square electrode segments of the AE have a size of $7 \times 7\,mm^2$. They are surrounded by four larger segments to fit the circular geometry of the planar electrode (see also the left panel of Fig. 16.19).

All 105 electrode pixels can be biased individually or in groups by an external $\pm$100 V dc-voltage or ac-voltage (sinus, square, and triangle shape) up to a frequency of 50 Hz and any phase. In addition, for three segments of the AE also an rf-power supply (13.56 MHz) up to 4 W is possible. The whole ensemble of electrode segments is surrounded by a ring electrode and a ground shield.

From other experiments, it is well known that the electric field in the plasma sheath indicates a basically linear decrease with distance from the wall toward the bulk plasma [73]. As the plasma potential is positive with respect to grounded surfaces, the electric field is directed toward the wall. Applying a local bias to some pixels of the AE influences the potential structure in the sheath and, thus, locally changes the direction and magnitude of the electric field. In this way, we can tailor confinement potentials for particles, which are levitating above the AE. The particles will adjust their position that gravitation and electrostatic force in the

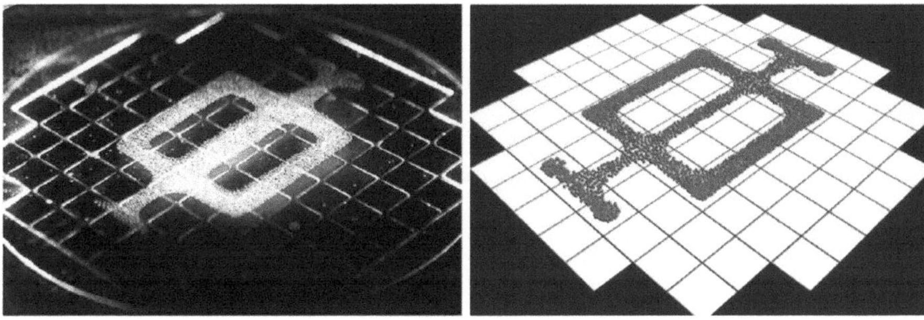

Fig. 16.19 *Left*: pattern of particles (melamine formaldehyde, MF, $d = 9.6\,\mu\text{m}$) which are charged in an rf-plasma and levitated in front of the adaptive electrode due to different pixel biasing. *Right*: simulation results for the same system using 3,000 particles

vertical direction balance each other, defining an equilibrium position. Around this equilibrium plane, the particles may oscillate (harmonically), but will be more or less confined to this region. It is clear that this equilibrium position will depend on the actual value of the bias potential, which is applied to the corresponding pixel of the AE. The potential differences at the surface induce additional forces in the horizontal direction, which are strongest at the interface between two pixels of different bias. Using a cloud of probe particles, which will arrange in a shape that their potential energy is minimized, the spatial variation of the potential can be mapped. The shape of this equipotential surface is influenced by several factors. Despite step-like potential differences between adjacent pixels, the potential distribution above the AE will be smooth due to the mediating effect of the plasma. Depending on the plasma parameters, this smoothing effect will be more or less pronounced, as for the shielding of the particle charge.

For a more detailed examination of the electric field structure in front of a wall, here the AE, we need to know the charge of the probe particles. In the literature, several methods for determining the particle charge can be found (for a summary see, e.g., [88, 89]). One of the most common methods is the excitation of trapped probe particles to oscillate around their equilibrium positions by applying an external low-frequency voltage [23, 74, 88].

In the experimental setup, a single MF particle is confined above the center pixel of the AE and its equilibrium position z_0 is measured. An additional sinusoidal voltage applied to the center pixel causes the particle to oscillate around z_0. Recording the oscillation amplitudes for different driving frequencies allows for a determination of its resonance frequency ω_0.

In Fig. 16.20, results for the dependence of the equilibrium position and resonance frequency on the discharge conditions, for example, neutral gas pressures, are shown. Using particles of different sizes, the corresponding equilibrium positions cover a wide range of the sheath and allow for a thorough characterization of the electric field in the sheath and the particle charge.

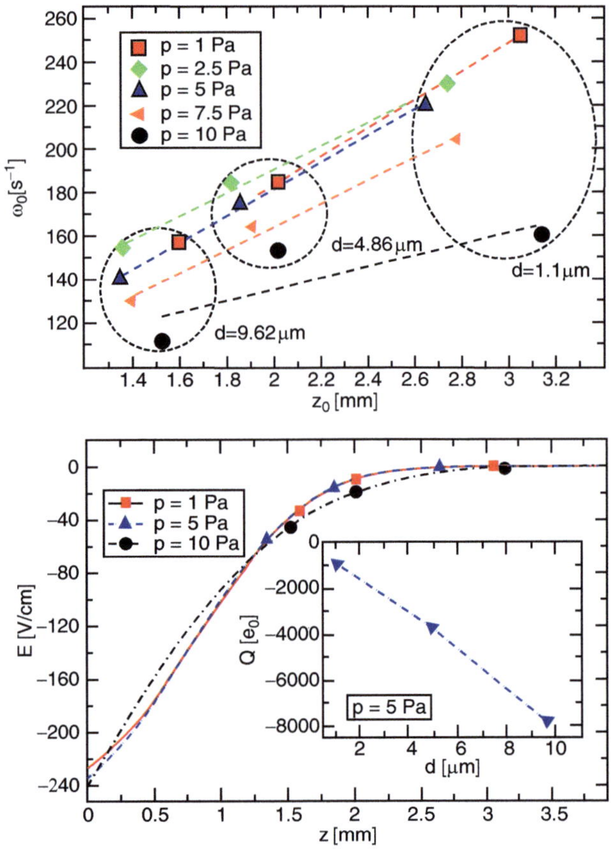

Fig. 16.20 *Top*: experimental results for probe particles in the sheath in front of the adaptive electrode: relation between the resonance frequency ω_0 and the equilibrium position z_0 for different particle diameters d and neutral gas pressures p. *Bottom*: calculated electric field E as a function of distance z from the adaptive electrode for different neutral gas pressures p. *Inset*: particle charge Q as a function of particle diameter d for $p = 5\,\text{Pa}$. These results are deduced from the data in the *left panel* [70]

We may exploit the obtained relation between ω_0 and z_0 to determine the particle charge and electric field structure in the sheath. To this end first, both drag forces and phoresis effects are neglected and the system is described as a driven harmonic oscillator. The equilibrium position z_0 of the (negatively) charged dust particle (charge $Q(z) = -Z(z)e_0$) is determined by $-Q(z_0)\boldsymbol{E}(z_0) = m\boldsymbol{g}$, where the number $Z(z_0)$ of elementary charges e_0 on the particle will depend on its vertical position z, m is the mass of the particle. $\boldsymbol{E}(z) = E(z)\boldsymbol{e}_z$ and $\boldsymbol{g}(z) = g(z)\boldsymbol{e}_z$ give $-e_0Z(z_0)E(z_0) = mg$. Upon applying a low-frequency voltage, the oscillations of the particle around z_0 are harmonic [81] provided the amplitudes are not too large. Within the harmonic oscillator model, the dust charge can thus be considered constant in the vicinity of the equilibrium position.

The resonance frequency of the particle at position z_0 is given by

$$\omega_0^2(z_0) = Z(z_0)\,\frac{e_0}{m}\frac{\mathrm{d}E(z)}{\mathrm{d}z}\bigg|_{z_0} = -\frac{g}{E(z_0)}\frac{\mathrm{d}E(z)}{\mathrm{d}z}\bigg|_{z_0}, \tag{16.4}$$

where the equilibrium condition is used to eliminate m in the second expression. This differential equation for the electric field can be solved by separation. Formal integration yields

$$E(z) = E(0)\exp\left(-\frac{1}{g}\int_0^z \omega_0^2(\xi)\mathrm{d}\xi\right). \tag{16.5}$$

Equating the (negative) integral over the electric field across the sheath with the sheath voltage fixes the value of $E(0)$ at the surface of the AE. For a further evaluation, we need to know the relation between the resonance frequency and the equilibrium position, $\omega_0(z)$, throughout the whole sheath. Experimental data (Fig. 16.20, top) suggest a linear behavior over a wide range of the sheath for low and moderate pressures ($p < 7.5$ Pa). For $p = 10$ Pa, the available data and the assumption of a linear relation between ω_0 and z agree only poorly. Unfortunately, we are not able to obtain experimental data in the very close vicinity of the AE due to surface attachment of the particles. Using a linear ansatz, $\omega_0 = a_0 + a_1 z$ in (16.5) we obtain the electric field,

$$E(z) = E(0)\exp\left(-\frac{a_1^2}{3g}z^3 - \frac{a_0 a_1}{g}z^2 - \frac{a_0^2}{g}z\right), \tag{16.6}$$

in the sheath in front of the AE, which we show in the right panel of Fig. 16.20 for different pressures. Utilizing the knowledge about the electric field, we can directly obtain the particle charge at its equilibrium position (inset in the bottom panel of Fig. 16.20).

Overall, the experimental results indicate that in front of grounded or additionally biased surfaces, we may experimentally determine the electric field structure of the sheath by means of charged microparticle probes.

The construction of the AE also allows for experiments on the sinking and oscillatory behavior of individual particles caused by a change of the local bias potential as already proposed by other authors [74, 90, 91]. To this end, a single particle (MF, $d = 9.6\,\mu$m) is trapped above the center pixel (E5) of the AE by the confining potential shown in Fig. 16.21.

The net force on the particle in vertical direction is given by

$$F(z) = F_{\mathrm{el}}(z) + F_{\mathrm{n}}(z) - F_{\mathrm{g}} - F_{\mathrm{ion}}(z), \tag{16.7}$$

with the electrostatic force F_{el}, the neutral drag force F_{n}, the gravitational force F_{g}, and the ion drag force F_{ion}. The net force vanishes at the equilibrium position z_0.

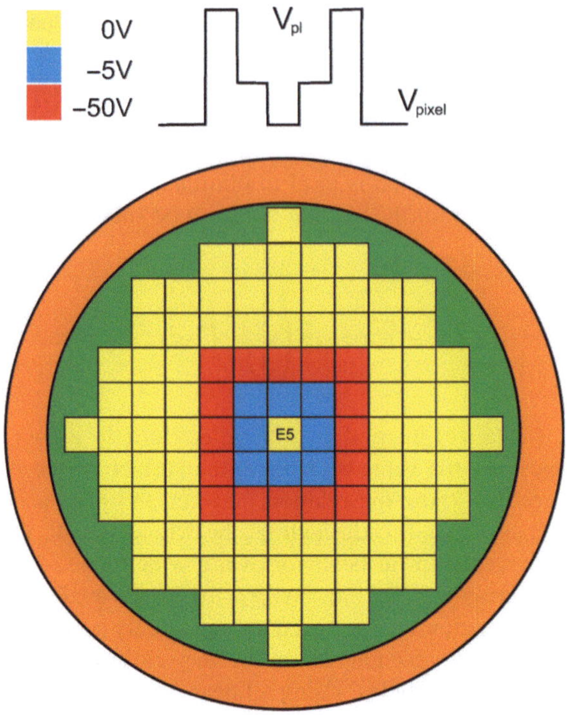

Fig. 16.21 Structure of the adaptive electrode. The center pixel (E5) can be biased to influence the sheath for particle trapping; here, it is at ground. The surrounding pixels (UG and IR) are biased as indicated for the particle confinement in the potential trap. The surrounding segments (*yellow*, *red*, and *green*) are at ground potential

If the confining plasma is switched off, and the pixel E5 is at ground potential ($V_{\text{bias}} = 0\,\text{V}$), the forces F_{el} and F_{ion} become zero. In the vacuum case, the free-fall condition for the particle is valid. However, taking into account the neutral drag force $F_{\text{n}} = -\beta\dot{z}$. the force balance is given by

$$F(z) = m\ddot{z} = -mg - \beta\dot{z}. \tag{16.8}$$

Solution of this differential equation results in

$$z(t) = z_0 - \frac{mg}{\beta}\left(t - \frac{m}{\beta}\left(1 - \mathrm{e}^{-(\beta/m)t}\right)\right). \tag{16.9}$$

The diagrams for the free fall and the real fall of the particles after switching off the plasma are plotted in Fig. 16.22.

The motion begins at the equilibrium position z_0, where the particle is trapped. The sheath width in front of the AE (pixel E5) is about 3 mm and the equilibrium position is at about 1.8 mm. From the experimental data shown in Fig. 16.22 (squares)

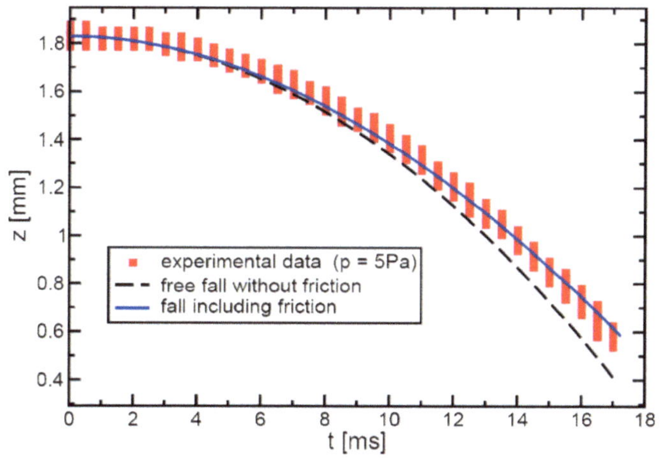

Fig. 16.22 Particle behavior after switching off the plasma

the damping constant β by the gas friction can be estimated to about 2×10^{-11} kg s^{-1} (solid line). For comparison, the dashed line in Fig. 16.22 shows a particle trajectory in the absence of any friction force (free fall).

Keeping the plasma operational, we may also investigate the particle behavior caused by sudden changes in the sheath due to additional bias voltages on some pixels. Compared with the previous experiment, the description here is more involved, but it is also promising in view of gaining information on particle charge and field distribution. We start again from the experimental configuration shown in Fig. 16.21 and confine a single particle above pixel E5 with no additional bias on this pixel. Then applying a negative bias voltage V_{bias} causes the particle to leave its initial equilibrium position. Depending on V_{bias}, we may distinguish two cases: up to some critical value of the bias potential, the particle is pushed away from the electrode, and reaches another equilibrium position to which it relaxes after some oscillations around it. Above a critical V_{bias} (about -30 V), the behavior changes fundamentally (see Fig. 16.23). In this case, the particle falls downwards and finally hits the AE. To explain these different behavior, we consider the following: the equilibrium position of a particle is determined by the balance of gravitation and electrical force, which depends on the local electrical field as well as on the actual particle charge . A more negative bias voltage broadens the plasma sheath within a few rf-cycles, reducing both the electron and ion densities. This effect is drastically more pronounced for the electrons than for the ions. By construction, the biasing of the AE should only have a negligible effect on the bulk plasma, leaving the plasma potential unaltered. In the sheath, the electric field is enhanced due to the increased potential difference, which surpasses the effect of the sheath broadening. An increase of the electric field strength implies that a particle rests at its initial position if its previous charge is decreased. More highly charged, the particle will move toward the bulk plasma as the electric force dominates the gravitation (in the opposite case it will move toward

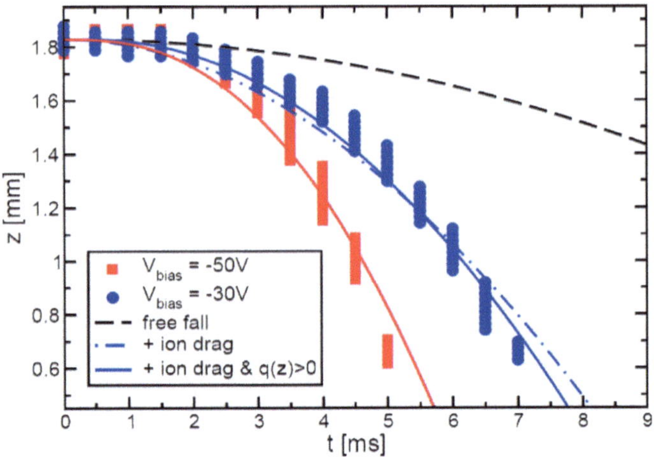

Fig. 16.23 Accelerated fall of a test particle after sudden application of a bias voltage of −30 and −50 V. For comparison, the *dashed line* indicates the trajectory for a free fall. For the *dashed-dotted and solid lines*, the ion drag force is also included. Whereas for the *dashed-dotted line* the negative particle charge may only be reduced, for the *solid line* positive charging is also possible

the AE). Depending on the electron and ion densities in the vicinity of the particle, the particle charge will adapt according to its local surroundings.

As compared with the particle dynamics, those recharging processes take place on a much faster time scale. This drastic change leads to such a fast loss of particle charge that the electric force is never able to compensate gravitation and the particle drops onto the AE.

The experimental data indicate even an additional acceleration toward the AE, as compared with the free fall of an uncharged particle (Fig. 16.23). For an explanation of this additional downward acceleration, the most obvious candidate is the electric force, provided we allow the particle to acquire a positive charge.

It is clear that the particle charge reflects the local balance of electron and ion density. Applying a negative bias, only electrons in the high-energy tail of the electron energy distribution function may penetrate into the enlarged sheath in front of the biased pixel. This leads to a further reduction of the local electron density, while the global plasma parameters of the discharge and the global charge balance are unaffected. Locally, the sheath in the close vicinity of the biased pixel is completely deprived of electrons. In such a region, a dust particle is exclusively surrounded by ions. Ion-dust collisions will reduce the negative dust charge and eventually also a positive charging of the dust is possible. Positive particle charges in the afterglow of a plasma discharge have been reported in the literature [92, 93]. However, to explain the observed behavior exclusively by electric forces, unrealistically high charges would be necessary ($10^4 e_0$ for $V_{\text{bias}} = -30\,\text{V}$ and $1.7 \times 10^4 e_0$ for $V_{\text{bias}} = -50\,\text{V}$) from the beginning on. Despite a strongly reduced averaged electron density at the starting position of the particle, some fraction of the rf-period electrons still reach this region. Due to their higher mobility, this is sufficient to keep the particle charge negative, though reduced as compared with the initial value. This means that a positive particle charging is only possible closer to the AE, in regions where the electrons are completely absent. Whether charges as high as mentioned earlier are possible is however questionable. For the early stages of the fall, surely another explanation is necessary. Here the (up to now neglected) ion drag force presents itself. The large negative bias enhances the ion current toward the AE pixel and also increases the ion energies. By collisions, the ions transfer part of their momentum to the dust particle, explaining the additional acceleration (Fig. 16.23). Closer to the electrode, the (more energetic) ions may transfer more momentum, but their density is reduced.

In total, the combination of the two forces allows for an qualitative explanation of the observed falling curves without the necessity of unrealistically large permanent positive dust charges. For lower bias, the effect of the ion drag is also present, but as long as the ion and electron densities are still alike as before, it is dominated by the upward electric force.

In addition to the main power supply by the PE, some pixels of the AE (there the central pixel (E5) again, see Fig. 16.21) can be driven by an additional rf-voltage with the same frequency (13.56 MHz) and phase to achieve a local enhancement of the plasma. An example of such a plasma bubble or dome obtained in an argon plasma at a pressure of 7.5 Pa is shown in Fig. 16.24. The figure illustrates a pronounced local enhancement of the light emission on top of the pixel E5. The lower part of the bright light represents a reflection of the light on the electrode surface. The size of the light bubble grows with increasing power, which is supplied through the pixel E5 up to 4 W.

A very interesting phenomenon was observed by injecting test particles, which interact with the dome-like plasma of the rf-biased center pixel E5. The particle behavior can be seen in Fig. 16.25.

Some MF particles were confined in front of pixel E5 and then the plasma rf-power at this pixel was increased from 0 to 4 W. The potential conditions of the surrounding pixels are not changed. At the beginning, with increasing power the particle cloud moves slightly in the direction of the biased pixel. Obviously, the particles are pulled into the plasma bubble, which is also attributed by a brighter glow of the small plasma bubble if the power increases. At higher power (e.g., $\approx$4 W) a very pronounced dome-like glow is formed and the particles are invisible due to

Fig. 16.24 Enhanced light emission (plasma bubble) above the pixel E5 (E5: rf-power = 2.5 W, UG: $V_{DC} = -15$ V, and all other pixels: $V_{DC} = -50$ V)

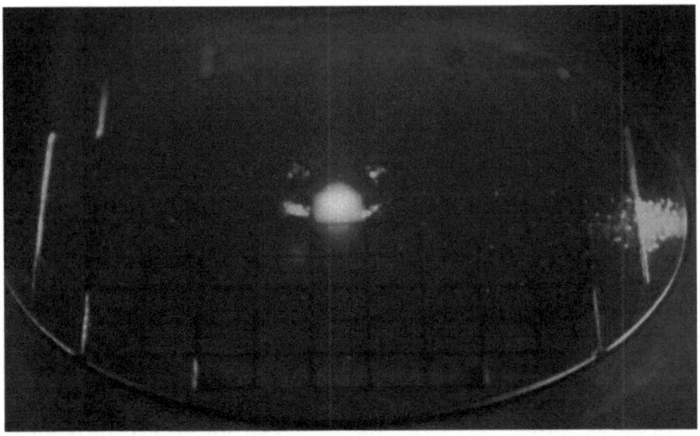

Fig. 16.25 MF-probe particles which are injected into the rf-plasma bubble are pushed outwards from the bubble and collected in the corners of the pixel

the bright plasma. But, suddenly, one after the other the particles jump outwards from the bubble to the corners of the pixel where they are collected. Now, the particles are obviously pushed outside the small plasma dome. This "funny" motion of the particles can qualitatively be explained by a dramatic change (reversal) of the electric field in front of the E5 pixel during rf-plasma operation with respect to the surrounding field. Vice versa, a decreasing power causes the particles to move back into the bubble. The interaction of dust with a plasma ball by rf-manipulation has also been studied to a certain extent by Annaratone et al. [94].

The major challenge for a numerical simulation of the particle motion in the vicinity of the light dome is the large range of involved time and length scales. These range from several ns for the rf-frequency over μs for the particle charging processes [83, 95] to the actual dynamics of the dust particles which takes place on a time scale of 0.1–100 s. Also to cover all length scales inherent to the system (μm dust radius, mm sheath thickness, and several cm vessel dimensions), within one single simulation run seems not to be practicable due to the numerical effort. But fortunately, this is also not necessary as we can separate the effects on the different scales by using a hierarchical set of models, moderating the numerical requirements. The basic idea behind this approach is to break up the simulation into a part concentrating on the fast dynamics on small-length scales, including the effects from larger scales only as fixed input. The results obtained that way then enter the simulation on the coarser time and space grid only in an averaged sense. Depending on the actual time and length scales and the physics involved, different numerical techniques have to be combined.

16.4.2 Interaction Between Dust Particles and Ion Beams

As described earlier in Sect. 16.4.1, if dust particles are injected into a plasma, they become negatively charged by the currents toward the particles and can be confined in the discharge. The spatial distribution and movement of the levitated dust particles in a low-temperature plasma is a consequence of several forces acting on the particles. Among the different forces the ion drag force is an important issue under investigation [23, 96]. Especially in astrophysical environment such as in comet tails [97, 98] and planetary magnetospheres [99], as well as from disturbing side effects in industrial plasmas in semiconductor processing [1, 24] the ion drag becomes important.

However, in common dusty plasma experiments the ion drag force cannot easily be changed without changing other plasma parameters. Therefore, to simulate independently the ion effect the influence of an external ion beam (additional ion drag) supplied by an ECR ion beam source has been investigated [100]. The superposition of the electrostatic field force in front of the rf-electrode, the gravity, and the ion beam results in typical particle arrangement, whereas the effect of the ion beam is threefold:

1. Change in the sheath structure and the electric field in particle trapping region
2. Recharging of the dust particles due to additional positive ion supply
3. Variation in the ion drag force by Coulomb interaction and momentum transfer

The different effects can be distinguished by variation in the gas pressure, the power of the ECR source, the ion beam voltage, and the particle size. Ion drag pushes, for example, the particles toward the electrode surface, contrary to the electric field force. However, the ionic force which changes the shape of the levitated particle cloud or even removes the particles from the glow has another dependence on the particle size than the electrostatic force or gravity.

The ion beam profile can be visualized by the interaction of the ions with the microdisperse particle cloud, too [100]. By this method, the thrust effect of the ions (momentum transfer, ion drag) as well as inhomogeneities in the beam can be really observed and estimated. Furthermore, studies on the ion effect of the additional ion beam might also help to clarify the questions coming up with void formation in dusty plasma clouds under microgravity conditions [58], since in the present investigations we have the possibility of an independent and external tuning of the ion influence.

To monitor the interaction of confined powder particles with the surrounding plasma and the external ion beam a common asymmetric, capacitively coupled rf-discharge was employed [101]. The experiments have been performed in a plasma reactor, which is schematically drawn in Fig. 16.26.

The plasma glow is located in the region between the planar aluminum rf-electrode ($d = 130$ mm) and the upper part of the cylindrically shaped reactor vessel ($d = 400$ mm), which serves as grounded electrode (see Fig. 16.26). A copper ring was placed on the electrode to confine the injected dust particles (SiO_2, 0.8 μm) by a parabolic potential trap. However, in some cases, we used agglomerates of particles in the order of 50–300 μm for observation. The 13.56 MHz rf-power (10 W) is supplied by a rf-generator in combination with an automatic matching network. Depending on the gas pressure, the rf-plasma induced a self-bias of 60–300 V at the

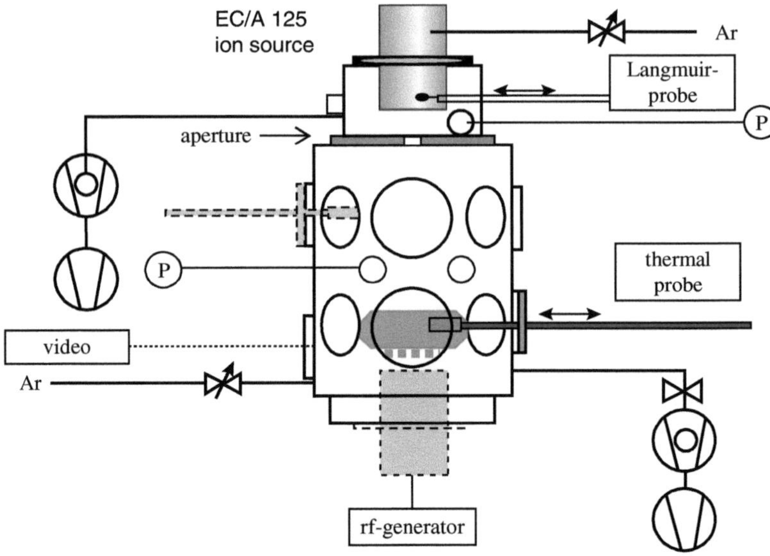

Fig. 16.26 Experimental setup for the investigation of an external ion beam with a particle cloud confined in an rf-plasma [101]

bottom electrode. The turbopump which allows for a base pressure of 5×10^{-5} Pa was connected to the vessel by a butterfly valve; the argon gas pressure was varied between 0.5 and 6 Pa using the valve and a flow controller.

For the determination of the plasma parameters, the experiments were carried out both with and without dust particles as well as with and without ion beam operation. The injected powder particles are charged and confined in the rf-plasma near the sheath edge ($\approx$10 mm) where they can be observed by light scattering of an illuminating laser fan (532 nm). A video camera at 125 frames per second with a filter at the laser wavelength was taken to observe the location and movement of the confined particles.

The ion beam source [102] is mounted on top of the vessel opposite to the rf-electrode (Fig. 16.27). Power of about 120 W was supplied by a generator via a microwave antenna. At 87.5 mT and 2.4 GHz, the electrons are strongly accelerated by the electron cyclotron resonance and ionize efficiently the argon atoms. The generated ions are extracted by a molybdenum grid system (diameter 125 mm) and accelerated by the beam voltage which was varied between 400 and 1,400 V. The second turbopump at the ion source allows for base pressure of 10^{-4} Pa; during ion beam operation, the gas pressure was 6×10^{-2} by using another flow controller. The corresponding gas flow is 8 sccm.

The distance of the levitated particle cloud from the extraction grid system of the ion source was about 640 mm. To get a good separation of the ion beam source (which is at lower pressure) and the plasma–particle interaction region in front of the rf-electrode which is at higher pressure, a tube of 75 mm diameter has been used

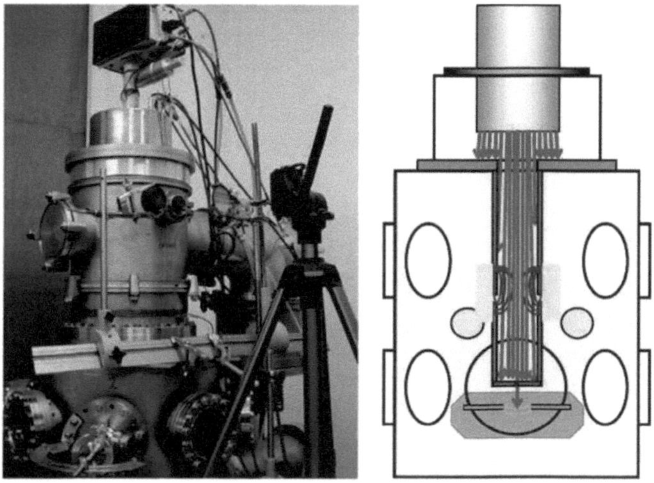

Fig. 16.27 Photograph (*left*) and schematic (*right*) of the ion beam operation. The ion beam source is on top of the vessel and the "beam tube" (*red*) is shown

(see Fig. 16.27). At the bottom of the "beam tube" is a hole ($d = 5\,\mathrm{mm}$) where the ions leave the tube for interaction with the confined particles. By this method, thermalization of the ions on their way from the extraction grid to the particles can be minimized.

In addition to Langmuir-probe measurements and optical emission spectroscopy [101], the integral energy flux from the ion beam toward the particles was measured by means of a thermal probe which has been described elsewhere [103]. The probe (copper, diameter 5 mm) is mounted on a manipulator arm to allow for radial scans along the beam diameter below the hole. The heat flux measurements are carried out by monitoring the rate of temperature change $\mathrm{d}T_S/\mathrm{d}t$ during "beam on" and "beam off." The radial profile of the energy influx (Fig. 16.28) reflects the profile of the escaping ion beam through the hole and its divergence due to the interaction with the rf-plasma at higher pressure. The maximum energy influx of $0.06\,\mathrm{J\,cm^{-2}\,s^{-1}}$ in the center of the beam corresponds at a beam voltage of 800 V and a pressure of 3 Pa to an ion current density of $75\,\mu\mathrm{A\,cm^{-2}}$.

The experimental studies have been conducted under static and dynamic conditions, respectively. Static conditions means the particle interaction with a constant ion beam and dynamic conditions means the interaction with a switched ion beam.

Static Conditions

The deformation of a levitated particle 2D dust cloud under broad beam ion operation has been observed and described elsewhere [100, 101, 104]. If the beam is switched on, the shape of the particle cloud changes in a characteristic manner due to inhomogeneity and divergence in the beam. The acting forces onto the particles can easily be seen in Fig. 16.29.

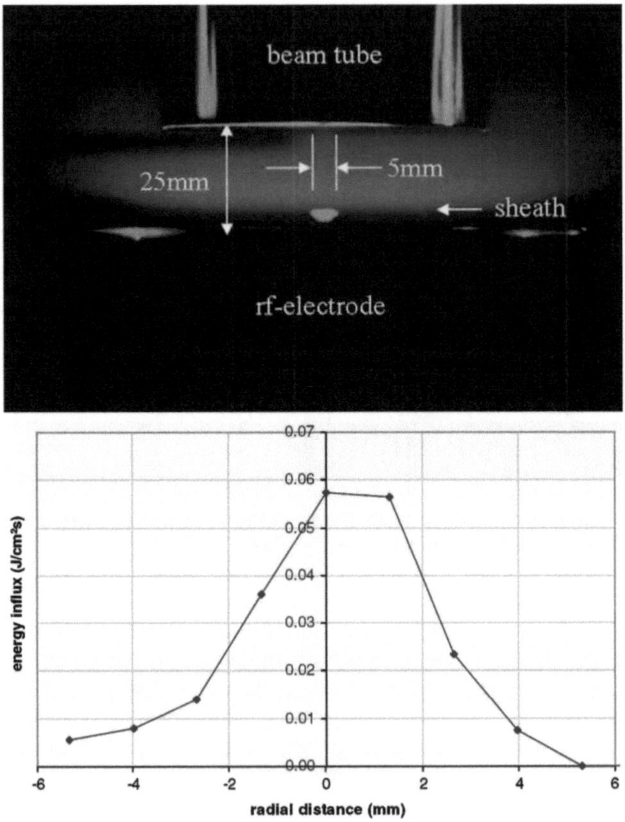

Fig. 16.28 Escaping ion beam through the hole (*top*) and the measured profile of the energy influx (*bottom*). The small particle cloud can be recognized underneath the hole

When the hole of the beam tube is closed by a shutter (Fig. 16.29a), the common situation of a complex plasma is realized. The particles are levitated due to the force balance between gravity F_g and electrostatic force F_{el} by the electric field in front of the rf-electrode. If the shutter is removed and, thus, the hole is open there exists a strong pressure gradient between the rf-plasma region (3 Pa) and the beam region inside the tube (0.1 Pa). The result is a neutral drag F_n by the gas flow which changes the shape of the originally flat dust cloud into a dome-like structure (see Fig. 16.29b). The additional force F_n acts in the radial as well as vertical direction due to the pressure gradient and the dust cloud structure can be explained by the flow patterns which can be simulated. Finally, if the ion beam is switched on the dome is distorted again by the pushing ion drag force F_{ion} (Fig. 16.29c). Since the electric field in the rf-sheath varies strongly with varying distance z from the electrode, the upper layers of the particle cloud are more likely influenced by the ion beam. Therefore, the displacement of the particle dome (e.g., top of the cloud) which looks like an indentation has been taken as measured quality.

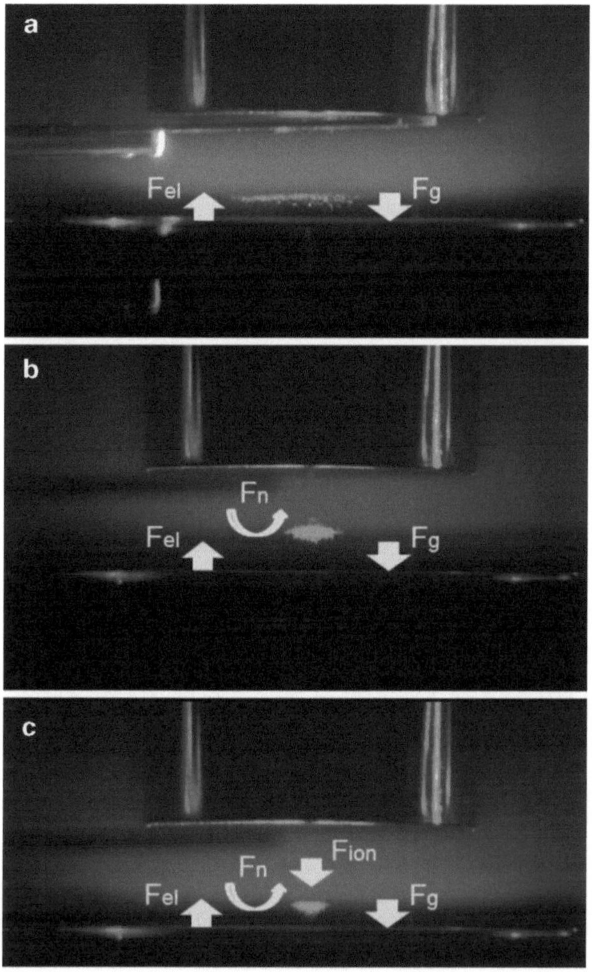

Fig. 16.29 Interaction of the particles confined in the plasma and the ion beam at $p = 3\,\mathrm{Pa}$ and $V_{\mathrm{bias}} = 146\,\mathrm{V}$. (**a**) Beam tube closed, (**b**) beam tube open, ion beam off, and (**c**) beam tube open, ion beam ($V_{\mathrm{beam}} = 1{,}300\ V$) on

For levitated particles of mass m_{d} and charge Q_{d} in steady state, the force balance in vertical z-direction can be written as

$$\begin{aligned} F(z) &= F_{\mathrm{el}}(z) + F_{\mathrm{n}}(z) - F_{\mathrm{g}} - F_{\mathrm{ion}}(z) \\ &\approx Q_{\mathrm{d}}(z)E(z) + F_{\mathrm{n}}(z) - m_{\mathrm{dg}} - F_{\mathrm{ion}}(z) = 0. \end{aligned} \qquad (16.10)$$

The ion drag force consists of two components: the orbital force and the collection force [23, 60]. The orbital force corresponds to the momentum transfer due to Coulomb scattering and the collection force is a consequence of direct collisions between the beam ions and the particle. Since the directed kinetic energy of the supersonic ions is much larger than the potential of the dust particles, only the collection force is of importance and the particle cross section πr_d^2 can be regarded as collection impact parameter. Then the ion drag force becomes

$$F_\mathrm{ion} = n_\mathrm{i} m_\mathrm{i} v_\mathrm{i}^2 \pi r_\mathrm{d}^2 = \pi r_\mathrm{d}^2 j_\mathrm{i} \sqrt{\frac{2m_\mathrm{i}}{e_0}} \sqrt{V_\mathrm{beam}}, \tag{16.11}$$

where n_i is the density of the ions of mass m_i and v_i is the directed velocity which can also be written in terms of the ion flux density j_i and the beam voltage V_beam. For typical experimental conditions as given earlier, the ion drag force is in the order of 1–5×10^{-14} N, depending on beam voltage.

On the other hand, the ion drag can be estimated from the force balance at the levitating position. Supposing that there is only a weak change in the electric field by the relatively low ion beam flux density in front of the rf-electrode and almost no influence on the particle charge during beam operation; the change in the position of the particles can only be caused by the ion drag (thrust). Under these assumptions, the sum of gravitation and ion drag has to compensate the sum of the electrostatic force and vertical neutral drag. The electrostatic force F_el is given by the product of the particle charge Q_d times the electric field strength $E(z_0)$ at trapping position z_0. The field is in the order of $2.5 \times 10^3\,\mathrm{V\,m^{-1}}$ and the charge is about $10^3 e_0$ which results in an electrostatic field force of 1–10×10^{14} N depending on the levitation height. The gravitational force F_g is for SiO_2 particles ($m_\mathrm{d} \approx 6 \times 10^{-16}$ kg) in the order of 6×10^{-15} N, for example, about 10% of the opposite electrostatic force. Considering the uncertainties in the determination of the particle charge and the field strength, the agreement between the estimation by the force balance and the calculation by (16.11) is rather satisfactory. The field strength at position $z = z_0$ has been obtained by a linear extrapolation along the sheath in front of the rf-electrode by measuring the bias voltage V_bias and the sheath thickness. Obviously, also the particle charge Q_d during ion beam operation is smaller than it is for the pure rf-plasma. Moreover, the ion drag by the rf-plasma which is certainly small in comparison with the ion beam drag has been neglected. The displacement of the particles by the ion beam in dependence on the beam voltage is shown in Fig. 16.30.

For small beam voltages ($V_\mathrm{beam} < 600$ V), there is almost no change. When the beam voltage increases, the height of the dust particles decreases. This means the originally dome-like dust cloud forms a more flat shape and moves in the direction of the rf-electrode. For higher beam voltage, the ion drag becomes stronger and the particles can deeply penetrate into the sheath and can compensate the influence of the electrostatic field more efficiently.

Dynamic Conditions

In addition to the static experiments, oscillations of particles have been excited by switching the ion beam at different frequencies. In contrast to the static experiments

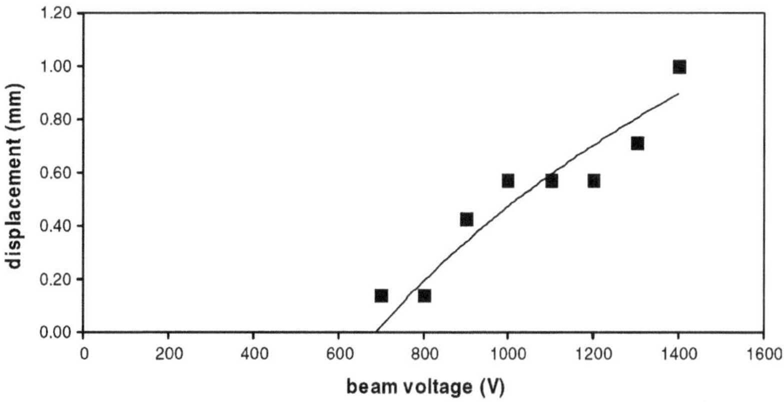

Fig. 16.30 Displacement of the particles from their original position in the sheath of the rf-electrode at ion beam operation ($p = 4\,\text{Pa}$ and $V_{\text{bias}} = 100\,\text{V}$)

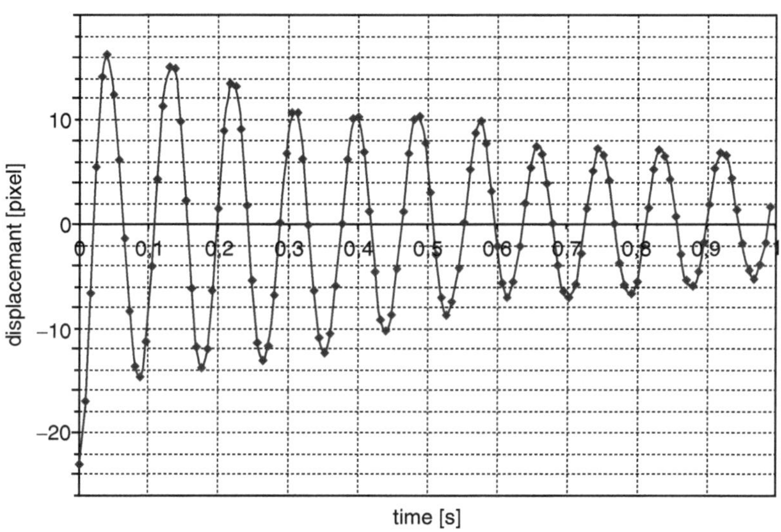

Fig. 16.31 Oscillations of the particles around a new equilibrium position after switching on the ion beam [86]

where it has been waited for the equilibrium of the particles, now the oscillations around the equilibrium position at switching on and switching off of the ion beam are under consideration.

After switching on the beam, it needs a few seconds for stabilization of the particle position. During this time, the particles (or clouds) perform oscillations with a frequency of about 14 Hz (see Fig. 16.31).

The oscillation in z-direction can be simply described by

$$m_{\text{d}}\ddot{z} + \gamma\dot{z} + \omega_0^2 z = F(z) = F_0 \cos \Omega t, \tag{16.12}$$

where γ is the damping constant and ω_0 is the resonance frequency of the system. The driving force $F(z)$ is in our case the ion drag by the beam operation. In the experiments performed here, a rectangular beam on/off profile has been used.

The damping behavior due to the neutral drag depends on the gas pressure; for typical conditions, γ is in the order of about 1 s^{-1}. The oscillation can be described as a periodic movement of the particles in the potential trap of the sheath. As long as the displacement of the particles by hitting the ions is smaller than the characteristic trapping length in the sheath, the particles will come to rest by damping the background gas.

16.5 Particles as Thermal Probes

In addition to the use of particles as electrostatic probes, it is also possible to employ particles as thermal probes. In situ thermometry could actually turn the suspended particles into microprobes for the particle energy balance, and indirectly also to probes for the balance between the several plasma–surface interaction mechanisms. For this purpose, special fluorescent particles were used as thermal probes [66, 105]. Especially, temperature-sensitive features of particular phosphors were utilized for measuring the temperature of the microparticles, confined in the sheath of an rf-plasma. The experiments were performed under variation of argon pressure and rf-power of the process plasma [105]. Parametric measurements of grain temperature have been accomplished in argon, showing quite promising results and match the expectations regarding variations in discharge power and gas pressure. The utilization of emission from rare-earth-activated phosphor particles offers a robust technique, overcoming the disadvantages of former experiments [105]. The particle temperature has been determined by evaluation of characteristic fluorescent lines. Furthermore, the influence of the background gas is assumed to play an important role for the cooling of the particles.

The surface temperature of plasma-treated objects is crucial in processes such as etching or thin film deposition [106–108]. However, in the case of powders or microscopic objects, no proper technique for the measurement of the surface temperature was available yet. On the other hand, it is well known, the possibility for obtaining the temperature of microparticles, levitated inside a plasma, would give access to the energetic conditions at their surface due to the balance of several contributions of energy gain and loss [103, 109]. Thus, the development of a temperature diagnostic for microparticles is not only valuable for the improvement of technical plasmas but could also improve the understanding of the physics of plasma–particle interactions.

Daugherty and Graves [110] measured particle temperatures in a pulsed rf-discharge by the detection of the decay time of the particle's fluorescence in the afterglow of the argon plasma. They reported particle temperatures of 410 ± 10 K ($137 \pm 10°$)C in an argon plasma of $P_{rf} = 50$ W and an argon pressure $p = 40$ Pa. A different approach was used by Swinkels et al. [111, 112]. They demonstrated a technique, utilizing temperature-dependent behavior of emission intensity and line broadening of melamine formaldehyde particles, dyed with rhodamine B, in argon and oxygen plasmas, respectively. This technique does not rely on pulsed operation, but suffers from oxygen, temperature, and UV bleachings of the phosphor (rhodamine B). Measured particle temperatures ranged from $T_p \approx 100°C$ at $P_{rf} = 5\,W$ to $T_p \approx 190°C$ at $P_{rf} = 50\,W$ for $p = 20\,Pa$, respectively.

Now results from the adoption of emission characteristics of inorganic phosphors, offering temperature-dependent features, for the temperature measurement on floating microparticles in a plasma will be shown. Especially rare-earth-activated phosphors, excited by UV illumination, are feasible for this purpose, showing strong line-type emission without temperature, oxygen, or UV bleachings under the operated conditions [113–115]. This family of phosphors has already been used for temperature measurements over a wide temperature range and for many applications such as projection thermographs and fiber optic thermometers [116, 117]. They have also been used for temperature measurements in gas turbines, rotors, or on projectiles, and for heat flux measurement through surfaces [115, 118, 119].

Our experiments are performed in the plasma reactor PULVA-INP with a capacitively coupled rf-discharge and adaptive electrode, which is particularly suited for the investigation of dusty plasmas [66, 87]. The experimental setup is described in Sect. 16.4.1 and elsewhere [70].

As a phosphor, polydisperse YVO_4:Eu particles ($d = 15$–$20\,\mu m$) are confined above the center of the segmented electrode. The phosphor shows strong temperature dependence in the emission spectrum over the desired temperature range from room temperature to 200°C. A mercury arc lamp is used to excite the phosphor. Its UV light is sent trough a 313 nm filter and focused onto the particles (see Fig. 16.32). The UV illumination can also be blocked to take records of background spectra. Additionally, the optics are set up on a translational stage, to track the particle's vertical position, which is affected by the plasma conditions such as power, pressure, and bias voltage. Phosphor luminescence is observed by an imaging spectrograph ($f = 500\,mm$) and a charge coupled device CCD camera. Again, the optics include a vertical translation stage to follow the position of the observed particles.

Each spectrum of phosphor particles inside the plasma, excited by UV illumination at $\lambda_{exc} = 313\,nm$, is a superposition of plasma emission and luminescent phosphor emission. To obtain the pure phosphor emission, a second background spectrum without UV illumination has to be taken. By computing the difference of these spectra, the phosphor emission can be determined (Fig. 16.33). To get rid of errors due to fluctuations in the plasma spectrum, multiple repetitions are performed and averaged, after background subtraction has taken place. From the standard deviation of the background spectra, information about the spectral accuracy is obtained and errors are identified and rejected.

The resulting data are then compared to a series of fluorescent calibration spectra of the phosphor, which have been recorded in a special oven, each at a certain known temperature (Fig. 16.34). In the oven, a small amount of particles has been placed in an object holder inside a heated copper box, increasing the temperature

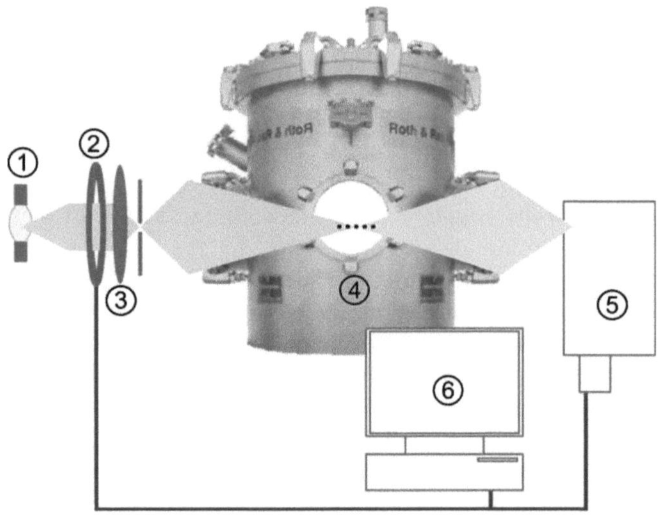

Fig. 16.32 Phosphor particles are confined in the plasma of an rf-discharge (4). Radiation of a Hg arc lamp (1) is filtered by a 313 nm filter (3) and focused onto the particles, to excite luminescence, or blocked by a shutter (2). Phosphor emission is collected and analyzed by an imaging spectrograph and a CCD camera (5). A PC (6) is used for data processing and to synchronize signal excitation and detection

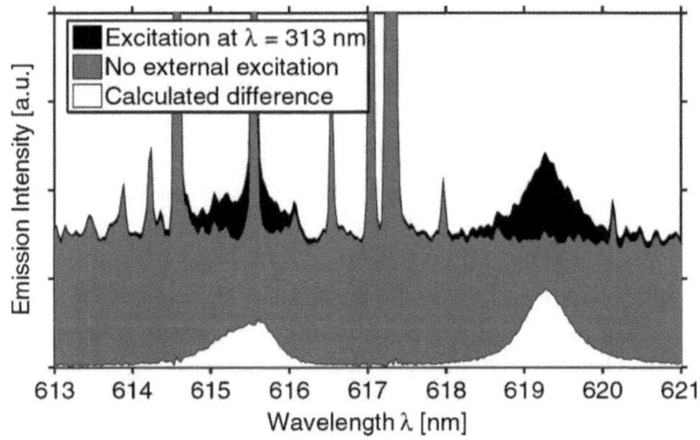

Fig. 16.33 Emission spectra of phosphor particles in plasma under UV excitation (*black*) and without excitation (*gray*). Pure phosphor emission (*white*) is obtained by computing the difference of both spectra

by less than 0.14°C min^{-1} during the measurements. Temperature was monitored by a platinum temperature sensor, fixed at the copper box. The spectra were then accumulated within 0.1°C intervals.

To determine the particle temperature T_p, a least squares fit is performed, scaling each calibration spectrum to fit the recorded data under plasma conditions. For each comparison, a residual χ is obtained. The minimum of $\chi(T)$ is finally determined

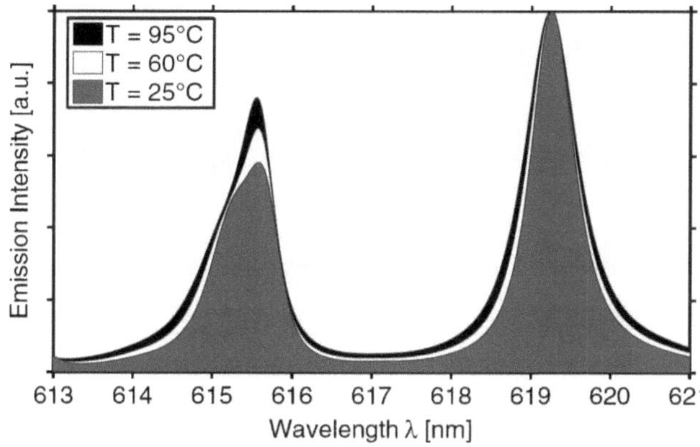

Fig. 16.34 Phosphor emission spectra, recorded in a special calibration oven, at different temperatures (normalized to emission maximum near 619 nm). For the presented results, calibration spectra were recorded with a temperature resolution better than 0.4°C

by a local polynomial fit, and the corresponding temperature is identified as the particle temperature T_p. As the determination of particle temperatures does not rely on absolute measurements, this technique is independent from the amount of emitting phosphor material. Experiments have been performed in argon at pressure $p = 10$, 20, 30, and 50 Pa and rf-power between 5 and 100 W. Before starting the measurements, the phosphor particles are confined in front of the AE at 5 Pa and power of 5 W. Then the desired gas pressure for the measurement series is adjusted. Measurements are always performed from low to high rf-power. If a new series was recorded, first the rf-power was reduced to 5 W and the argon pressure was adjusted. During one series, each change in rf-power is followed by an alignment of excitation and detection optics to the particle position, which is governed by the altered plasma sheath. Each spectrum was taken with an exposure time of $t_{exp} = 5$ s and averaged over 20 cycles. Thus, one measurement took approximately 7 min to perform, during which the discharge parameters were kept constant.

The results of measuring the grain temperature in the plasma are shown in Fig. 16.35a–d. The particles were confined in front of the central segment of the AE. As expected, T_p is rising with increasing plasma power. The higher the gas pressure, the smaller is the increase in T_p. In Fig. 16.35d ($p = 50$ Pa), the slope is very weak and a dependence of particle temperature on the rf-power is barely observable. This observation could be explained by enhanced heat loss of the particles with rising argon pressures in the low-pressure regime (assuming a flowing Knudsen gas to be an appropriate description, as reported by Daugherty and Graves [110] and Swinkels [111, 112], and references therein).

At the same time, also the electron temperatures are dropping with increasing rf-power [87] at $p = 50$ Pa. Thus, even if the electron density increases with the rf-power, the energy influx toward the particle at 50 Pa is less strong dependent on the rf-power than at lower pressures.

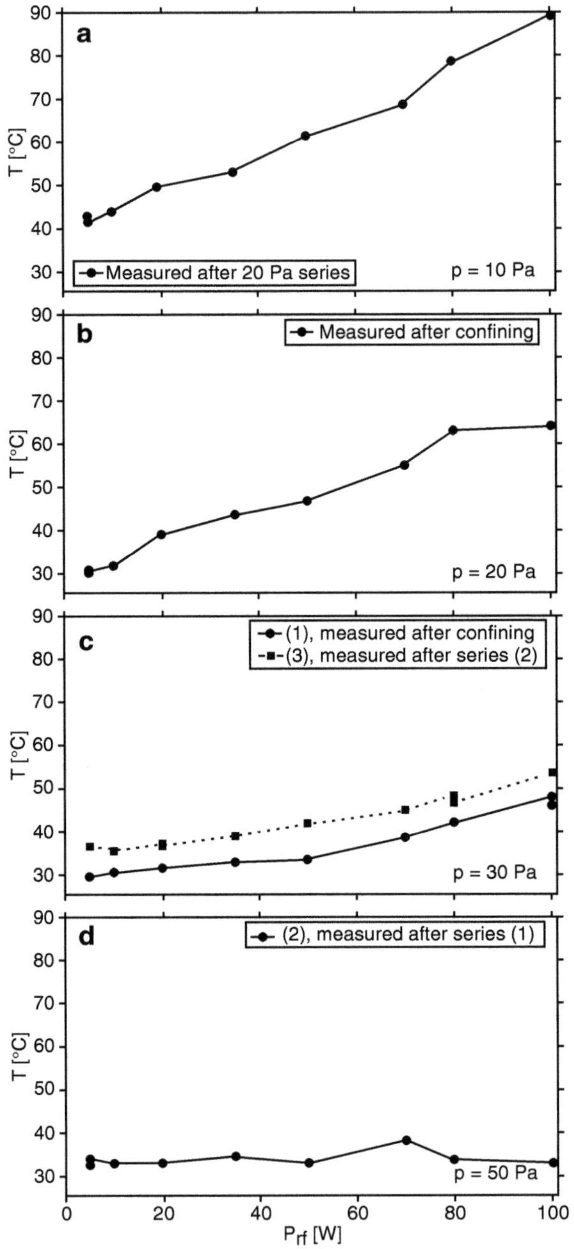

Fig. 16.35 Results for the particle temperature T_p in the plasma at parametric variation of rf-power at argon pressures of (**a**) 10 Pa, (**b**) 20 Pa, (**c**) 30 Pa, and (**d**) 50 Pa. For better comparison, also the previous history of plasma treatment is given in the figures [105]

However, the particle temperature should also depend on the argon gas temperature [103, 110, 112] which has not been monitored in our measurements yet. A comparison of two series, performed at the same day and using the same discharge parameters, indicates an influence of the gas temperature (Fig. 16.35c, d): The first measurement series at $p = 30$ Pa in Fig. 16.35c, marked with (1), has been performed directly after confining the phosphor particles. After completion, another series at $p = 50$ Pa has been recorded, assigned with (2) in Fig. 16.35d, and finally a second series at $p = 30$ Pa was accomplished (3) in Fig. 16.35c. The results for the particle temperature are increased by $\approx 5°$C, compared to series (1). This can be explained by increasing gas temperature due to heating of the vessel by higher plasma power, resulting in higher particle temperatures in the latter measurement series. For this reason, the previous history of plasma treatment for every measurement series has been mentioned in Fig. 16.35.

The particle temperatures presented in Fig. 16.35 show lower values than the measurements by Daugherty and Graves and Swinkels [111, 112]. However, Swinkels et al. [112] measured in front of the driven electrode in a gaseous electronic conference (GEC) reference cell, where the particle temperature should be expected to be remarkably higher. The plasma device, used by Daugherty and Graves, was rather small compared to this experimental setup, particularly the electrode distance of 25 mm is comparatively small. Thus, the plasma conditions near the lower electrode are not comparable to those in our device.

Acknowledgements The experimental studies were supported by the Deutsche Forschungsgemeinschaft under SFB TR 24/B4 and the Deutsches Zentrum für Luft- und Raumfahrt (DLR) under grant 50 JR 0644. This chapter is a summary and overview on results which have been obtained in collaboration with many coworkers and students. The authors would like to thank R. Basner, G. Schubert, H. Fehske, G. Thieme, H. Neumann, H. Wulff, M. Quaas, R. Hippler, K.D. Weltmann, H. Maurer, R. Wiese, G.M.W. Kroesen, W.W. Stoffels, E. Stoffels, R. Schneider, H. Deutsch, F. Scholze, M. Zeuner, and M. Tartz for their collaboration and contributions.

References

1. A. Bouchoule, *Dusty Plasmas: Physics, Chemistry and Technological Impacts in Plasma Processing* (Wiley, Chichester, 1999)
2. A. Bouchoule, Phys. World **6**, 47 (1993)
3. A. Bouchoule, L. Boufendi, J. Hermann, A. Plain, T. Hbid, G.M.W. Kroesen, E. Stoffels, W.W. Stoffels, Pure Appl. Chem. **68**, 1121 (1996)
4. G.S. Selwyn, Plasma Sour. Sci. Technol. **3**, 340 (1994)
5. G. Lapenta, J.U. Brackbill, Plasma Sour. Sci. Technol. **6**, 61 (1997)
6. W. Böhme, W. Köhler, M. Römheld, S. Veprek, IEEE Trans. Plasma Sci. **22**(2), 110 (1994)
7. Ch. Hollenstein, W. Schwarzenbach, A.A. Howling, C. Courteille, J. Vac. Sci. Technol. A **14**(2), 535 (1996)
8. E. Stoffels, W.W. Stoffels, G.M.W. Kroesen, F.J. Dehoog, J. Vac. Sci. Technol. A **14**, 556 (1996)
9. C. Hollenstein, J.L. Dorier, J. Dutta, L. Sansonnens, A.A. Howling, Plasma Sour. Sci. Technol. **3**, 278 (1994)
10. L. Boufendi, J. Hermann, A. Bouchoule, B. Dubreuil, E. Stoffels, W.W. Stoffels, M.L. De Giorgi, J. Appl. Phys. **76**, 148 (1994)

11. T. Fukuzawa, K. Obata, H. Kawasaki, M. Shiratani, Y. Watanabe, J. Appl. Phys. **80**, 3202 (1996)
12. W.W. Stoffels, E. Stoffels, G. Swinkels, M. Boufnichel, Phys. Rev. E **59**(2), 2302 (1999)
13. H. Kersten, H. Deutsch, M. Otte, G.H.P.M. Swinkels, G.M.W. Kroesen, Thin Solid Films **377**, 530 (2000)
14. M. Klick, J. Appl. Phys. **79**, 3445 (1996)
15. H.H. Anderson, Plasma Sour. Sci. Technol. **3**, 302 (1994)
16. W.W. Stoffels, E. Stoffels, G.M.W. Kroesen, M. Haverlag, J.H. Den Boer, F.J. De Hoog, Plasma Sour. Sci. Technol. **3**, 329 (1994)
17. A. Garscadden, B.N. Ganguly, P.D. Haaland, J. Williams, Plasma Sour. Sci. Technol. **3**, 239 (1994)
18. G. Praburam, J. Goree, J. Vac. Sci. Technol. A **12**(6), 3137 (1994)
19. M. Keidar, I.I. Beilis, R.L. Boxman, IEEE Trans. Plasma Sci. **23**(6), 902 (1995)
20. I.I. Beilis, M. Keidar, R.L. Boxman, S. Goldsmith, J. Appl. Phys. **85**(3), 1358 (1999)
21. B. Schultrich, Contrib. Plasma Phys. **39**, 463 (1999)
22. A.C. Xenoulis, G. Doukellis, P. Tsouris, A. Karydas, C. Potiriadis, A.A. Katsanos, Th. Tsakalakos, Vacuum **51**, 357 (1998)
23. S.V. Vladimirov, K. Ostrikov, A.A. Samarian, *Physics and Applications of Complex Plasmas* (Imperial College, London, 2005)
24. H. Kersten, H. Deutsch, E. Stoffels, W.W. Stoffels, G.M.W. Kroesen, R. Hippler, Contrib. Plasma Phys. **41**(6), 598 (2001)
25. J. Winter, Plasma Phys. Control. Fusion **40**, 1201 (1998)
26. J. Winter, G. Gebauer, J. Nucl. Mater. **266–269**, 228 (1999)
27. K. Narihara, K. Toi, Y. Hamada, Nucl. Fusion **37**, 1177 (1996)
28. M. Dalvie, G.S. Selwyn, M. Surendra, C.R. Guarnieri, J.J. Mcgill, Appl. Phys. Lett. **63**, 3279 (1993)
29. C.K. Yeon, K.W. Whang, J. Vac. Sci. Technol. A **15**, 66 (1997)
30. Y. Watanabe, M. Shiratani, H. Makoni, Appl. Phys. Lett. **57**, 1616 (1990)
31. G. Uchida, S. Iizuka, N. Sato, IEEE Trans. Plasma Sci. **29**, 274 (2001)
32. M. Onozuka, Y. Ueda, K. Takahashi, Y. Seki, U. Ueda, I. Aoki, Vacuum **47**(6–8), 541 (1996)
33. A. Brattli, O. Havnes, J. Vac. Sci. Technol. **14**(2), 644 (1996)
34. J. Perrin, P. Molinas-Mata, P. Belenguer, J. Phys. D **27**, 2499 (1994)
35. M. Klindworth, A. Melzer, A. Piel, Phys. Rev. B **61**(12), 8404 (2000)
36. M. Wolter, A. Melzer, Phys. Rev. E **71**, 036414 (2005)
37. U. Kogelschatz, B. Eliasson, W. Egli, Pure Appl. Chem. **71**(10), 1819 (1999)
38. N. Rao, S. Girshick, J. Heberlein, P. Mcmurry, S. Jones, D. Hansen, B. Micheel, Plasma Chem. Plasma Proc. **15**, 581 (1995)
39. R. Scholl, G. Natour, in *Proceedings of ICPIG XXII*, Hoboken, NJ, vol. 363, 1995
40. T. Ishigaki, T. Sato, Y. Moriyoshi, M.I. Boulos, J. Mater. Sci. Lett. **14**, 694 (1995)
41. I. Roca, P. Cabarrocas, M. Kunst, C. Clerc, H. Bernas, J. Vac. Sci. Technol. A **16**(2), 436 (1998)
42. S. Veprek, Pure Appl. Chem. **68**, 1023 (1996)
43. B. Schultrich, H.J. Scheibe, H. Mai, Adv. Eng. Mater. **2**, 419 (2000)
44. E. Stoffels, W.W. Stoffels, G. Ceccone, R. Hasnaoui, J. Appl. Phys. **86**(6), 3442 (1999)
45. I. Herrmann, V. Brüser, S. Fiechter, H. Kersten, P. Bogdanoff, J. Electrochem. Soc. **152**, A2179 (2005)
46. S. Yan, H. Maeda, J.I. Kusakabe, K. Morooka, T. Okubo, J. Mater. Sci. **28**, 1892 (1993)
47. H. Kersten, P. Schmetz, G.M.W. Kroesen, Surf. Coat. Technol. **507**, 108 (1998)
48. K. Tachibana, Y. Hayashi, Pure Appl. Chem. **68**, 1107 (1996)
49. F. Vivet, A. Bouchoule, L. Boufendi, in *Proceedings of ICPIG XXIII*, Toulouse, France, vol. I/200, 1997
50. U. Schürmann, W. Hartung, H. Takele, V. Zaporojtchenko, F. Faupel, Nanotechnology **16**, 1078 (2005)
51. K. Kitamura, S. Akutsu, S. Ito, K. Akashi, in *Proceedings of the ICRP-3*, Nara, Japan, vol. 357, 1997

52. G. Thieme, M. Quaas, H. Kersten, H. Wulff, R. Hippler, in *Proceedings of the BFPT-10*, Greifswald, Germany, vol. 68, 2001
53. C. Arpagaus, A. Sonnenfeld, P.R. von Rohr, Chem. Eng. Technol. **28**, 87 (2005)
54. H. Kersten, G. Thieme, M. Fröhlich, D. Bojic, D.H. Tung, M. Quaas, H. Wulff, R. Hippler, Pure Appl. Chem. **77**, 415 (2005)
55. F. Vollertsen, S. Vogler, *Werkstoffeigenschaften und Mikrostruktur* (Hanser, Muenchen, 1989)
56. U. Konopka, D. Samsonov, A.V. Ivlev, J. Goree, Phys. Rev. E **61**(2), 1890 (2000)
57. K. Matyash, M. Fröhlich, H. Kersten, G. Thieme, R. Schneider, M. Hannemann, R. Hippler, J. Phys. D **37**, 2703 (2004)
58. M. Wolter, A. Melzer, O. Arp, M. Klindworth, M. Kroll, A. Piel, IEEE Trans. Plasma Sci. **35**, 266 (2007)
59. P.S. Epstein, Phys. Rev. **23**, 710 (1924)
60. M.S. Barnes, J.K. Keller, J.C. Forster, J.A. O'Neiil, Phys. Rev. Lett. **68**(3), 313 (1992)
61. M.D. Kilgore, J.E. Daugherty, R.K. Porteous, D.B. Graves, J. Appl. Phys. **73**, 7195 (1993)
62. H. Kersten, G.M.W. Kroesen, R. Hippler, Thin Solid Films **332**, 282 (1998)
63. H.S. Shin, D.G. Goodwin, Mater. Lett. **19**, 119 (1994)
64. M. Karches, Ch. Bayer, P.R. von Rohr, Surf. Coat. Technol. **116–119**, 879 (1999)
65. R. Symes, R.M. Sayer, J. Reid, Phys. Chem. Phys. **6**, 474 (2004)
66. G. Thieme, R. Basner, R. Wiese, H. Kersten, Faraday Discuss. **137**, 157 (2008)
67. G. Mie, Ann. Phys. **25**, 337 (1908)
68. R.J. Hopkins, R. Symes, R.M. Sayer, J.P. Reid, Chem. Phys. Lett. **380**, 665 (2003)
69. M. Wolter, T. Ockenga, M. Haass, J. Blazek, H. Kersten, Plasma Process. Polym. **6**(Suppl. 1), 620 (2009). DOI 10.1002/ppap.200931603
70. R. Basner, F. Sigeneger, D. Loffhagen, G. Schubert, H. Fehske, H. Kersten, New J. Phys. **11**, 013041 (2008)
71. R. Basner, H. Fehske, H. Kersten, S. Kosse, G. Schubert, Vak. Forsch. Prax. **17**(5), 259 (2005)
72. B.M. Annaratone, T. Antonova, H.M. Thomas, G.E. Morfill, Phys. Rev. Lett. **93**(18), 185001 (2004)
73. A.A. Samarian, B.W. James, Phys. Lett. A **287**(1–2), 125 (2001)
74. C. Zafiu, A. Melzer, A. Piel, Phys. Plasmas **9**, 4794 (2002)
75. J. Goree, Plasma Sour. Sci. Technol. **3**, 400 (1994)
76. Y.N. Wang, J.R. Hou, Acta Phys. Sin. **52**, 434 (2003)
77. P.K. Shukla, Phys. Plasmas **8**(5), 1791 (2001)
78. H. Kersten, R. Wiese, G. Thieme, M. Fröhlich, A. Kapitov, D. Bojic, F. Scholze, H. Neumann, M. Quaas, H. Wulff, New J. Phys. **5**, 93.1 (2003)
79. T. Cao, J. Matsoukas, J. Appl. Phys. **92**(5), 2916 (2002)
80. H. Kersten, H. Deutsch, G.M.W. Kroesen, Int. J. Mass Spectrom. **233**, 51 (2004)
81. E.B. Tomme, B.M. Annaratone, J.E. Allen, Plasma Sour. Sci. Technol. **9**, 87 (2000)
82. E.B. Tomme, D.A. Law, B.M. Annaratone, J.E. Allen, Phys. Rev. Lett. **85**(12), 2518 (2000)
83. F.X. Bronold, H. Fehske, H. Kersten, H. Deutsch, Phys. Rev. Lett. **101**, 175002 (2008)
84. G. Thieme, R. Wiese, D. Gorbov, H. Kersten, R. Hippler, AIP Conf. Proc. **649**, 333 (2002)
85. B.M. Annaratone, M. Glier, T. Stuffler, M. Raif, H.M. Thomas, G.E. Morfill, New J. Phys. **5**, 92 (2003)
86. H. Kersten, R. Wiese, H. Neumann, R. Hippler, Plasma Phys. Control. Fusion **48**, B105 (2006)
87. M. Tatanova, G. Thieme, R. Basner, M. Hannemann, Y.B. Golubovskii, H. Kersten, Plasma Sour. Sci. Technol. **15**(3), 507 (2006)
88. A.A. Samarian, S.V. Vladimirov, Phys. Rev. E **67**(6), 066404 (2003)
89. A. Barkan, N. D'Angelo, R.L. Merlino, Phys. Rev. Lett. **73**(23), 3093 (1994)
90. C. Zafiu, A. Melzer, A. Piel, Phys. Plasmas **10**(5), 1278 (2003)
91. P.W. Smith, C.M. Ticos, A. Dyson, Plasma Sour. Sci. Technol. **13**, 395 (2004)
92. L. Couedel, M. Mikikian, L. Boufendi, A.A. Samarian, Phys. Rev. E **74**(2), 026403 (2006)
93. S.V. Vladimirov, N.F. Cramer, Phys. Rev. E **62**(2), 2754 (2000)
94. B.M. Annaratone, T. Antonova, D.D. Goldbeck, H.M. Thomas, G.E. Morfill, Plasma Phys. Control. Fusion **46**, B495 (2004)

95. K. Matyash, R. Schneider, J. Plasma Phys. **72**(6), 809 (2006)
96. S.A. Khrapak, A.V. Ivlev, S.K. Zhdanov, G.E. Morfill, Phys. Plasmas **12**(4), 042308 (2005)
97. L. Spitzer, *Physical Processes in the Interstellar Medium* (Wiley, New York, 1982)
98. E. Gruen, G.E. Morfill, D.A. Mendis, *Planetary Rings* (University Of Arizona Press, Tucson, 1984)
99. C.K. Goertz, Rev. Geophys. **27**, 271 (1989)
100. H. Kersten, R. Wiese, D. Gorbov, A. Kapitov, F. Scholze, H. Neumann, Surf. Coat. Technol. **174**, 918 (2003)
101. H. Kersten, R. Wiese, M. Hannemann, A. Kapitov, F. Scholze, H. Neumann, R. Hippler, Surf. Coat. Technol. **200**(1–4), 809 (2005)
102. M. Zeuner, H. Neumann, F. Scholze, D. Flamm, M. Tartz, F. Bigl, Plasma Sour. Sci. Technol. **7**, 252–267 (1998)
103. H. Kersten, H. Deutsch, H. Steffen, G.M. Kroesen, R. Hippler, Vaccum **63**, 385–431 (2001)
104. H. Neumann, H. Kersten, R. Wiese, A. Kapitov, F. Scholze, in *ESA Conference on Space Propulsion*, Cagliari, Italy, p. 61, 2004
105. H. Maurer, R. Basner, H. Kersten, Rev. Sci. Instrum. **79**, 093508 (2008)
106. S.D. Bernstein, T.Y. Wong, R.W. Tustison, J. Vac. Sci. Technol. **17**(2), 571 (1999)
107. C. Cardinaud, M.C. Peignon, P.Y. Tessier, Appl. Surf. Sci. **164**(1–4), 72 (2000)
108. H. Deutsch, H. Kersten, A. Rutscher, Contrib. Plasma Phys. **29**, 263 (1989)
109. H. Kersten, D. Rohde, H. Steffen, H. Deutsch, R. Hippler, G. Swinkels, G.M.W. Kroesen, Appl. Phys. A **72**(5), 531 (2001)
110. J.E. Daugherty, D.B. Graves, J. Vac. Sci. Technol. A **11**(4), 1126 (1993)
111. G. Swinkels, Ph.D. Thesis, Technische Universiteit Eindhoven (1999)
112. G. Swinkels, H. Kersten, H. Deutsch, G.M.W. Kroesen, J. Appl. Phys. **88**, 1747 (2000)
113. L.P. Goss, A.A. Smith, M.E. Post, Rev. Sci. Instrum. **60**(12), 3702 (1989)
114. K. Cates, D. Smith, in *Oak Ridge National Laboratory Technical Report*, ORNL/ATD-40, 1990
115. S.W. Allison, G.T. Gillies, Rev. Sci. Instrum. **68**(7), 2615 (1997)
116. F. Urbach, Technical Report, U.S. Patent 2,551,650 (1951)
117. S.A. Wade, S.F. Collins, G.W. Baxter, J. Appl. Phys. **94**(8), 4743 (2003)
118. J.P. Feist, A.L. Heyes, in *Proceedings of the 10th International Symposium on Application of Laser Techniques to Fluid Mechanics*, Lisbon, Portugal, 2000
119. K. Tobin, G. Capps, J. Muhs, D. Smith, M. Cates, in *Oak Ridge National Laboratory Technical Report*, ORNL/ATD-43, 1990

Index

activation energy, E_A, 364
adaptive electrode, 414, 417–425, 435, 437
alumina, Al_2O_3, 347, 358
Andersen thermostat, 241
annihilation operator, 69, 94
antisymmetrization operator, 90
approximation
  dipole, 88
  finite difference, 82
  local density, 191–197
  mean-field, 180–187, 190
  random phase, 90, 111–127
asymmetric Bragg case, 347, 349
atomic force microscopy (AFM), 306, 309
atomic units, *see* system of units

BBGKY hierarchy, 67
Bohmian quantum mechanics, 72–73
Bohr radius, 43
  effective, 44
Boltzmann equation, 206, 217
Boltzmann factor, 55, 100
boltzmannon, 104
Bose–Einstein condensate, 60–61
boson, 58–60, 104
bottom-up approach, 80, 233
boundary conditions, 83–84, 88
  absorbing, 83–84
  Dirichlet, 83, 84
  periodic, 247–249
Brownian motion, 161, 171
Brueckner parameter, 6, *see* coupling parameter, Coulomb

canonical ensemble, 55, 177
capacitively coupled rf-discharge, 415, 416, 435
capacitively coupled rf-plasma, 417
cavity enhanced spectroscopy, 413
cavity-enhanced spectroscopy, 410
Cayley's form, 82
chaining mesh technique, 247
chemical potential, 92
cluster, 299, 300, 302–313
cluster deposition, 311
cluster evaporation, 312, 313
cluster film, 308, 309
cluster source, 305
coating, 399–402, 406, 407, 410
color gradient method, 142
complex fluids, 7
configuration space, 97
correlation, 4
correlation energy, 109
correlations, 79, 93, 94, 179, 195–197
Coulomb
  energy, 44
  regularized potential, 94
Coulomb cluster, 155
Coulomb crystal, 231, 234
Coulomb potential, 109, 110, 113, 114, 129
coupling parameter, 5, 95
  Coulomb, 44–45, 156
  harmonic trap, 60
Crank–Nicolson, 81–83, 88
creation operator, 69, 94
critical angle for total reflection, 346, 349
cutoff radius, *see* potential, truncation

DC magnetron, 350
Debye length, 235
Debye screening, 254–255
Debye–Hückel potential, *see* Yukawa potential
Debye–Thomas–Fermi screening, 115, 118, 120

degeneracy parameter, 5, 44, 45
density matrix, 90, 92, 101, 102
  convolution property, 101
  discrete-time path-integral representation, 103
  high-temperature, 102
  low-temperature, 101, 102
density operator, 54, 67, 73, 100
  (anti)symmetric, 59
density profile, 183
deposition, 298, 300, 302, 305, 306, 309, 311, 313, 443
deposition rate, 350, 352, 354
depth of field, 149
detailed balance condition, 98
dielectric function, 111, 112, 116, 117, 234
  Lindhard, 114
  of Graphene, 123, 126
dielectric response, 109, 111
  of Graphene, 127–129
diffractometry, 306–308
diffration pattern, 351, 354
diffusion coefficient, 361–363
digital holography, 146
Dirichlet boundary conditions, *see* boundary conditions, Dirichlet, 84
dislocation density, 356, 357
dust particle, 395–398, 401, 404, 415–417, 420, 424–428, 432
dynamically screened pair potential, 234, 262
Dyson equation, 94

Einstein relation, 251
electron scattering, 88–89, 218
electron trap, 301, 302
electrostatic probe, 396, 413, 415, 434
energy (in)flux, 395, 429, 430, 437
equation
  Boltzmann, 206, 217
  Dyson, 94
  Keldysh/Kadanoff–Baym, 93
  Liouville, 67
  Roothaan–Hall, 92
  Vlasov–Boltzmann, 113, 114, 118
  von Neumann, 55
Euler integration scheme, *see* numerical integration, Euler
Ewald summation, 246
exchange potential, 91
expansion coefficients, 86
explicit propagation scheme, *see* Schrödinger equation, explicit propagation

Fermi distribution, 112, 114
Fermi energy, 44, 116, 119, 120
Fermi liquid, 128
Fermi–Dirac distribution, 92
fermion, 58–60, 104
  neutrino, 123
  relativistic, 123
  sign problem, 105
field operator, 69, 93
finite difference approximation, 82
floating potential, 415, 416, 418
fluctuation–dissipation theorem, 243
Fock potential, 91
force
  electric field, 157
  gravitation, 157
  thermophoretic, 157, 163
free particle dispersion relation, 88
Fresnel zone plate, 146
Fresnel–Kirchhoff integral, 147
friction, 255
Friedel oscillations, 114, 115, 117
Friedel–Kohn "wiggle", 120, 121
fugacity, 55

Gaussian filter, 140
Gaussian wave packet, 86, 88
grand canonical ensemble, 55
Graphene, 121–130
  band structure, 123
grazing incidence X-ray diffractometry, GIXD, 345–350, 352, 359, 365
Green's function, 113, 123
  for Graphene, 123, 124
  free phonon, 128
  Matsubara, 93
  nonequilibrium, 90, 91
  of Helmholtz operator, 182
  retarded, 112, 123, 124
  thermodynamic, 112, 113
Green–Kubo relation, 251, 258

Haas–van Alphen oscillations, 119, 120
Hamilton matrix, 86
Hamiltonian, 80, 110, 176
  of Graphene, 123
  plasma, 54
  trapped plasma, 60
Hartree potential, 91
Hartree–Fock, 79, 86, 89–92
  multiconfigurational time-dependent, 91
Helmholtz operator, 182
holography, 164

imaginary time, 103
propagation, *see* Schrödinger equation, imaginary time propagation
slice, 103
implicit propagation scheme, *see* Schrödinger equation, implicit propagation
in situ high temperature diffractometry, 352
indium tin oxide, ITO, 347, 350, 351
inductively coupled plasma, 396
information depth, 347–349
initial conditions, 84–87, 91, 94
eigenstate, 84
Gaussian wave packet, 84
interaction, *see* potential
ion beam source, 427–429
isokinetic thermostat, *see* thermostat, isokinetic
ITP, *see* Schrödinger equation, imaginary time propagation

KBE, *see* Keldysh/Kadanoff–Baym equations
Keldysh contour, *see* Schwinger/Keldysh contour
Keldysh/Kadanoff–Baym equations, 93
kinetic equations
generalized, 93

Landau quantization, 117–121, 130
Langevin dynamics, 12, *see* simulations, Langevin dynamics
laser field, 88
lattice defects, 355
LDA, *see* approximation, local density
Least quadratic kernel method, 140
Li-algorithm, 243
Lindhard dielectric function, 114
Liouville equation, 67
low-temperature plasmas, 9

macrolenses, 138
magic configurations, 252
magnetic field, 117–121
magnetron discharge, 299–302, 305, 402–406, 417
magnetron sputtering, 402
Markov
chain, 98, 99
process, 98, 99
Matsubara Green's function, 93
MC, *see* simulations, Monte Carlo
MD, *see* simulations, molecular dynamics
mean-field
approximation, 180–187, 190
potential, 91
mean-squared displacement, 251, 258
melamine formaldehyde, 413, 414, 417, 419, 421, 425, 426, 435
memory effects, 94
Mesh–Ewald algorithm, 246
Particle–Particle–Particle–Mesh, 246
metastable configurations, 167
Metropolis algorithm, 98–99
microstrain, 346, 356, 357
Mie theory, 137, 145
minimum image convention, 248
molecular dynamic simulations, 12
molecular dynamics, *see* simulations, molecular dynamics
moment method, 139
Monte Carlo, *see* simulations, Monte Carlo
quantum, 79
Mott effect, 50, 52

nanoparticles, 203
density, 218, 225
in discharge plasma, 204
NEGF, *see* nonequilibrium Green's functions
neighbor list method, 246
nonequilibrium Green's functions, 70–71, 79, 90–97
normal modes, 161
normalization, 82, 181, 183, 193
Nosé–Hoover thermostat, *see* thermostat, Nosé–Hoover
numerical integration
Euler, 236–237
Runge–Kutta, 239–240
velocity Verlet, 238–239

operator
annihilation/creation, 69
antisymmetrization, 90
density, *see* density operator
dipole moment, 129
field, 69, 93
Helmholtz, 182
particle number, 69
permutation, 58
self-energy, 91
time evolution, 87
optical potential, 84

Pade expansion, 82
pair correlation function, 178

pair distribution function, 249
particle charge, 419–421, 423–425, 432
particle growth, 396, 397, 400
particle number operator, 69
particle size, 346, 356, 357
particle treatment, 399
Particle–Particle–Particle–Mesh, *see* Mesh–Ewald algorithm, Particle–Particle–Particle–Mesh
partition function, 55, 177
path-integral
  discrete-time, 103
path-integral Monte Carlo, *see* simulations, path-integral Monte Carlo
Pauli principle, 59, 64, 91, 114
PBC, *see* boundary conditions, periodic
PECA, 149
periodic boundary conditions, *see* boundary conditions, periodic
periodic table of clusters, 160, 165
permutation operator, 58
phosphor particles, 407–409, 434–437, 439
PIC–MCC method, *see* simulations, combined PIC–MCC method
PIMC, *see* simulations, path-integral Monte Carlo
pixel locking, 139
plasma
  astrophysical, 3, 48
  bosonic, 60–64
  chemically reacting, 56–58
  crystal, 8
  dense, 48–49
  discharge, 217
  discharge parameters, 219, 223
  dusty, 156
  electron–hole, 49–50
  equations of motion, 236–237
  fermionic, 64–66
  laser, 49
  magnetized, 117–121, 130
  natural, 3
  nonequilibrium, 66
  quantum, 5, 53–75, 110
  quark–gluon, 6, 52
  solid state, 112–130
  technological, 3, 10
  ultra-dense, 52–53
  weakly coupled, 56
plasma coupling parameter, 5
plasma-enhanced chemical vapor deposition, 397, 401, 407, 409
plasmon, 110
  “Bernstein”-type resonances, 120
  dispersion relation, 118, 119, 121
  free, 110
  magneto, 118
  of Graphene, 127, 128, 130
  spectrum, 114, 117, 121
potential
  Coulomb, 43, 109, 110, 113, 114, 129
  electron–plasmon, 110
  exchange, 91
  Fock, 91
  Hartree, 91
  mean-field, 91
  optical, 84
  regularized Coulomb, 94
  screened, 110, 113, 128
  truncation, 245
  van der Waals, 129, 130
  Yukawa, 176, 235
powder particle, 396, 398, 399, 402, 406, 417, 427, 428
pressure ionization, *see* Mott effect

quantum breathing motion, 97
Quantum Hall effect, 122
quantum hydrodynamics, 73–75
quantum well, 116
quantum wire, 116, 117, 130

radial pair distribution function, *see* pair distribution function
radiation pressure, 138
random phase approximation, 90, 111–127
rapid-thermal annealing (RTA), 308, 309, 311
Roothaan–Hall equations, 92
RPA, *see* random phase approximation
Runge–Kutta integration scheme, *see* numerical integration, Runge–Kutta

scanning electron microscopy (SEM), 309, 310
Scanning Video Microscope, 141
Schmidt orthogonalization, 85
Schrödinger equation, 54
  stationary, 54
Schrödinger equation, 100
  basis representation, 86–88
  explicit propagation, 82
  imaginary time propagation, 84–86, 91
  implicit propagation, 82–83
  time-dependent, 79–89, 91
Schwinger/Keldysh contour, 93

screening, *see* Debye screening
  Debye–Thomas–Fermi, 115, 118, 120
  dynamic, 110
screening function, 111, 114, 115, 128
screening strength, 158
second quantization, 68–70
self-diffusion coefficient, 251
self-energy, 92–94, 128
self-energy operator, 91
semiconductor heterostructure, 116
shadowed particle, 146
shell model, 198–200
shielding, *see* screening
shooting algorithm, 86
simulated annealing, 189, 252
simulations
  combined PIC–MCC method, 206–216
  Langevin dynamics, 242–244
  molecular dynamics, 232–262
  Monte Carlo, 97–105, 187
  path-integral Monte Carlo, 60, 100–105
singular value decomposition, 161
Slater determinant, 90
soft matter, 7
spectral power density, 162, 171
specular reflection, 348, 349
sputter rate, 361–363
static structure factor, 249
statistical ensemble
  canonical, 55, 177
  grand canonical, 55
  grandcanonical, 92
stereoscopy, 143, 164
substrate bias voltage, 347, 350, 354, 355, 357
superdiffusion, 258
superfluidity, 61–64
surface effects, 247
symplectic low-order algorithm, 243
system of units, 43–46
  atomic units, 81
  dimensionless, 244–245

TDHF, *see* Hartree–Fock
technological plasma applications, 9
thermal average, 55, 101
thermal probe, 396, 429, 434
thermal wavelength, 43, 103
thermostat
  Andersen, 241
  isokinetic, 241
  Nosé–Hoover, 242
thin film deposition, 397, 398, 408, 434
threshold method, 139
time evolution operator, 81, 82, 87–88
transparent conductive oxide, TCO, 350
Trotter formula, 102
Trotter identity, 239
two-particle integrals, 92
two-time correlation functions, *see* nonequilibrium Green's functions

unitary time evolution, 82

van der Waals potential, 129, 130
variational problem, 179
velocity scaling, *see* thermostat, isokinetic
velocity Verlet integration scheme, *see* numerical integration, velocity Verlet
video microscope, 141
Vlasov–Boltzmann equation, 113, 114, 118
von Neumann boundary conditions, *see* boundary conditions, von Neumann
von Neumann equation, 55

warm dense matter, 4
Wigner crystal, 8, 231
Wigner function, 71

X-ray reflectometry, XR, 345, 346, 349, 352, 353, 359, 365

Yukawa balls, 166, 175, 232
Yukawa potential, 158, 176, 235

zigzag transition, 159

MIX
Papier aus verantwortungsvollen Quellen
Paper from responsible sources
FSC® C105338

If you have any concerns about our products,
you can contact us on
ProductSafety@springernature.com

In case Publisher is established outside the EU,
the EU authorized representative is:
Springer Nature Customer Service Center GmbH
Europaplatz 3, 69115 Heidelberg, Germany

Printed by Libri Plureos GmbH
in Hamburg, Germany